T0335494

THE ELEMENTS OF FINANCIAL ECONOMETRICS

Financial econometrics is an interdisciplinary subject that uses statistical methods and economic theory to address a variety of quantitative problems in finance. This compact, master's-level textbook focuses on methodology and includes real financial data illustrations throughout. The mathematical level is purposely kept moderate, allowing the power of the quantitative methods to be understood without too much technical detail. Wherever possible, the authors indicate where to find the relevant R codes to implement the various methods.

This book grew out of the course at Princeton University which is one of the world's flagship programs in computational finance and financial engineering. It will therefore be useful for those with an economics and finance background who are looking to sharpen their quantitative skills, and also for those with strong quantitative skills who want to learn how to apply them to finance.

THE ELEMENTS OF FINANCIAL ECONOMETRICS

JIANQING FAN
Princeton University, New Jersey

QIWEI YAO
London School of Economics and Political Science

CAMBRIDGE
UNIVERSITY PRESS

University Printing House, Cambridge CB2 8BS, United Kingdom

One Liberty Plaza, 20th Floor, New York, NY 10006, USA

477 Williamstown Road, Port Melbourne, VIC 3207, Australia

314-321, 3rd Floor, Plot 3, Splendor Forum, Jasola District Centre, New Delhi - 110025, India

79 Anson Road, #06-04/06, Singapore 079906

Cambridge University Press is part of the University of Cambridge.

It furthers the University's mission by disseminating knowledge in the pursuit of education, learning and research at the highest international levels of excellence.

www.cambridge.org
Information on this title: www.cambridge.org/9781107191174
DOI: 10.1017/9781108120616

First published 2017

A catalogue record for this publication is available from the British Library

ISBN 978-1-107-19117-4 Hardback

Additional resources for this publication at www.cambridge.org/fanyao

Contents

Preface

This is an introductory textbook for financial econometrics at the master's level. Readers are assumed to have some background in calculus, linear algebra, statistics, and probability, all at undergraduate level. Knowledge in economics and finance is beneficial but not essential.

This book grew out of the lecture notes for the "Financial Econometrics" course taught by Jianqing Fan for Master in Finance students at Princeton University since 2003 and for Master in Financial Engineering students at Fudan University since 2011. The audiences are always very broad with diverse backgrounds in mathematics, physics, computer science, economics, or finance. Those talented students wish to learn fundamentals of financial econometrics in order to work in the financial industry or to pursue a Ph.D. degree in related fields. The challenge is to give all of them sufficient financial econometrics knowledge upon the completion of the class. This textbook is written with the aim to achieve such an ambitious goal.

We trust that the book will be of interest to those coming to the area for the first time, to readers who already have an economics and finance background but would like to further sharpen their quantitative skills, and also to those who have strong quantitative skills and would like to learn how to apply them to finance. Application-oriented analysts will also find this book useful, as it focuses on methodology and includes numerous case studies with real data sets. We purposely keep the level of mathematics moderate, as the power of the quantitative methods can be understood without sophisticated technical details. We also avoid the cook book style of writing, as a good understanding of statistical and econometric principles in finance enables readers to apply the knowledge far beyond the problems stated in the book. Numerical illustration with real financial data is included throughout the book. We also indicate, whenever possible, where to find the relevant `R` codes to implement the various methods. Due to the nature of the subject, it is inevitable that we occasionally step into more sophisticated techniques which rely on more advanced mathematics; such sections are marked with "$*$" and can be ignored by beginners. Most technical arguments are collected in a "Complements" section at the end of some chapters, but key ideas are left within the main body of the text.

What is financial econometrics? Broadly speaking, it is an interdisciplinary subject that uses statistical methods and economic theory to address a variety of quantitative problems in finance. These include building financial models, testing financial economics theory, simulating financial systems, volatility estimation, risk management, capital asset pricing, derivative pricing, portfolio allocation, proprietary trading, portfolio and derivative hedging, among others. Financial econometrics is an active field of integration of finance, economics, probability, statistics, and applied mathematics. Financial activities generate many new problems and products, economics provides useful theoretical foundation and guidance, and quantitative methods such as statistics, probability and applied mathematics are essential tools for solving quantitative problems in finance. Professionals in finance now routinely use sophisticated statistical techniques and modern computation power in portfolio management, proprietary trading, derivative pricing, financial consulting, securities regulation, and risk management.

When the class was first taught in 2003, there were very few books on financial econometrics. The books that had a strong impact on our preparation of lecture notes were Campbell et al. (1997) and Fan and Yao (2003). With one semester of teaching, we can only cover the important elements of financial econometrics. We use this name as the title of the book, as it also reflects the modular aspect of the book. This allows us to expand this textbook to cover other fundamental materials in the future. For example, "Simulation methods in finance" and "Econometrics of continuous time finance" are taught at Princeton, but are not included in this book nor in Fudan's financial engineering class due to time constraints. Another important topic that we wish to cover is the analysis of high-frequency financial data.

The book consists of two integrated parts: The first four chapters are on time series aspects of financial econometrics while the last five chapters focus on cross-sectional aspects. The introduction in Chapter 1 sets the scene for the book: using two financial price time series we illustrate the stylized features in financial returns. The efficient markets hypothesis is deliberated together with statistical tests for random walks and white noise. A compact view of linear time series models is given in Chapter 2, including ARMA models, random walks, and inference with trends. We also include a brief introduction to the exponential smoothing based forecasting techniques for trends and momentum, which are widely used in the financial industry. Chapter 3 introduces various heteroscedastic volatility models. A compact introduction to state space models including techniques such as Kalman filter and particle filters is included as an appendix. Chapter 4 contains some selective topics in multivariate time series analysis. Within the context of vector autoregressive models, we also introduce topics such as Granger causality, impulse response functions, and cointegration, as they play important roles in economics and finance.

The second part begins with Chapter 5, which introduces portfolio theory and derives the celebrated capital asset model. We also introduce statistical techniques to test such a celebrated model and provide extensive empirical studies. Chapter 6 extends the capital asset pricing model to a multi-factor pricing model. The applications of the factor models and econometrics tests on the validity of such pricing models are introduced. In addition, principal component analysis and factor analysis are briefly discussed. Chapter 7 touches on several practical aspects of portfolio allocation and risk management. The highlights of this chapter include risk assessments of large portfolios, portfolio allocation under gross-exposure constraints, and large volatility matrix estimation using factor models and covariance regularization. Chapter 8 derives the capital asset pricing model from a consumption, investment, and saving point of view. This gives students a different perspective on where the financial prices come from and a chance to appreciate how this differs from pricing financial derivatives. Chapter 9 calculates the prices implied by the models of returns. It gives us an idea of what the fundamental price of a stock is and how the prices are related to the dividend payments and short-term interest rates.

Many people have been of great help to our work on this book. Early drafts of this book have been taught to roughly five hundred students and there are also our enthusiastic readers. In particular, we are grateful to Yingying Fan, Yue Niu, Jingjing Zhang, Feng Yang, Weijie Gu, Xin Tong, Wei Dai, Jiawei Yao, Xiaofeng Shi, and Weichen Wang for their gracious assistance teaching the Financial Econometrics class at Princeton. Réne Carmona provided us with his course outlines on Financial Econometrics which helped us in the selection of the topics of the course. Many treatments of ARIMA models are inspired by the lecture notes of George Tiao. Alex Furger and Michael Lachans spent a great amount of their precious time proof-reading the final version of the book. We are very grateful for their contributions and generosity. We would like to thank Yacine Ait-Sahalia and Yazhen Wang for stimulating discussions on the topic and Shaojun Guo for formatting the references of the book. We are indebted to Jiaan Yan for his encouragement and support to publish this book in China and to Xiongwen Lu for inviting us to teach the course in Fudan University. We would also like to thank Yuzuo Chen for providing various editorial assistance.

Jianqing Fan's research was generously supported by the National Science Foundation and National Institutes of Health of the USA, and the Academy of Mathematics and System Science and the National Center for Mathematics and Interdisciplinary Sciences, Chinese Academy of Sciences.

1

Asset Returns

The primary goal of investing in a financial market is to make profits without taking excessive risks. Most common investments involve purchasing financial assets such as stocks, bonds or bank deposits, and holding them for certain periods. Positive revenue is generated if the price of a holding asset at the end of holding period is higher than that at the time of purchase (for the time being we ignore transaction charges). Obviously the size of the revenue depends on three factors: (i) the initial capital (i.e. the number of assets purchased), (ii) the length of holding period, and (iii) the changes of the asset price over the holding period. A successful investment pursues the maximum revenue with a given initial capital, which may be measured explicitly in terms of the so-called *return*. A return is a percentage defined as the change of price expressed as a fraction of the initial price. It turns out that asset returns exhibit more attractive statistical properties than asset prices themselves. Therefore it also makes more statistical sense to analyze return data rather than price series.

1.1 Returns

Let P_t denote the price of an asset at time t. First we introduce various definitions for the returns for the asset.

1.1.1 One-period simple returns and gross returns

Holding an asset from time $t-1$ to t, the value of the asset changes from P_{t-1} to P_t. Assuming that no dividends paid are over the period. Then the *one-period simple return* is defined as

$$R_t = (P_t - P_{t-1})/P_{t-1}. \tag{1.1}$$

It is the profit rate of holding the asset from time $t-1$ to t. Often we write $R_t = 100R_t\%$, as $100R_t$ is the percentage of the gain with respect to the initial capital P_{t-1}. This is particularly useful when the time unit is small (such as a day or an hour); in such cases R_t typically takes very small values. The returns for less risky assets such as bonds can be even smaller in a short period and are often quoted in *basis points*, which is $10,000R_t$.

The *one-period gross return* is defined as $P_t/P_{t-1} = R_t + 1$. It is the ratio of the new market value at the end of the holding period over the initial market value.

1.1.2 Multiperiod returns

The holding period for an investment may be more than one time unit. For any integer $k \geqslant 1$, the returns for over k periods may be defined in a similar manner. For example, the *k-period simple return* from time $t-k$ to t is

$$R_t(k) = (P_t - P_{t-k})/P_{t-k},$$

and the *k-period gross return* is $P_t/P_{t-k} = R_t(k) + 1$. It is easy to see that the multiperiod returns may be expressed in terms of one-period returns as follows:

$$\frac{P_t}{P_{t-k}} = \frac{P_t}{P_{t-1}}\frac{P_{t-1}}{P_{t-2}}\cdots\frac{P_{t-k+1}}{P_{t-k}}, \tag{1.2}$$

$$R_t(k) = \frac{P_t}{P_{t-k}} - 1 = (R_t + 1)(R_{t-1} + 1)\cdots(R_{t-k+1} + 1) - 1. \tag{1.3}$$

If all one-period returns $R_t, \ldots, R_{t-k+1}$ are small, (1.3) implies an approximation

$$R_t(k) \approx R_t + R_{t-1} + \cdots + R_{t-k+1}. \tag{1.4}$$

This is a useful approximation when the time unit is small (such as a day, an hour or a minute).

1.1.3 Log returns and continuously compounding

In addition to the simple return R_t, the commonly used *one-period log return* is defined as

$$r_t = \log P_t - \log P_{t-1} = \log(P_t/P_{t-1}) = \log(1 + R_t). \tag{1.5}$$

Note that a log return is the logarithm (with the natural base) of a gross return and $\log P_t$ is called the log price. One immediate convenience in using

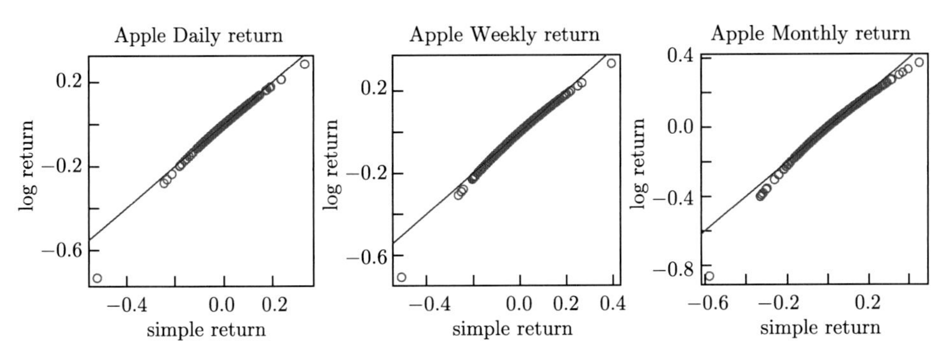

Figure 1.1 Plots of log returns against simple returns of the Apple Inc share prices in January 1985 to February 2011. The blue straight lines mark the positions where the two returns are identical.

log returns is that the additivity in multiperiod log returns, i.e. the *k-period log return* $r_t(k) \equiv \log(P_t/P_{t-k})$ is the sum of the k one-period log returns:

$$r_t(k) = r_t + r_{t-1} + \cdots + r_{t-k+1}. \tag{1.6}$$

An investment at time $t-k$ with initial capital A yields at time t the capital

$$A\exp\{r_t(k)\} = A\exp(r_t + r_{t-1} + \cdots + r_{t-k+1}) = Ae^{k\bar{r}},$$

where $\bar{r} = (r_t + r_{t-1} + \cdots + r_{t-k+1})/k$ is the average one-period log returns. In this book *returns* refer to log returns unless specified otherwise.

Note that the identity (1.6) is in contrast with the approximation (1.4) which is only valid when the time unit is small. Indeed when the values are small, the two returns are approximately the same:

$$r_t = \log(1 + R_t) \approx R_t.$$

However, $r_t < R_t$. Figure 1.1 plots the log returns against the simple returns for the Apple Inc share prices in the period of January 1985 to February 2011. The returns are calculated based on the daily close prices for the three holding periods: a day, a week and a month. The figure shows that the two definitions result almost the same daily returns, especially for those with the values between −0.2 and 0.2. However when the holding period increases to a week or a month, the discrepancy between the two definitions is more apparent with a simple return always greater than the corresponding log return.

The log return r_t is also called *continuously compounded return* due to its close link with the concept of compound rates or interest rates. For a bank deposit account, the quoted interest rate often refers to as 'simple interest'. For example, an interest rate of 5% payable every six months will be quoted as a simple interest of 10% per annum in the market. However if an account with the initial capital $1 is held for 12 months and interest rate remains unchanged, it follows from (1.2) that the gross return for the two periods is

$$1 \times (1 + 0.05)^2 = 1.1025,$$

i.e. the annual simple return is $1.1025 - 1 = 10.25\%$, which is called the *compound return* and is greater than the quoted annual rate of 10%. This is due to the earning from 'interest-on-interest' in the second six-month period.

Now suppose that the quoted simple interest rate per annum is r and is unchanged, and the earnings are paid more frequently, say, m times per annum (at the rate r/m each time of course). For example, the account holder is paid every quarter when $m = 4$, every month when $m = 12$, and every day when $m = 365$. Suppose m continues to increase, and the earnings are paid continuously eventually. Then the gross return at the end of one year is

$$\lim_{m \to \infty} (1 + r/m)^m = e^r.$$

More generally, if the initial capital is C, invested in a bond that compounds continuously the interest at annual rate r, then the value of the investment at time t is

$$C \exp(rt).$$

Hence the log return per annum is r, which is the logarithm of the gross return. This indicates that the simple annual interest rate r quoted in the market is in fact the annual log return if the interest is compounded continuously. Note that if the interest is only paid once at the end of the year, the simple return will be r, and the log return will be $\log(1 + r)$ which is always smaller than r.

In summary, a simple annual interest rate quoted in the market has two interpretations: it is the simple annual return if the interest is only paid once at the end of the year, and it is the annual log return if the interest is compounded continuously.

1.1.4 Adjustment for dividends

Many assets, for example some bluechip stocks, pay dividends to their shareholders from time to time. A dividend is typically allocated as a fixed amount of cash per share. Therefore adjustments must then be made in computing returns to account for the contribution towards the earnings from dividends. Let D_t denote the dividend payment between time $t - 1$ and t. Then the returns are now defined as follows:

$$R_t = (P_t + D_t)/P_{t-1} - 1, \quad r_t = \log(P_t + D_t) - \log P_{t-1},$$

$$R_t(k) = \big(P_t + D_t + \cdots + D_{t-k+1}\big)/P_{t-k} - 1,$$

$$r_t(k) = r_t + \cdots + r_{t-k+1} = \sum_{j=0}^{k-1} \log\left(\frac{P_{t-j} + D_{t-j}}{P_{t-j-1}}\right).$$

The above definitions are based on the assumption that all dividends are cashed out and are not re-invested in the asset.

1.1.5 Bond yields and prices

Bonds are quoted in annualized yields. A so-called zero-coupon bond is a bond bought at a price lower than its face value (also called par value or principal), with the face value repaid at the time of maturity. It does not make periodic interest payments (i.e. coupons), hence the term 'zero-coupon'. Now we consider a zero-coupon bond with the face value \$1. If the current yield is r_t and the remaining duration is D units of time, with continuous compounding, its current price B_t should satisfy the condition

$$B_t \exp(Dr_t) = \$1,$$

i.e. the price is $B_t = \exp(-Dr_t)$ dollars. Thus, the annualized log-return of the bond is

$$\log(B_{t+1}/B_t) = D(r_t - r_{t+1}). \tag{1.7}$$

Here, we ignore the fact that B_{t+1} has one unit of time shorter maturity than B_t.

Suppose that we have two baskets of high-yield bonds and investment-grade bonds (i.e. the bonds with relatively low risk of default) with an average duration of 4.4 years each. Their yields spread (i.e. the difference) over the Treasury bond with similar maturity are quoted and plotted in Figure 1.2. The daily returns of bonds can then be deduced from (1.7), which is the change of yields multiplied by the duration. The daily changes of treasury bonds are typically much smaller. Hence, the changes of yield spreads can directly be used as proxies of the changes of yields. As expected, the high-yield bonds have higher yields than the investment grade bonds, but have higher volatility too (about 3 times). The yield spreads widened significantly in a period after the financial crisis following Lehman Brothers filing bankrupt protection on September 15, 2008, reflecting higher default risks in corporate bonds.

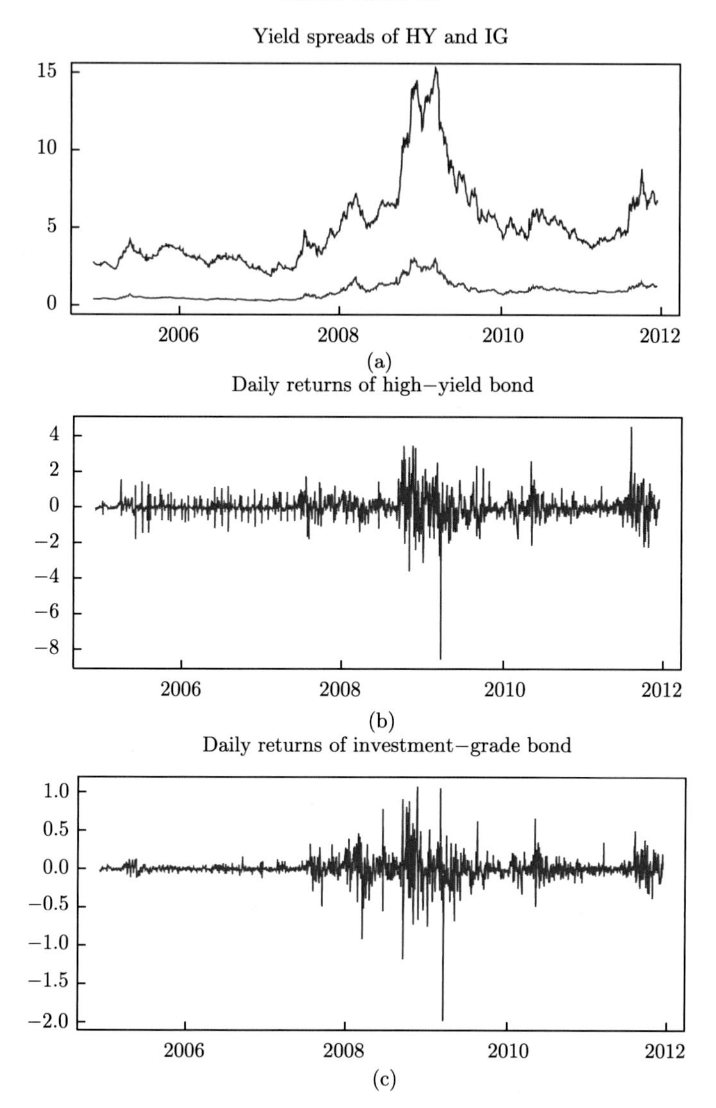

Figure 1.2 Time series of the yield spreads (the top panel) of high-yield bonds (blue curve) and investment-grade bonds (red curve), and their associated daily returns (the 2nd and 3rd panels) in November 29, 2004 to December 10, 2014.

1.1.6 Excess returns

In many applications, it is convenient to use an *excess return*, which is defined in the form $r_t - r_t^\star$, where $r_t^\star$ is a reference rate. The commonly used reference rates are, for example, bank interest rates, *LIBOR* rates (London Interbank Offered Rate: the average interest rate that leading banks in London charge when lending to other banks), log returns of a riskless asset (e.g., yields of short-term government bonds such as the 3-month US treasury bills) or market portfolio (e.g. the S&P 500 index or CRSP value-weighted index, which is the value-weighted index of all stocks traded in three major stock exchanges, created by the Center for Research in Security Prices of University of Chicago).

For bonds, *yield spread* is an excess yield defined as the difference between the yield of a bond and the yield of a reference bond such as a US treasury bill with a similar maturity.

1.2 Behavior of financial return data

In order to build useful statistical models for financial returns, we collect some empirical evidence first. To this end, we look into the daily closing indices of the S&P 500 and the daily closing share prices (in US dollars) of the Apple Inc in the period of January 1985 to February 2011. The data were adjusted for all splits and dividends, and were downloaded from *Yahoo!Finance*.

The S&P 500 is a value-weighted index of the prices of the 500 large-cap common stocks actively traded in the United States. Its present form has been published since 1957, but its history dates back to 1923 when it was a value-weighted index based on 90 stocks. It is regarded as a bellwether for the American economy. Many mutual funds, exchange-traded funds, pension funds etc are designed to track the performance of S&P 500. The first panel in Figure 1.3 is the time series plot for the daily closing indices of S&P 500. It shows clearly that there was a slow and steady increase momentum in 1985–1987. The index then reached an all-time high on March 24, 2000 during the dotcom bubble, and consequently lost about 50% of its value during the stock market downturn in 2002. It peaked again on October 9, 2007 before suffering during the credit crisis stemming from subprime mortgage lending in 2008–2010. The other three panels in Figure 1.3 show the daily, the weekly and the monthly log returns of the index. Although the profiles of the three plots are similar, the monthly return curve is a 'smoothed' version of, for example, the daily return curve which exhibits higher volatile fluctuations. In particular, the high volatilities during the 2008-2010 are more vividly depicted in the daily return plot. In contrast to the prices, the returns oscillate around a constant level close to 0. Furthermore, high oscillations tend to cluster together, reflecting more volatile market periods. Those features on return data are also apparent in the Apple stock displayed in Figure 1.4. The share prices of the Apple Inc are also non-stationary in time in the sense that the price movements over different time periods are different. For example in 1985–1998, the prices almost stayed at a low level. Then it experienced a steady increase until September 29, 2000 when the Apple's value sliced in half due to the earning warning in the last quarter of the year. The more recent surge of the price increase was largely due to Apple's success in the mobile consumer electronics market via its products the iPod, iPhone and iPad, in spite of its fluctuations during the subprime mortgage credit crisis.

We plot the normalized histograms of the daily, the weekly and the monthly log returns of the S&P 500 index in Figure 1.5. For each histogram, we also superimpose the normal density function with the same mean and variance. Also plotted in Figure 1.5 are the quantile–quantile plots for the three returns. (See an introduction to Q–Q plots in Section 1.5.) It is clear that the returns within the given holding periods are not normally distributed. Especially the tails of the return distributions are heavier than those of the normal distribution, which is highlighted explicitly in the Q–Q plots: the left tail (red circles) is below (negatively larger) the blue line, and the right tail (red circles) is above (larger) the blue line. We have also noticed that when the holding period increases from a day, a week to a month, the tails of the distributions become lighter. In particular the upper tail of the distribution for the monthly returns is about equally heavy as that of a normal distribution (red circles and blue line are about the same). All the distributions are skewed to the left due to a few large negative returns. The histograms also show that the distribution for the monthly returns is closer to a normal distribution than those for the weekly returns and the daily returns. The similar patterns are also observed in the Apple return data; see Figure 1.6.

Figures 1.7 and 1.8 plot the sample autocorrelation function (ACF) $\widehat{\rho}_k$ against the time lag k for the log returns, the squared log returns and the absolute log returns. Given a return series $r_1, \ldots, r_T$, the sample autocorrelation function is defined as $\widehat{\rho}_k = \widehat{\gamma}_k / \widehat{\gamma}_0$, where

$$\widehat{\gamma}_k = \frac{1}{T}\sum_{t=1}^{T-k}(r_t - \bar{r})(r_{t+k} - \bar{r}), \quad \bar{r} = \frac{1}{T}\sum_{t=1}^{T} r_t. \tag{1.8}$$

$\widehat{\gamma}_k$ is the sample autocovariance at lag k. It is (about) the same as the sample correlation coefficient of the paired observations $\{(r_t, r_{t+k})\}_{t=1}^{T-k}$ (the difference is in the definition of the sample mean in the calculation of the sample covariance). The sample autocorrelation functions for the squared and the absolute returns are defined in the same manner but with r_t replaced by, respectively, r_t^2 and $|r_t|$. For each ACF plot in Figures 1.7 and 1.8,

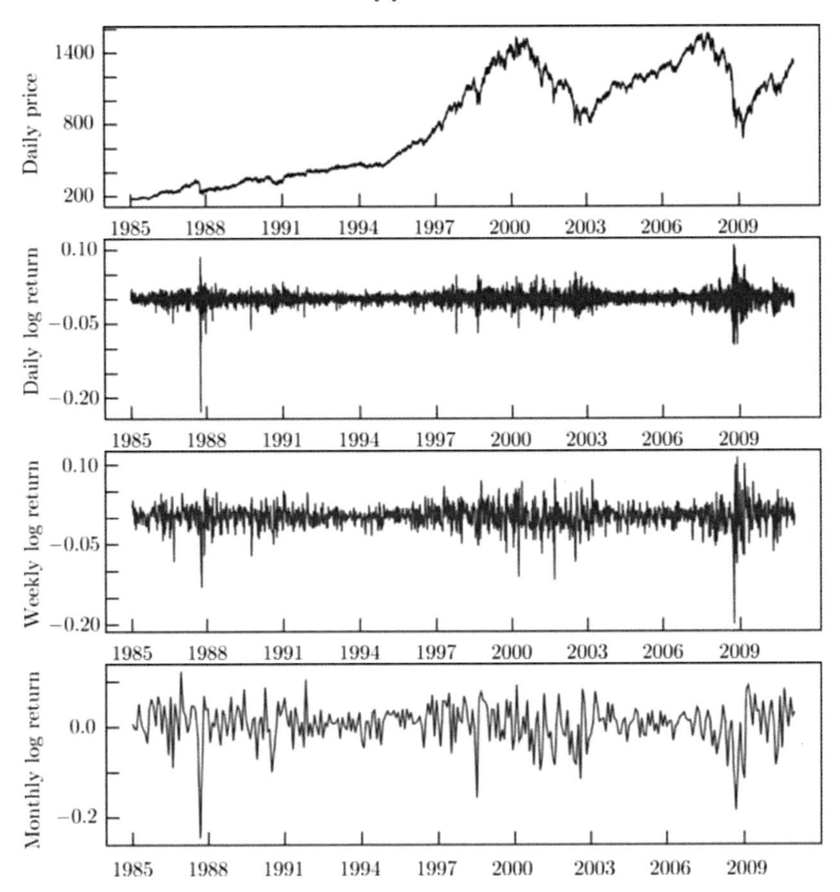

Figure 1.3 Time series plots of the daily indices, the daily log returns, the weekly log returns, and the monthly log returns of S&P 500 index in January 1985 to February 2011.

the two dashed horizontal lines, which are $\pm 1.96/\sqrt{T}$, are the bounds for the 95% confidence interval for ρ_k if the true value is $\rho_k = 0$. Hence ρ_k would be viewed as not significantly different from 0 if its estimator $\widehat{\rho}_k$ is between those two lines. It is clear from Figures 1.7 and 1.8 that all the daily, weekly and monthly returns for both S&P 500 and the Apple stock exhibit no significant autocorrelation, supporting the hypothesis that the returns of a financial asset are uncorrelated across time. However there are some small but significant autocorrelations in the squared returns and more in the absolute returns.

Furthermore the autocorrelations are more pronounced and more persistent in the daily data than in weekly and monthly data. Since the correlation coefficient is a measure of linear dependence, the above empirical evidence

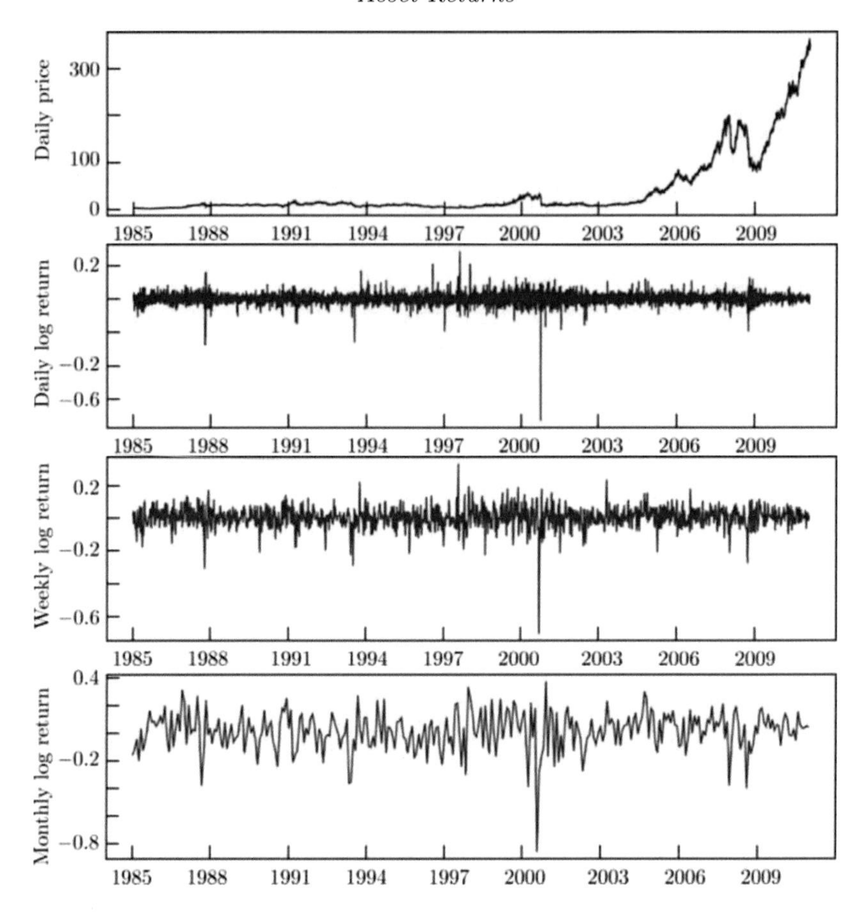

Figure 1.4 Time series plots of the daily prices, the daily log returns, the weekly log returns, and the monthly log returns of the Apple stock in January 1985 to February 2011.

indicates that the returns of a financial asset are linearly independent with each other, although there exist nonlinear dependencies among the returns at different lags. Especially the daily absolute returns exhibit significant and persistent autocorrelations – a characteristic of so-called long memory processes.

1.2.1 Stylized features of financial returns

The above findings from the two real data sets are in line with the so-called stylized features in financial returns series, which are observed across different kinds of assets including stocks, portfolios, bonds and currencies. See, e.g. Rydberg (2000). We summarize these features below.

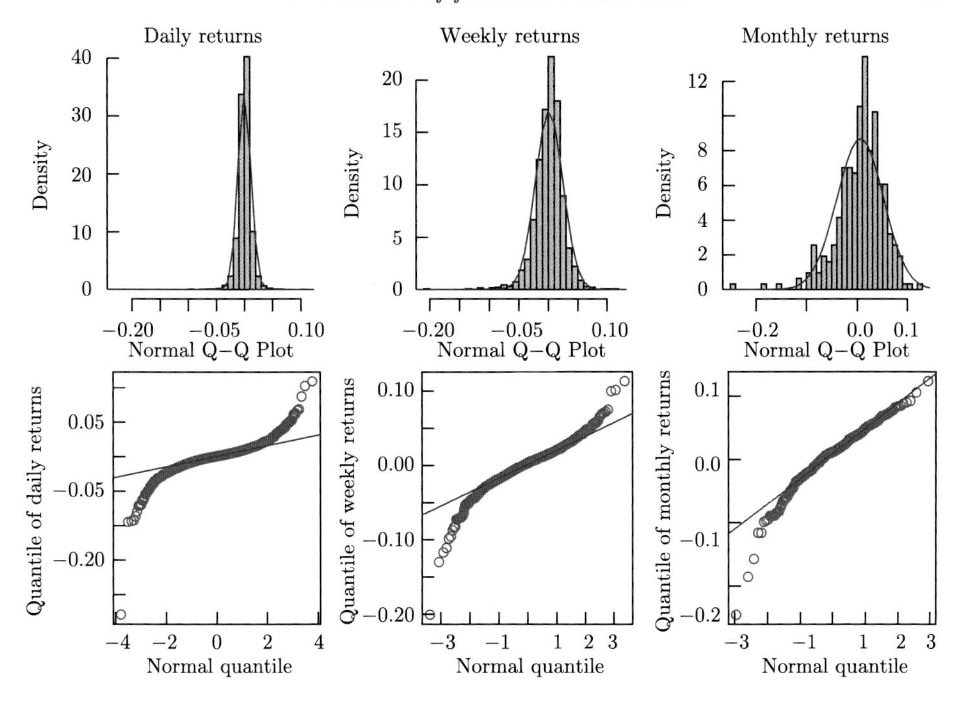

Figure 1.5 Histograms (the top panels) and Q–Q plots (the bottom panels) of the daily, weekly, and monthly log returns of S&P 500 in January 1985 to February 2011. The normal density with the same mean and variance are superimposed on the histogram plots.

(i) *Stationarity*. The prices of an asset recorded over times are often not stationary due to, for example, the steady expansion of economy, the increase of productivity resulting from technology innovation, and economic recessions or financial crisis. However their returns, denoted by r_t for $t \geqslant 1$, typically fluctuates around a constant level, suggesting a constant mean over time. See Figures 1.3 and 1.4. In fact most return sequences can be modeled as a stochastic processes with at least time-invariant first two moments (i.e. the weak stationarity; see 2.1). A simple (and perhaps over-simplistic) approach is to assume that all the finite dimensional distributions of a return sequence are time-invariant.

(ii) *Heavy tails*. The probability distribution of return r_t often exhibits heavier tails than those of a normal distribution. Figures 1.5 and 1.6 provide the *quantile–quantile plot* or *Q–Q plot* for graphical checking of normality. See Section 1.5 for detail. A frequently used statistic for checking the normality (including tail-heaviness) is the Jarque–Bera test presented in Section 1.5. Nevertheless r_t is assumed typically to have at least two finite

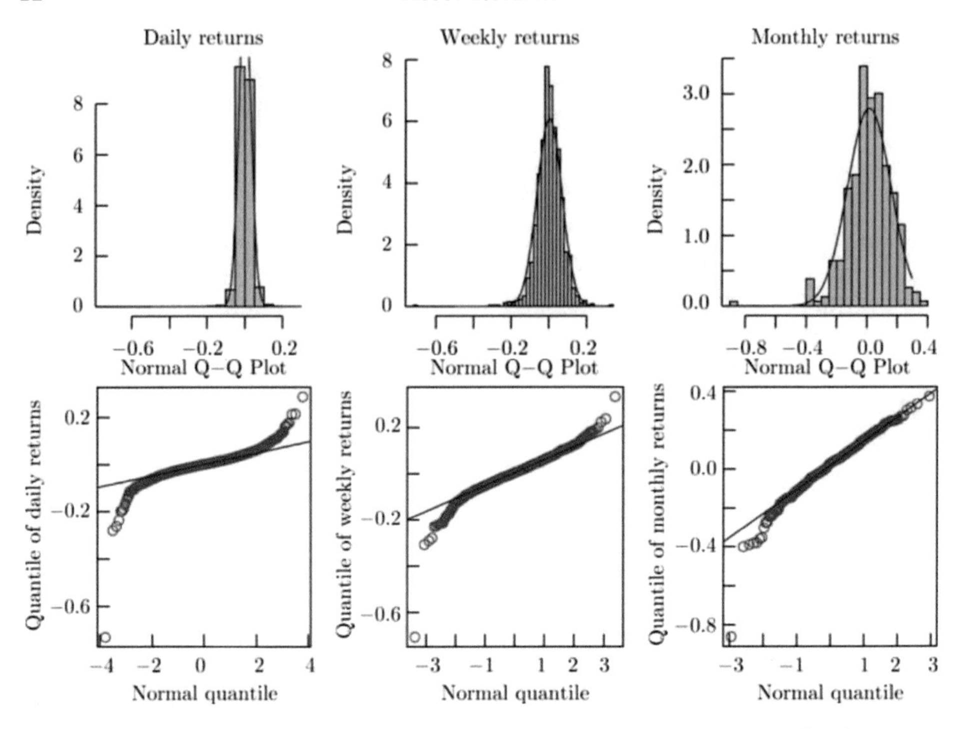

Figure 1.6 Histograms (the top panels) and Q–Q plots (the bottom panels) of the daily, weekly, and monthly log returns of the Apple stock in January 1985 to February 2011. The normal density with the same mean and variance are superimposed on the histogram plots.

moments (i.e. $E(r_t^2) < \infty$), although it is debatable how many moments actually exist for a given asset.

The density of *t-distribution* with degree of freedom ν is given by

$$f_\nu(x) = d_\nu^{-1}\left(1 + \frac{x^2}{\nu}\right)^{-(\nu+1)/2}, \tag{1.9}$$

where $d_\nu = B(0.5, 0.5\nu)\sqrt{\nu}$ is the normalization constant and B is a beta function. This distribution is often denoted by $t(\nu)$ or t_ν. Its tails are of polynomial order $f_\nu(x) \asymp |x|^{-(\nu+1)}$ (as $|x| \to \infty$), which are heavier than the normal density. Note that for any random variable $X \sim t(\nu)$, $E\{|X|^\nu\} = \infty$ and $E\{|X|^{\nu-\delta}\} < \infty$ for any $\delta \in (0, \nu]$.

When ν is large, $t(v)$ is close to a normal distribution. In fact, based on a sample of size 2500 (approximately 10-year daily data), one can not differentiate $t(10)$ from a normal distribution based on, for example, the Kolmogorov–Smirnov test (the function `KS.test` in R). However their tail behaviors are very different: A 5-standard-deviation (SD) event occurs once

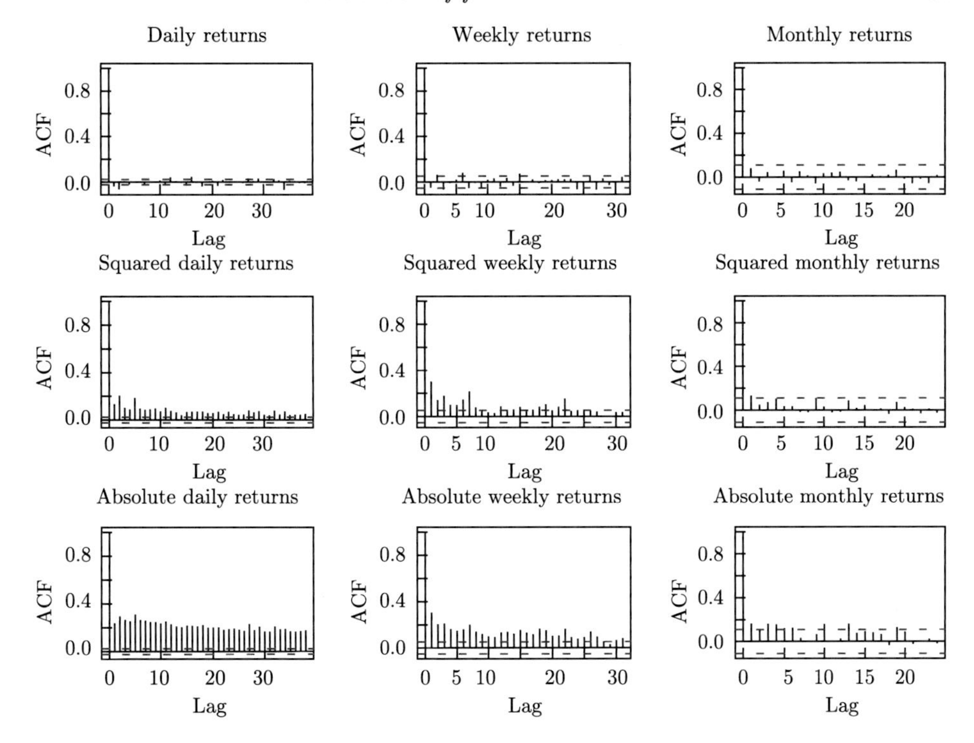

Figure 1.7 Autocorrelations of the daily, weekly, and monthly log returns, the squared daily, weekly, and monthly log returns, and the absolute daily, weekly, and monthly log returns of S&P 500 in January 1985 to February 2011.

in every 14000 years under a normal distribution, once in every 15 years under t_{10}, and once in every 1.5 years under $t_{4.5}$. The calculation goes as follows. The probability of getting a -5 SD daily shock or worse under the normal distribution is 2.8665×10^{-7} (which is $P(Z < -5)$ for $Z \sim N(0, 1)$), or 1 in 3488575 days. Dividing this by approximately 252 trading days per year yields the result of 13844 years. A similar calculation can be done with different kinds of t-distributions. If the tails of stock returns behave like $t_{4.5}$, the left tail of typical daily S&P 500 returns, the occurrence of -5 SD event is more often than what we would conceive.

Figure 1.9 plots the quantiles of the S&P 500 returns in the period January 1985 to February 2011 against the quantiles of $t(\nu)$ distributions with $\nu = 2, 3, \ldots, 7$. It is clear that the tails of the S&P 500 return are heavier than the tails of $t(5)$, and are thinner than the tails of $t(2)$ (and perhaps also $t(3)$). Hence it is reasonable to assume that the second moment of the return of the S&P 500 is finite while the 5th moment should be infinity.

(iii) *Asymmetry*. The distribution of return r_t is often negatively skewed

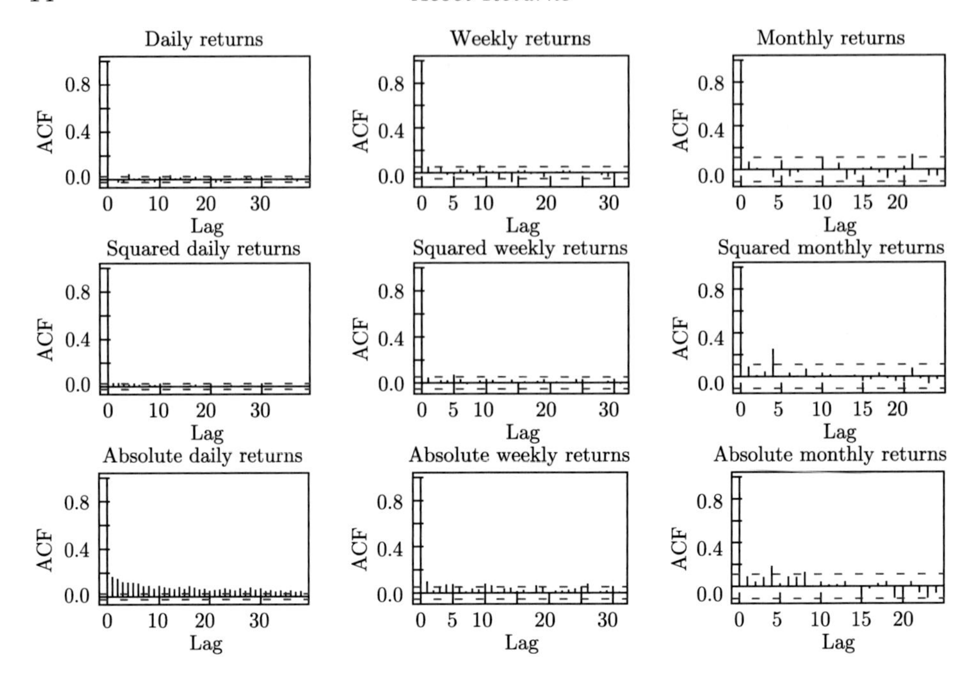

Figure 1.8 Autocorrelations of the daily, weekly, and monthly log returns, the squared daily, weekly, and monthly log returns, and the absolute daily, weekly, and monthly log returns of the Apple stock in January 1985 to February 2011.

(Figures 1.5 and 1.6), reflecting the fact that the downturns of financial markets are often much steeper than the recoveries. Investors tend to react more strongly to negative news than to positive news.

(iv) *Volatility clustering.* This term refers to the fact that large price changes (i.e. returns with large absolute values) occur in clusters. See Figures 1.3 and 1.4. Indeed, large price changes tend to be followed by large price changes, and periods of tranquility alternate with periods of high volatility. The time varying volatility can easily be see in Figures 1.3. The standard deviation of S&P 500 returns from November 29, 2004 to December 31, 2007 (3 months before JP Morgan Chase offered to acquire Bear Stearns at a price of $2 per share on March 17, 2008) is 0.78%, and the standard deviation of the returns since 2008 is 1.83%. The volatility of S&P 500 in 2005 and 2006 is merely 0.64%, whereas the volatility at the height of the 2008 financial crisis (September 15, 2008 to March 16, 2009, i.e. from a month before to 6 months after Lehmann Brother's fall) is 3.44%.

(v) *Aggregational Gaussianity.* Note that a return over k days is simply the aggregation of k daily returns; see (1.6). When the time horizon increases,

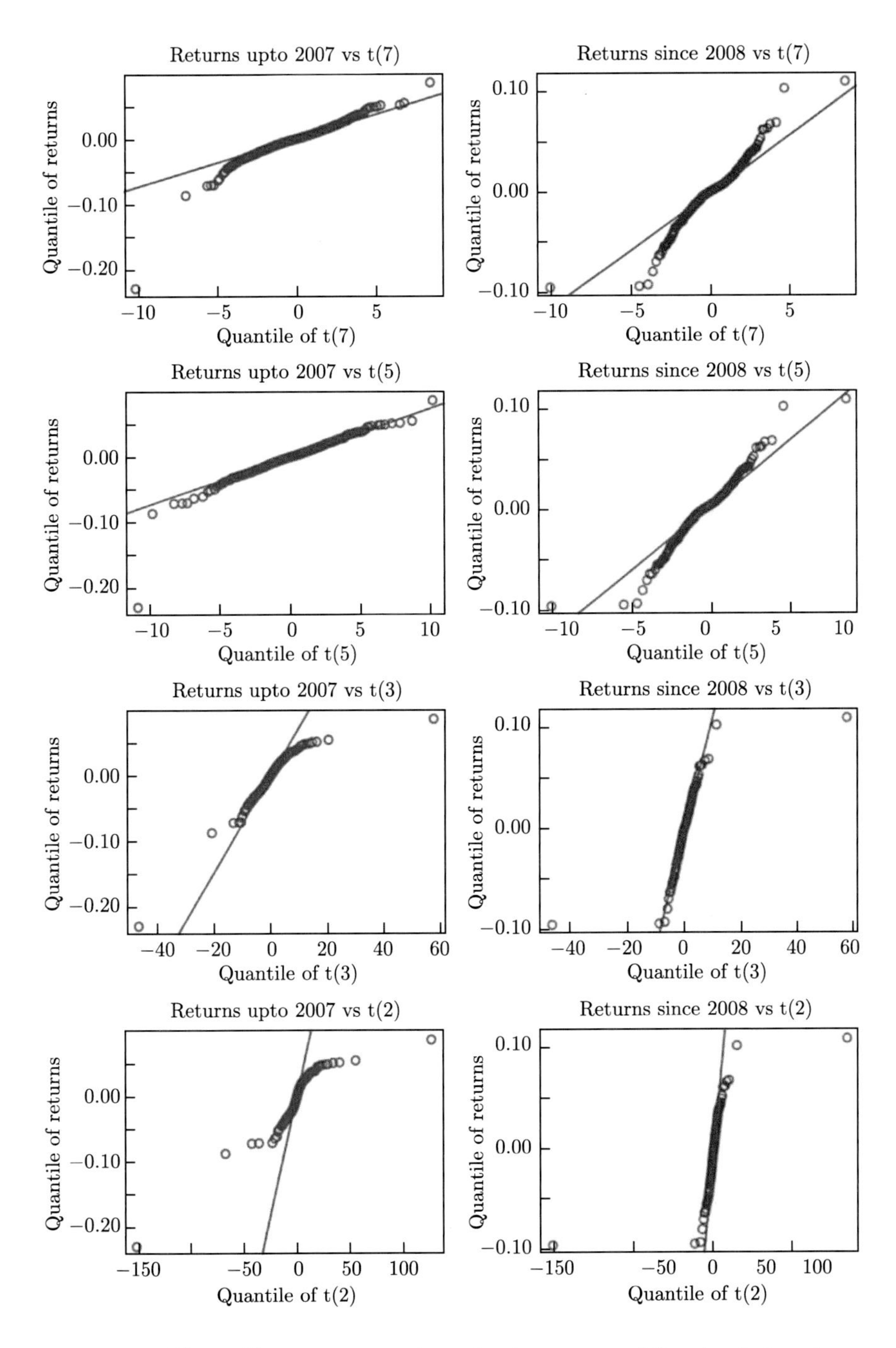

Figure 1.9 Plots of log returns against simple returns of the Apple Inc share prices in January 1985 to February 2011. The blue straight lines mark the positions where the two returns are identical.

the central limit law sets in and the distribution of the returns over a long time-horizon (such as a month) tends toward a normal distribution. See Figures 1.5 and 1.6.

(vi) *Long range dependence.* The returns themselves hardly show any serial correlation, which, however, does not mean that they are independent. In fact, both daily squared and absolute returns often exhibit small and significant autocorrelations. Those autocorrelations are persistent for absolute returns, indicating possible long-memory properties. It is also noticeable that those autocorrelations become weaker and less persistent when the sampling interval is increased from a day, to a week to a month. See Figure 1.7.

(vii) *Leverage effect.* Asset returns are negatively correlated with the changes of their volatilities (Black 1976, Christie 1982). As asset prices decline, companies become more leveraged (debt to equity ratios increase) and riskier, and hence their stock prices become more volatile. On the other hand, when stock prices become more volatile, investors demand high returns and hence stock prices go down. Volatilities caused by price decline are typically larger than the appreciations due to declined volatilities. To examine the leverage effect, we use the VIX, which is a proxy of the implied volatility of a basket of at-money S&P 500 options maturity in one month, as the proxy of volatility of the S&P 500 index. Figure 1.10 shows the time series of VIX and S&P 500 index (left panel) and the returns of S&P 500 index against the change of volatilities (right panel). The leverage effect is strong, albeit VIX is not a prefect measure of the volatility of the S&P 500 index, involving the volatility risk premium (Ait-Sahalia, Fan, and Li, 2013)

1.3 Efficient markets hypothesis and statistical models for returns

The efficient markets hypothesis (EMH) in finance assumes that asset prices are fair, information is accessible for everybody and is assimilated rapidly to adjust prices, and people (including traders) are rational. Therefore price P_t incorporates all relevant information up to time t, individuals do not have comparative advantages in the acquisition of information. The price change $P_t - P_{t-1}$ is *only* due to the arrival of "news" between t and $t+1$. Hence individuals have no opportunities for making an investment with return greater than a fair payment for undertaking riskiness of the asset. A shorthand for the EHM: *the price is right*, and *there exist no arbitrage opportunities*.

The above describes the strong form of the EMH: security prices of traded assets reflect instantly all available information, public or private. A semi-

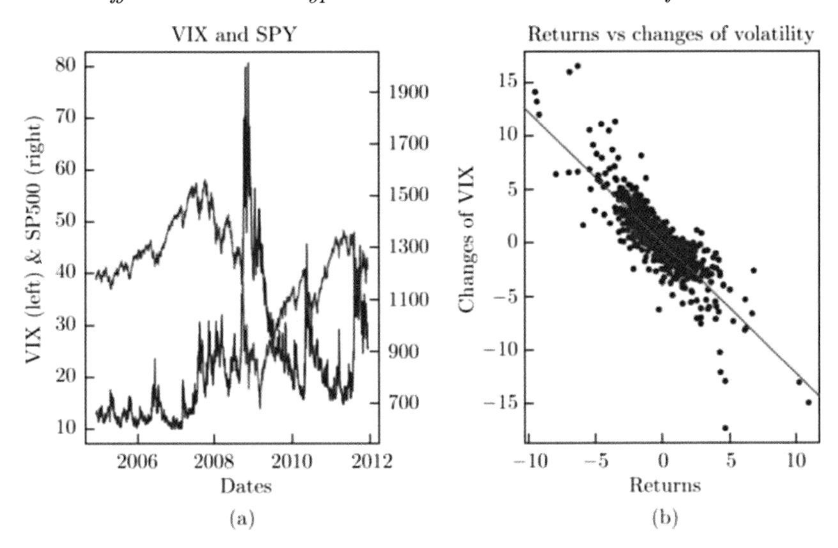

Figure 1.10 Time series plot of VIX (blue) and the S&P 500 index (red) in Nov. 29, 2004 to Dec. 14, 2011 (the left panel), and the plot of the daily S&P 500 returns (in percent) against the changes of VIX (the right panel).

strong form states that security prices reflect efficiently all public information, leaving rooms for the value of private information. The weak form merely assumes security prices reflect all past publicly available information.

Under the EHM, an asset return process may be expressed as

$$r_t = \mu_t + \varepsilon_t, \quad \varepsilon_t \sim (0, \sigma_t^2), \tag{1.10}$$

where μ_t is the *rational expectation* of r_t at time $t-1$, and ε_t represents the return due to unpredictable "news" which arrives between time $t-1$ and t. In this sense, ε_t is an *innovation* – a term used very often in time series literature. We use the notation $X \sim (\mu, \nu)$ to denote that a random variable X has mean μ and variance ν, respectively. The assumption that $E\varepsilon_t = 0$ reflects the belief that on average the actual change of log price equals the expectation μ_t.

By combining the EHM model (1.10) with some stylized features outlined in Section 1.2, most frequently used statistical models for financial returns admit the form

$$r_t = \mu + \varepsilon_t, \quad \varepsilon_t \sim \mathrm{WN}(0, \sigma^2), \tag{1.11}$$

where $\mu = Er_t$ is the expected return, which is assumed to be a constant. The notation $\varepsilon_t \sim \mathrm{WN}(0, \sigma^2)$ denotes that $\varepsilon_1, \varepsilon_2, \ldots$ form a white noise process with $E\varepsilon_t = 0$ and $\mathrm{var}(\varepsilon_t) = \sigma^2$; see (i) below. Here we assume that $\mathrm{var}(\varepsilon_t) = \mathrm{var}(r_t) = \sigma^2$ is a finite positive constant, noting that *most* varying volatilities in the return plots in Figures 1.3 and 1.4 can be represented by conditional heteroscadasticity under the martingale difference assumption (ii) below. The assumption (1.11) is reasonable, supported by the empirical analysis in Section 1.2.

There are three different types of assumptions about the innovations $\{\varepsilon_t\}$ in (1.11), from the weakest to the strongest.

(i) *White noise innovations*: $\{\varepsilon_t\}$ are white noise, written $\varepsilon_t \sim \mathrm{WN}(0, \sigma^2)$. Under this assumption, $\mathrm{Corr}(\varepsilon_t, \varepsilon_s) = 0$ for all $t \neq s$.

(ii) *Martingale difference innovations*: ε_t form a martingale difference sequence in the sense that for any t

$$E(\varepsilon_t | r_{t-1}, r_{t-2}, \ldots) = E(\varepsilon_t | \varepsilon_{t-1}, \varepsilon_{t-2}, \ldots) = 0. \tag{1.12}$$

One of the most frequently used format for martingale difference innovations is of the form

$$\varepsilon_t = \sigma_t \eta_t, \tag{1.13}$$

where $\eta_t \sim \mathrm{IID}(0, 1)$ (see (iii) below), and σ_t is a predictable volatility process, known at time $t - 1$, satisfying the condition

$$E(\sigma_t | r_{t-1}, r_{t-2}, \ldots) = \sigma_t.$$

Note that ARCH and GARCH processes are special cases of (1.13).

(iii) *IID innovations*: ε_t are independent and identically distributed, denoted as $\varepsilon_t \sim \mathrm{IID}(0, \sigma^2)$.

The assumption of IID innovations is the strongest. It implies that the innovations are martingale differences. On the other hand, if $\{\varepsilon_t\}$ satisfies (1.12), it holds that for any $t > s$,

$$\begin{aligned} \mathrm{cov}(\varepsilon_t, \varepsilon_s) &= E(\varepsilon_t \varepsilon_s) = E\{E(\varepsilon_t \varepsilon_s | \varepsilon_{t-1}, \varepsilon_{t-2}, \ldots)\} \\ &= E\{\varepsilon_s E(\varepsilon_t | \varepsilon_{t-1}, \varepsilon_{t-2}, \ldots)\} = 0. \end{aligned}$$

Hence, $\{\varepsilon_t\}$ is a white noise series. Therefore the relationship among the three types of innovations is as follows:

$$\text{IID} \Rightarrow \text{Martingale differences} \Rightarrow \text{White noise}.$$

The relationship is summarized in Figure 1.11.

The white noise assumption is widely observed in financial return data. It is consistent with the stylized features presented in Figures 1.7 and 1.8. It is implied by the EMH, as the existence of the non-zero correlation between ε_{t+1} and its lagged values leads to an improvement on the prediction for r_{t+1} over the rational expectation μ. This violates the hypothesis that ε_{t+1} is unpredictable at time t. To illustrate this point, suppose $\mathrm{Corr}(\varepsilon_{t+1}, \varepsilon_s) =$

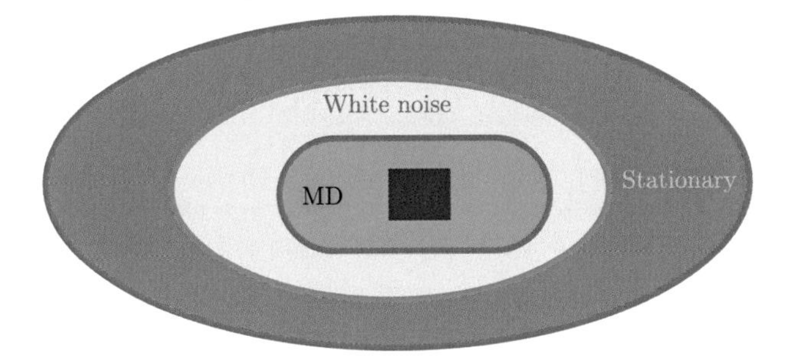

Figure 1.11 Relationship among different processes: Stationary processes are the largest set, followed by white noise, martingale difference (MD), and IID processes. There are many useful processes between stationary processes and white noise processes, to be detailed in Chapters 2 and 3.

$\rho \neq 0$ for some $s \leqslant t$. Then $\widetilde{r}_{t+1} = \mu + \rho(r_s - \mu)$ is a legitimate predictor for r_{t+1} at time t. Under the EMH, the fair predictor for r_{t+1} at time t is $\widehat{r}_{t+1} = \mu$. Then it is easy to see that

$$E\{(\widehat{r}_{t+1} - r_{t+1})^2\} = \text{var}(\varepsilon_{t+1}) = \sigma^2,$$

whereas

$$E\{(\widetilde{r}_{t+1} - r_{t+1})^2\} = E\{(\rho\varepsilon_s - \varepsilon_{t+1})^2\} = (1 - \rho^2)\sigma^2 < \sigma^2.$$

i.e. the mean squared predictive error of $\widetilde{r}_{t+1}$ is smaller. Hence the white noise assumption is appropriate and arguably necessary under the EMH. It states merely that the asset returns cannot be predicted by any linear rules. However it says nothing beyond the first two moments and remains silent on the question whether the traded asset returns can be predicted by nonlinear rules or by other complicated strategies.

On the other hand, the empirical evidence reported in Section 1.2 indicates that the IID assumption is too strong and too restrictive to be true in general. For example, the squared and the absolute returns of both S&P 500 index and the Apply stock exhibit significant serial correlations, indicating that $r_1, r_2, \ldots$, therefore also $\varepsilon_1, \varepsilon_2, \ldots$, are not independent with each other; see Figures 1.7 & 1.8.

Note that $r_t = \log(P_t/P_{t-1})$. It follows from (1.11) that

$$\log P_t = \mu + \log P_{t-1} + \varepsilon_t. \tag{1.14}$$

Hence under the assumption that the innovations ε_t are IID, the log prices $\log P_t$, $t = 1, 2, \ldots$ form a *random walk*, and the prices P_t, $t = 0, 1, 2, \ldots$, are a geometric random walk. Since the future is independent of the present and the past, the EMH holds in the most strict sense and nothing in the future can be predicted based on the available information up to the present. If we further assume ε_t to be normal, P_t follows a log normal distribution. Then the price process P_t, $t = 0, 1, 2, \ldots$, is a log normal geometric random walk. As the length of time unit shrinks to zero, the number of periods goes to infinity and the appropriately normalized random walk $\log P_t$ converges to a Brownian motion, and the geometric random walk P_t converges to a geometric Brownian motion under which the celebrated Black–Scholes formula is derived. The concept that stock market prices evolve according to a random walk can be traced back at least to French mathematician Louis Bachelier in his PhD dissertation in 1900.

A weaker form of random walks relaxes ε_t to be, for example, martingale differences. The martingale difference assumption offers a middle ground between white noise and IID. While retaining the white noise (i.e. the linear independence) property, it does not rule out the possibility of some nonlinear dependence, i.e. $\{r_t\}$ are uncorrelated but $\{r_t^2\}$ or $\{|r_t|\}$ may be dependent with each other. Under this assumption model (1.11) may accommodate conditional heteroscadasticity as in (1.13). In fact, many volatility models including ARCH, GARCH and stochastic volatility models are special cases of (1.11) and (1.13) with ε_t being martingale differences.

The martingale difference assumption retains the hypothesis that the innovation ε_{t+1} is unpredictable at time t at least as far as the point prediction is concerned. (Later we will learn that the interval predictions for ε_{t+1}, or more precisely, the risks ε_{t+1} may be better predicted by incorporating the information from its lagged values.) The best point predictor for r_{t+1} based on $r_t, r_{t-1}, \ldots$ is the conditional expectation

$$\widehat{r}_{t+1} = E(r_{t+1}|r_t, r_{t-1}, \ldots) = \mu + E(\varepsilon_{t+1}|\varepsilon_t, \varepsilon_{t-1}, \ldots) = \mu,$$

which is the fair expectation of r_{t+1} under the EMH. The last equality in the above expression is guaranteed by (1.12). We call $\widehat{r}_{t+1}$ as the best in the sense that it minimizes the mean squared predictive errors among all the point predictors based on $r_t, r_{t-1}, \ldots$. See Section 2.9.1 for additional details.

In summary, the *martingale hypothesis*, which postulates model (1.11) with martingale difference sequence $\{\varepsilon_t\}$, assures that the returns of assets cannot be predicted by any rules, but allow volatility to be predictable. It is the most appropriate mathematical form of the efficient market hypothesis.

1.4 Tests related to efficient markets hypothesis

One fundamental question in financial econometrics is if the efficient markets hypothesis is consistent with empirical data. One way to verify this hypothesis is to test if returns are predictable. In the sequel we will introduce two statistical tests which address this issue from different angles.

1.4.1 Tests for white noise

From the discussion in the previous section, we have learned that if returns are unpredictable, they should be at least white noise. On the other hand, the assumption that returns are IID is obviously too strong. The autocorrelations in squared and absolute returns shown in Figures 1.7 & 1.8 clearly indicate that returns at different times are not independent of each other. In spite of a large body of statistical tests for IID (e.g. see the rank-based test of Hallin and Puri (1988), and also see Section 2.2 of Campbell et al. (1997), we focus on testing white noise hypothesis, i.e. returns are linearly independent but may depend on each other in some nonlinear manners. The test for white noise is one of the oldest and the most important tests in statistics, as many testing problems in linear modelling may be transformed into a white noise test. There exist quite a few testing methods; see Section 7.4 of Fan and Yao (2003) and the references within. We introduce below a simple and frequently used omnibus test, i.e. *Ljung–Box portmanteau test*.

The linear dependence between r_t and r_{t-k} is comprehensively depicted by the correlation function between r_t and r_{t-k}:

$$\rho_k \equiv \text{Corr}(r_t, r_{t-k}) = \frac{\text{cov}(r_t, r_{t-k})}{\sqrt{\text{var}(r_t)\text{var}(r_{t-k})}}.$$

In fact if $\rho_k = 0$, r_t and r_{t-k} are linearly independent, and $\rho_k = \pm 1$ if and only if $r_t = a + br_{t-k}$ for some constants a and b. When $\{r_t\}$ is a white noise sequence, $\rho_k = 0$ for all $k \neq 0$. In practice, we do not know ρ_k. Based on observed returns $r_1, \ldots, r_T$, we use the estimator $\widehat{\rho}_k = \widehat{\gamma}_k/\widehat{\gamma}_0$ instead, where $\widehat{\gamma}_k$ is defined in (1.8). The Ljung–Box Q_m-statistic is defined as

$$Q_m = T(T+2)\sum_{j=1}^{m} \frac{1}{T-j}\widehat{\rho}_j^2, \tag{1.15}$$

where $m \geqslant 1$ is a prescribed integer. Note that Q_m is essentially a weighted sum of the squared sample ACF over the first m lags, though the weights are approximately the same when $T \gg m$. Intuitively we reject the white noise hypothesis for large values of Q_m. How large is large depends on the

Table 1.1 *P-values based on the Ljung–Box test for the S&P 500 data*

	m	1	6	12	24
returns	Q_m	2.101	5.149	8.958	14.080
	P-value	0.147	0.525	0.707	0.945
squared returns	Q_m	5.517	12.292	16.964	23.474
	P-value	0.019	0.056	0.151	0.492
absolute returns	Q_m	8.687	39.283	49.721	76.446
	P-value	0.003	0.000	0.000	0.000

theoretical distribution of Q_m under the null hypothesis, which turns out to be problematic; see below. In practice a chi-square approximation is used: For $\alpha \in (0,1)$, let $\chi^2_{\alpha,m}$ denote the top αth percentile of the χ^2-distribution with m degrees of freedom.

The Ljung–Box portmanteau test: Reject the hypothesis that $\{r_t\}$ is a white noise at the significance level α if $Q_m > \chi^2_{\alpha,m}$ or its P-value, computed as $P(Q > Q_m)$ with $Q \sim \chi^2_m$, is smaller than α.

In practice one needs to choose m in (1.15). Conventional wisdom suggests to use small m, as serial correlation is often at its strongest at small lags. This also avoids the large estimation errors in sample ACF at large lags and error accumulation issue in summation (1.15). However using too small m may miss the autocorrelations beyond the lag m. It is not uncommon in practice that the Ljung–Box test is performed simultaneously with different values of m.

The *R*-function to perform the Ljung–Box test is `Box.test(x, lag=m, type= "Ljung")`, where `x` is a data vector. The command `Box.test(x, lag=m, type= "Box")` performs the Box–Pierce test with the statistic Q^*_m.

We apply the Ljung–Box test to the monthly log returns of the S&P 500 index displayed in the bottom panel of Figure 1.3. The sample size is $T = 313$. The testing results are shown in Table 1.1. We cannot reject the white noise hypothesis for the log return data, as the tests with $m = 1, 6, 12$ and 24 are all not significant. In contrast, the tests for the absolute returns are statistically significant with the P-values not greater than 0.3%, indicating that the absolute returns are not white noise. The tests for the squared returns are less clearly cut with the smallest P-value 0.019 for $m = 1$ and the largest P-value 0.492 for $m = 24$. Indeed, the monthly data exhibits much weaker correlation than daily. See also the three panels on the right in Figure 1.7.

A different but related approach is to consider the normalized Q_m^*-statistic:

$$\frac{1}{\sqrt{2m}}\Big\{T\sum_{j=1}^{m}\widehat{\rho}(j)^2 - m\Big\}.$$

The asymptotic normality of this statistic under the condition that $m \to \infty$ and $m/T \to 0$ has been established for IID data by Hong (1996), for martingale differences by Hong and Lee (2003) (see also Durlauf (1991) and Deo (2000)), and for other non-IID white noise processes by Shao (2011), and Xiao and Wu (2011). However, those convergences are typically slow or very slow, resulting in the size distortation of the tests based on the asymptotic normality. In addition, as pointed out above, when j is large, ρ_j tends to be small. Therefore, including those terms $\hat{\rho}_j^2$ adds noises to the test statistic without increasing signals. How to choose a relevant m adds a further complication in using this approach. Horowitz et al.(2006) proposed a double blockwise bootstrap method to perform the tests with the statistic Q_m^* for non-IID white noise.

1.4.2 Remarks on the Ljung–Box test*

The chi-square approximation for the null distribution of Ljung–Box test statistic is based on the fact that when $\{r_t\}$ is an IID sequence, $\widehat{\rho}_1, \dots, \widehat{\rho}_m$ are asymptotically independent, and each of them has an asymptotic distribution $N(0, 1/T)$ (Theorem 2.8(iii) of Fan and Yao (2003)). Hence

$$Q_m^* \equiv T\sum_{j=1}^{m}\widehat{\rho}_j^2 \sim \chi_m^2 \quad \text{approximately for large } T.$$

Now, it is easy to see that Q_m is approximately the same as Q_m^* when T is large, since $(T+2)/(T-j) \approx 1$. Hence, it also follows χ_m^2-distribution under the null hypothesis. In fact Q_m^* is the test statistic proposed by Box and Pierce (1970). However Ljung and Box (1978) subsequently discovered that the χ^2-approximation to the distribution of Q_m^* is not always adequate even for T as large as 100. They suggest to use the statistic Q_m instead as its distribution is closer to χ_m^2. See also Davies, Triggs and Newbold (1977).

A more fundamental problem in applying the Ljung–Box test is that the statistic itself is defined to detecting the departure from white noise, but the asymptotic χ^2-distribution can only be justified under the IID assumption. Therefore, as formulated above, it should not be used to test the hypothesis that the returns are white noise but not IID, as then the asymptotic null distributions of $\widehat{\rho}(k)$ depend on the high moments of the underlying distribution of r_t. These asymptotic null distributions may typically be too complicated to be directly useful in the sense that the asymptotic null distributions of Q_m or Q^*_m may then not be of the known forms for fixed m; see, e.g. Romano and Thombs (1996). Unfortunately this problem also applies to most (if not all) other omnibus white noise tests.

One alternative is to impose an explicit assumption on the structure of white noise process (such as a GARCH structure), then some resampling methods may be employed to simulate the null distribution of Q_m. Furthermore if one is also willing to impose some assumptions on the parametric form of a possible departure from white noise, a likelihood ratio test can be employed, which is often more powerful than a omnibus nonparametric test, as the latter tries to detect the departure (from white noise) to all different possibilities. The analogy is that an all-purpose tool is typically less powerful on a particular task than a customized tool. An example of the customized tool is the Dicky–Fuller test in the next section. However it itself is a challenge to find relevant assumptions. This is why omnibus tests such as the Ljung–Box test are often used in practice, in spite of their potential problems in mispecifying significance levels.

1.4.3 Tests for random walks

Another way to test the EMH is to look at the random walk model (1.14) for log prices $X_t \equiv \log P_t$. In general we may impose an autoregressive model for the log prices:

$$X_t = \mu + \alpha X_{t-1} + \varepsilon_t. \tag{1.16}$$

To test the validity of model (1.14) is equivalent to testing the hypothesis $H_0 : \alpha = 1$ in the above model. This is a special case of the unit-root test which we will revisit again later. We introduce here the Dickey–Fuller test which in fact deals with three different cases: (i) the model (1.16) with a drift μ, (ii) the model without drift

$$X_t = \alpha X_{t-1} + \varepsilon_t, \tag{1.17}$$

and (iii) the model with both drift and a linear trend

$$X_t = \mu + \beta t + \alpha X_{t-1} + \varepsilon_t. \tag{1.18}$$

Based on observations $X_1, \ldots, X_T$, let $\widehat{\alpha}$ be the least squares estimator for α, and $\mathrm{SE}(\widehat{\alpha})$ be the standard error of $\widehat{\alpha}$. These can easily be obtained from any least-squares package. Then the Dickey–Fuller statistic is defined as

$$W = (\widehat{\alpha} - 1)/\mathrm{SE}(\widehat{\alpha}). \tag{1.19}$$

We reject $H_0 : \alpha = 1$ if W is smaller than a critical value determined by the significance level of the test and the distribution of W under H_0. The intuition behind this one-sided test may be understood as follows. This random walk test is only relevant when the evidence for $\alpha < 1$ is overwhelming. Then we reject $H_0 : \alpha = 1$ only if the statistical evidence is in favor of $H_1 : \alpha < 1$. The hypothesis H_1 implies that X_t is a stationary and causal process (see Section 2.2.2 below) for models (1.16) and (1.17), and, furthermore, the changes $\{X_t - X_{t-1}\}$ is an auto-correlated process. In the context of model (1.14), this implies that the returns $r_t = \log P_t - \log P_{t-1}$ are auto-correlated and, therefore, are not white noise. When $\alpha > 1$, the process X_t is explosive, which implies $r_t = \mu + \gamma \log P_{t-1} + \varepsilon_t$ for some positive constant γ. The latter equation has little bearing in modelling real financial prices except that it can be used as a tool for modeling financial bubbles; see Phillips and Yu (2011).

The least squares estimators $\widehat{\alpha}$ may easily be evaluated explicitly. For example, under (1.16), the least-squares estimate is

$$\widehat{\alpha} = \sum_{t=2}^{T} (X_t - \bar{X}_T)(X_{t-1} - \bar{X}_{T-1}) \Big/ \sum_{t=2}^{T} (X_{t-1} - \bar{X}_{T-1})^2,$$

where

$$\bar{X}_T = \frac{1}{T-1} \sum_{t=2}^{T} X_t, \qquad \bar{X}_{T-1} = \frac{1}{T-1} \sum_{t=2}^{T} X_{t-1}.$$

Furthermore let $\widehat{\mu}$ be the least squares estimator for μ in (1.16). Then

$$\mathrm{SE}(\widehat{\alpha})^2 = \frac{1}{T-3} \sum_{t=2}^{T} (X_t - \widehat{\mu} - \widehat{\alpha} X_{t-1})^2 \Big/ \sum_{t=2}^{T} (X_{t-1} - \bar{X}_{T-1})^2.$$

Under model (1.17),

$$\widehat{\alpha} = \frac{\sum_{t=2}^{T} X_t X_{t-1}}{\sum_{t=2}^{T} X_{t-1}^2}, \qquad \mathrm{SE}(\widehat{\alpha}) = \frac{\sum_{t=2}^{T} (X_t - \widehat{\alpha} X_{t-1})^2}{(T-2) \sum_{t=2}^{T} X_{t-1}^2}.$$

There also exists the Dickey–Fuller coefficient test, which is based on the test statistic $T(\hat{\alpha} - 1)$. The asymptotic null distributions are complicated, but can be tabulated. At significant level $\alpha = 0.05$, the critical values are -8.347 and -13.96 respectively for testing model (1.17) (without drift) and model (1.16) (with drift).

Table 1.2 *The critical values of the (augmented) Dickey–Fuller test*

Model	Significance level		
	10%	5%	1%
(1.17) or (2.66): no drift, no trend	−1.61	−1.95	−2.60
(1.16) or (2.67): drift, no trend	−2.25	−2.89	−3.51
(1.18) or (2.68): drift & trend	−3.15	−3.45	−4.04

Although the Dickey–Fuller statistic is of the form of a t-statistic (see (1.19)), t-distributions cannot be used for this test as all the three models under H_0 are nonstationary (see Section 2.1 below). In fact the Dickey–Fuller test statistic admits certain non-standard asymptotic null distributions and those distributions under models (1.16)–(1.18) are different from each other. Fortunately the quantiles or critical values of those distributions have been tabulated in many places; see, e.g. Fuller (1996). Table 1.2 lists the most frequently used critical values, evaluated by simulation with the sample size $T = 100$. Larger sample sizes will result in critical values that are slightly smaller in absolute value and smaller sample sizes will result in somewhat larger critical values.

The *R*-code "aDF.test.r" defines a function `aDF.test` which implements the (augmented) Dickey–Fuller test: `aDF.test(x, kind=i, k=0)`, where `x` is a data vector, and `i` should be set at 2 for model (1.16), 1 for model (1.17), and 3 for model (1.18).

We now apply the Dickey–Fuller test to the log daily, weekly, and monthly prices displayed in Figure 1.3. Since the returns (i.e. the differenced log prices) fluctuate around 0 and show no linear trend, we tend to carry out the test based on either model (1.16) or model (1.17). But for the illustration purpose, we also report the tests based on model (1.18). The P-values of the test with the three models for the daily, weekly, and monthly prices are listed below.

Model used	(1.17)	(1.16)	(1.18)
daily	> 0.9	0.392	0.646
weekly	> 0.9	0.336	0.698
monthly	> 0.9	0.413	0.791

Since none of those tests are statistically significant, we cannot reject the hypothesis that the log prices for the S&P 500 are random walk. This applies to daily, weekly, and also monthly data. We also repeat the above exercise for the daily, weekly and monthly returns (i.e. the differenced log prices),

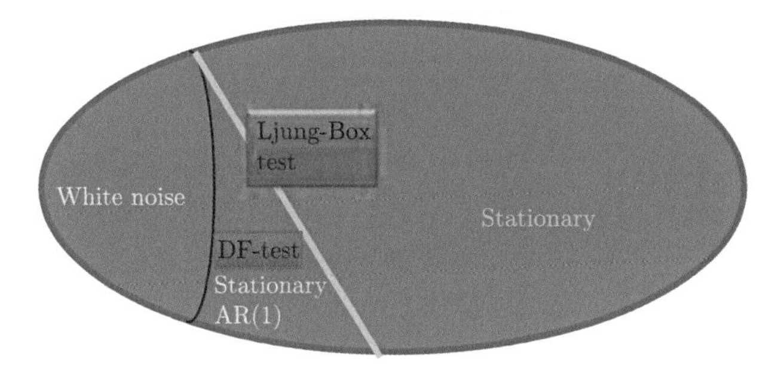

Figure 1.12 In terms of returns, the null hypotheses of both Ljung–Box and Dickey–Fuller tests are the same. However, the alternative of Ljung–Box is larger.

obtaining the P-values smaller than 0.01 for all the cases. This shows that the returns are not random walks across difference frequencies.

The Dickey–Fuller test was originally proposed in Dickey and Fuller (1979). It has been further adapted in handling the situations when there are some autoregressive terms in models(1.16)–(1.18); see Section 2.8.2 below.

1.4.4 Ljung–Box test and Dickey–Fuller test

Both Ljung–Box and Dickey–Fuller tests can be used to validate some aspects of efficient markets hypothesis. First of all, the input of Ljung–Box is the returns of assets, whereas the Dickey–Fuller test utilizes the log-prices. Secondly, the alternative hypothesis of Ljung–Box is nonparametric, which merely requires the stationary correlated processes, whereas the Dickey–Fuller test is designed to test against the parametric alternative hypothesis, which is a stationary AR(1) process. Thus, the Ljung–Box test is more omnibus whereas the Dickey–Fuller is more specific. Putting both in terms of asset returns, the null hypotheses of both problems are the same: returns behaves like uncorrelated white noise. However, the alternatives of the Ljung–Box is larger, as shown in Figure 1.12.

1.5 Appendix: Q–Q plot and Jarque–Bera test

1.5.1 Q–Q plot

A Q–Q plot is a graphical method for comparing two probability distribution functions based on their quantiles. It is particularly effective to reveal the differences in the tail-heaviness of the two distributions.

For any probability distribution function F and $\alpha \in (0, 1)$, the αth quantile of F is defined as

$$F^{-1}(\alpha) = \max\{x : F(x) \leqslant \alpha\}. \tag{1.20}$$

For any two probability distribution functions F and G, the quantile–quantile plot, or simply the *Q–Q plot*, of F and Q is a curve on a two-dimensional plane obtained by plotting $F^{-1}(\alpha)$ against $G^{-1}(\alpha)$ for $0 < \alpha < 1$.

It can be shown that if one distribution is a location-scale transformation of the other, i.e.

$$F(x) = \sigma^{-1}G\Big(\frac{x-\mu}{\sigma}\Big)$$

for some constant μ and $\sigma > 0$, their Q–Q plot is a straightline. (In fact the inverse is also true.) Hence it is useful to draw a straight line passing through the two inter-quarters of the quantiles to highlight the differences in the tails of the two distributions.

We illustrate the usefulness of a Q–Q plot by an example using the daily S&P 500 returns. The lower-left panel in Figure 1.5 is the Q–Q plot of F and G, where F is the standard normal distribution and G is the empirical distribution of the daily returns of S&P 500 index. It gives basically the scatter plot of pairs

$$\Big(F^{-1}\Big(\frac{i-0.5}{n}\Big), x_{(i)}\Big), \qquad i = 1, \ldots, n,$$

where $x_{(i)}$ is the ith smallest value of the data $\{x_i\}_{i=1}^n$, representing the empirical i/n-quantile, and $F^{-1}((i-0.5)/n)$ is its corresponding theoretical quantile modulus a location-scale transform. We do not use $F^{-1}(i/n)$ as its theoretical quantile to avoid $F^{-1}(i/n) = \infty$ for $i = n$. Different software has slightly different modifications from what is presented above, but the key idea of comparing the empirical quantiles with those of their referenced distribution remains the same. The blue straight line marks the position if the two distribution are identical under a location-scale transformation. The points on the left in the graph are the lower quantiles, corresponding to α close to 0 (see (1.20) above). Since those points are below the blue line, the lower quantiles of G (empirical quantiles) are smaller (i.e. negatively larger) than their expected values if G is a location-scale transformation of F. This means that the left tail of G is heavier than that of F, namely the daily returns of S&P 500 index have a heavier left tail than the normal distribution. Similarly, it can be concluded that the daily returns of S&P 500 index have a heavier right tail than the normal distribution. However, the daily returns of S&P 500 index have lighter tails than the t-distribution with degree of freedom 3, as shown in Figure 1.9.

Q–Q plots may be produced using the R-functions `qqplot`, `qqline` or `qqnorm`.

1.5.2 Jarque–Bera test

Q–Q plots check normality assumptions by informal graphical inspection. They are particularly powerful in revealing tail behavior of data. Formal tests can also be constructed. For example, one can employ the Kolmogorov–Smirnov test for testing normality. A popular test for normality is the Jarque–Bera test, which is defined as

$$\mathrm{JB} = \frac{n}{6}[S^2 + (K-3)^2/4]$$

where for a given sequence of data $\{x_i\}_{i=1}^n$,

$$S = \frac{\hat{\mu}_3}{\hat{\sigma}^3} = \frac{\sum_{i=1}^{n}(x_i - \bar{x})^3/n}{\left(\sum_{i=1}^{n}(x_i - \bar{x})^2/n\right)^{3/2}}$$

is the *sample skewness* and

$$K = \frac{\hat{\mu}_4}{\hat{\sigma}^4} = \frac{\sum_{i=1}^{n}(x_i - \bar{x})^4/n}{\left(\sum_{i=1}^{n}(x_i - \bar{x})^2/n\right)^{2}}$$

is the *sample kurtosis*. Therefore, the JB-statistic validates really only the skewness and kurtosis of normal distributions.

Under the null hypothesis that the data are drawn independently from a normal distribution, the asymptotic null distribution of the JB-statistic follows approximately χ_2^2-distribution. Therefore, the P-value can easily be computed by using χ_2^2-distribution.

1.6 Further reading and software implementation

The books that have strong impact on our writing are Fan and Yao (2003), and Campbell et al. (1997). The former emphasizes advanced theory and methods on nonlinear time series and has influenced our writing on the time series aspect. The latter emphasizes on the economic interpretation of econometric results of financial markets and has shaped our writing on the cross-sectional aspect of the book. Since preparing the first draft of the lecture notes on Financial Econometrics taught at Princeton University in 2004, a number of books on the subject have been published. For an introduction to financial statistics, see Ruppert (2004, 2010), Carmona (2004, 2013), and Franke et al. (2015). Tsay (2010, 2013) provide an excellent and comprehensive account on the analysis of financial time series. Gourieroux and Jasiak (2001) provide an excellent introduction to financial econometrics for those who are already familiar with econometric theory. For the financial econometrics with emphasis on investments, see Rachev et al. (2013).

Most of the computation in this book was carried out using the software package R, which is free and publicly available. See Section 2.10 for an introduction and installation. The books by Ruppert (2010), Carmona (2013), and Tsay (2013) are also implemented in R.

It is our hope that readers will be stimulated to use the methods described in this book for their own applications and research. Our aim is to provide information in sufficient detail so that readers can produce their own implementations. This will be a valuable exercise for students and readers who are new to the area. To assist this endeavor, we have placed all of the data sets and codes used in this book on the following web site:

`http://orfe.princeton.edu/~jqfan/fan/FinEcon.html`

1.7 Exercises

1.1 Download the daily, weekly and monthly prices for the Nasdaq index and the IBM stock from *Yahoo!Finance*. Reproduce Figures 1.3–1.8 using the Nasdaq index and the IBM stock data instead.

1.2 Consider a path dependent payoff function $Y_t = a_1 r_{t+1} + \cdots + a_k r_{t+k}$ where $\{a_i\}_{i=1}^k$ are given weights. If the return time series is weak stationary in the sense that $\text{cov}(r_t, r_{t+j}) = \gamma(j)$. Show that

$$\text{var}(Y_t) = \sum_{i=1}^{k} \sum_{j=1}^{k} a_i a_j \gamma_{i-j}.$$

A natural estimate of this variance is the following substitution estimator:

$$\hat{\text{var}}(Y_t) = \sum_{i=1}^{k} \sum_{j=1}^{k} a_i a_j \hat{\gamma}_{i-j},$$

where $\hat{\gamma}_{i-j}$ is defined by (1.8). Show that $\hat{\text{var}}(Y_t) \geqslant 0$.

1.3 Consider the following quote from Eugene Fama who was Myron Scholes' thesis adviser:

If the population of prices changes is strictly normal, on the average for any stock ... an observation more than five standard deviations from the mean

should be observed about once every 7000 years. In fact such observations seem to occur about once every three or four years.
(See Lowenstein, 2000, page 71.) For $X \sim N(\mu, \sigma^2)$, $P(|X - \mu| > 5\sigma) = 5.733 \times 10^{-7}$, deduce how many observations per year Fama was implicitly assuming to be made. If a year is defined as 252 trading days and daily returns are normal, how many years is it expected to take to get a 5 standard deviation event? How does the answer to the last question change when the daily returns follow the t-distribution with 4 degrees of freedom.

1.4 Is the (marginal) distribution of log-returns over a long time horizon (e.g. monthly or quarterly) close to normal? Explain briefly.

1.5 Generate a random sample of size 1000 from the t-distribution with ν degrees of freedom and another random sample of size 1000 from the standard normal distribution. Apply the Kolmogorov–Smirnov test to check if they come from the same distribution. Report the results for $\nu = 5$, 10, 15 and 20.

1.6 Report the P-values for appying the Jarque–Bera test to the data given in Exercise 1.1. What can you conclude based on these P-values?

1.7 Generate a random sample of size 100 from the t-distribution with ν degrees of freedom for $\nu = 5$, 15 and ∞ (i.e. normal distribution). Apply the Jarque–Bera test to check the normality and report the P-values.

1.8 According to the efficient market hypothesis, is the return of a portfolio predictable? Is the volatility of a portfolio predictable? State the most appropriate mathematical form of the efficient market hypothesis.

1.9 If the Ljung–Box test is employed to test the efficient market hypothesis, what null hypothesis is to be tested? If the autocorrelation for the first 4 lags of the monthly log-returns of the S&P 500 is

$$\hat{\rho}(1) = 0.2, \hat{\rho}(2) = -0.15, \hat{\rho}(3) = 0.25, \hat{\rho}(4) = 0.12$$

based on past 5 years data, is the efficient market hypothesis reasonable?

1.10 Generate 400 time series from the independent Gaussian white noise $\{r_t\}_{t=1}^T$ with $T = 100$. Compute

$$Z = \sqrt{T}\hat{\rho}(1), \quad Q_m, \quad Q_m^*$$

for $m = 3, 6$, and 12. Plot the histograms of Z, Q_3, Q_3^* and Q_6 and compare it with their asymptotic distributions. Report the first, fifth and tenth percentiles of the statistic $|Z_1|$, Q_3, Q_3^*, Q_6, Q_6^*, Q_{12} and Q_{12}^*, among 400 simulations and compare them with their theoretical (asymptotic) percentiles.

1.11 Repeat the experiment in Exercise 1.10 when $T = 400$ and r_t is generated from the t-distribution with degree of freedom 5.

1.12 What is the alternative hypothesis of the Dickey–Fuller test for the random walk? Suppose that based on last 120 quarterly data on the US GDP, it was computed that $\hat{\alpha} = 0.95$. Does the US GDP follow a random walk with a drift? Answer the question at 5% significance level (the critical value is –13.96) using the Dickey–Fuller coefficient test for the model with drift.

1.13 (*Implication of martingale hypothesis*). Let S_t be the price of an asset at time t. One version of the EMH assumes that the prices of any asset form a martingale process in the sense that

$$E(S_{t+1}|S_t, S_{t-1}, \ldots) = S_t, \quad \text{for all } t.$$

To understand the implication of this assumption, we consider the following simple investment strategy. With initial capital C_0 dollars, at the time t we hold α_t dollars in cash and β_t shares of an asset at the price S_t. Hence the value of our investment at time t is $C_t = \alpha_t + \beta_t S_t$. Suppose that our investment is self-financing in the sense that

$$C_{t+1} = \alpha_t + \beta_t S_{t+1} = \alpha_{t+1} + \beta_{t+1} S_{t+1},$$

and our investment strategy $(\alpha_{t+1}, \beta_{t+1})$ is entirely determined by the asset prices up to the time t. Show that if $\{S_t\}$ is a martingale process, there exist no strategies such that $C_{t+1} > C_t$ with probability 1.

2

Linear Time Series Models

Data obtained from observations collected sequentially over time are common in this information age. For example, we have collections on daily stock prices, weekly interest rates, monthly sales figures, quarterly consumer price indices (CPI) and annual gross domestic product (GDP) figures. Those data collected over time are called time series. The purpose of analyzing time series data is in general two-fold: to understand the stochastic mechanism that generates the data, and to predict or forecast the future values of a time series. This chapter introduces a class of linear time series models, or more precisely, a class of models which depict the linear features (including linear dependence) of time series. Those linear models and associated inference techniques provide the basic framework for the study of the linear dynamic structure of financial time series and for forecasting future values based on linear dependence structures.

2.1 Stationarity

One of the important aspects of time series analysis is to use the data collected in the past to forecast the future. How can historical data be useful for forecasting a future event? This is through the assumption of stationarity which refers to some time-invariance properties of the underlying process. For example, we may assume that the correlation between the returns of tomorrow and today is the same as those between any two successive days in the past. This enables us to aggregate the information from the data in the past to learn about the correlation. This correlation invariance over time is a typical characteristic of the so called *weak stationarity* or *covariance stationarity*. It facilitates linear prediction which is essentially based on the correlation between a predicated variable and its predictor (such as in linear regression). A stronger time-invariant assumption is that the joint distribution of the returns in a week in the future is the same as that in any weeks in the past. In other words, prediction is always based on some invariance properties over time, although the invariance may refer to some characteristics of the probability distribution of the process, or to the law governing the change of the distribution. We introduce the concept of stationarity more formally below.

A sequence of random variables $\{X_t,\ t = 0, \pm 1, \pm 2, \ldots\}$ is called a stochastic process and is served as a model for a set of observed time series data. It is convenient to refer to $\{X_t\}$ itself as a time series. It is known that the complete probability structure of $\{X_t\}$ is determined by the set of all the finite-dimensional distributions of $\{X_t\}$. Fortunately most linear features concerned depend on the first two moments of $\{X_t\}$, which are the main objects depicted in linear time series models. Of course if $\{X_t\}$ is a Gaussian process in the sense that all its finite-dimensional distributions are normal, the first two moments then determine the the probability structure of $\{X_t\}$ completely and $\{X_t\}$ is a linear process.

A time series $\{X_t\}$ is said to be *weakly stationary* (or *second-order stationary* or *covariance stationary*) if $E(X_t^2) < \infty$ and, for any integer k, neither EX_t nor $\mathrm{cov}(X_t, X_{t+k})$ depend on t.

For weakly stationary time series $\{X_t\}$, let $\mu = EX_t$ denote its common mean. We define the *autocovariance function* (ACVF) as

$$\gamma(k) = \mathrm{cov}(X_t, X_{t+k}) = E\{(X_t - \mu)(X_{t+k} - \mu)\}, \tag{2.1}$$

and the *autocorrelation function* (ACF) as

$$\rho(k) = \mathrm{Corr}(X_t, X_{t+k}) = \gamma(k)/\gamma(0) \tag{2.2}$$

for $k = 0, \pm 1, \pm 2, \ldots$. Note that $\gamma(0) = \mathrm{var}(X_t)$ is independent of t. For simplicity, we drop the adverb "weakly" and call $\{X_t\}$ *stationary* if it is weakly stationary, i.e. $\{X_t\}$ has finite and time-invariant first two moments. It is easy to see that $\rho(0) = 1$ and $\rho(k) = \rho(-k)$ for any stationary processes, and that the variance–covariance matrix of the vector $(X_t, \ldots, X_{t+k})$ is

$$\mathrm{var}(X_t, \ldots, X_{t+k}) = \begin{pmatrix} \gamma(0) & \gamma(1) & \gamma(2) & \cdots & \gamma(k-1) \\ \gamma(1) & \gamma(0) & \gamma(1) & \cdots & \gamma(k-2) \\ \vdots & \vdots & \vdots & & \vdots \\ \gamma(k-2) & \gamma(k-3) & \gamma(k-4) & \cdots & \gamma(1) \\ \gamma(k-1) & \gamma(k-2) & \gamma(k-3) & \cdots & \gamma(0) \end{pmatrix}.$$

Therefore, for any linear combinations,

$$\begin{aligned}\text{var}(\sum_{i=1}^{k} a_i X_{t+i}) &= \sum_{i=1}^{k}\sum_{j=1}^{k} a_i a_j \text{cov}(X_{t+i}, X_{t+k}) \\ &= \sum_{i=1}^{k}\sum_{j=1}^{k} a_i a_j \gamma(i-j) \geqslant 0.\end{aligned} \tag{2.3}$$

As such, the function $\gamma(\cdot)$ is referred to as the *semi-positive definite* function.

A very specific class of processes that plays a similar role to zero in the number theory is the *white noise*. When $\rho(k) = 0$ for any $k \neq 0$, $\{X_t\}$ is called a *white noise*, and is denoted by $X_t \sim \text{WN}(\mu, \sigma^2)$, where $\sigma^2 = \gamma(0) = \text{var}(X_t)$. In other words, a white noise is a sequence of uncorrelated random variables with the same mean and the same variance. White noise processes are building blocks for constructing general stationary processes.

In practice we use an observed sample $X_1, \ldots, X_T$ to estimate ACVF and ACF by the *sample ACVF* and *sample ACF*. They are basically the sample covariance and sample correlation of the lagged pairs $\{(X_{t-k}, X_k)\}_{t=k+1}^{T}$. Formally, they are defined as follows

$$\widehat{\gamma}(k) = \frac{1}{T}\sum_{t=k+1}^{T}(X_t - \bar{X})(X_{t-k} - \bar{X}), \qquad \widehat{\rho}(k) = \widehat{\gamma}(k)/\widehat{\gamma}(0), \tag{2.4}$$

where $\bar{X} = T^{-1}\sum_{1\leqslant t\leqslant T} X_t$. In the estimator $\widehat{\gamma}(k)$, we use the divisor T instead of $T-k$. This is a common practice adopted by almost all statistical packages. It ensures that the function $\hat{\gamma}(\cdot)$ is semi-positive definite (Exercise 2.2), a property given by (2.3). See Fan and Yao (2003) p. 42 for further discussion on this choice.

Weak stationarity is indeed a very weak notion of stationarity. For example, if $\{X_t\}$ is weakly stationary, it does not imply that $\{X_t^2\}$ is weakly stationary. Yet, the latter time series has very strong connections with the volatility of financial returns. Therefore, we need a stronger version of stationarity as follows.

A time series $\{X_t,\ t = 0, \pm1, \pm2, \ldots\}$ is said to be *strongly stationary* or *strictly stationary*if the k-dimensional distribution of $(X_1, \ldots, X_k)$ is the same as that of $(X_{t+1}, \ldots, X_{t+k})$ for any $k \geqslant 1$ and t.

This assumption is needed in the context of nonlinear prediction. Obviously the strict stationarity implies the (weak) stationarity provided $E(X_t^2) < \infty$. In addition, the strong stationarity of $\{X_t,\ t = 0, \pm1, \pm2, \ldots\}$ implies that of the time series $\{g(X_t),\ t = 0, \pm 1, \pm 2, \ldots\}$ is also strongly stationary for any function g.

2.2 Stationary ARMA models

One of the most frequently used time series models is the stationary autoregressive moving average (ARMA) model. It is frequently used in modeling the dynamics on returns of financial assets and other time series.

2.2.1 Moving average processes

Perhaps the simplest stationary time series are moving average (MA) processes. They also facilitate the computation of the autocovariance function easily. A simple example of this is the k-period return (1.6).

Let $\varepsilon_t \sim \text{WN}(0, \sigma^2)$. For a fixed integer $q \geqslant 1$, we write $X_t \sim \text{MA}(q)$ if X_t is defined as a moving average of q successive ε_t as follows:

$$X_t = \mu + \varepsilon_t + a_1 \varepsilon_{t-1} + \cdots + a_q \varepsilon_{t-q}, \tag{2.5}$$

where $\mu, a_1, \ldots, a_q$ are constant coefficients. In the above definition, ε_t stands for the innovation at time t, and the innovations $\varepsilon_t, \varepsilon_{t-1}, \ldots$ are unobservable. The intuition behind the moving average equation (2.5) may be understood as follows: innovation ε_t stands for the shock to the market at time t and X_t is the impact on the return from the innovations up to time t. The coefficient a_k is regarded as a "discount" factor on the k-lagged innovation ε_{t-k}. For example, for $a_k = b^k$ and $|b| < 1$, the impact of ε_t fades away exponentially over time.

In fact X_t defined by (2.5) is always stationary with $EX_t = \mu$, as the coefficients $a_1, \ldots, a_q$ do not vary over time. We first use a simple example to illustrate how to calculate ACVF and ACF for MA processes.

Example 2.1 For MA(1) model $X_t = \mu + \varepsilon_t + a\varepsilon_{t-1}$, it holds

$$\gamma(0) = \text{var}(X_t) = \text{var}(\varepsilon_t) + \text{var}(a\varepsilon_{t-1}) + 2\text{cov}(\varepsilon_t, a\varepsilon_{t-1}) = (1 + a^2)\sigma^2.$$

Similarly,

$$\gamma(1) = \text{cov}(X_t, X_{t-1}) = \text{cov}(\varepsilon_t + a\varepsilon_{t-1}, \varepsilon_{t-1} + a\varepsilon_{t-2}) = a\sigma^2,$$

since there is only one common term, ε_{t-1} in both X_t and X_{t-1}. Now for the ACVF of lag two, we have

$$X_t = \mu + \varepsilon_t + a\varepsilon_{t-1},$$

which does not share the same set of innovations $\{\varepsilon_t\}$ as

$$X_{t-2} = \mu + \varepsilon_{t-2} + a\varepsilon_{t-3}.$$

Therefore $\gamma(2) = 0$. Similarly, $\gamma(k) = 0$ for any $k \geqslant 2$. Consequently,

$$\rho(1) = a/(1 + a^2), \quad \rho(k) = 0 \text{ for any } |k| > 1. \tag{2.6}$$

Since $2|a| < 1 + a^2$, $|\rho(1)| \leqslant 0.5$ for any MA(1) process.

The above calculation can easily be extended to more general cases. For an MA(q) process defined as (2.5),

$$\begin{aligned} \text{var}(X_t) &= E\{(X_t - \mu)^2\} = E\{(\varepsilon_t + a_1\varepsilon_{t-1} + \cdots + a_q\varepsilon_{t-q})^2\} \\ &= \sigma^2(1 + a_1^2 + \cdots + a_q^2), \end{aligned} \tag{2.7}$$

which is independent of t. Furthermore, noticing the common white noise terms in both X_{t+1} and X_t are $\varepsilon_t, \ldots, \varepsilon_{t+1-q}$, we obtain

$$\begin{aligned} &\text{cov}(X_{t+1}, X_t) \\ =&E\{(\varepsilon_{t+1} + a_1\varepsilon_t + \cdots + a_q\varepsilon_{t-q+1})(\varepsilon_t + a_1\varepsilon_{t-1} + \cdots + a_q\varepsilon_{t-q})\} \\ =&\sigma^2(a_1 + a_2a_1 + \cdots + a_qa_{q-1}), \end{aligned}$$

which is also independent of t. In general, for any $1 \leqslant k \leqslant q$, the common white noise terms in both X_{t+k} and X_t are $\varepsilon_t, \ldots, \varepsilon_{t+k-q}$. Hence,

$$\text{cov}(X_{t+k}, X_t) = \sigma^2(a_k + a_{k+1}a_1 + \cdots + a_qa_{q-k}), \quad 0 \leqslant k \leqslant q. \tag{2.8}$$

We extend the above expression for $k = 0$ by adopting the notation $a_0 = 1$. For $k > q$, X_t and X_{t+k} are defined as the moving averages of two non-overlapping subsets of $\{\varepsilon_t\}$. Hence $\text{cov}(X_{t+k}, X_t) = 0$. Since the RHS of (2.8) is independent of t, X_t is always stationary. Hence, we conclude that any MA(q) process is stationary with ACF of the form

$$\rho(k) = \begin{cases} \dfrac{a_{|k|} + a_{|k|+1}a_1 + \cdots + a_qa_{q-|k|}}{1 + a_1^2 + \cdots + a_q^2} & 0 \leqslant |k| \leqslant q, \\ 0 & |k| > q. \end{cases} \tag{2.9}$$

The expression for $0 \leqslant |k| \leqslant q$ in (2.9) follows from (2.8), (2.7) and the symmetry of ACF. It is a trade-mark property for any MA(q) processes that their ACF cuts off at q, indicating that the (linear) memory of an MA(q) only lasts q time units.

For any MA(q) process, the ACF cuts off at q, i.e. $\rho(k) = 0$ for any $|k| > q$.

It is worth pointing out that the ACF depends on the coefficients a_j only, and is independent of the variance of white noise σ^2. An increase in the value of σ^2 leads to the increase of $\text{Var}(X_t)$, but will not change the correlations between X_t and its lagged values.

Figure 2.1 displays the plots of the four MA processes together with their sample ACF. For the two MA(1) models with $a = \pm 0.7$, the sample ACF $\widehat{\rho}(1)$, defined as in (2.6), are very close to the true values

$$\rho(1) = \pm 0.7/(1 + 0.49) = \pm 0.47$$

with the sample size $T = 100$; see the first two panels on the left in Figure 2.1. However $\widehat{\rho}(k) \neq 0$, in spite of $\rho(k) = 0$ for all $k \geqslant 2$. Note also that the sample ACF for the MA(2) and MA(4) in Figure 2.1 do not exhibit the clear cut-off property stated above. This is due to estimating errors caused by random fluctuations in finite samples. For an MA(q) process defined in (2.5) with $\varepsilon_t \sim \text{IID}(0, \sigma^2)$ and $E(\varepsilon_t^4) < \infty$, it holds that as $T \to \infty$,

$$\sqrt{T}\,\widehat{\rho}(k) \xrightarrow{D} N\big(0,\ 1 + 2\sum_{j=1}^{q} \rho(j)^2\big), \quad \text{for all } k > q. \tag{2.10}$$

See Theorem 2.8(iii) of Fan and Yao (2003). In the sample ACF plots in Figure 2.1, we superimpose two confidence bounds

$$\pm 1.96\{1 + 2\sum_{j=1}^{k-1} \rho(j)^2\}^{1/2}/\sqrt{T} \tag{2.11}$$

against k. Then for an MA(q) process, we expect all $\widehat{\rho}(k)$, for $k > q$, will be sandwiched by those two bounds (with approximate probability 0.95). See §2.10 at the end of this chapter on how to simulate those time series data and how to produce those figures using R.

Let us revisit a simple MA(1) process

$$X_t = \varepsilon_t + a\varepsilon_{t-1}. \tag{2.12}$$

Now we know that $\gamma(0) = \sigma^2(1 + a^2)$, $\gamma(1) = \sigma^2 a$ and $\gamma(k) = 0$ for all $k \geqslant 1$. To make this model identifiable in terms of its ACVF, we require $|a| \leqslant 1$ in (2.12). This rules out the model

$$Y_t = e_t + a^{-1}e_{t-1}, \quad e_t \sim \text{WN}(0, a^2\sigma^2)$$

when $|a| < 1$. It is easy to see that the ACVF of Y_t is exactly the same as that of X_t defined in (2.12). For an MA(q) defined in (2.5), we often in practice impose the following invertibility condition under which $a_1, \ldots, a_q$ are identifiable in terms of its ACF. Under the invertibility condition, we can recover the innovations $\varepsilon_t, \varepsilon_{t-1}, \ldots$ from the observations $X_t, X_{t-1}, \ldots$,

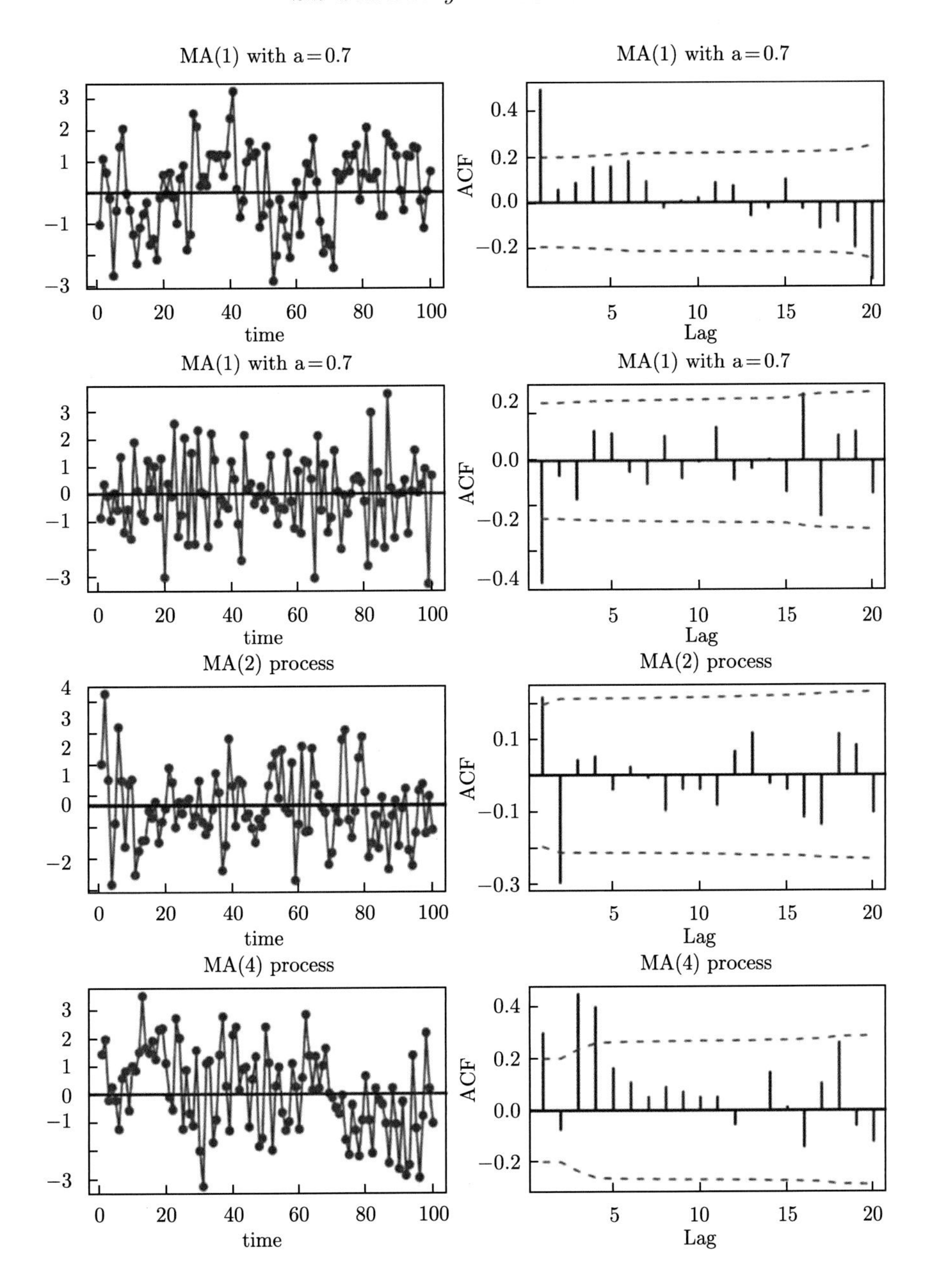

Figure 2.1 Time series plot and sample ACF plot of four moving-average processes of different orders (from top to bottom): MA(1): $X_t = \varepsilon_t + 0.7\varepsilon_{t-1}$, MA(1): $X_t = \varepsilon_t - 0.7\varepsilon_{t-1}$, MA(2): $X_t = \varepsilon_t + 0.7\varepsilon_{t-1} - 0.4\varepsilon_{t-2}$, and MA(4): $X_t = \varepsilon_t + 0.7\varepsilon_{t-1} - 0.4\varepsilon_{t-2} + 0.6\varepsilon_{t-3} + 0.8\varepsilon_{t-4}$, where ε_t are independent $N(0, 1)$. The horizontal line in each time series plot marks the position of mean EX_t. Two dashed curves in each ACF plot are the two confidence bounds (2.11) at lag k.

hence the name 'invertibility'. See also the discussion on the stationarity condition in Section 2.2.3.

The MA(q) process defined by (2.5) is said to be *invertible* if the q roots of the characteristic equation $1 + a_1 x + \cdots + a_q x^q = 0$ are outside of the unit circle.

In general we may consider an MA(∞) model

$$X_t = \mu + \varepsilon_t + \sum_{j=1}^{\infty} a_j \varepsilon_{t-j}, \tag{2.13}$$

where the coefficients a_j satisfy the condition $\sum_j a_j^2 < \infty$. (Then the infinite sum on the RHS of (2.13) is taken as the mean squared limit of the partial sums.) Under this condition, $EX_t = \mu$ and the ACVF of X_t admits the expression

$$\gamma(k) = \sigma^2 \sum_{j=0}^{\infty} a_j a_{j+|k|}, \quad k = 0, \pm 1, \pm 2, \ldots, \tag{2.14}$$

where $a_0 = 1$, which is derived in the same way as that leading to (2.8) by observing the common white noise terms. Indeed, it is an extension of (2.8) with $q = \infty$.

2.2.2 Autoregressive processes

A more intuitive model is the autoregressive (AR) model, which expresses explicitly the present value X_t as a linear regression of its lagged values with innovation ε_t as noise:

$$X_t = c + b_1 X_{t-1} + \cdots + b_p X_{t-p} + \varepsilon_t, \tag{2.15}$$

where $\varepsilon_t \sim \mathrm{WN}(0, \sigma^2)$, and $c, b_1, \ldots, b_p$ are parameters. We write $X_t \sim \mathrm{AR}(p)$. Obviously a predicted value for X_t based on its past is then $c + b_1 X_{t-1} + \cdots + b_p X_{t-p}$. This facilitates the task of forecasting. However, not all autoregressive models are stationary. For example, the AR(1) model $X_t = 2X_{t-1} + \varepsilon_t$ exhibits explosive behavior and is not expected to be stationary. When it is stationary, by taking expectation on both sides of (2.15), it is easy to see that

$$\mu \equiv EX_t = c/(1 - b_1 - \cdots - b_p).$$

Now (2.15) can be written as

$$X_t - \mu = b_1(X_{t-1} - \mu) + \cdots + b_p(X_{t-p} - \mu) + \varepsilon_t. \tag{2.16}$$

Below, we consider simple AR(1) model first, which also reveals the essence of the condition for stationarity AR processes.

Example 2.2 For an AR(1) model

$$X_t = bX_{t-1} + \varepsilon_t, \quad t = 0, \pm 1, \pm 2, \ldots, \tag{2.17}$$

where ε_t is assumed to be uncorrelated with X_{t-k} for all $k \geqslant 1$. This is intuitively reasonable, as innovation ε_t only enters the system at time t. Thus, by unfoiling the squared term on the right hand side of (2.17) and taking the expectation, we obtain

$$E(X_t^2) = b^2 E(X_{t-1}^2) + \sigma^2.$$

If $E(X_t^2) = E(X_{t-1}^2)$, a necessary condition for weak stationarity, then

$$E(X_t^2) = \sigma^2/(1 - b^2)$$

or $|b| < 1$. This shows that X_t can only be stationary if $|b| < 1$. Now we show that $|b| < 1$ is also the sufficient condition for (2.17) defining a stationary X_t.

When $|b| < 1$, the MA(∞) process $X_t = \sum_{j=0}^{\infty} b^j \varepsilon_{t-j}$ is stationary and satisfies (2.17), hence it is a stationary solution of (2.17). On the other hand, for any stationary X_t defined by (2.17), it follows from recursive substitution that

$$\begin{aligned} X_t &= \varepsilon_t + bX_{t-1} = \varepsilon_t + b(\varepsilon_{t-1} + bX_{t-2}) = \varepsilon_t + b\varepsilon_{t-1} + b^2 X_{t-2} \\ &= \varepsilon_t + b\varepsilon_{t-1} + \cdots + b^k \varepsilon_{t-k} + b^{k+1} X_{t-k-1} \to \sum_{j=0}^{\infty} b^j \varepsilon_{t-j} \end{aligned}$$

in mean square, since

$$\lim_{k\to\infty} E\Big\{\Big(X_t - \sum_{j=0}^{k} b^j \varepsilon_{t-j}\Big)^2\Big\} = \lim_{k\to\infty} |b|^{2(k+1)} E(X_{t-k-1}^2) = 0.$$

This shows that a stationary $AR(1)$ process is effectively MA(∞), hence its ACF does not cut off at a finite lag; see (2.14). This is an important characteristic distinguishing AR from MA processes.

The recursive substitution in Example 2.2 can be compactly represented by the backshift operator B defined as

$$BX_t = X_{t-1}, \quad B^k X_t = X_{t-k} \quad \text{for } k = \pm 1, \pm 2, \ldots.$$

Then model (2.17) can be written as $(1 - bB)X_t = \varepsilon_t$. As $(1 - bx)^{-1}$ admits the infinite series expansion

$$(1 - bx)^{-1} = 1 + bx + b^2x^2 + \cdots, \tag{2.18}$$

we may formally define $(1 - bB)^{-1} = 1 + bB + b^2B^2 + \cdots$. Then it holds that

$$(1 - bB)^{-1}(1 - bB) = 1.$$

Therefore,

$$\begin{aligned} X_t &= (1 - bB)^{-1}\varepsilon_t \\ &= (1 + bB + b^2B^2 + \cdots)\varepsilon_t \\ &= \varepsilon_t + b\varepsilon_{t-1} + b^2\varepsilon_{t-2} + \cdots, \end{aligned} \tag{2.19}$$

which is the stationary solution of (2.17).

Example 2.3 The stationary AR(1) exhibits the *mean-reversion* feature, which is very important for modeling interest rate dynamic and corresponds to the Vasicek (1977) model observed at discrete time. The mean is given $\mu = c/(1 - b_1)$ by (2.16). The AR(1) model can now be expressed as

$$X_t - X_{t-1} = -\kappa(X_{t-1} - \mu) + \varepsilon_t$$

where $\kappa = 1 - b_1 > 0$. Whenever X_{t-1} exceeds its mean, there is a negative force to drive it down (the change of the values are expected to be negative); whereas when X_{t-1} is below its mean μ, there is a positive force driving it up. Hence, the AR(1) model is appropriate for modeling the yield of bonds with mean reversion features.

The above idea can be extended to high-order AR processes. Using the backshift operator, AR(p) model (2.15) can be written as

$$b(B)X_t = c + \varepsilon_t, \tag{2.20}$$

where the characteristic polynomial $b(\cdot)$ is defined as

$$b(x) = 1 - b_1x - \cdots - b_px^p.$$

Let $\alpha_1^{-1}, \ldots, \alpha_p^{-1}$ be the roots of the equation $b(x) = 0$, i.e.

$$b(x) = (1 - \alpha_1x)\cdots(1 - \alpha_px).$$

Suppose $|\alpha_j| < 1$ for all $1 \leqslant j \leqslant p$. Then, using (2.18), $b(x)^{-1}$ may be expressed as

$$b(x)^{-1} = \prod_{j=1}^{p}(1-\alpha_j x)^{-1} = \prod_{j=1}^{p}(1+\alpha_j x+\alpha_j^2 x^2+\alpha_j^3 x^3+\cdots).$$

The above expression can further be expressed as

$$b(x)^{-1} = 1+\sum_{k=1}^{\infty} a_k x^k,$$

where a_k's are determined by $\alpha_1, \ldots, \alpha_p$, and $\sum_k |a_k| < \infty$. In fact

$$|a_k| \leqslant \big(\max_{1\leqslant j\leqslant p}|\alpha_j|\big)^k, \quad k=1,2,\ldots. \tag{2.21}$$

Now it follows from (2.20) that

$$\begin{aligned} X_t &= b(B)^{-1}(c+\varepsilon_t) \\ &= c+b(B)^{-1}\varepsilon_t = c+\varepsilon_t+\sum_{k=1}^{\infty} a_k\varepsilon_{t-k}, \end{aligned} \tag{2.22}$$

i.e. $X_t \sim$ MA(∞). Hence X_t is stationary. Furthermore, an AR(p) model with small p may offer a parsimonious representation for processes with memories last much longer than p.

The MA(∞) representation for a stationary AR process indicates an important fact that X_t depends on ε_t and its lagged values, and is uncorrelated with the future white noise terms. Such a process is also called a *causal process*. By symmetry, any invertible MA(q) is a stationary AR(∞).

To calculate the ACVF and ACF for a stationary AR process, we may use formula (2.14) based on its MA(∞) representation above. Although it is cumbersome to evaluate explicitly the moving average coefficients a_k from the autoregressive coefficients b_k, we can see from (2.14) and (2.21) that $\gamma(k)$ and, therefore, also $\rho(k)$ converge to 0 at an exponential (or geometric) rate as $k \to \infty$. In other words, AR processes can only be used for modeling *short memory* data. This is an important character of stationary AR processes, and can also be shown directly (see Exercise 2.5).

The AR(p) process defined by (2.15) is stationary if the p roots of the characteristic equation $1-b_1x-\cdots-b_px^p=0$ are outside of the unit circle (the modulus of all complex roots are bounded by 1). Furthermore the ACF of a stationary AR(p) process decays at an exponential rate, i.e. $\rho(k)=O(a^k)$ as $k\to\infty$ for some constant $a\in(0,1)$.

Example 2.4 Let us consider the following AR(2) model

$$X_t = 0.26 + 0.5X_{t-1} + 0.24X_{t-2} + \varepsilon_t, \quad \varepsilon_t \sim \text{WN}(0, \sigma^2). \tag{2.23}$$

Its characteristic function is

$$1 - 0.5x - 0.24x^2 = (1 - 0.8x)(1 + 0.3x),$$

which has both roots $1/0.8$ and $-1/0.3$, outside the unit circle. Hence the above equation defines a stationary and causal process. Taking the expectation on the both sides of the equation in (2.23), we obtain

$$\mu = EX_t = 0.26/(1 - 0.5 - 0.24) = 1.$$

We now calculate the ACVF of the process. Since the future noise and the past data are uncorrelated, i.e., $\text{cov}(X_{t-k}, \varepsilon_t) = 0$ for any $k \geqslant 1$, by (2.23), we have

$$\text{cov}(X_{t-k}, X_t - 0.26 - 0.5X_{t-1} - 0.24X_{t-2}) = 0.$$

This implies that

$$\gamma(k) - 0.5\gamma(k-1) - 0.24\gamma(k-2) = 0, \quad \text{for } k \geqslant 1, \tag{2.24}$$

which is the famous Yule–Walker equation. The initial value of the above iterative equations can be obtained as follows. For $k = 0$, $\text{cov}(X_t, \varepsilon_t) = \sigma^2$, we have

$$\gamma(0) - 0.5\gamma(-1) - 0.24\gamma(-2) = \sigma^2, \tag{2.25}$$

Recall that $\gamma(-k) = \gamma(k)$. Using (2.25) and (2.24) with $k = 1$ and 2, we obtain

$$\begin{cases} \gamma(0) = 0.5\gamma(1) + 0.24\gamma(2) + \sigma^2, \\ \gamma(1) = 0.5\gamma(0) + 0.24\gamma(1), \\ \gamma(2) = 0.5\gamma(1) + 0.24\gamma(0). \end{cases} \tag{2.26}$$

Solving the above three linear equations, we obtain

$$\gamma(0) = 1.871\sigma^2, \quad \gamma(1) = 1.231\sigma^2, \quad \gamma(2) = 1.064\sigma^2.$$

For $k \geqslant 3$, $\gamma(k)$ may be calculated recursively using (2.24).

The above example indicates an alternative way to compute ACVFs based on the *Yule–Walker equation*:

$$\gamma(k) = b_1\gamma(k-1) + \cdots + b_p\gamma(k-p), \quad k \geqslant 1. \tag{2.27}$$

This is obtained similarly to (2.24). Note that for $k = 0$, an extra σ^2 occurs on the RHS of the equation, which is analogous to (2.25). The initial $p + 1$

values of $\gamma(\cdot)$ can be obtained by a system of $(p+1)$ linear equations, similar to (2.26). The values $\gamma(k)$ for $k > p+1$ are then calculated recursively using Yule–Walker equation (2.27). The same equation to (2.27) holds for the ACF function. For example for AR(1) process,

$$X_t = bX_{t-1} + \varepsilon_t,$$

(2.27) reduces to equation

$$\rho(k) = b\rho(k-1) = b^k \rho(0) = b^k.$$

Hence, as calculated in Example 2.2, $\text{var}(X_t) = \sigma^2/(1-b^2)$.

The order of an MA process is characterized by its ACF in the sense that the ACF cuts off at q for an MA(q) process. The *partial autocorrelation function* (PACF) introduced below plays the same role for AR processes. Basically, the PACF at lag k, denoted by $\pi(k)$, is the conditional correlation between X_1 and X_{k+1} given the intermediate variables $X_2, \ldots, X_k$. The precise definition is somewhat technical, see, for example, §2.2.3 of Fan and Yao (2003). Here we present a characterization result which also paves a way for estimating PACF from data. For each given k, run the linear regression by minimizing

$$E(X_{k+1} - \beta_0 - \beta_1 X_k - \cdots - \beta_k X_1)^2 \tag{2.28}$$

with respect to $\boldsymbol{\beta}$. The regression coefficients depend on k and are denoted by b_{k0}, ..., b_{kk}. The PACF is then $\pi(k) = b_{kk}$, the last regression coefficient in the fit.

For the AR(p) model (2.15), when $k > p$, the best fit to (2.28) is obviously (see Exercise 2.6)

$$b_{k0} = c, b_{kj} = b_1, \ldots, b_{kp} = b_p, b_{k,p+1} = 0, \ldots, b_{kk} = 0.$$

Hence, $\pi(k) = 0$.

For a stationary AR(p) process, the PACF cuts off at p, i.e. $\pi(k) = 0$ for any $k > p$.

The *sample PACF* is defined by least squares estimation: $\widehat{\pi}(k) = \widehat{b}_{kk}$, where $(\widehat{b}_{k1}, \ldots, \widehat{b}_{kk})$ minimizes the sum

$$\sum_{t=k+1}^{T} (X_t - b_1 X_{t-1} - \cdots - b_k X_{t-k})^2.$$

This is clearly a sample version of (2.28). However, the sample PACF will not admit exactly zero cut-off. For a stationary AR(p) process defined by (2.15) with $\varepsilon_t \sim \text{IID}(0, \sigma^2)$ and $E(\varepsilon_t^4) < \infty$, it holds that as $T \to \infty$,

$$\sqrt{T}\,\widehat{\pi}(k) \xrightarrow{D} N(0,1), \quad \text{for any } k > p. \tag{2.29}$$

See Proposition 3.1 of Fan and Yao (2003).

Figure 2.2 displays the time series plots of four stationary AR series (with length $T = 100$) together with their sample PACF plots. For each PACF plot, we also superimpose the confidence bounds $\pm 1.96/\sqrt{T}$. By (2.29), we expect that for an AR(p) process, $\widehat{\pi}(k)$ for most $k > p$ are within those two bounds. Figure 2.2 shows that this criterion is adhered for all the four time series considered here.

Example 2.5 For the monthly returns of the Center for Research in Security Prices (CRSP) value-weighted index in January 1926 to December 1997($T = 864$), it can be computed that

Lag	1	2	3	4	5	6	7	8	9	10
PACF	0.11	−0.02	−0.12	0.04	0.07	−0.06	0.02	0.06	0.06	−0.01

Since the confidence bounds are $1.96/\sqrt{T} = 0.067$, we may select the order 5, since the $|\hat{\pi}(k)| < 0.066$ for $k > 5$. Note that $\hat{\pi}(5)$ is slightly larger than 0.66, we may also choose the order $p = 3$, resulting in a more parsimonious model.

2.2.3 *Autoregressive and moving average processes*

Combing AR and MA together, a general *autoregressive and moving average (ARMA) model* with the order (p, q) admits the form

$$X_t = c + b_1 X_{t-1} + \cdots + b_p X_{t-p} + \varepsilon_t + a_1 \varepsilon_{t-1} + \cdots + a_q \varepsilon_{t-q}, \tag{2.30}$$

where $\varepsilon_t \sim \text{WN}(0, \sigma^2)$, $c, b_1, \ldots, b_p, a_1, \ldots, a_q$ are parameters. Let

$$a(x) = 1 + a_1 x + \cdots + a_q x^q \quad \text{and} \quad b(x) = 1 - b_1 x - \cdots - b_p x^p.$$

Then, model (2.30) can be more compactly written as

$$b(B) X_t = a(B) \varepsilon_t.$$

To ensure that (p, q) are the genuine orders of the model, we assume that the two equations $a(x) = 0$ and $b(x) = 0$ do not have common roots. Otherwise, we can cancel the common factors from the above equation. Based on the discussions in Sections 2.2.1 and 2.2.2, the following assertions hold.

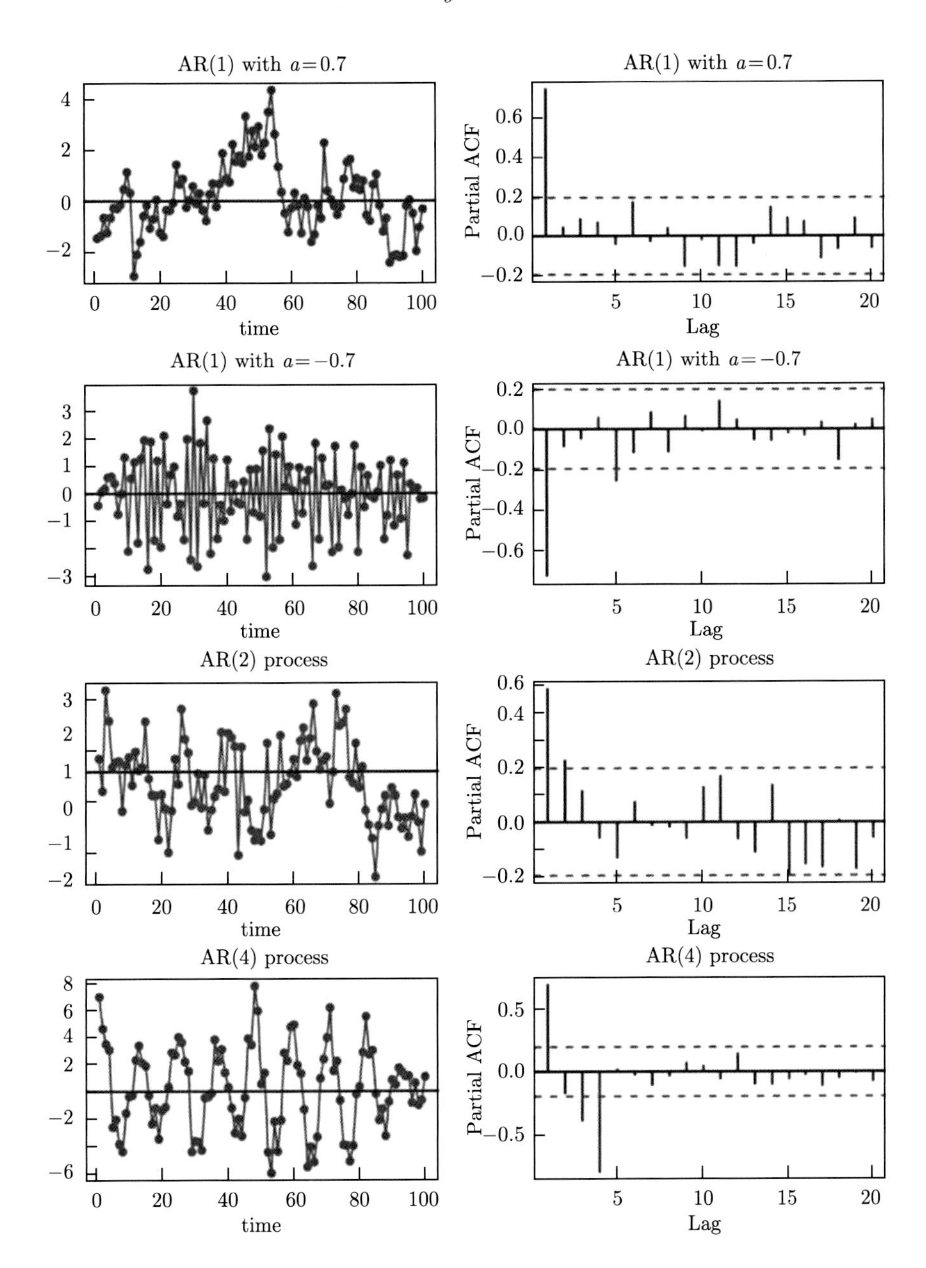

Figure 2.2 Time series plot and sample PACF plot of (from top to bottom) 4 autoregressive processes. AR(1): $X_t = 0.7X_{t-1} + \varepsilon_t$, AR(1): $X_t = -0.7X_{t-1} + \varepsilon_t$, AR(2): $X_t = 0.26 + 0.5X_{t-1} + 0.24X_{t-2} + \varepsilon_t$, and AR(4): $X_t = 0.5X_{t-1} + 0.24X_{t-2} + 0.2X_{t-3} - 0.8X_{t-4} + \varepsilon_t$, where ε_t are independent $N(0, 1)$ random variables. The horizontal line in each time series plot marks the position of the mean EX_t. The two dashed lines in each PACF plot are the confidence bounds $\pm 1.96/\sqrt{T}$.

Stationarity When the p roots of the AR characteristic equation $b(x) = 0$ are outside of the unit cycle, equation (2.30) defines a stationary process $\{X_t, t = 0, \pm 1, \pm 2, \ldots\}$ which admits an MA(∞) representation (see also (2.22) above.)

$$X_t = b(B)^{-1} a(B)\varepsilon_t \sim \text{MA}(\infty),$$

i.e. such a stationary solution is also causal. Furthermore $EX_t = c/(1 - b_1 - \cdots - b_p)$, and the ACF $\rho(k) \to 0$ at an exponential rate as $k \to \infty$. In other words, ARMA models are suitable for short memory processes.

Yule–Walker equation For the stationary process defined by (2.30), by multiplying $(X_{t-k} - \mu)$ on both sides and then taking the expectation with the use of $\gamma(k) = EX_t(X_{t-k} - \mu)$, we have

$$\gamma(k) = b_1\gamma(k-1) + \cdots + b_p\gamma(k-p), \quad \text{for any } k > q.$$

Here, the condition $k > q$ is imposed so that X_{t-k} is even more past than $\varepsilon_t, \ldots, \varepsilon_{t-q}$ so that they are uncorrelated. Dividing both sides by $\gamma(0)$, we have a similar equation for the ACF:

$$\rho(k) = b_1\rho(k-1) + \cdots + b_p\rho(k-p), \quad \text{for any } k > q. \tag{2.31}$$

From this equation, it can also be shown that the autocovariance function decays exponentially fast, $\rho(k) = O(a^k)$ where a is the reciprocal of the minimum modulus of the root of the characteristic function. See Exercise 2.5.

Invertibility When the q roots of the MA characteristic equation $a(x) = 0$ are outside of the unit cycle, the process $\{X_t\}$ defined by (2.30) is invertible in the sense that $X_t \sim \text{AR}(\infty)$.

Figure 2.3 presents the time plots for five stationary time series (with the length $T = 100$ for each of them). The top panel is for normal white noise $X_t \sim_{iid} N(1, 1.81)$. The other four are taken from the following ARMA(1,1) model

$$X_t - \mu = b(X_t - \mu) + \varepsilon_t + a\varepsilon_{t-1}, \quad \varepsilon_t \sim_{iid} N(0, 1),$$

with $(b, a) = (0, 0.9)$ for MA(1), $(b, a) = (0.669, 0)$ for AR(1), $(b, a) = (-0.669, 0)$ for another AR(1), and $(b, a) = (0.5, 0.279)$ for ARMA(1,1). We always set mean $\mu = E(X_t) = 1$. The values of a and b are specified such that the marginal distributions of X_t under the five different settings are the same. We add a horizontal line at 1 in each panel to mark the mean position of the series.

The top panel is patternless as the points fluctuate randomly around the

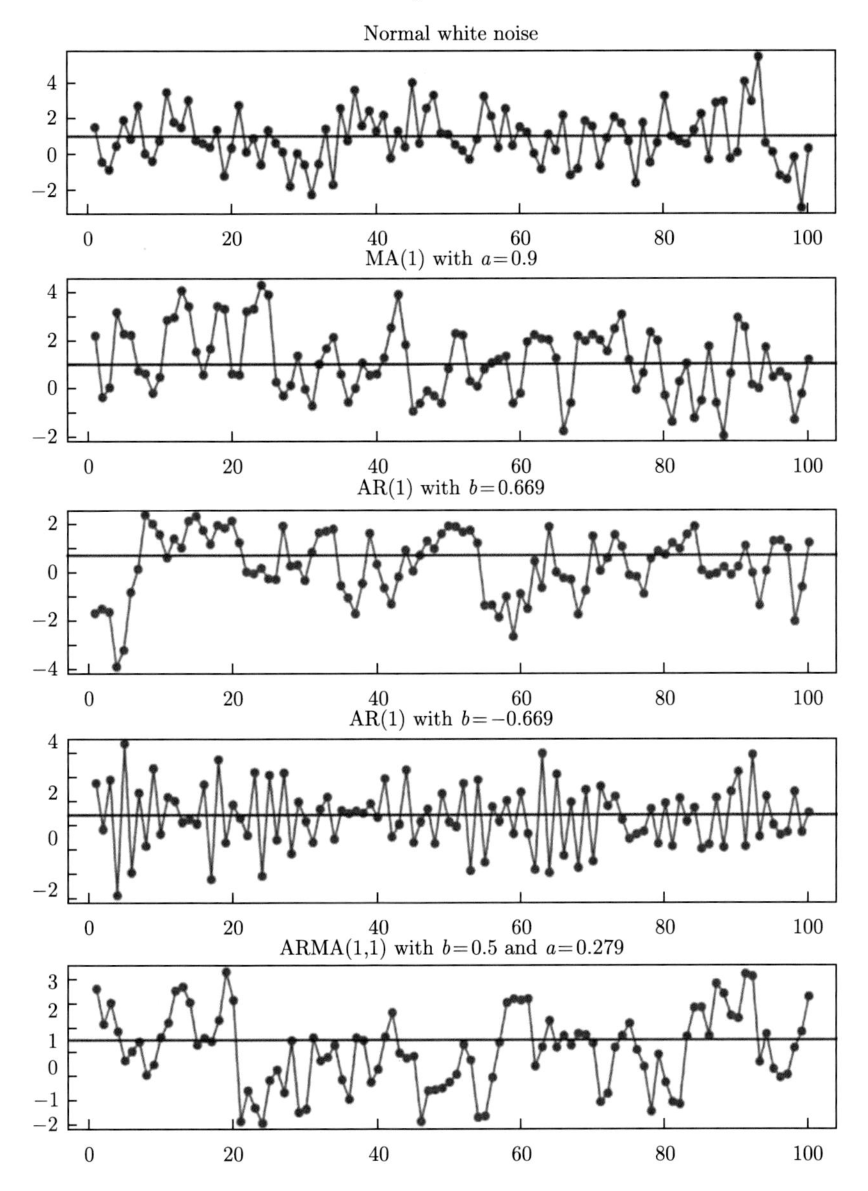

Figure 2.3 Plots of X_t against t for five stationary time series with the same marginal distribution $X_t \sim N(1, 1.81)$.

mean. In contrast the two successive points of MA(1) process in the second panel is positively correlated as $\rho(1) = 0.9/(1 + 0.9^2) = 0.497$, and there is the tendency that X_t and X_{t-1} stay on the same side of the mean. This tendency is much more pronounced for the AR(1) process with $b = 0.669$ in the 3rd panel and the ARMA(1, 1) process in the bottom panel. In both cases

X_t is effectively an MA(∞) process with positive coefficients, and it tends to take long excursions before it returns to the mean — a clear indication of serial dependence in the data. The 4th panel shows a different pattern of AR(1): since $b = -0.669$, X_t oscillates around the mean in such a way that X_t and X_{t-1} tend to take opposite sites of the mean.

If the evidence for the serial dependence in time plots is subtle, such evidence is much more apparent with both ACF and PACF plots. Figure 2.4 presents the sample ACF and sample PACF based on five time series data sets plotted in Figure 2.3.

An ACF plot plots $\widehat{\rho}(k)$ against k with two added varying confidence bounds at

$$\pm\frac{1.96}{\sqrt{T}}\left\{1+2\sum_{j=1}^{k-1}\widehat{\rho}(j)^2\right\}^{1/2}, \quad k=1,2,\ldots. \tag{2.32}$$

For each k, $\widehat{\rho}(k)$ should be sandwiched by the above two bounds with approximate probability 0.95 if $X_t \sim \text{MA}(k-1)$; see (2.10). Hence if indeed $X_t \sim \text{MA}(q)$, we expect that most $\widehat{\rho}(k)$ are between the above two bounds among 95% of $k > q$. Figure 2.4 shows that this is indeed the case. For MA(1), $\widehat{\rho}(k)$ is beyond the bounds only at lag $k = 1$; showing off the property that the ACF for MA(1) cuts off at lag 1. The sample ACFs for the two AR(1) and the ARMA(1,1) processes delay exponentially.

A PACF plot plots $\widehat{\pi}(k)$ against k with the two added constant confidence bounds at $\pm 1.96/\sqrt{T}$. When $X_t \sim \text{AR}(p)$, we expect that most $\widehat{\pi}(k)$ are sandwiched by those two bounds 95% of the time among all $k > p$; see (2.29). Figure 2.4 shows that for two AR(1) processes, $\widehat{\pi}(1)$ is beyond the bounds; indicating the property that the PACF for AR(1) cuts off at lag 1.

Example 2.6 We compute the autocovariance function for the ARMA(1,1) model. For $k > 1$, we have

$$\gamma(k) = b_1\gamma(k-1) = \cdots = b_1^{k-1}\gamma(1),$$

which decays exponentially. To determine the value $\gamma(0)$ and $\gamma(1)$, we use the fact that the present noise and the past data are uncorrelated to obtain

$$\gamma(1) = \text{Cov}(X_t, X_{t-1}) = \text{Cov}(b_1X_{t-1} + a_1\varepsilon_{t-1}, X_{t-1}) = b_1\gamma(0) + a_1\sigma^2,$$

noticing X_{t-1} contains a term ε_{t-1}. A similar calculation yields

$$\gamma(0) = \text{var}(X_t) = b_1^2\gamma(0) + (1+a_1^2)\sigma^2 + 2a_1b_1\sigma^2.$$

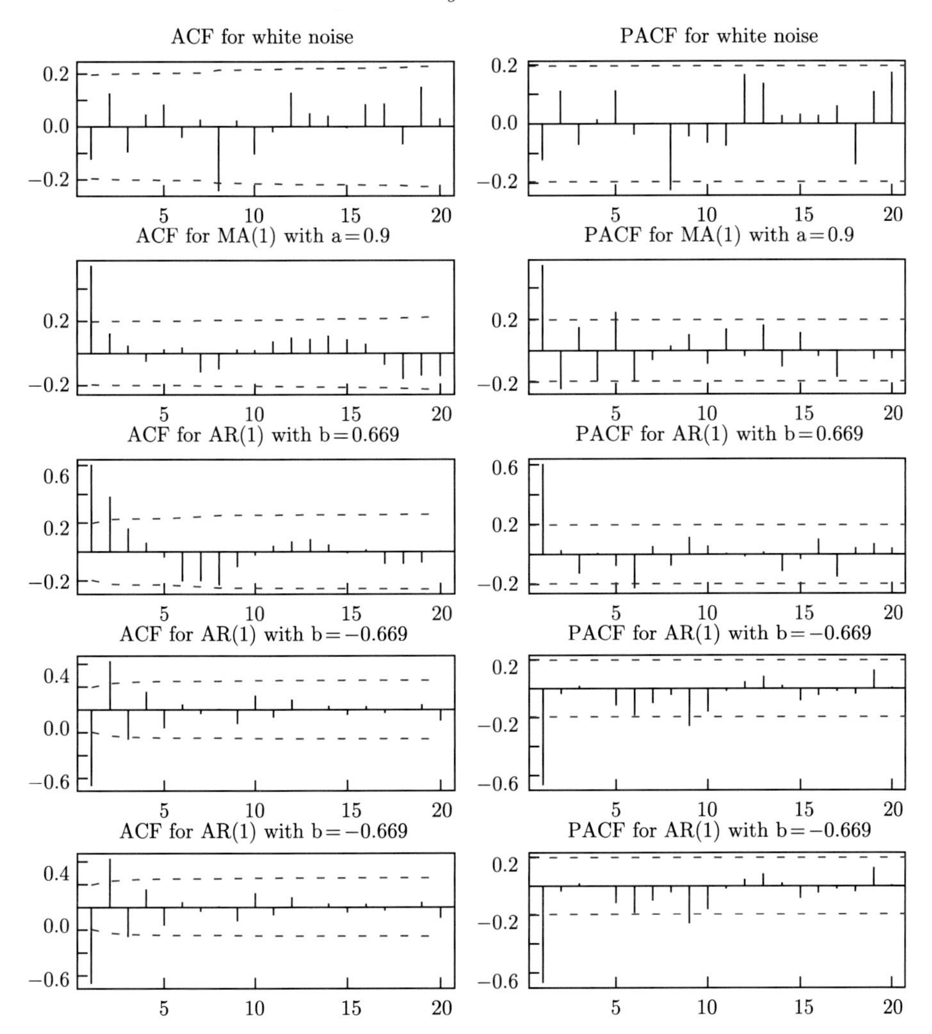

Figure 2.4 Sample ACF and PACF for five stationary time series plotted in Figure 2.3. Two dashed curves in each ACF plot are the confidence bounds (2.32), and in each PACF plot are the confidence bounds $\pm 1.96/\sqrt{T}$.

Solving the above two equations yield

$$\gamma(0) = \frac{1 + a_1^2 + 2a_1 b_1}{1 - b_1^2}\sigma^2, \qquad \gamma(1) = \frac{a_1 + b_1(1 + a_1^2 + a_1 b_1)}{1 - b_1^2}\sigma^2, \tag{2.33}$$

and

$$\gamma(k) = \gamma(1) b_1^{k-1}, \text{ for } k \geqslant 1. \tag{2.34}$$

Example 2.7 To gain further insight on ARMA models, let us express

ARMA(1,1) model in terms of AR(∞) model. Assuming $|a| < 1$, we have

$$a(B) = (1 - aB)^{-1} = 1 + aB + a^2B^2 + \cdots .$$

Using this and ARMA(1,1) equation (2.30), we have

$$(1 + aB + a^2B^2 + \cdots)(X_t - b_1X_{t-1} - c) = \varepsilon_t.$$

In other words,

$$X_t + aX_{t-1} + a^2X_{t-2} + \cdots - b_1(X_{t-1} + aX_{t-2} + \cdots) - \frac{c}{1-a} = \varepsilon_t.$$

Simplifying the above equation, we have

$$X_t = \frac{c}{1-a} + (b_1 - a)(X_{t-1} + aX_{t-2} + a^2X_{t-3} + \cdots) + \varepsilon_t.$$

In other words, the present value is the weighted average of the past values, with weights that decay exponentially. This kind of weighted average is referred to as the exponential smoothing.

*Stationarity, Causality and Stability**. In this book, we call an ARMA process, defined by (2.30), stationary if the roots of the corresponding AR characteristic equation $b(x) = 0$ are outside of the unit circle. In this case, the process is MA(∞), i.e. X_t is a linear combination of $\varepsilon_t, \varepsilon_{t-1}, \ldots$. Such a process is called *causal* as X_t is caused by the innovations at time t and its lagged values. In fact as long as no roots of equation $b(x) = 0$ are on the unit cycle, there exists a unique stationary solution $\{X_t,\, t = 0, \pm 1, \pm 2, \ldots\}$ from model (2.30) (see §2.1.2 of Fan and Yao (2003)), although we deal with *causal stationary processes* only in this book. The processes with unit roots (such as random walks and ARIMA models) are nonstationary, which will be discussed further in Section 2.8.1 below. Note that the aforementioned stationary solution is the process which satisfies model (2.30) for all $-\infty < t < \infty$. It should not be confused with a process starting with p arbitrary initial values $X_1, \ldots, X_p$. For the latter, $\{X_t,\, t > p\}$ defined by (2.30) is *stable* only if all the roots of $b(x) = 0$ are outside of the unit cycle. This stable process will merge together with the stationary process as t increases. On the other hand, if there are no unit roots but at least one root inside the the unit cycle, $\{X_t,\, t > p\}$ defined by (2.30) with arbitrary initial values $X_1, \ldots, X_p$ is *explosive*, and it will not converge to the stationary solution of (2.30).

So far, we have introduced a number of stationary models. To help us comprehend them, Figure 2.5 depicts their relationship. Stationary time series models are a large class, containing all processes with time-invariant

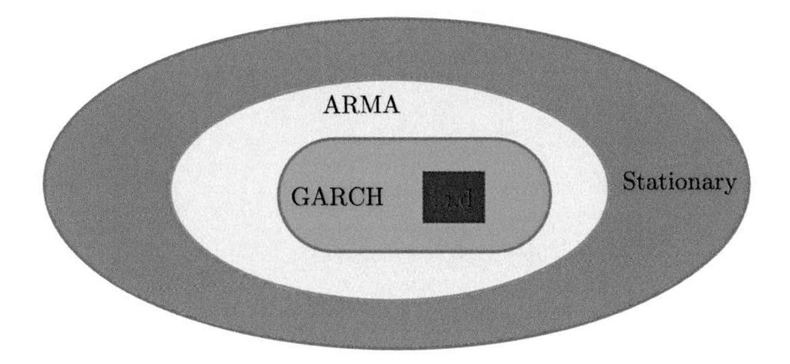

Figure 2.5 Relationship among different processes: Stationary processes are the largest set, followed by ARMA, ARMA-GARCH, and i.i.d. white noises processes.

second moments. We specify a subclass of useful models, namely stationary ARMA models. Within the context of ARMA models, no attempts are made to depict the structure of a white noise process beyond its first two moments. However GARCH processes introduced in the next chapter are white noise. The conditional variances of GARCH processes play important roles in understanding the volatility of financial time series. Of course all i.i.d. sequences form a subset of GARCH processes, which exhibit no serial dependence at all.

2.3 Nonstationary and long memory ARMA processes

The stationarity requires some time-invariable properties for time series, which is not always observed in practice. For example, according to the efficient market hypothesis, the logarithm of the price processes behaves like a random walk.

2.3.1 Random walks

A random walk without drift [see also (1.17)] is defined as

$$X_t = X_{t-1} + \varepsilon_t, \quad t = 1, 2, \ldots, \tag{2.35}$$

where $\varepsilon_t \sim \text{WN}(0, \sigma^2)$. Hence a random walk is an AR(1) model with the AR coefficient equal to 1, whose characteristic equation $b(x) = 1 - x$ has a unit root, i.e. a root on the unit cycle. We may assume that the process starts at an arbitrary point $X_0 = c$, then

$$\text{var}(X_t) = \text{var}(X_{t-1} + \varepsilon_t) = \text{var}(\varepsilon_1 + \cdots + \varepsilon_t) = t\sigma^2,$$

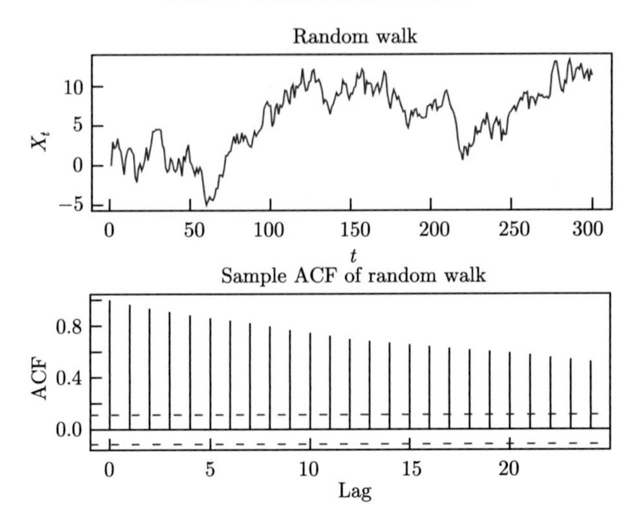

Figure 2.6 Time series plot and ACF of a random walk.

which increases linearly with time t. Therefore X_t is not stationary. A random walk is a mathematical model for the position of a 'drunk' who walks a random step on a line with size ε_t at time t. When $\{\varepsilon_t\}$ are independent and normally distributed, it is a discrete version of Brownian motion. Although the structure of random walks is extremely simple, it is one of the most frequently used model for some nonstationary components of time series with complex structures.

Figure 2.6 displays a random walk of length 300 generated from (2.35) with standard normal ε_t, together with its sample ACF. The time series plot resembles patterns of financial asset prices. See, for example, the top panel in Figures 1.3 and 1.3. As X_t is not stationary, the autocorrelation $\text{Corr}(X_t, X_{t+k})$ depends on both t and k. Since $X_t = c + \sum_{1 \leqslant j \leqslant t} \varepsilon_t$,

$$\text{cov}(X_t, X_{t+k}) = \text{var}(X_t) = t\sigma^2, \qquad \text{for all } k \geqslant 0.$$

Hence, it is easy to see for $k \geqslant 1$ that

$$\text{Corr}(X_t, X_{t+k}) = t\sigma^2 / \sqrt{t\sigma^2\,(t+k)\sigma^2} = \{t/(t+k)\}^{1/2}, \tag{2.36}$$

which is close to 1 for small k and large t. This explains why the sample ACF, defined as in (2.3), is close to 1 at small time lags, and decays slowly when the lag increases; see the second panel in Figure 2.6. This is in fact a common feature for nonstationary processes with unit roots. Note that differenced random walk $\nabla X_t = X_t - X_{t-1}$ is white noise, where ∇ denotes the difference operator, and for any integer $d \geqslant 2$, $\nabla^d X_t \equiv \nabla(\nabla^{d-1} X_t)$.

2.3.2 ARIMA model and exponential smoothing

For a nonstationary time series such as the log-prices of financial assets, one often takes the difference and hopes the resulting time series is stationary. *Autoregressive integrated moving average (ARIMA) models* form a useful class of nonstationary models. We say $X_t \sim \text{ARIMA}(p, d, q)$ if $\nabla^d X_t$ is stationary ARMA(p, q). Of course the models with small p, d, q are practically more useful. One such an example is ARIMA(0,1,1) model

$$X_t - X_{t-1} = \varepsilon_t - \theta\varepsilon_{t-1}, \quad \varepsilon_t \sim \text{WN}(0, \sigma^2), \tag{2.37}$$

where $|\theta| < 1$. Although differenced X_t is MA(1) which has only one time lag memory, the sample ACF of X_t exhibits the similar pattern as in the second panel of Figure 2.6. Note (2.37) can be written as

$$\varepsilon_t = (1 - \theta B)^{-1}(1 - B)X_t = \{1 - (1 - \theta)(B + \theta B^2 + \theta^2 B^3 + \cdots)\}X_t.$$

Hence,

$$X_t = (1 - \theta)(X_{t-1} + \theta X_{t-2} + \theta^2 X_{t-3} + \cdots) + \varepsilon_t. \tag{2.38}$$

This is in the form of an AR(∞) with coefficients $(1 - \theta)\theta^k$ decaying exponentially. Hence at time t, the "best" predictor for the future value X_{t+1} is

$$\widehat{X}_{t+1} = (1 - \theta)(X_t + \theta X_{t-1} + \theta^2 X_{t-2} + \theta_3^3 X_{t-3} + \cdots), \tag{2.39}$$

i.e. $\widehat{X}_{t+1}$ is the weighted average of the past values $X_t, X_{t-1}, \ldots$ with larger weights on more recent values. Note $\sum_{j\geqslant 0}(1 - \theta)\theta^j = 1$, i.e. all the weights sum up to 1. Predictor (2.39) is the celebrated *exponential smoothing*. It has been widely used in practice. We discuss its application in forecasting asset prices in Section 2.9.2 below.

Figure 2.7 plots the daily yields of a basket of high-yield (HY) bonds with average duration 4.4 years. The yields themselves do not look stationary. Their sample ACF is close to 1 up to lags over 30 and decays very slowly. The differenced log yields may (or may not) be stationary. Their sample ACF shows a small but significant negative correlation at lag 1; suggesting ARIMA(0,1,1) model the yields. Note that the negative correlation at lag 1 may be caused by the differencing, as both ∇X_t and ∇X_{t-1} contain X_{t-1} with opposite signs.

2.3.3 FARIMA model and long memory processes*

For stationary ARMA models, its autocorrelation decays exponentially fast. There also exists a class of stationary time series whose autocorrelation

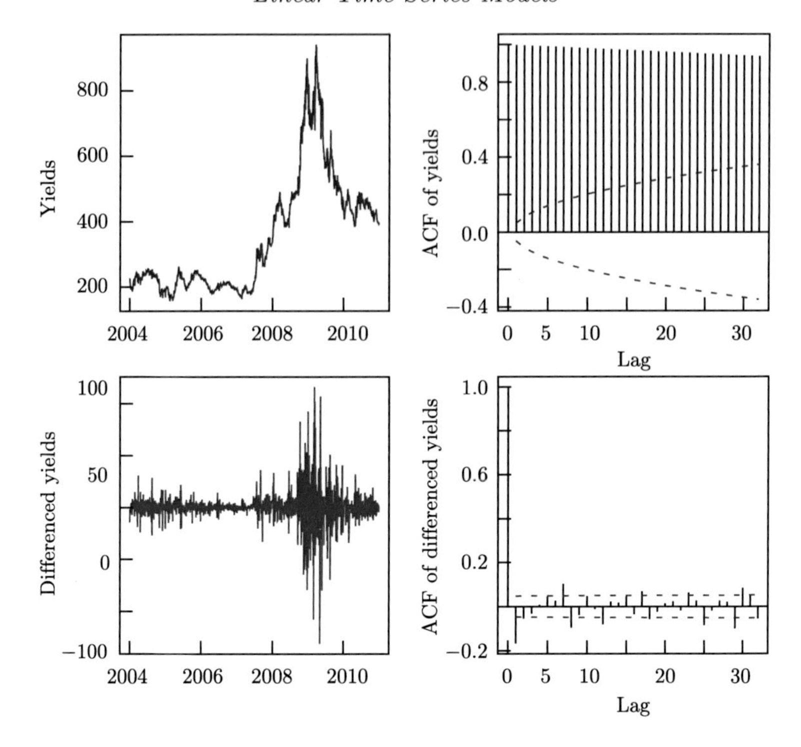

Figure 2.7 The daily yields of a basket HY bonds in 2004–2010 and their differences: time series and sample ACF plots. Two dashed curves in each sample ACF plots are the confidence bounds (2.32).

function decays algebraically:

$$\rho(k) \sim Ck^{2d-1}, \quad \text{as} \quad k \to \infty, \quad d < 0.5.$$

For $d \in (0, 0.5), \sum |\rho(k)| = \infty$. Such a process is called a *long memory process*. As shown in Figures 1.7 and 1.8, the autocorrelation functions of the absolute daily returns of S&P 500 and Apple stock exhibit long-memory features. The same holds true for the log-prices. See for example Figure 2.16.

A family of models with $\rho(k) \sim Ck^{2d-1}$ is given by the *fractional difference*

$$(1 - B)^d X_t = \varepsilon_t, \quad -0.5 < d < 0.5,$$

where the fractional difference is defined through the infinite series expansion:

$$(1 - B)^d = 1 - dB + \frac{d(1-d)}{2!}B^2 - \cdots - (-1)^k \frac{d(1-d)\cdots(k-1-d)}{k!}B^k - \cdots.$$

Thus, X_t admits an AR(∞) representation

$$X_t - dX_{t-1} + \cdots + (-1)^k \frac{d(1-d)\cdots(k-1-d)}{k!} X_{t-k} - \cdots = \varepsilon_t.$$

For the process defined above, it has the following properties (see Hosking, 1981):

(i) ACF: $\rho(k) = \dfrac{d(1+d)\cdots(k-1+d)}{(1-d)(2-d)\cdots(k-d)} \sim Ck^{2d-1}, \quad k \to \infty.$
(ii) PACF: $\quad \pi(k) = d/(k-d).$
(iii) Spectral density: The Fourier transform of the ACF admits

$$f(\omega) \sim w^{-2d}, \quad \text{as} \quad \omega \to 0.$$

Therefore, d can be estimated from log-periodogram with small ω.

The first property confirms the long memory feature induced by the fractional difference. In general, if $\{X_t\}$ satisfies

$$b(B)(1-B)^d X_t = a(B)\varepsilon_t,$$

then it is called a FARIMA(p,d,q), fractional ARIMA model. Note that

$$b(B)X_t = a(B)\eta_t,$$

where $\eta_t = (1-B)^{-d}\varepsilon_t$ is a long-memory process. Thus, FARIMA can be viewed as an ARMA model driven by a long memory noise.

2.3.4 Summary of time series models

Models for time series can be classified into stationary and nonstationary ones. Both classes are large. In terms of time series modelling, we typically use some interesting subsets from those classes. Figure 2.8 gives a schematic overview of different models.

A useful specification of stationary time series models is the family of ARMA models. They are used to model time series data with short memory. FARIMA models are used to model stationary time series with long memories.

There are also various specifications of nonstationary time series models. An example of this is ARIMA models, which include random walks as a specific example. Other examples are given in Section 2.8. Time series with seasonality and time trend can not be stationary. See Section 2.8 for the techniques to remove time and seasonal trends.

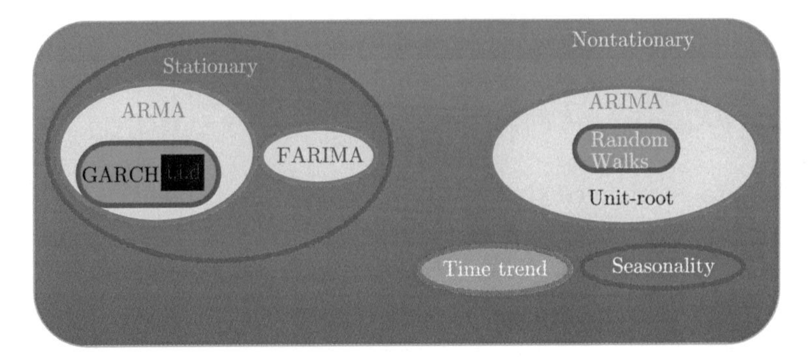

Figure 2.8 A schematic overview of time series models.

2.4 Model selection using ACF, PACF and EACF*

From the previous section, we have learned that both ACF and PACF provide effective tools in identifying pure MA(q) or pure AR(p) models. We introduce below the *extended autocorrelation function* (EACF) which plays a similar role in determining the orders of ARMA(p, q) models.

The EACF was proposed by Tsay and Tiao (1984, 1985). Its basic idea is simple. Let X_t be a stationary ARMA(p, q) process. For the simplicity, we assume that $EX_t = 0$. Then

$$Z_t \equiv X_t - b_1 X_{t-1} - \cdots - b_p X_{t-p} \sim \text{MA}(q).$$

Hence the ACF of "residual" process Z_t cuts off at q. The autoregressive coefficients $b_1, \ldots, b_p$ can be estimated by Yule–Walker equation (2.31). Since p is unknown, we try out with the AR-order $\ell = 0, 1, 2, \ldots$ in order. Consequently an EACF is defined for each non-negative integer ℓ, and the ℓth EACF is denoted by $\rho(\cdot\,; \ell)$, in which $\rho(k; \ell)$ is the ACF function of the residuals after fitting an AR model with order ℓ. Obviously $\rho(\cdot\,; 0) \equiv \rho(\cdot)$ is the original ACF.

Now we define the ℓth EACF $\rho(\cdot\,; \ell)$ for a fixed $\ell \geqslant 0$. It is a function of the ACF $\rho(\cdot)$. Since the value of q is unknown, we have also to try different potential values of k in order to properly use (2.31). For $k = 0, 1, 2, \ldots$, let $b_1^{(k)}, \ldots, b_\ell^{(k)}$ be the solutions of the following ℓ Yule–Walker equations:

$$\rho(k+j) = b_1^{(k)} \rho(k+j-1) + \cdots + b_\ell^{(k)} \rho(k+j-\ell), \quad j = 1, \ldots, \ell \tag{2.40}$$

(the solution depends also on ℓ, but its dependence is suppressed) and

$$Z_t^{(k)} = X_t - b_1^{(k)} X_{t-1} - \cdots - b_\ell^{(k)} X_{t-\ell}, \quad t = 0, \pm 1, \pm 2, \ldots \tag{2.41}$$

be the residuals after the AR(ℓ) fit. If $\ell = p$ the true AR-order of the model, it follows from (2.31) that

$$b_i^{(k)} = b_i, \quad \text{for } i = 1, \ldots, p, \text{ for all } k > q.$$

Hence $\{Z_t^{(k)}\}$ is then an MA(q) process for all $k > q$. Now the ℓth *extended autocorrelation function* (EACF) at lag k is defined as

$$\rho(k; \ell) = \text{ACF of } \{Z_t^{(k)},\ t = 0, \pm 1, \pm 2, \ldots\} \text{ at lag } k. \tag{2.42}$$

Then the assertion below holds.

The pth EACF $\rho(\cdot; p)$ cuts off at q for stationary ARMA(p, q) processes, i.e.

$$\rho(k; p) = 0 \text{ for all } k > q.$$

One might expect that the above property could be extended to $\rho(k; \ell) = 0$ for all $k > q$ when $\ell > p$, since an ARMA(p, q) model may be viewed as an ARMA(ℓ, q) model with the last $\ell - p$ autoregressive coefficients equal to 0. Unfortunately, or fortunately, this is not the case, due to the fact that the system of equations (2.40) is under determined, and its sample version does not necessarily produce a consistent estimate. Instead the ℓth EACF cuts off at $q + (\ell - p)$ for ARMA(p, q) models, for $\ell > p$. This property is also illustrated below with a simple AR(1) model.

Example 2.8 Consider $AR(1)$ model: $X_t = bX_{t-1} + \varepsilon_t$, where $|b| < 1$ and $\varepsilon_t \sim \text{WN}(0, \sigma^2)$. Then $\rho(k; 0) = \rho(k) = b^k$ for all $k \geqslant 0$.

To compute the 1st EACF $\rho(k; 1)$, let

$$Z_t^{(k)} = X_t - b_1^{(k)} X_{t-1},$$

where $b_1^{(k)}$ is the solution of (2.40) with $\ell = 1$, i.e. $b_1^{(k)} = \rho(k+1)/\rho(k) = b$. Thus $Z_t^{(k)} = X_t - bX_{t-1} = \varepsilon_t$ is a white noise. Therefore

$$\rho(k; 1) = \text{ACF of } \{\varepsilon_t\} \text{ at lag } k,$$

which is equal to 0 for all $k \geqslant 1$. Thus the 1st EACF of AR(1)=ARMA(1, 0) cuts off at q=0.

To find the 2nd EACF $\rho(k; 2)$, let

$$Z_t^{(k)} = X_t - b_1^{(k)} X_{t-1} - b_2^{(k)} X_{t-2},$$

where $(b_1^{(k)}, b_2^{(k)})$ is the solution of equations (2.40) with $\ell = 2$. Then, using the definition of X_t and X_{t-1}

$$\begin{aligned} Z_t^{(k)} &= \varepsilon_t + (b - b_1^{(k)})X_{t-1} - b_2^{(k)} X_{t-2} \\ &= \varepsilon_t + (b - b_1^{(k)})\varepsilon_{t-1} + (b^2 - b_1^{(k)} b - b_2^{(k)})X_{t-2}. \end{aligned} \tag{2.43}$$

Note that now $(b_1^{(k)}, b_2^{(k)})$ is under-determined by (2.40), as the two equations in (2.40) collapse to a single equation

$$b^2 = b_1^{(k)} b + b_2^{(k)}.$$

Thus the last term on the RHS of (2.43) is 0, and

$$Z_t^{(k)} = \varepsilon_t + (b - b_1^{(k)})\varepsilon_{t-1} \sim \text{MA}(1),$$

where $b_1^{(k)}$ is an arbitrary constant. As $\rho(k;2)$ is the ACF of $Z_t^{(k)}$ at lag k, $\rho(k;2) = 0$ for all $k > 1$, i.e. the 2nd EACF for AR(1)=ARMA(1, 0) processes cuts off at $q + (2-1) = 1$ instead of $q = 0$. Only when $b_1^{(1)} = b_1$ (consistent estimate), $\rho(1;2) = 0$.

It can be shown that the ℓth EACF for this AR(1) process cuts off at $(\ell - 1)$.

In practice we replace $\rho(\cdot)$ in (2.40) by the sample ACF $\widehat{\rho}(\cdot)$, which leads to the estimators $\widehat{b}_j^{(k)}$. Replacing $b_j^{(k)}$ in (2.41) by $\widehat{b}_j^{(k)}$, we obtain $\widehat{Z}_t^{(k)}$. The sample ACF of $\widehat{Z}_t^{(k)}$ at lag k is taken as the ℓth sample EACF at lag k and is denoted by $\widehat{\rho}(k;\ell)$. Tsay and Tiao (1984) suggested to list the sample EACF in a table of the following form. Note that the entry at the (p,q)th place is $\widehat{\rho}(q+1;p)$. Also $\widehat{\rho}(k;0) = \widehat{\rho}(k)$ is the sample ACF.

		MA				
		0	1	2	3	$\cdots$
AR	0	$\widehat{\rho}(1)$	$\widehat{\rho}(2)$	$\widehat{\rho}(3)$	$\widehat{\rho}(4)$	$\cdots$
	1	$\widehat{\rho}(1;1)$	$\widehat{\rho}(2;1)$	$\widehat{\rho}(3;1)$	$\widehat{\rho}(4;1)$	$\cdots$
	2	$\widehat{\rho}(1;2)$	$\widehat{\rho}(2;2)$	$\widehat{\rho}(3;2)$	$\widehat{\rho}(4;2)$	$\cdots$
	3	$\widehat{\rho}(1;3)$	$\widehat{\rho}(2;3)$	$\widehat{\rho}(3;3)$	$\widehat{\rho}(4;3)$	$\cdots$
	$\vdots$	$\vdots$	$\vdots$	$\vdots$	$\vdots$	$\vdots$

The *R*-function `eacf` in the package TSA (see §2.10.3 below) computes sample EACF automatically and prints out a coded table in the above form with $\widehat{\rho}(i,j)$ replaced by symbol "x" if $|\widehat{\rho}(i,j)| > 1.96/\sqrt{T-i-j}$, and "o" otherwise. See Table 2.1 for the "ideal" pattern of the sample EACF for an ARMA(p,q) processes, which contains a triangle of "o" with the upper-left vertex at the positive (p,q). This triangular shape is implied by the property

Table 2.1 *The ideal pattern of the sample EACF for ARMA(p, q) processes*

	0	1	$\cdots$	q	$q+1$	$q+2$	$q+3$	$\cdots$
0	x	x	$\cdots$	x	x	x	x	$\cdots$
1	x	x	$\cdots$	x	x	x	x	$\cdots$
			$\cdots$		$\cdots$			
$p-1$	x	x	$\cdots$	x	x	x	x	$\cdots$
p	x	x	$\cdots$	o	o	o	o	$\cdots$
$p+1$	x	x	$\cdots$	x	o	o	o	$\cdots$
$p+2$	x	x	$\cdots$	x	x	o	o	$\cdots$
$p+3$	x	x	$\cdots$	x	x	x	o	$\cdots$
			$\cdots$		$\cdots$			

that the ℓth EACF of an stationary ARAM(p, q) process cuts off at $q+\ell-p$ for $\ell = p, p+1, \ldots$.

Below we summarize some useful guidelines for model identification based on sample ACF, PACF and EACF.

---

1. If the sample ACF $\widehat{\rho}(k)$ are close to 1 for all small k and decay slowly, difference the data first.
2. If the sample ACF $\widehat{\rho}(k)$ is bounded by the bounds (2.32) for any $k > q$, fit an MA(q) model to the data.
3. If the sample PACF $\widehat{\pi}(k)$ is bounded by $\pm 1.96/\sqrt{T}$ for any $k > p$, fit an AR(p) model to the data.
4. If the sample PACF $\widehat{\rho}(q+j; p+i)$ is bounded by $\pm 1.96/\sqrt{T-i-j}$ for any $i \geqslant 0$ and $j > i$, fit an ARMA(p, q) model to the data.

Among choices that satisfy simultaneous Criteria 2, 3 and 4, we choose the one with the most parsimonious representation. The sample ACF and PACF plots in Figure 2.4 conform well with the above rules. For example, the sample ACF and PACF for the white noise lie between the two confidence bounds at all (but one) lags. For the MA(1) process with the MA coefficient $a = 0.9$, the ACF plot suggests MA(1) while the PACF plot indicates AR(6); both make perfect sense as this is an invertible process which admits AR(∞) representation with the AR coefficients $b_k = (0.9)^k$. For the two AR(1) processes with the AR coefficient $b = \pm 0.669$, the two PACF plots indicate AR(1) process.

For the ARMA(1,1) model with $b = 0.5$ and $a = 0.279$, the ACF plot in Figure 2.4 suggests an MA(5) model while the PACF plot suggests an

AR(3) model. The EACF calculated by R below indicates clearly that the most parsimonious model for this data is ARMA(1,1). This is identified from the EACF table as follows: we look for a triangle, similar to the blue triangle in Table 2.1, such that most of the elements within this triangle are "o". The upper-left vertex of the triangle is the order (p, q) of the identified model, the upper side of the triangle is always horizontal, the vertical edge of the EACF table forms another side of the triangle, and the other side is always parallel to the main diagonal line. We try to identify such a triangle with the smallest values of p and q. For this example, the EACF table below contains such a triangle with the upper-left vertex at (1,1).

```
> x <- arima.sim(n=100, list(ar=0.5, ma=0.279))+1
> eacf(x)
AR/MA
  0 1 2 3 4 5 6 7 8 9 10 11 12 13
0 x x o o o o o o o x x  x  o  o
1 x o o o o o o o o o o  o  o  o
2 o x o o o o o o o o o  o  o  o
3 o x x o o o o o o o o  o  o  o
4 x o o x x o o o o o o  o  o  o
5 o x o o o o o o o o o  o  o  o
6 o x o o o o o o o o o  o  o  o
7 x x o o o o o o o o o  o  o  o
```

2.5 Fitting ARMA models: MLE and LSE

Fitting an ARMA model to an observed time series data set is typically done nowadays using computers. For example, the R-function `arima` performs such a fitting with various options. However it is important to understand the methodology upon which most computer implementations are based upon. We introduce below the two most frequently used methods for estimating the parameters in ARMA(p, q) models with prescribed p and q.

2.5.1 Least squares estimation

We consider purely AR processes first. Let $X_1, \ldots, X_T$ be the observations from AR(p) model

$$X_t = c + b_1 X_{t-1} + \cdots + b_p X_{t-p} + \varepsilon_t, \qquad \varepsilon_t \sim \text{WN}(0, \sigma^2).$$

The *least squares estimator* (LSE) for $(c, \boldsymbol{b}) \equiv (c, b_1, \ldots, b_p)$ is defined as

$$\begin{aligned}(\widehat{c},\, \widehat{\boldsymbol{b}}) &\equiv (\widehat{c},\, \widehat{b}_1,\, \ldots,\, \widehat{b}_p) \\ &= \arg\min_{c,\boldsymbol{b}} \sum_{t=p+1}^{T} (X_t - c - b_1 X_{t-1} - \cdots - b_p X_{t-p})^2.\end{aligned}$$

This is the same as the LSE for linear regression models, and the estimator $(\widehat{c},\, \widehat{\boldsymbol{b}})$ admits an explicit expression. Consequently the LSE for σ^2 may be defined as

$$\widehat{\sigma}^2 = \frac{1}{T-2p-1} \sum_{t=p+1}^{T} (X_t - \widehat{c} - \widehat{b}_1 X_{t-1} - \cdots - \widehat{b}_p X_{t-p})^2.$$

We divide the sum by $T-2p-1$ above, as the effective sample size is $T-p$ and the number of parameters estimated is $p+1$.

For the ARMA(p, q) model

$$X_t = c + b_1 X_{t-1} + \cdots + b_p X_{t-p} + \varepsilon_t + a_1 \varepsilon_{t-1} + \cdots + a_q \varepsilon_{t-q}, \tag{2.44}$$

the innovations $\varepsilon_1, \ldots, \varepsilon_T$ are unobservable. We assume that $\varepsilon_{p+1-k} \equiv 0$ for all $1 \leqslant k \leqslant q$. This assumption has limited impact on our estimation when the sample size T is large. With these initial values, and given parameters $\boldsymbol{a} = (a_1, \ldots, a_q)'$, $\boldsymbol{b} = (b_1, \ldots, b_p)'$ and c, we now compute recursively for $t = p+1, \ldots, T$ that

$$\varepsilon_t(c, \boldsymbol{a}, \boldsymbol{b}) = X_t - c - \sum_{j=1}^{p} b_j X_{t-i} - \sum_{i=1}^{q} a_i \varepsilon_{t-i}(c, \boldsymbol{a}, \boldsymbol{b}) \tag{2.45}$$

The LSE for $(c, \boldsymbol{a}, \boldsymbol{b})$ is defined as

$$(\widehat{c},\, \widehat{\boldsymbol{a}},\, \widehat{\boldsymbol{b}}) = \arg\min_{c, \boldsymbol{a}, \boldsymbol{b}} \sum_{t=p+1}^{T} \{\varepsilon_t(c, \boldsymbol{a}, \boldsymbol{b})\}^2. \tag{2.46}$$

Unfortunately this is a nonlinear optimization problem. One possible solution is an iterative linear approximation: starting with an initial estimator $(\widehat{c}_0, \widehat{\boldsymbol{a}}_0, \widehat{\boldsymbol{b}}_0)$, we define the kth iterative estimator as

$$\begin{aligned}(\widehat{c}_k, \widehat{\boldsymbol{a}}_k, \widehat{\boldsymbol{b}}_k) = \arg\min_{c,\boldsymbol{a},\boldsymbol{b}} \sum_{t=p+1}^{T} \Big\{ X_t - c - \sum_{j=1}^{p} b_j X_{t-i} \\ - \sum_{i=1}^{q} a_i \varepsilon_{t-i}(\widehat{c}_{k-1}, \widehat{\boldsymbol{a}}_{k-1}, \widehat{\boldsymbol{b}}_{k-1}) \Big\}^2\end{aligned} \tag{2.47}$$

for $k = 1, 2, \ldots$, where $\varepsilon_t(\cdot)$ is defined as in (2.45). We stop the iteration when the two successive estimators differ by a small amount. Note that $(\widehat{c}_k, \widehat{\boldsymbol{a}}_k, \widehat{\boldsymbol{b}}_k)$ admits the same explicit expression as the LSE for a linear regression model.

2.5.2 Gaussian maximum likelihood estimation

The least squares estimation method introduced above is relatively easy to compute. However this may not be a most efficient statistical estimator. It makes no use of the information on the underlying distribution.

When the distribution of ε_t is known, the maximum likelihood estimation is more efficient. The most common assumption in (linear) time series analysis is that ε_t is independent and $N(0, \sigma^2)$ in model (2.44). For simplicity, we assume $c = 0$ in (2.44), i.e. $\mu = EX_t = 0$. This implies that we should center the data first before the fitting, as we can hardly estimate the mean μ better than the sample mean. Under those assumptions, $\boldsymbol{X}_T \equiv (X_1, \ldots, X_T)$ are jointly normal with the common mean 0 and covariance matrix $\boldsymbol{\Sigma}$, whose (i,j) element is $\gamma(i-j)$ that depends on the model parameters $(\boldsymbol{a}, \boldsymbol{b}, \sigma^2)$ and is written as $\boldsymbol{\Sigma}(\boldsymbol{a}, \boldsymbol{b}, \sigma^2)$. See Example 2.6 for the specification of ARMA(1,1).

To derive the maximum likelihood estimator (see Section 2.5.4 for an overview), we need to find the density of $\boldsymbol{X}_T$. From the multivariate analysis (see, e.g., Anderson, 2003), the density of $\boldsymbol{X}_T \sim N(\boldsymbol{\mu}, \boldsymbol{\Sigma})$ is given by

$$f(\boldsymbol{X}_T; \boldsymbol{\theta}) = (2\pi)^{-T/2} |\boldsymbol{\Sigma}|^{-1/2} \exp\big(-(\boldsymbol{X}_T - \boldsymbol{\mu})^{\mathrm{T}} \boldsymbol{\Sigma}^{-1} (\boldsymbol{X}_T - \boldsymbol{\mu})/2\big), \quad (2.48)$$

where $\boldsymbol{\theta}$ are the parameters, $\boldsymbol{\mu}$ and $\boldsymbol{\Sigma}$. Regarding $f(\boldsymbol{X}_T; \boldsymbol{\theta})$ as a function of $\boldsymbol{\theta}$ results in the likelihood function $L(\boldsymbol{\theta}) \equiv f(\boldsymbol{X}_T; \boldsymbol{\theta})$ with the log-likelihood function $\ell(\boldsymbol{\theta}) = \log L(\boldsymbol{\theta})$. Dropping the constant factors in (2.48),

$$-2\ell(\boldsymbol{\theta}) = \log |\boldsymbol{\Sigma}| + (\boldsymbol{X}_T - \boldsymbol{\mu})^{\mathrm{T}} \boldsymbol{\Sigma}^{-1} (\boldsymbol{X}_T - \boldsymbol{\mu}).$$

In our application, $\boldsymbol{\mu} = 0$ and $\boldsymbol{\Sigma} = \boldsymbol{\Sigma}(\boldsymbol{a}, \boldsymbol{b}, \sigma^2)$ so that the unknown parameters are $\boldsymbol{a}$, $\boldsymbol{b}$ and σ^2. Thus, the *maximum likelihood estimator* (MLE) for $(\boldsymbol{a}, \boldsymbol{b}, \sigma^2)$ is then defined as

$$(\widehat{\boldsymbol{a}}, \widehat{\boldsymbol{b}}, \widehat{\sigma}^2) = \arg \min_{\boldsymbol{a}, \boldsymbol{b}, \sigma^2} \big\{\boldsymbol{X}_T' \{\boldsymbol{\Sigma}(\boldsymbol{a}, \boldsymbol{b}, \sigma^2)\}^{-1} \boldsymbol{X}_T + \log \big|\boldsymbol{\Sigma}(\boldsymbol{a}, \boldsymbol{b}, \sigma^2)\big|\big\}. \quad (2.49)$$

This is a nonlinear optimization problem involving computing both the inverse and the determinant of the $T \times T$ matrix $\boldsymbol{\Sigma}(\boldsymbol{a}, \boldsymbol{b}, \sigma^2)$. There are some efficient algorithms for calculating the likelihood function, including, for example, the innovation algorithm via prewhitening which avoids the direct calculation for both the inverse and the determinant of $\boldsymbol{\Sigma}(\boldsymbol{a}, \boldsymbol{b}, \sigma^2)$ (see §3.2

of Fan and Yao 2003), and the Kalman filter via the state space representation for ARMA models (see (3.13) in Section 3.5.2 below). Coupled with such an algorithm, the MLE may be calculated in the Newton–Raphson iterative manner; see, e.g. §3.3.1 of Fan and Yao (2003). In fact the calculation of the MLE based on Gaussian likelihood, which we simply called *Gaussian MLE*, has been implemented in most statistical packages.

The Gaussian MLE $(\widehat{\boldsymbol{a}}, \widehat{\boldsymbol{b}}, \widehat{\sigma}^2)$ defined in (2.49) is asymptotically unbiased, normal with the limit covariance matrix determined by the ACF of X_t, provided that X_t is causal and invertible, while ε_t is not necessarily normal (Theorem 3.2 of Fan and Yao (2003)). This covariance can also be produced by the computer software (see also Section 2.5.4). Statistical tests and confidence intervals for $\boldsymbol{a}$ and/or $\boldsymbol{b}$ are constructed based on the asymptotic normality.

The Gaussian MLE is often used when ε_t is not normal. The resulting estimator is in fact a *quasi-MLE*.

Example 2.9 For the monthly returns of the CRSP index in 1926–1997 ($T = 864$), we fit an AR(3) model following the suggestion from Example 2.5. The fitted model is

$$\begin{array}{ccccccccc} r_t & = & 0.0103 & + & 0.104\, r_{t-1} & - & 0.010\, r_{t-2} & - & 0.120\, r_{t-3} & + & \varepsilon_t, \\ & & (0.002) & & (0.034) & & (0.034) & & (0.034) & & \end{array}$$

where the numbers in parentheses are the standard errors of the estimated coefficients. The estimated intercept is $\widehat{c} = 0.0103$ which seems very small. To test the hypothesis $H_0 : c = 0$, the t-statistic is

$$T = \widehat{c}/\mathrm{SE}(\widehat{c}) = 0.0103/0.002 = 5.15.$$

Hence the p-value for the test against the one-sided alternative $H_1 : c > 0$ is $1 - \Phi(5.15) = 0$, and that against the two-sided alternative $H_1 : c \neq 0$ is $2(1 - \Phi(5.15)) = 0$. Therefore the evidence against the hypothesis $c = 0$ is overwhelming. We conclude that $c > 0$. By (2.16), the estimate for the expected monthly return $\mu \equiv Er_t$ is

$$\widehat{\mu} = \frac{\widehat{c}}{1 - \widehat{b}_1 - \widehat{b}_2 - \widehat{b}_3} = \frac{0.0103}{1 - 0.104 + 0.010 + 0.120} = 0.01, \tag{2.50}$$

which is small but statistically significantly positive. The implied annualized expected return is

$$(1 + 0.01)^{12} - 1 = 12.68\%,$$

while the actual annualized return over the whole period of 1/1926 to 12/1997 is

$$\left\{\prod_{t=1}^{864}(1+R_t)\right\}^{12/864} - 1 = 10.53\%.$$

We have a point estimate for the expected monthly return in (2.50). A natural question is how to find out the confidence interval for the expected return. To this end, we need to evaluate the standard error of $\widehat{\mu}$. By (2.50), we may write $\widehat{\mu} = f(\widehat{\boldsymbol{\theta}})$ as a function of $\widehat{\boldsymbol{\theta}} = (\widehat{c}, \widehat{b}_1, \widehat{b}_2, \widehat{b}_3)'$. Using the Delta method below, we have

$$\text{var}(\widehat{\mu}) \approx \dot{f}(\boldsymbol{\theta})'\text{var}(\widehat{\boldsymbol{\theta}})\dot{f}(\boldsymbol{\theta}), \tag{2.51}$$

where $\dot{f}(\boldsymbol{\theta})$ is the gradient vector of the function f. For this particular application, it is a four-component vector, consisting of the partial derivatives of f with respect to c, b_1, b_2, and b_3, respectively. Since $\dot{f}(\boldsymbol{\theta})$ evaluated at $\hat{\boldsymbol{\theta}}$ is $\dot{f}(\boldsymbol{\theta})|_{\boldsymbol{\theta}=\hat{\boldsymbol{\theta}}} = (0.9746, 0.0098, 0.0098, 0.0098)'$, and

$$\text{var}(\widehat{\boldsymbol{\theta}}) = \frac{1}{1000^2}\begin{pmatrix} 2^2 & 34 & 0 & 0 \\ 34 & 34^2 & 0 & 0 \\ 0 & 0 & 34^2 & 0 \\ 0 & 0 & 0 & 34^2 \end{pmatrix},$$

which is returned by a statistical software such as R, we obtain $\text{var}(\widehat{\mu}) = 4.7804 \times 10^{-6}$. Hence the standard error $\text{SE}(\widehat{\mu}) = \sqrt{4.7804 \times 10^{-6}} = 0.2186\%$. Consequently an approximate 95% confidence interval for the expected monthly return is

$$1\% \pm 1.96 \times 0.2186\% = [0.37\%,\ 1.43\%].$$

The Delta Method. Suppose that k-dimensional random vector $\widehat{\boldsymbol{\theta}}$ is asymptotically normal with mean $\boldsymbol{\theta}$ and covariance $\boldsymbol{\Sigma}$. Let $\mu = f(\boldsymbol{\theta})$ be a smooth function and the vector $\dot{f}(\boldsymbol{\theta}) = \partial f/\partial \boldsymbol{\theta}$ does not vanish. Then $\widehat{\mu} = f(\widehat{\boldsymbol{\theta}})$ is asymptotically normal with mean $f(\boldsymbol{\theta})$ and variance $\dot{f}(\boldsymbol{\theta})'\boldsymbol{\Sigma}\dot{f}(\boldsymbol{\theta})$.

The Delta method is one of the standard methods for deriving the asymptotic distributions. Intuitively it follows a simple Taylor expansion

$$\widehat{\mu} = f(\boldsymbol{\theta}) + \dot{f}(\boldsymbol{\theta})'(\widehat{\boldsymbol{\theta}} - \boldsymbol{\theta})\{1 + o_P(1)\}.$$

Hence, the asymptotic variance of $\hat{\mu}$ is approximately the same as that of $\dot{f}(\boldsymbol{\theta})'(\widehat{\boldsymbol{\theta}} - \boldsymbol{\theta})$, which is $\dot{f}(\boldsymbol{\theta})'\boldsymbol{\Sigma}\dot{f}(\boldsymbol{\theta})$. For a formal proof, we refer to, for example, Theorem 5.4.6 of Lehmann (1999).

2.5.3 Illustration with gold prices

As an illustration, we fit an ARIMA model to the daily close prices (in US-dollars) of the SPDR gold shares in January 1–November 23, 2011. This example was completed on 25 November 2011 using all the available prices to that day. The gold prices were extremely volatile in the second half of 2011 due to the financial turbulence and led to, among others, the Euro debt crisis. There are $T = 188$ data points in this period. The data were downloaded from *Yahoo!Finance*.

Figure 2.9 depicts the time series plot, the ACF plot and the PACF plot for the daily prices and their differences taken at lag 1. The dynamic of the prices somehow resembles that of random walks; see the first panels in both Figure 2.9 and Figure 2.6. The ACF of the prices decay slowly, suggesting that we should take differences of the data. Note that the ACF of the differences is not significant at lag 1; indicating further the appropriateness of the differencing. The price series is clearly non-stationary while it is debatable whether the series of the differenced prices is stationary. Since the differenced prices exhibit significant ACF and PACF at, for example, lag 9, pure AR or MA models with small orders are not appropriate for this data set.

The EACF below suggests ARMA(3,2) model for the changes of prices, i.e. ARIMA(3,1,2) for the original daily prices.

```
AR/MA
  0 1 2 3 4 5 6 7 8 9 10 11 12 13
0 x x x x x x x x x x x  x  x  x
1 o o o o o o o o x o o  o  o  o
2 o o o o o o o o x x o  o  o  o
3 o x o o o o o o x o o  o  o  o
4 x x o o o o o o x o o  o  o  o
5 x x o o o o o o x o o  o  o  o
6 x x o o o o o o o o x  o  o  o
7 o x x x o x o o o o x  o  o  o
```

We fit this model to the data using R. Suppose the daily prices have been imported into R as `goldPrices`. The following line command returns the fitted model as an object `arima312`:

```
> arima312 = arima(goldPrices, order=c(3,1,2))
```

To use the LSE method in §2.5.1 or the MLE in §2.5.2 to estimate the parameters, one needs to add in the above command the option `method="CSS"` or `method="ML"` respectively. Without those options, R uses the LSE to obtain an initial value and then apply the MLE to obtain the final estimates. The

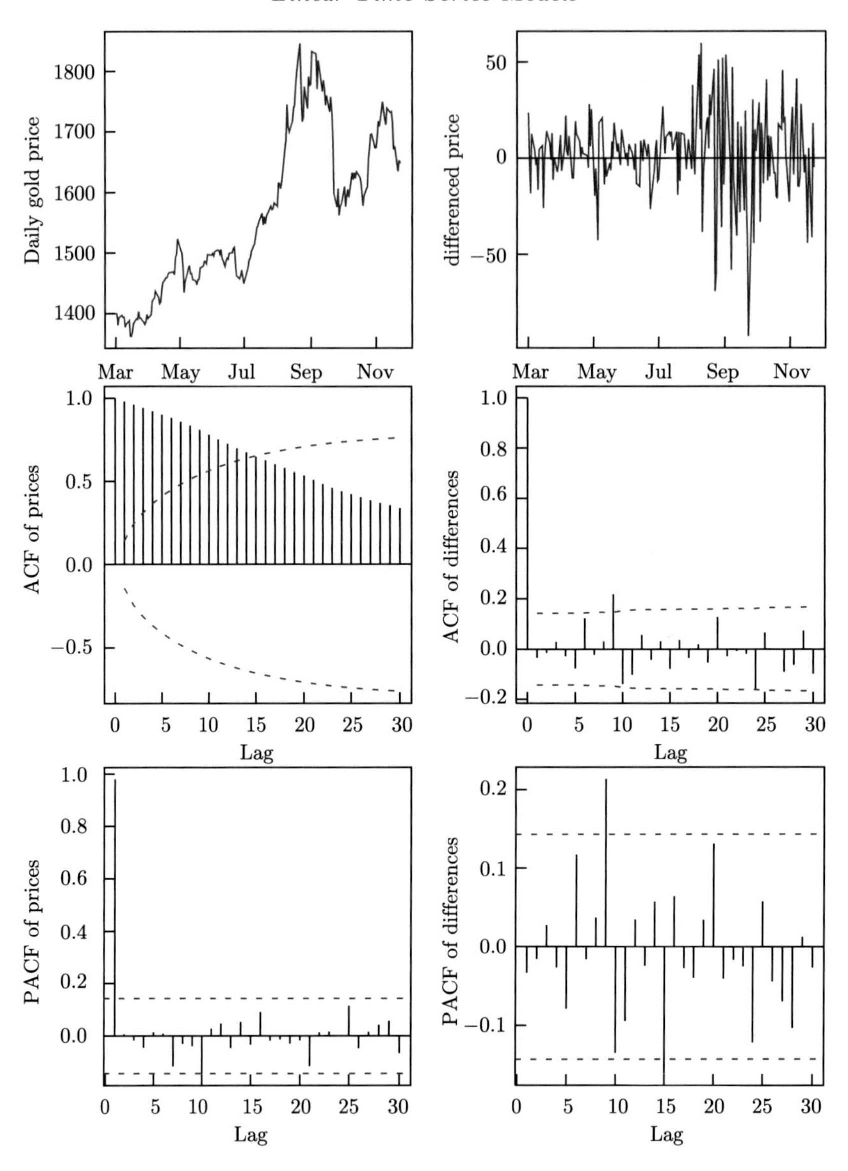

Figure 2.9 Plots of log returns against simple returns of the Apple Inc share prices in January 1985 to February 2011. The blue straight lines mark the positions where the two returns are identical.

estimated coefficients and the corresponding standard errors for the fitted ARMA(3,2) model for the changes of prices are

```
          ar1      ar2      ar3     ma1     ma2
coef.  -1.4054  -1.0316  -0.1109  1.4371  1.0000
```

```
s.e.    0.0734   0.1002   0.0735  0.0395  0.0525
```

The estimated variance for the white noise is $\widehat{\sigma}^2 = 442.5$.

To gain some appreciation of how good the model is, we consider the out-sample prediction of the 17 prices in November. More precisely for $j = 17, 16, \ldots, 1$, we fit the first $T = 188 - j$ daily prices with an ARIMA(3,1,2) model and we use this fitted model to predict the next day's price. We repeat the exercise with the random walk model (i.e. ARIMA(0,1,0)), as well as the exponential smoothing (i.e. ARIMA(0,1,1)); see §2.3.1. For the detailed description on how to predict future values based on ARIMA models, we refer to Section 2.9.1 below. Figure 2.10 displays those 17 true prices together with their predicted values based on the three different models. We also present in Figure 2.10 the approximate 95% predictive intervals:

$$\text{Predictive value} \pm 2 \times \text{Standard predictive error}$$

Note that the random walk model simply predicts the price next day by the price today. The predictive values and their standard errors can be obtained by using the *R*-command `predict(arima312, n.ahead=1)`, where `arima312` is an output from a fitting using `arima`; see above. Three models provide very similar performances. They all predict the prices on the 1st, 4th, 10th, 14th and 23nd of November well. All the three methods contain all 17 true values in their approximate 95% predictive intervals. The mean absolute predictive errors are $17.08 for ARIMA(3,1,2) models, $17.74 for random walks, and $17.54 for exponential smoothing. This shows that ARIMA(3,1,2) provides the best post-sample predictions among the three methods.

It is worth pointing out that using a more complex model will not necessarily lead to an improvement in post-sample prediction. For example, the mean absolute predictive error for the above data with ARIMA(2,1,2) models is $18.44, which is greater than those with the simple random walk ARIMA(0,1,0) model and the exponential smoothing ARIMA(0,1,1) model.

2.5.4 A snapshot of maximum likelihood methods*

The *maximum likelihood estimate (MLE)* is one of mostly widely used techniques for statistical inference. Suppose that we have data $\boldsymbol{X}_T$ that can be regarded as a random sample from a certain population with a density that depends on unknown parameters $\boldsymbol{\theta}$. Let us denote it as $f(\boldsymbol{X}_T; \boldsymbol{\theta})$. The density and the *likelihood function* are really two different sides of the same coin. When it is regarded as a function of $\boldsymbol{X}_T$, it is called the density function, which is used in the probability theory to describe the likelihood of

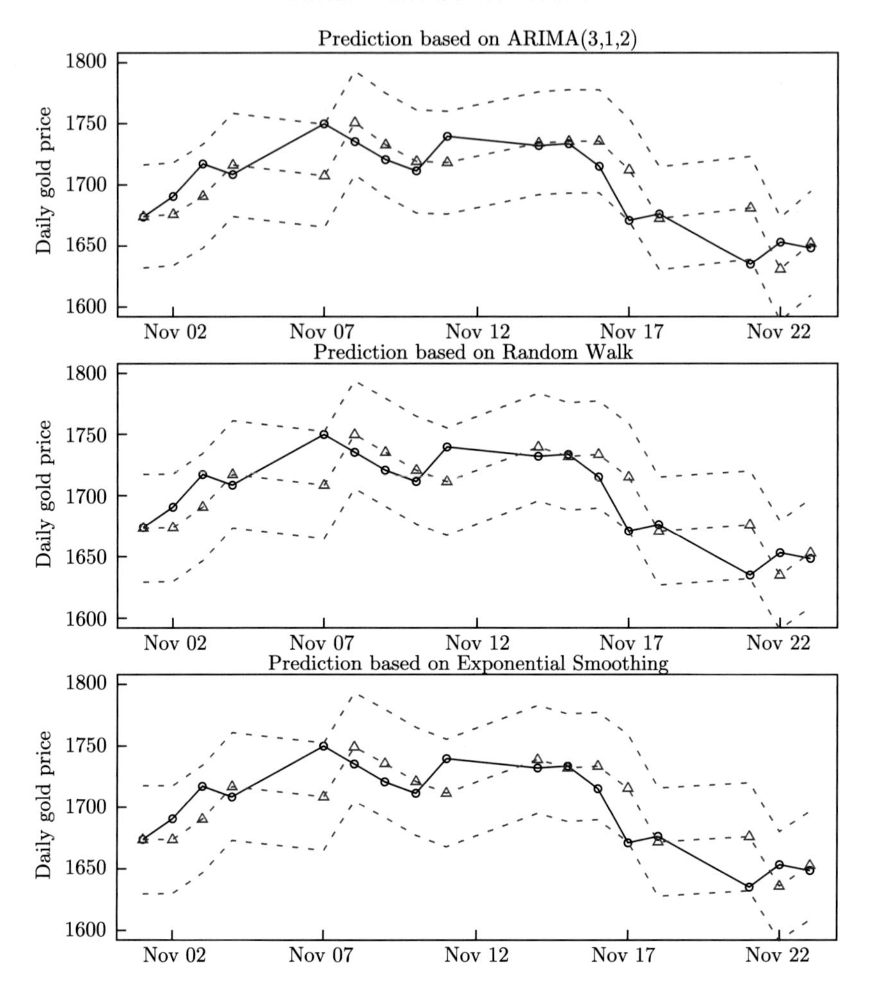

Figure 2.10 The true daily gold prices (solid-dot lines) in November 2011, and the one-step-ahead predicted prices (dashed-triangle lines) based on 3 models: ARIMA(3,1,2), random walk and exponential smoothing. Two dashed lines are the boundaries of the approximate 95% predictive intervals.

getting the data around a neighborhood of $\boldsymbol{X}_T$. In statistical data analysis, however, the data $\boldsymbol{X}_T$ are already given and our task is reverse engineering: estimating the parameters of the distribution under which the data were generated. When $f(\boldsymbol{X}_T; \boldsymbol{\theta})$ is regarded as a function of parameters $\boldsymbol{\theta}$, it shows the likelihood of getting the data $\boldsymbol{X}_T$ for each given parameter value $\boldsymbol{\theta}$. In other words,

$$L(\boldsymbol{\theta}) = f(\boldsymbol{X}_T; \boldsymbol{\theta})$$

is the likelihood function. Maximizing the likelihood function with respect to $\boldsymbol{\theta}$ amounts to finding the parameter $\hat{\boldsymbol{\theta}}$ that is most likely to produce the data $\boldsymbol{X}_T$. It is equivalent to maximizing the log-likelihood $\ell(\boldsymbol{\theta}) = \log L(\boldsymbol{\theta})$ or minimizing $-\ell(\boldsymbol{\theta})$. The resulting maximizer is called the maximum likelihood estimator (MLE).

To obtain the MLE, we first need to derive the probability density function under the assumptions on the data generating process. An example of this is given in (2.48) for the ARMA process with Gaussian innovation. With a given log-likelihood, we solve the likelihood equation

$$\dot{\ell}(\widehat{\boldsymbol{\theta}}) = 0.$$

This usually solves by the Newton Raphason method: Given an initial value $\boldsymbol{\theta}_0$, the next iteration is to solve, by Taylor's expansion,

$$\dot{\ell}(\widehat{\boldsymbol{\theta}}) \approx \dot{\ell}(\boldsymbol{\theta}_0) + \ddot{\ell}(\boldsymbol{\theta}_0)(\widehat{\boldsymbol{\theta}} - \boldsymbol{\theta}_0) = 0,$$

where $\ddot{\ell}(\boldsymbol{\theta}_0)$ is the Hessian matrix of the log-likelihood function $\ell(\boldsymbol{\theta})$ and is assumed to be invertible. This is equivalent to compute the next iteration

$$\widehat{\boldsymbol{\theta}} = \boldsymbol{\theta}_0 + \ddot{\ell}(\boldsymbol{\theta}_0)^{-1}\dot{\ell}(\boldsymbol{\theta}_0). \tag{2.52}$$

One can now regard the newly computed $\widehat{\boldsymbol{\theta}}$ as the initial value and use (2.52) to compute a new update. Continue this iterative process until convergence.

Under fairly general conditions (see Lehmann, 1999), the maximum likelihood estimator is asymptotically unbiased and statistically efficient. It is asymptotically normal with mean $\boldsymbol{\theta}$, the true parameter, and covariance being the *Fisher information* matrix. Such a covariance matrix can be consistently estimated as

$$\widehat{\mathrm{var}}(\widehat{\boldsymbol{\theta}}) = -\ddot{\ell}(\widehat{\boldsymbol{\theta}})^{-1}, \tag{2.53}$$

the negative inverse of Hessian matrix. This matrix is typically reported by statistical software packages. Note that when $\ell'(\boldsymbol{\theta})$ is maximized at $\hat{\boldsymbol{\theta}}$, the matrix $-\ddot{\ell}(\widehat{\boldsymbol{\theta}})$ is positively definite and hence is a covariance matrix. The square roots of the diagonal elements are the standard errors of components of the MLE $\hat{\boldsymbol{\theta}}$. See Example 2.9. The delta-method (2.51) continues to apply with the asymptotic variance given by (2.53).

The asymptotic mean and variance are useful for statistical inference. For example, to test the null hypothesis

$$H_0 : \boldsymbol{\theta} = \boldsymbol{\theta}_0, \tag{2.54}$$

the Wald test statistic is

$$W = (\widehat{\boldsymbol{\theta}} - \boldsymbol{\theta}_0)'\widehat{\mathrm{var}}(\widehat{\boldsymbol{\theta}})^{-1}(\widehat{\boldsymbol{\theta}} - \boldsymbol{\theta}_0). \tag{2.55}$$

We may also replace $\widehat{\text{var}}(\widehat{\boldsymbol{\theta}})$ in the above expression by $\widehat{\text{var}}(\boldsymbol{\theta}_0) = -\ddot{\ell}(\boldsymbol{\theta}_0)^{-1}$. Under the null hypothesis (2.54), the asymptotic distribution of W is a χ^2-distribution with p degrees of freedom, denoted by

$$W \overset{a}{\sim}_{H_0} \chi^2_p,$$

where p is the number of parameters, and "$\overset{a}{\sim}_{H_0}$" can be read as "distributed approximately under the null hypothesis as".

Related to the Wald test is the *maximum likelihood ratio (MLR) test*:

$$\text{MLR} = 2\{\ell(\widehat{\boldsymbol{\theta}}) - \ell(\boldsymbol{\theta}_0)\} \overset{a}{\sim}_{H_0} \chi^2_p. \tag{2.56}$$

This test is also referred to as the Wilk test, which compares the likelihood ratios or equivalently the difference of the log-likelihood of two models: Full model without restrictions on the parameters $\boldsymbol{\theta}$ and the restricted model with underlying parameters $\boldsymbol{\theta} = \boldsymbol{\theta}_0$. If the difference is small, the evidence against the null hypothesis is weak. P-value can be computed under the asymptotic null distribution χ^2_p.

The MLR test can further be extended to test the composite hypothesis

$$H_0 : \boldsymbol{\theta} \in \boldsymbol{\Theta}_0 \quad \text{vs.} \quad H_1 : \boldsymbol{\theta} \notin \boldsymbol{\Theta}_0, \tag{2.57}$$

where $\boldsymbol{\Theta}_0$ is a p_0-dimensional subspace in p-dimensional Euclidean space. Let $\widehat{\boldsymbol{\theta}}_0$ be the maximum likelihood estimator under the null hypothesis. The MLR test statistic is then generalized as

$$\text{MLR} = 2\{\ell(\widehat{\boldsymbol{\theta}}) - \ell(\widehat{\boldsymbol{\theta}}_0)\} = 2\{\max_{\boldsymbol{\theta}} \ell(\boldsymbol{\theta}) - \max_{\boldsymbol{\theta}\in\boldsymbol{\Theta}_0} \ell(\boldsymbol{\theta})\} \overset{a}{\sim}_{H_0} \chi^2_{p-p_0}. \tag{2.58}$$

This again compares the likelihood ratios or equivalently the difference of the log-likelihood of two models: restricted parameter spaces versus unrestricted parameter spaces.

To keep this section brief, we do not use examples here to illustrate the maximum likelihood method. They will be illustrated in the context where it is needed.

2.6 Model diagnostics: residual analysis

Data analysis involves both arts and sciences. The formulation of ARMA models is a statistical assumption for unknown data generating processes. Whether it is appropriate for a given financial time series requires a validation. Model diagnostics is one of the important steps in building a time series model. It checks the goodness-of-fit with the data. The key idea is simple: if ARMA model (2.44) is adequate for a given data set, the residuals

$$\widehat{\varepsilon}_t = X_t - \sum_{j=1}^{p} \widehat{b}_j X_{t-j} - \sum_{i=1}^{q} \widehat{a}_i \widehat{\varepsilon}_{t-i}, \qquad t = p+1, \ldots, T, \tag{2.59}$$

should behave like white noise. We may calculate the above recursion with the initial values $\widehat{\varepsilon}_{p+1-i} = 0$ for $i = 1, \ldots, q$. Most computer packages revert to the AR(∞) representation for an invertible model to calculate the residuals. There are two classes of the methods for residual analysis: visual diagnostics, and the tests for white noise.

2.6.1 Residual plots

It is useful to plot the residuals $\widehat{\varepsilon}_t$ against time t. Such a plot allows us to examine whether there is any time trend or strong serial correlation in the residuals. Such a plot should be patternless if the residuals behave like white noise. Figure 2.11 displays some good and bad residual patterns. An adequate fitting leads to a residual plot similar to the top-left panel in which the residuals fluctuate around zero randomly. The other three panels exhibit some systematic patterns, indicating inadequacies of the fitted models. We often plot the standardized residuals, i.e. the residuals divided by their standard errors. If ε_t are normal, approximately 95% of points in a standardized residual plot should be within the range $[-2, 2]$.

On the same token, we may also plot $\widehat{\varepsilon}_t$ against X_t or X_{t-k}, or even the fitted values $\widehat{X}_t$, where

$$\widehat{X}_t = \sum_{j=1}^{p} \widehat{b}_j X_{t-j} + \sum_{i=1}^{q} \widehat{a}_i \widehat{\varepsilon}_{t-i}, \qquad t = p+1, \ldots, T.$$

The evidence that all those plots are patternless is indicative for the appropriateness of the fitted model.

We may also look into the ACF, PACF and EACF of the residuals. All of them should show no significant non-zero correlations at non-zero lags. The *R*-function `tsdiag` produces the three diagnostic plots for each fitted model. For example, `arima312` is an *R* object produced by fitting ARIMA(3,1,2) model to the gold prices in 2011; see Section 2.5.3 above. To produce the plots in Figure 2.12, we call the *R* function:

```
> tsdiag(arima312, gof.lag=15)
```

There are 3 panels in Figure 2.12. The time series plot of the standardized residuals indicates that the residuals tend to be larger in the second half of

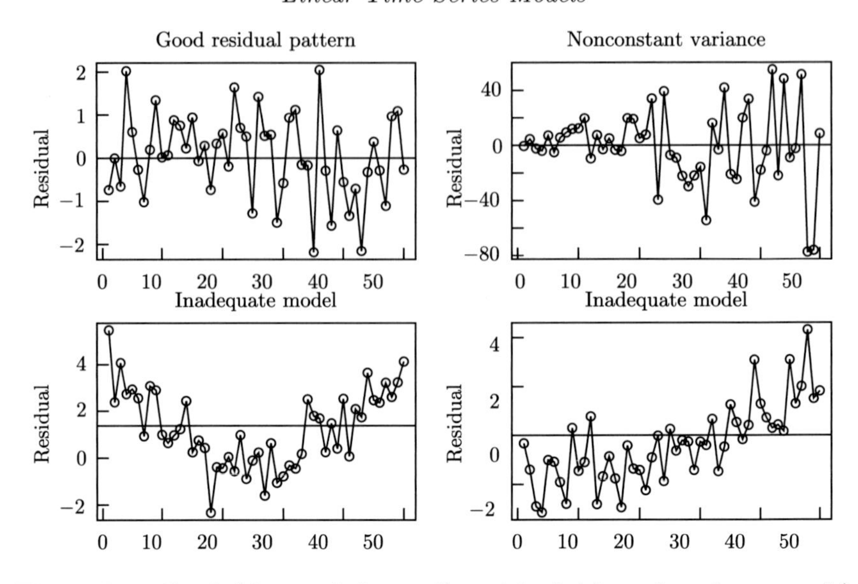

Figure 2.11 Good (the top-left panel) and bad (the other three panels) residual patterns.

the period than those in the first half of period, although most of them are within the range of [-2, 2]. The second panel in the figure shows no significant correlation among all non-zero lags; indicating that the residuals behave like a white noise process. The third panel depicts the P-values of the Ljung–Box test statistics Q_m defined by (1.15) as a function of m. It confirms the white noise pattern observed in the first and the second panel. Figure 2.13 shows the same plots for the residuals that result from fitting the same data set with ARIMA(0,1,1) model, namely the exponential smoothing. Since the model is simpler, the range of the residuals is greater and there are a few non-zero time lags at which the autocorrelation is beyond the 95% significance bounds.

Q–Q plots introduced in Section 1.5 are also often used for diagnosing whether the assumption of normal white noise is appropriate or not. Figure 2.14 displays the Q–Q plots of the residuals of the fitted ARIMA(3,1,2) model and those of the fitted ARIMA(0,1,1) model for the gold prices in 2011 against normal distributions. The tails of the residual distributions for both the fitted models are heavier than those of normal distributions. This is very common for financial data.

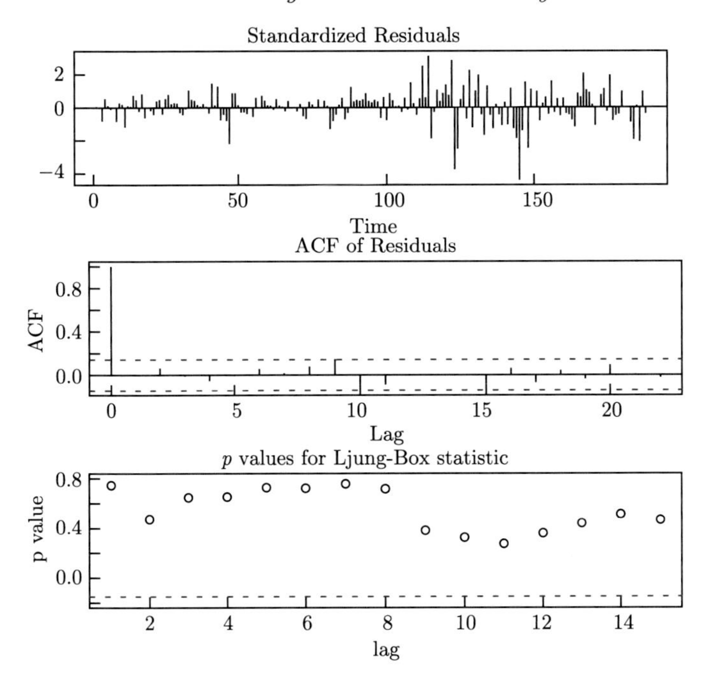

Figure 2.12 Diagnostic plots for the fitted ARIMA(3,1,2) model for the daily gold prices in January to November 2011.

2.6.2 Goodness-of-fit tests for residuals

In principle, the methods for testing white noise described in §1.4.1 may apply to testing if residuals are white noise. However one complication here is that those residuals resulting from a fitted model are not raw data. The fact that the data have been used to estimate the parameters in the model leads to the reduction of the "freedom" in the tests. For example, when the Ljung–Box statistic Q_m defined in (1.15) is applied for testing the residuals from fitting a stationary ARMA(p, q) model, the degree of freedom needs to be adjusted from χ^2_m to χ^2_{m-p-q}. Namely, the reference distribution for computing the P-value should now be χ^2-distribution with $m-p-q$ degrees of freedom. Note that this adjustment is similar to (2.58). In practice we often apply the Ljung–Box tests with different values of $m (> p+q)$.

The third panel in the diagnostic plot produced by the R-function `tsdiag` plots the p-value of the Ljung–Box test for the residuals against the values of $m-p-q$; see Figures 2.12 and 2.13. It is interesting to note that for the gold prices in January to November 2011, the Ljung–Box test for the residuals from the fitted ARIMA(3,1,2) model is not significant at all lags,

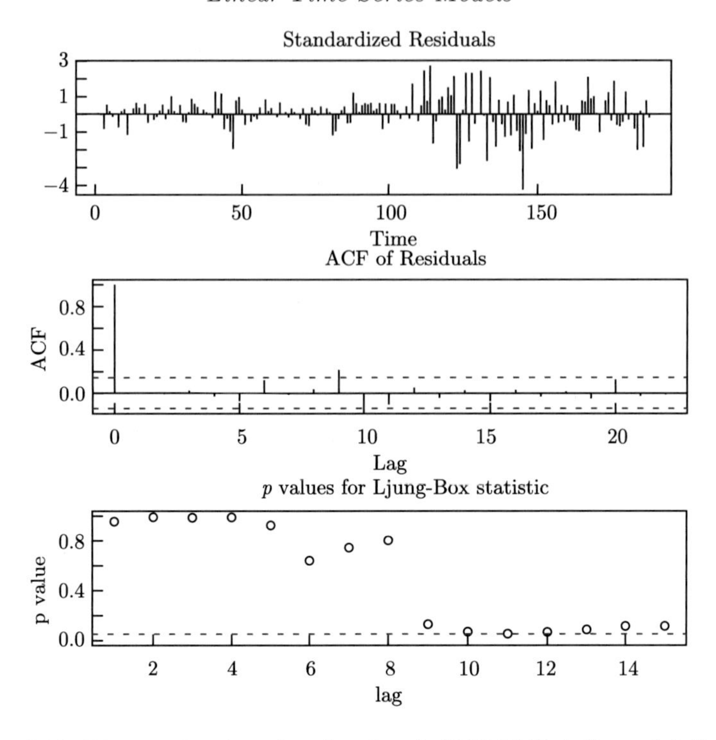

Figure 2.13 Diagnostic plots for the fitted ARIMA(0,1,1) model (i.e. exponential smoothing) for the daily gold prices in January to November 2011.

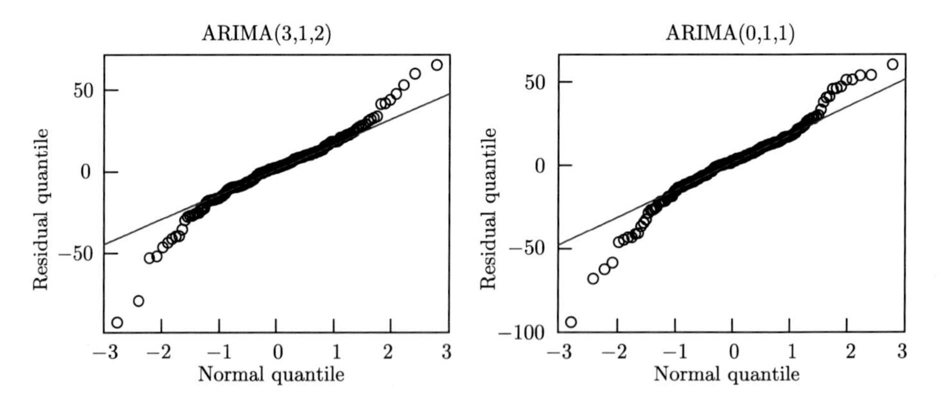

Figure 2.14 The Q–Q plot of the residuals from fitted ARIMA models for the daily gold prices in January to November 2011 against normal distributions.

while the same test for the residuals from the fitted ARIMA(0,1,1) model leads to very small p-values when $m \geqslant 10$ (i.e. $m - p - q = m - 1 \geqslant 9$).

2.7 Model identification based on information criteria

Although the model diagnostics methods based on residuals are effective and widely used in practice beyond time series modeling, they suffer from an obvious drawback that they are powerless in detecting overfitting. Overfitting often leads to an unnecessarily complicated model with some redundant parameters, and hence increases the errors in the estimated parameters. It also makes the interpretation of the fitted model more difficult. A good strategy to build a model may be to combine the consideration on both the goodness of the fit and the simplicity of the model. Akaike's information criterion(*AIC*) (Akaike 1973) and its numerous variations are proposed to select an optimal model based on the trade-off between those two factors.

For a fitted model by the maximum likelihood method, the AIC is defined as

$$\begin{aligned} \text{AIC} = & -2(\text{maximized log-likelihood}) \\ & +2(\text{No. of estimated parameters}). \end{aligned} \tag{2.60}$$

For a stationary ARMA(p, q) model with normal innovations, it reduces to

$$\text{AIC}(p, q) = T \log(\widehat{\sigma}^2) + 2(p + q + 1), \tag{2.61}$$

where $\widehat{\sigma}^2$ is the Gaussian MLE for the variance of innovations defined in (2.49).

The first term on the RHS of (2.61) reflects the goodness of fit of the model. An increase in the value of p and q leads to a decrease in $\widehat{\sigma}^2$. The second term on the RHS of (2.61) is a penalty for the complexity of the model. It increases when p or q increases. We select the order (p, q) which minimizes AIC(p, q).

The AIC-criterion is resulted from the consideration of minimizing the Kullback–Leibler information distance between the unknown true model and the fitted model. A gentle illustration of this derivation may be found in §3.4.1 of Fan and Yao (2003). We cite Kitagawa (2010) for a more systematic treatment of the AIC and the related topics. In particular, AIC does not lead to a consistent order selection (Akaike 1970; Shibata 1980; Woodroofe 1982).

As the AIC tends to overestimate the orders, a popular alternative is the *Bayesian information criterion* (BIC, Schwarz, 1978) given by

$$\text{BIC}(p, q) = T \log(\widehat{\sigma}^2) + (p + q + 1) \log T. \tag{2.62}$$

Since BIC penalizes the order more severely than AIC (when $\log T > 2$), it selects orders of ARMA models that are no larger than those by the AIC. Indeed, it can result in oversimplified ARMA models.

Since Akaike's pioneering work on AIC, various information criteria have been developed; see Choi (1992) for a survey. For example, Hurvich and Tsai (1989) proposed to increase the penalty by inserting a factor $T/(T-p-q-2)$. This leads to the following AICC criterion

$$\text{AICC}(p,q) = T\log(\widehat{\sigma}^2) + \frac{2(p+q+1)T}{T-p-q-2}. \tag{2.63}$$

Still, AIC and BIC are among the most frequently used information criteria in practice. Empirical experience suggests that AIC is a good starting point.

Since the Gaussian MLE for the coefficients of an ARMA model depends on the ACF of the time series only [see (2.48) and §3.2 and §3.3 of Fan and Yao (2003)], there may exist quite a few different ARMA models that provide almost equally good approximations of the sample ACF of the observed data set. Therefore, we may consider the models with AIC values within a small distance from the minimum AIC value as competitive candidates. Selection among the competitive models may be based on interpretation, simplicity, diagnostic checking or other considerations. If we prefer a simple model that reflects the main and interpretable features, we may also try BIC, for example. On the other hand, forecasting based on an AR model with a slightly overestimated order does little harm. In fact Shibata (1980) showed that AIC is asymptotically efficient, while BIC is not. The asymptotic efficiency is a desirable property defined in terms of the one-step mean square prediction error achieved by the fitted model. The AIC was not designed to be consistent, nor is its inconsistency necessarily a defect (Hannan 1986).

For the daily gold prices in January to November 2011 analyzed before, the table below list the values AIC, AICC and BIC (all substracted by 1600 to facilitate reading) for the three fitted models, namely ARIMA(3,1,2), exponential smoothing model ARIMA(0,1,1), and random walk model ARIMA(0,1,0).

	ARIMA(3,1,2)	ARIMA(0,1,1)	ARIMA(0,1,0)
AIC	86.63	91.39	89.54
AICC	87.09	91.46	89.56
BIC	106.05	97.86	92.78

Among those three models, both AIC and AICC are in favor of ARIMA(3,1,2) which passes the Ljung–Box goodness of fit test (see Figure 2.12), and provides the best post-sample forecasting (see §2.5.3). On the other hand, BIC chooses the random walk model ARIMA(0,1,0), and prefers the exponential smoothing ARIMA(0,1,1) to ARIMA(3,1,2). Note the Ljung–Box test indicates that there exist some autocorrelations in the residuals (see Figure 2.13).

If we enlarge our search among all ARIMA$(p, 1, q)$ models with $0 \leqslant p, q \leqslant 5$, AIC choose ARIMA(5,1,4) (with AIC=1686.47) as the best model with ARIMA(3,1,2) as the close second best, AICC selects ARIMA(3,1,2) as the best model with ARIMA(2,1,2) as the second best, and BIC chooses ARIMA(0,1,0) and ARIMA(0,1,1) as the two best models. Combining the post-sample predication performance and the results from diagnostic tests, we recommend ARIMA(3,1,2) as an overall competitive model for this particular data set.

It is worth pointing out that the AIC values reported above were returned by the *R* function `arima`. *R* calculates AIC values based on formula (2.60) which differs from (2.61) by a common constant. This difference has no bearings in the model selection.

In applying AIC or BIC, we always assume that all candidate models are nested in a smooth parametric family. Furthermore, we assume that the time series is stationary and the true model is contained in the family concerned. The extension for nonstationary cases can be found in, e.g., Pötscher (1989) and Kim (1998). The information criteria for selecting the best approximation when the true model is not within the given family was investigated by, e.g., Konishi and Kitagawa (1996).

2.8 Stochastic and deterministic trends

Section 2.3.1 indicates that ACF plots decay slowly for random walk and ARIMA types of nonstationary processes. The same phenomenon may be caused by the existence of a deterministic trend. Figure 2.15 displays two time series generated from two different models, namely a random walk model

$$X_t = X_{t-1} + \varepsilon_t, \tag{2.64}$$

and a white noise with a deterministic trend

$$Y_t = 0.1t + \varepsilon_t, \tag{2.65}$$

where ε_t are independent $N(0, 1)$ random variables. The profiles of the sample ACFs for both the processes are very similar: close to 1 at small lags and decay slowly when the lag increases; see the two lower panels in Figure 2.15. For random walk defined in (2.64), X_t may be written as

$$X_t = X_0 + \sum_{j=1}^{t} \varepsilon_t = X_0 + v_t,$$

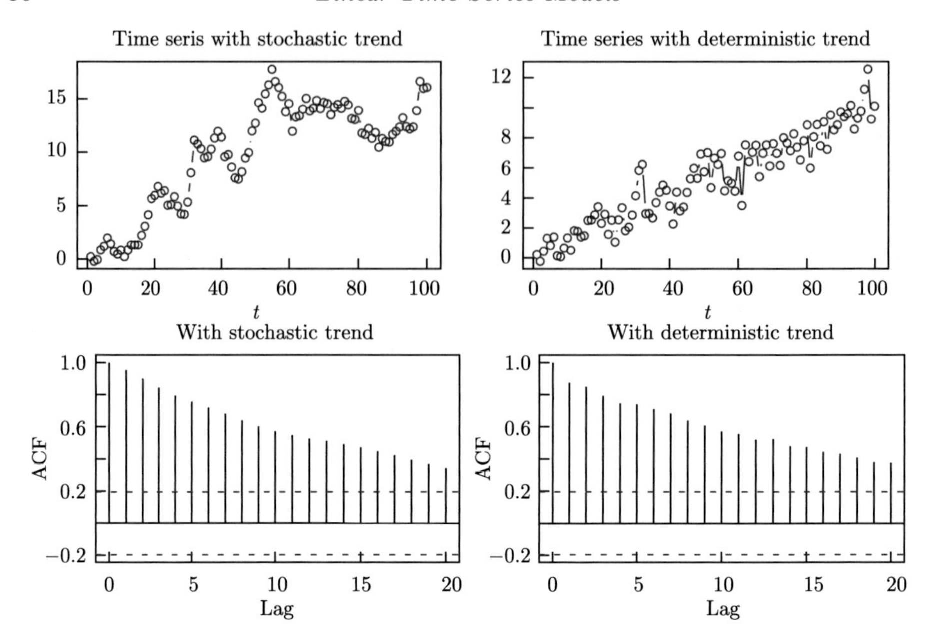

Figure 2.15 Top-left: time series plot of a random walk. Top-right: time series plot of white noise plus a linear trend $0.1t$. Bottom-left: sample ACF of the random walk. Bottom-right: sample ACF of the white noise with a linear trend.

where $v_t = \sum_{j=1}^{t} \varepsilon_t$ is called a stochastic trend. It follows from (2.36) that

$$\text{Corr}(X_t, X_{t+k}) = \sqrt{t/(t+k)} > 0, \qquad k = 0, 1, \ldots.$$

The increasing momentum exhibited in the top-left panel in Figure 2.15 is an artifact due to the strong positive correlations at the nearby time points and the increasing variances over the time. In fact different realizations from random walk model (2.64) may show completely different "trends" (hence the wording "stochastic trend"). This is in marked contrast to the increasing deterministic trend $0.1t$ in model (2.65), see also the top-left panel in Figure 2.15.

2.8.1 Trend removal

As we have discussed in Section 2.3.1, the time series models with stochastic trends are important in modelling nonstationary phenomena such as asset price fluctuations. An effective way to handle those stochastic trends is to use ARIMA models, i.e. to take differences and then to model the differenced data by stationary ARMA models. The same strategy can also be applied to the models with deterministic trends. For example for Y_t defined by (2.65), $\nabla Y_t = 0.1 + \varepsilon_t - \varepsilon_{t-1} \sim \text{MA}(1)$, and hence $Y_t \sim \text{ARIMA}(0,1,1)$. If the deterministic trend is a quadratic function such as $Y_t = at^2 + bt + c + \varepsilon_t$, then $\nabla^2 Y_t = 2a + \varepsilon_t - 2\varepsilon_{t-1} + \varepsilon_{t-2} \sim \text{MA}(2)$. In practice, a trend, whether stochastic or deterministic, may be removed by taking differences enough times. A drawback of differencing data several times is that each differencing is likely to make the dependence structure of the data more complicated. Therefore it is important in practice to avoid overdifferencing by looking carefully at the ACF of each differenced series in succession.

For the time series with deterministic trends, we may also use an appropriate regression model to remove trends. For example, a simple linear regression $Y_t = a+bt+\varepsilon_t$ will remove the linear trend from the data generated from (2.65) successfully. The advantage for doing so is that the residuals from this regression fitting are likely to retain the simple structure (e.g. white noise) in the original model. However we should only use a parametric regression model to remove a trend when we have good reasons for assuming such a concrete parametric form. For example, using a linear regression to remove the trend in the series displayed in the top-left panel of Figure 2.15 would be disastrous in spite of the perceived increasing trend, as the underlying dynamics is a random walk and thus fitting a deterministic linear trend is completely misleading. For the use of nonparametric regression techniques to remove deterministic trends, we refer to §6.2 of Fan and Yao (2003). On the other hand, it is safer to take differences, as the differencing can handle both deterministic and stochastic trends even without knowing which is which.

2.8.2 Augmented Dickey–Fuller test

The slow decay of the sample ACF is often taken as a symptom of nonstationary time series due to the existence of either dynamic or deterministic trend. Differencing is often adopted, which eventually leads to the fitting of the data with an appropriate ARIMA model. Nevertheless we would also like to apply some formal statistical tests to assess if differencing is necessary. The *augmented Dickey–Fuller (ADF) test* is one of such tests. Furthermore it can be used to identify some concrete parametric forms for the underlying deterministic trends. The ADF test is essentially the same as the Dickey–Fuller test presented in Section 1.4.2, except it applies to the following extended models: With $Z_t = \nabla X_t = X_t - X_{t-1}$,

$$X_t = \alpha X_{t-1} + b_1 Z_{t-1} + \cdots + b_p Z_{t-p} + \varepsilon_t, \tag{2.66}$$

$$X_t = \mu + \alpha X_{t-1} + b_1 Z_{t-1} + \cdots + b_p Z_{t-p} + \varepsilon_t, \tag{2.67}$$

$$X_t = \mu + \beta t + \alpha X_{t-1} + b_1 Z_{t-1} + \cdots + b_p Z_{t-p} + \varepsilon_t. \tag{2.68}$$

The extended models allow lags of differenced series, drift and linear trends in the model. The null hypothesis is $H_0 : \alpha = 1$ and the alternative hypothesis is $H_1 : \alpha < 1$. The former represents nonstationary process whereas the latter indicates stationary process. When $\alpha = 1$, the first two models in (2.66) states further that Z_t is an AR(p) model, whereas the third model is an AR(p) model with a linear trend.

Let $\widehat{\alpha}$ be the least squares estimator for α, and SE($\widehat{\alpha}$) denote its standard error. We reject H_0 when the ADF test statistic W is smaller than a critical value, where $W = (\widehat{\alpha} - 1)/\text{SE}(\widehat{\alpha})$ is in the same form as the Dickey–Fuller test statistic (1.19). The critical values for the ADF test are taken as the same as those for the Dickey–Fuller test, and are listed in Table 1.2 in Section 1.4.3. Further improvement from making the critical values depend on the autoregressive order p and the sample size n can be found in, e.g., Cheung and Lai (1995). The extension of the test for dealing with more complex structures of ε_t in models (2.66)–(2.68) was studied by Phillips and Perron (1988).

An implicit assumption in models (2.66)–(2.68) is that the autoregressive coefficients $b_1, \ldots, b_p$ satisfy the condition that all the roots of the equation $1 - b_1 x - \cdots - b_p x^p = 0$ are outside of the unit cycle. Hence if the null hypothesis $H_0 : \alpha = 1$ cannot be rejected within (2.66), the differenced process Z_t may be stationary AR(p). If H_0 cannot be rejected within (2.67), Z_t may be stationary AR(p) with a non-zero drift. In this case, the original process X_t may contain a deterministic trend $c + \mu t$ (and a stochastic trend as well). Note that the ADF test is incapable of distinguishing between a deterministic trend and a stochastic trend. If H_0 cannot be rejected within (2.68), it is possible that X_t contains a (deterministic) quadratic trend. Note that some real time series may contain both deterministic and stochastic trends. The settings in (2.66)–(2.68) do allow them to be present together.

The *R*-code "aDF.test.r" defines a function `aDF.test` which implements the ADF test: `aDF.test(x, kind=i, k=p)`, where `x` is a data vector, `p` is the autoregressive order in the model, and `i` is set at 1 for model (2.66), 2 for model (2.67), and 3 for model (2.68). We apply `aDF.test` to the two data sets displayed in the two top panels of Figure 2.15 with `k=0`, and list the obtained p-values below.

Model used	(2.66)	(2.67)	(2.68)
random walk data	> 0.90	0.59	0.64
data with trend $0.1t$	> 0.90	0.24	< 0.01

For the data generated by the random walk model (2.64), the ADF test cannot reject the null hypothesis $\alpha = 1$ in all the three models (2.66)–(2.68). This suggests that we may difference the data first. For the data generated by model (2.65) which has a linear deterministic trend $0.1t$ only, the ADF test correctly rejects the hypothesis $\alpha = 1$ in model (2.68) at the significance level 1%. This confirms that the differenced data do not contain a deterministic linear trend. The test cannot reject the null hypothesis $\alpha = 1$ under model (2.67) – this is the correct inference, as the differenced data contain a drift $\mu = 0.1$ now. The test also fails to reject the null hypothesis under model (2.66). This should not be taken for granted as a vindication for the hypothesized model. Instead the test only confirms that there exists no significant evidence against the hypothesized model, and we may take the outcome of "not rejecting H_0" as an indication for differencing the data. Although the p-values listed above were obtained with the option `k=0` in `aDF.test`. Using `k=1` or `k=2` produces the similar p-values, and will not alter the outcomes of the tests.

2.8.3 An illustration

We now consider the daily S&P 500 index prices in November 16, 2011 to April 5, 2012. We choose this period as the log prices exhibit a linear increasing trend; see Figure 2.16. The sample ACF decays linearly and very slowly; suggesting the existence of either stochastic or deterministic (linear) trend. By applying the ADF test to the log price series, the R-function `aDF.test(y, kind = 1, k = i)` for `i` between 0 and 6 return the p-values greater than 0.9. Hence we cannot reject the hypothesis $\alpha = 1$ with model (2.66). This suggests that we may difference the log prices first, which may lead to fitting an ARIMA model to the log price series, or equivalently, a stationary ARMA model to the return series. To carry out the test based on model (2.67), `aDF.test(y, kind = 2, k = i)` for `i` between 0 and 6 also return the p-values greater than 0.25. This suggests that the log prices in this period may contain a linear deterministic trend. As `aDF.test(y, kind = 3, k = i)` for `i` between 0 and 6 return the p-values smaller than 0.06, we reject model (2.68). We can safely conclude that there is no evidence indicating any (deterministic) quadratic trends in the prices.

The analysis based on the ADF test above suggests that there may exist a linear deterministic and/or stochastic trend in the S&P 500 index prices

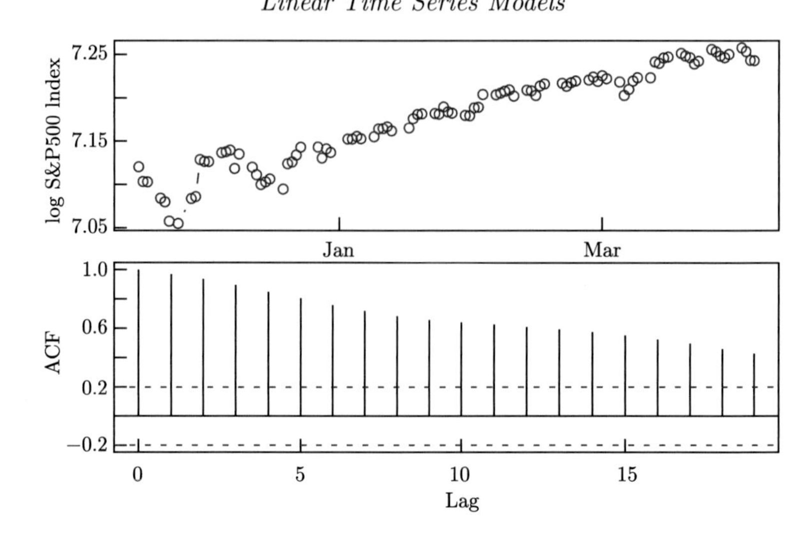

Figure 2.16 The plot of the log daily S&P 500 index prices in November 16, 2011 to April 5, 2012 (the top panel), and their sample ACF plot (the low panel).

in this period. Let Y_t denote the log price at time t. Regressing Y_t on time t, we obtained a fitted linear model:

$$
\begin{aligned}
Y_t &= 7.089 + 0.002t + X_t, \quad t = 1, \ldots, 97, \\
X_t &= 0.721X_{t-1} + 0.197X_{t-2} - 0.275X_{t-3} + \varepsilon_t,
\end{aligned}
\tag{2.69}
$$

where $\varepsilon_t \sim \mathrm{WN}(0, (0.009)^2)$. In the above expression, the residuals X_t from the linear regression were fitted with an AR(3) model. Note that the residuals X_t are plotted together with their sample ACF and PACF in the three panels on the left in Figure 2.17. The PACF plot indicates that AR(3) is an appropriate model which is also selected by the AIC. Also displayed in Figure 2.17 are the returns (i.e. the differenced log prices) together with their sample ACF and PACF. Now the PACF plot suggests an AR(8) model

```
AR/MA
  0 1 2 3 4 5 6 7 8 9 10 11 12 13
0 o o o o o x o x o o o  o  o  o
1 x o o o o x o x o o o  o  o  o
2 o x o o o o o o o o o  o  o  o
3 o x o o o o o o o o o  o  o  o
4 x x o o o o o o o o o  o  o  o
5 o x x o o o o o o o o  o  o  o
6 o x o o o o o o o o o  o  o  o
7 o x o o o o o o o o o  o  o  o
```

and the ACF plot suggests an MA(8) model for the return data. However the EACF above indicates ARMA(2,2) is a more parsimonious alternative, which in fact has a smaller AIC value.

The fitted ARMA(2,2) model for the return data is

$$Z_t = 0.040Z_{t-1} + 0.043Z_{t-2} + \varepsilon_t - 0.103\varepsilon_{t-1} + 0.194\varepsilon_{t-2}, \tag{2.70}$$

where $Z_t = Y_t - Y_{t-1}$, and $\varepsilon_t \sim \text{WN}(0, (0.009)^2)$.

For the log prices of the S&P 500 index in the selected period, both models (2.69) and (2.70) provide reasonable fitting for the increasing dynamics exhibited in the data. The preference between the two depends on the goal of analysis. The models with deterministic trends are relatively easier to interpret. However the trend function is not flexible and should only be used if we are certain that such a trend function is practically meaningful. On the other hand, ARIMA models such as (2.70) are flexible enough to catch up the direction-changes of the trend although they tend to be more complex. For example, if our goal is to predict the future prices of the S&P 500 index, model (2.70) would be a safer option unless we are certain that the increasing momentum exhibited in Figure 2.16 will be sustained for a while in the future. Further discussion on forecasting trends and momentum of financial markets will be presented in Section 2.9.2.

2.8.4 Seasonality

Some financial time series exhibit periodic behavior. A typical example is corporate quarterly earnings per share, which exhibits both an increasing trend and a season pattern. Figure 2.18 depicts the quarterly earning of IBM and Johnson and Johnson pharmaceutical company from 1984 to 2013. IBM experienced operation losses from the last quarter of 1992 to the third quarter of 1993 and hence their logarithms do not exist.

Let X_t be the time series of the logarithm of the quarterly earnings of the Johnson and Johnson pharmaceutical company. The company records a loss of five cents in the first quarter of 1986. To facilitate the illustration, we replace it by 0.07, the average of earning of the first quarter of 1985 and that of 1987.

Serial correlation is significantly weakened by a differencing: $\nabla X_t = X_t - X_{t-1}$. Comparing the ACF functions given in Figure 2.19(a) and (b). The oscillation patterns are clearly seen due to seasonality. To remove the *seasonality* effect, we typically employ a seasonal difference:

$$\nabla_4 X_t = (1 - B^4)X_t = X_t - X_{t-4}. \tag{2.71}$$

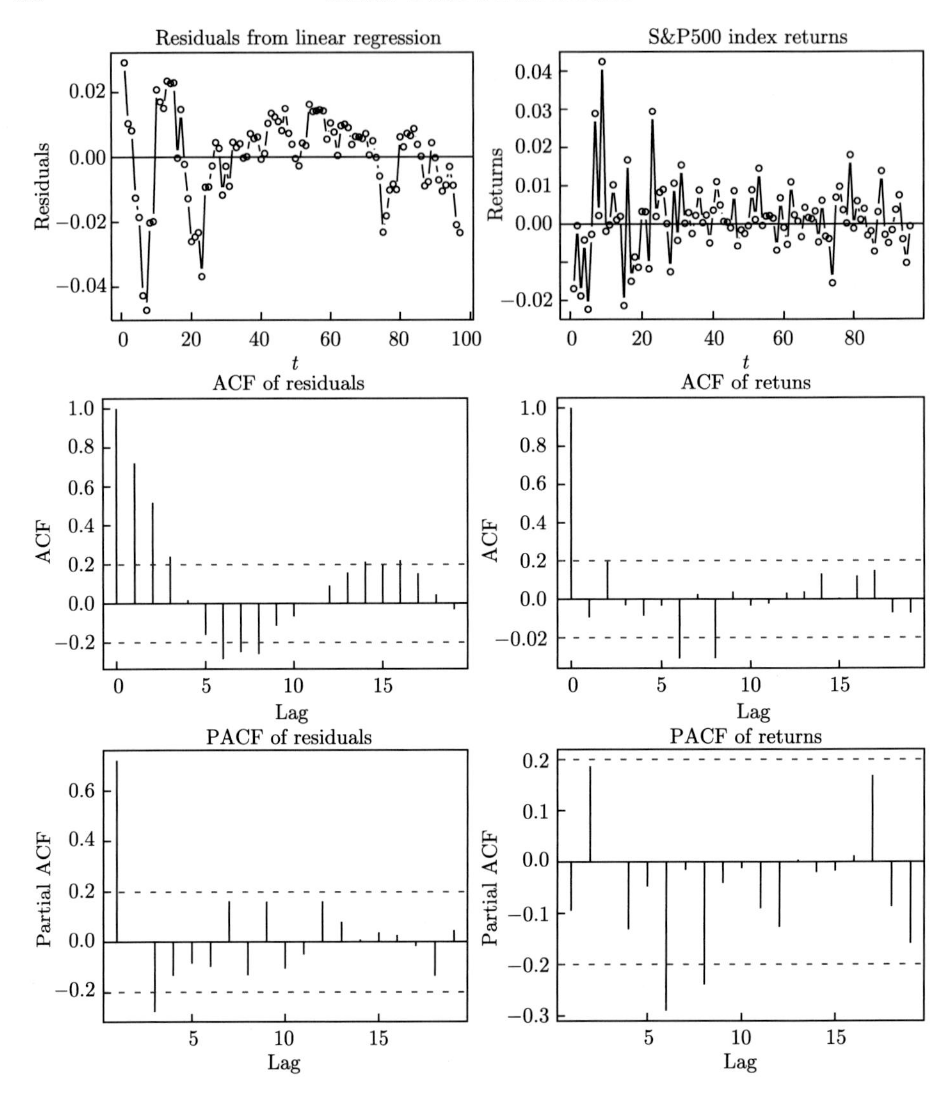

Figure 2.17 Time series plot of the residuals from fitting a straight line regression to the log daily S&P 500 prices from November 16, 2011 to April 5, 2012 and the their sample ACF and PACF plots (the 3 panels on the left). Time series plot of the returns of the S&P 500 index and their sample ACF and PACF plots (the 3 panels on the right).

This removes the seasonal pattern. The ACF of the resulting time series is given in Figure 2.19(c). In general, seasonal difference with periodicity m applies the operator $\nabla_m = (1-B^m)$. Serial correlation and seasonality can be handled simultaneously by

$$(1-B^4)(1-B)X_t = X_t - X_{t-1} - X_{t-4} + X_{t-5}. \tag{2.72}$$

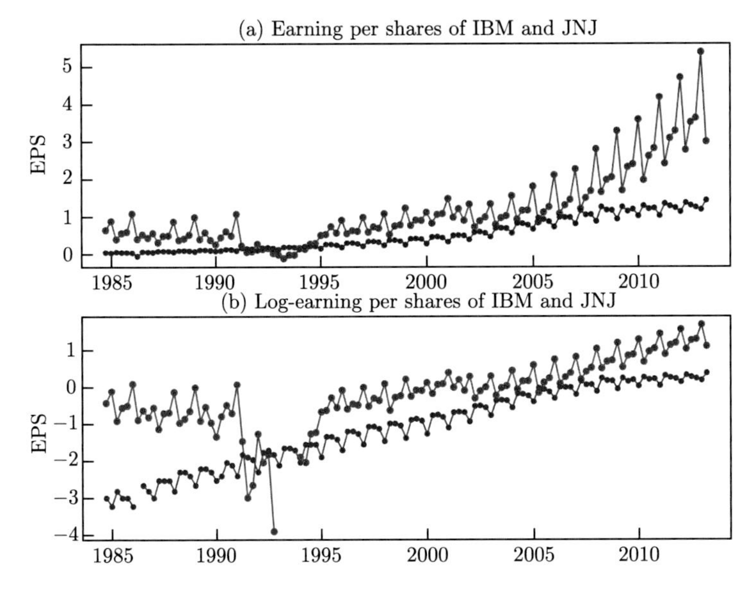

Figure 2.18 Time Quarterly earnings per share of IBM (red) and Johnson and Johnson (blue) from 1984–2013. Top panel: earning per share; bottom panel: logarithm of earings per share.

The resulting series has ACF given in Figure 2.19(d). With the de-trend and de-seasonalized data, we can now fit ARMA models. In fact, de-seasonalized data has already removed linear trend. Therefore, it can also be fitted directly to an ARMA model.

2.9 Forecasting

One of the primary goals in time series analysis is to forecast the future values. This is often done by building up an appropriate model (such as ARIMA) based on the historical data. Often for financial applications some empirical rules play important and effective roles in forecasting market trends and momentum. In this section, we first discuss the model-based forecasting. It applies to the forecasting with stationary time series, or at least the series of which the differences are stationary. We then give a brief introduction of the so-called *technical analysis* widely used in financial trading to detect trends and momentum of financial markets. The direct goal of the analysis is to forecast the directions of price movements rather than prices themselves. The exponential smoothing technique (i.e. the forecaster based on ARIMA(0,1,1) model) introduced in Section 2.3.2 is the key building block for the indicators used on the technical analysis.

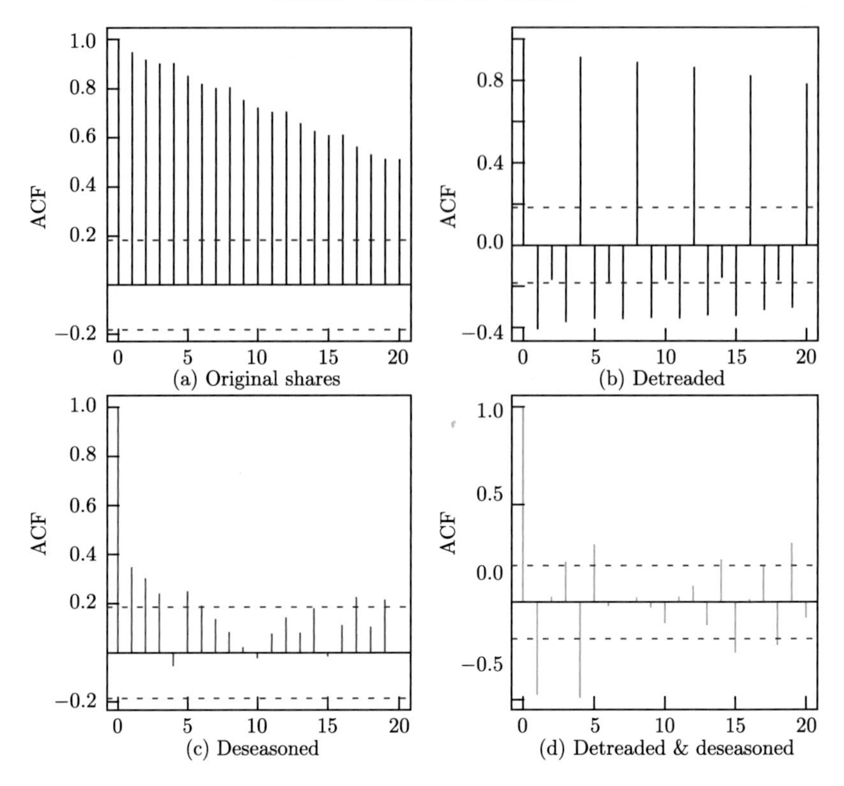

Figure 2.19 Autocorrelation functions for the logarithm of earning per share of Johnson and Johnson from 1984–2013. (a) ACF of the original time series $\{X_t\}$. (b) ACF of de-trend series $\{\nabla X_t\}$; (c) ACF of de-seasonalized series $\{\nabla_4 X_t\}$; (d) ACF of de-trend and de-seasonalized data $\{(1-B^4)(1-B)X_t.\}$

2.9.1 Forecasting ARMA processes

In this section we assume that the model is known completely, including all the parameters. Although such an assumption is never true, it simplifies the derivation substantially. In practice unknown parameters are replaced by their estimates.

Let $X_1, \ldots, X_T$ be the observations from a time series. Our goal is to forecast the future values X_{T+k} for $k \geqslant 1$. Assuming no extra information available, we forecast the future based on the observations $X_1, \ldots, X_T$ only. Denote by $X_T(k)$ the predictor for X_{T+k}. We may seek for the predictor $X_T(k) = f_0(X_1, \ldots, X_T)$ as a function of observed data such that the mean squared predictive error (MSPE)

$$E[\{X_{T+k} - f(X_1, \ldots, X_T)\}^2] \tag{2.73}$$

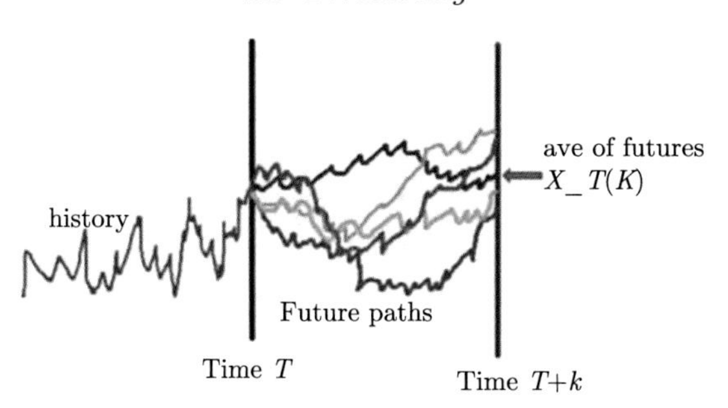

Figure 2.20 Illustration of the best prediction of X_{T+k} given $X_1, \ldots, X_T$, which is the average of possible values.

is minimized among all predictors f. This leads to the least squares predictor

$$X_T(k) = E(X_{T+k}|X_T, X_{T-1}, \ldots, X_1), \quad k \geqslant 1.$$

See Proposition 3.2 of Fan and Yao (2003). The intuition of this is very clear. The best predictor $X_T(k)$ is simply the average of all possible realizations X_{T+k} given the information given up to time point T. Figure 2.20 gives an illustration of this. Note that so far we have not imposed any condition on $\{X_t\}$. Technically it is more convenient to assume that we know the whole infinite past of time series. Consequently the above least square predictor can be adjusted to

$$X_T(k) = E(X_{T+k}|X_T, X_{T-1}, \ldots) = E_T X_{T+k}, \quad k \geqslant 1 \tag{2.74}$$

where and thereafter we use the notation E_T to denote the conditional expectation given the information up to time T. When the sample size T is large, it has a little effect on $X_T(k)$ to assume that we have also observed $X_0, X_{-1}, X_{-2}, \ldots$.

It is also instructive to consider the class of linear predictors:

$$\beta_0 + \beta_1 X_1 + \cdots + \beta_T X_T.$$

The *best linear predictor* is to find the coefficient vector $\boldsymbol{\beta}$ that minimizes

$$E[\{X_{T+k} - \beta_0 - \beta_1 X_1 - \cdots - \beta_T X_T\}^2] \tag{2.75}$$

By expanding the above quadratic function, the criterion function (2.75) depends only on the first two moments of the process, namely the mean and autocovariance functions. Therefore, the optimal linear predictors depends only on the mean and autocovariance functions. It depends also on k.

Clearly, the class of all predictors used in (2.73) is much bigger than that of linear predictors. Therefore, the least squares predictor should have smaller

MSPE than the best linear predictor. However, the least squares predictor is much harder to estimate from empirical data. The conditional expectation (2.74) is determined by the joint distribution of $(X_{T+k}, X_T, X_{T-1}, \ldots)$. It may be a nonlinear function of the past values and cannot be determined completely by the autocorrelation function. However, for a special class of ARMA model, the best prediction rule happens to be linear as illustrated by the following example.

Example 2.10 Let us consider $AR(1)$ model

$$X_t = bX_{t-1} + \varepsilon_t,$$

where $|b| < 1$, $\varepsilon_t \sim \mathrm{WN}(0, \sigma^2)$. Assume further the unpredictable condition:

$$E\{\varepsilon_t | X_{t-1}, X_{t-2}, \ldots\} = 0, \tag{2.76}$$

which is weaker that the condition $\varepsilon_t \sim \mathrm{IID}(0, \sigma^2)$. Then the one-step ahead predictor is

$$X_T(1) = E_T\{bX_T + \varepsilon_{T+1}\} = bX_T,$$

which is linear. Its mean squared predictive error is

$$\mathrm{MSPE}\{X_T(1)\} = E[\{X_{T+1} - X_T(1)\}^2] = E(\varepsilon_{T+1}^2) = \sigma^2.$$

To forecast X_{T+2}, note that

$$X_{T+2} = \varepsilon_{T+2} + bX_{T+1} = \varepsilon_{T+2} + b\varepsilon_{T+1} + b^2 X_T.$$

By (2.76), it holds that

$$X_T(2) = E_T X_{T+2} = b^2 X_T.$$

What is not predicted in X_{T+2} is the future noises $\varepsilon_{T+2} + b\varepsilon_{T+1}$. Hence,

$$\mathrm{MSPE}\{X_T(2)\} = E\{(\varepsilon_{T+2} + b\varepsilon_{T+1})^2\} = (1 + b^2)\sigma^2.$$

In general, for any $k \geqslant 1$,

$$X_{T+k} = \varepsilon_{T+k} + b\varepsilon_{T+k-1} + \cdots + b^{k-1}\varepsilon_{T+1} + b^k X_T.$$

Thus

$$X_T(k) = b^k X_T,$$

and

$$\mathrm{MSPE}\{X_T(k)\} = (1 + b^2 + \cdots + b^{2(k-1)})\sigma^2.$$

Hence a 95% predictive interval for X_{T+k} is given as

$$X_T(k) \pm 1.96\sigma\sqrt{1 + b^2 + \cdots + b^{2(k-1)}}.$$

When the forecast horizon k increases, the mean squared predictive error increases. When $k \to \infty$, $X_T(k) \xrightarrow{P} 0$, the expected value of $E(X_{T+k})$, and

$$\mathrm{MSPE}\{X_T(k)\} \to \frac{\sigma^2}{1-b^2} = \mathrm{var}(X_{T+k}). \tag{2.77}$$

This indicates that the long-term forecasting is difficult, as the best prediction is the mean of the time series. In fact, the prediction error approaches the unconditional variance.

We consider a causal and invertible ARMA(p,q) model:

$$X_t - b_1 X_{t-1} - \cdots - b_p X_{t-p} = c + \varepsilon_t + a_1\varepsilon_{t-1} + \cdots + a_q\varepsilon_{t-q}, \tag{2.78}$$

where $\varepsilon_t \sim \mathrm{WN}(0,\sigma^2)$ satisfying the unpredictable condition (2.76). An added complication in forecasting with an MA component is due to the fact that we do not observe directly the innovations ε_t. The invertible condition implies that X_t admits the following AR(∞) representation

$$X_t - \mu = \varepsilon_t + \sum_{\ell=1}^{\infty} \varphi_\ell (X_{t-\ell} - \mu), \tag{2.79}$$

where $\mu = c/(1-b_1-\cdots-b_p)$, φ_ℓ are determined by a_i and b_j in the original model (2.78); see §2.2.3. This indicates that we can recover the innovations $\varepsilon_t, \varepsilon_{t-1}, \ldots$ from the observations $X_t, X_{t-1}, \ldots$. Hence

$$\varepsilon_T(k) \equiv E_T(\varepsilon_{T+k}) = \begin{cases} 0 & \text{if } k \geqslant 1 \\ \varepsilon_{T+k} & \text{if } k \leqslant 0, \end{cases} \tag{2.80}$$

namely, the best prediction for the future noise is 0 and for the realized noise is itself. Consequently, noticing that

$$X_{T+1} = c + b_1 X_T + \cdots + b_p X_{T+1-p} + \varepsilon_{T+1} + a_1\varepsilon_T + \cdots + a_q\varepsilon_{T+1-q}$$

and taking E_T on both sides, the one-step ahead predictor admits the expressions:

$$X_T(1) = c + b_1 X_T + \cdots + b_p X_{T+1-p} + a_1\varepsilon_{T-1} + \cdots + a_q\varepsilon_{T+1-q}. \tag{2.81}$$

To use this formula in practice, we assume that $X_t = 0$ for all $t \leqslant 0$. We then recover the values $\varepsilon_1, \ldots, \varepsilon_T$ using (2.79).

More generally, from

$$X_{T+k} = c + b_1 X_{T+k-1} + \cdots + b_p X_{T+k-p} + \varepsilon_{T+k} + a_1\varepsilon_{T+k-1} + \cdots + a_q\varepsilon_{T+k-q},$$

by taking the operation E_T on both sides, we obtain the following recursive formula for multi-step prediction:

$$X_T(k) = c + b_1 X_T(k-1) + \cdots + b_p X_T(k-p) \qquad (2.82)$$
$$+ a_1 \varepsilon_T(k-1) + \cdots + a_q \varepsilon_T(k-q), \quad k \geqslant 1,$$

where $\varepsilon_T(k)$ is defined as in (2.80), and $X_T(j) = X_{T+j}$ for all $j \leqslant 0$.

Comparing (2.81) with (2.78), it is easy to see that

$$\text{MSPE}\{X_T(1)\} = \text{var}(\varepsilon_{T+1}) = \sigma^2.$$

To compute MSPE$\{X_T(k)\}$ for $k > 1$, we appeal to the causal property of model (2.78). Note that the causality implies that X_t admits an MA(∞) expression:

$$X_t - \mu = \varepsilon_t + \sum_{\ell=1}^{\infty} \psi_\ell \varepsilon_{t-\ell}, \qquad (2.83)$$

where ψ_ℓ are determined by a_i and b_j in (2.78); see §2.2.3. One immediate consequence of this MA-representation is a simple expression for the variance of X_t:

$$\text{var}(X_t) = \sigma^2 \Big\{ 1 + \sum_{\ell=1}^{\infty} \psi_\ell^2 \Big\}. \qquad (2.84)$$

For $k \geqslant 1$,

$$X_{T+k} = \mu + \varepsilon_{T+k} + \sum_{\ell=1}^{\infty} \psi_\ell \varepsilon_{T+k-\ell}.$$

Combining this with (2.74) and (2.80), we obtain that

$$X_T(k) = \mu + \sum_{\ell=k}^{\infty} \psi_\ell \varepsilon_{T+k-\ell}, \qquad (2.85)$$

namely the future noises are not predicted. Hence

$$\text{MSPE}\{X_T(k)\} = E[\{X_{T+k} - X_T(k)\}^2] = \sigma^2 \Big\{ 1 + \sum_{\ell=1}^{k-1} \psi_\ell^2 \Big\}, \qquad (2.86)$$

which increases as k increases; indicating that the prediction error increases when the forecast horizon increases.

The long-term forecasting is difficult. When $k \to \infty$, by (2.85), we have $X_T(k) \xrightarrow{P} \mu$. In other words, the best predictor is the mean of the time series itself. In fact, by (2.84), MSPE$\{X_T(k)\} \to \text{var}(X_{T+k})$, the same as unpredicted one.

For an ARIMA process, we may forecast the future values of an appropriately differenced process using the above techniques. For example, if $X_t \sim \text{ARIMA}(p, 1, q)$, the predicted value of $X_{T+k} - X_{T+k-1}$ is close to a constant μ for large k. Consequently the predicted increment from X_T to X_{T+k} is close to μk for large k, which is the linear trend in the process when $\mu \neq 0$.

The gold price data in Section 2.5.3 was used to illustrate the one-step ahead forecast based on ARIMA models; see also Figure 2.10. Now we use the same data to illustrate the multiple-step ahead prediction in *R*. Suppose that the gold prices in January to October 2011 are stored in the object `Gold` in *R*. We fit an ARIMA(3,1,2) model to this data set (see the justification of using this model in Section 2.5.3) and predict the 17 prices in November using the *R*-function `predict.arima`:

```
> fitGold <- arima(Gold, order=c(3,1,2))
> preGold <- predict(fitGold, n.ahead=17)
```

R returns the 17 predicted values in `preGold$pre` and their standard errors in `preGold$se`. A standard error is the square root of the MSPE defined in (2.86). We list the predicted values, the approximate 95% predictive intervals (i.e. predicted value $\pm 1.96 \times \text{SE}$) together with the true values in Table 2.2. The predictive intervals contain the true prices for all those 17 days, however they are much wider than those one-step ahead intervals plotted in Figure 2.10. Furthermore, the width of the interval increases as the prediction horizon increases. This reflects the increasing uncertainty in predicting further ahead into the future, as all the prediction is based on the data up to the end of October only.

2.9.2 Forecasting trends and momentum of financial markets

While the model-based prediction method presented in Section 2.9.1 is useful in forecasting stochastic phenomena exhibiting a certain degree of stationarity, it is typically too rigid to forecast the trends and the momentum of financial markets. Even without the impact of external forces such as wars or natural disasters, financial prices themselves are very nonstationary: the market trend and momentum change randomly with little warning, and each new change tends to be different from previous ones. While the returns of financial prices may be viewed as (weakly) stationary, there exists little serial correlation, as any tangible correlations would be quickly explored by keen market participants. The lack of the stationarity in financial prices and the lack of autocorrelations in their returns are the two fundamental

Table 2.2 *The multiple-step ahead prediction for the gold prices in November 2011*

Date	Predicted price	Predictive interval	True price
1	1671.2	(1631.0, 1711.4)	1673.8
2	1672.9	(1615.9, 1729.9)	1690.6
3	1672.5	(1600.6, 1744.3)	1717.2
4	1671.1	(1586.9, 1755.4)	1708.5
7	1672.4	(1578.7, 1766.1)	1749.8
8	1672.8	(1569.9, 1775.8)	1735.3
9	1671.3	(1559.2, 1783.5)	1720.7
10	1671.9	(1552.2, 1791.5)	1711.4
11	1672.9	(1546.1, 1799.6)	1739.6
14	1671.7	(1537.4, 1806.1)	1732.0
15	1671.5	(1530.7, 1812.4)	1733.6
16	1672.7	(1525.9, 1819.6)	1715.1
17	1672.1	(1518.9, 1825.4)	1671.0
18	1671.4	(1512.2, 1830.7)	1676.2
21	1672.4	(1507.9, 1837.0)	1635.0
22	1672.4	(1502.4, 1842.5)	1653.1
23	1671.4	(1495.9, 1847.1)	1648.3

reasons which prevent ARIMA models from being effective in predicting future market movements. On the other hand, the so called *technical analysis* developed and employed by the trading professionals gains increasing momentum and has proved to be more successful in forecasting market trends and momentum.

A *market trend* refers to a putative tendency of a financial market moving either upwards or downwards for a certain period. These moves are not necessarily strictly monotonic, and are typically coupled with small scale fluctuations. The length of a trend may vary from a few minutes to a few years. An upward trend is often called a bullish trend and a downward trend is referred to as a bearish trend. Furthermore a *momentum* is referred to as a tendency for rising prices to rise further, or falling prices to keep falling. Although it is difficult, if ever possible, to give a universally accepted definition for trend and momentum, they are widely used concepts in the financial market. Many *technical indicators* have been developed to detect market trends and momentums. We introduce two classes of indicators below.

MACD

Developed by Gerald Appel in late 1970s, the *moving average convergence–divergence* (MACD) is one of the simplest and the most often used trend and momentum indicators. See Appel (2009). Let P_t be the price at time t. For a given parameter k, the exponential moving average is defined as

$$\mathrm{EMA}_k(t) = \frac{2}{k+1} P_t + \frac{k-1}{k+1} \mathrm{EMA}_k(t-1). \tag{2.87}$$

It is easy to see that $\mathrm{EMA}_k(t)$ is an exponential smoothing forecast for P_{t+1}:

$$\mathrm{EMA}_k(t) = (1-\lambda) \sum_{j=0}^{\infty} \lambda^j P_{t-j}, \qquad \lambda = (k-1)/(k+1), \tag{2.88}$$

see (2.39) and (2.38). Thus the above EMA is the predictor for P_{t+1} based on an ARIMA(0,1,1) model for the price process P_t with the moving average coefficient $\lambda = (k-1)/(k+1)$. Even though the summation in (2.88) involves an infinite sum, the weighting coefficients in the sum decay exponentially fast. Therefore, there are effectively finitely many terms involved in computing (2.88). For an integer k, the coefficient of the $(k+1)$th is

$$\lambda^k = \left(1 - \frac{2}{k+1}\right)^k \to e^{-2} = 0.135 \quad \text{as } k \to \infty,$$

which is non-negligible, but its $(2k+1)$ coefficient is approximately $e^{-4} = 0.018$, which is negligible. For this reason, (2.88) is also referred to as the k-period exponential moving average.

We now use the exponential smoothing to define moving average convergence–divergence. For two given parameters $l > s \geqslant 1$, representing short-term and long-term moving averages, MACD is defined as

$$\mathrm{MACD}(t) \equiv \mathrm{MACD}_{s,l}(t) = \mathrm{EMA}_s(t) - \mathrm{EMA}_l(t),$$

i.e. it is the difference between a shorter exponential moving average with period s and a longer exponential moving average with period l. As the name suggested, the information in an MACD is reflected by the convergence and the divergence of those two exponential moving averages. The shorter moving average $\mathrm{EMA}_s(t)$ is faster and responsive to market movements while the longer moving average $\mathrm{EMA}_l(t)$ is slower and less reactive to price changes.

MACD oscillates above and below the zero line $y = 0$, which is called the centerline on an MACD chart. Positive MACD means that the short-term moving average is greater than the long-term moving average; indicating a bullish trend and the time to buy. Positive values increase as $\mathrm{EMA}_s(t)$ diverges further above $\mathrm{EMA}_l(t)$. This means that the upward momentum is increasing. Negative MACD means that the short-term moving average is smaller than the long-term moving average; indicating a bearish trend, and the time to sell. Negative values increase as $\text{EMA}_s(t)$ diverges further below $\text{EMA}_l(t)$. This means that downward momentum is increasing.

It is informative to superimpose a signal line in an MACD chart. A k-period signal is defined as

$$\text{Signal}(t) \equiv \text{Signal}_k(t) = \frac{2}{k+1}\text{MACD}(t) + \frac{k-1}{k+1}\text{Signal}_k(t-1),$$

i.e. Signal$(t;k)$ is an exponential smoothing of MACD(t) with $\lambda = \dfrac{(k-1)}{(k+1)}$; see (2.87) and (2.88). Signal line crossovers are the most common MACD signals. As a moving average of the MACD indicator, $\text{Signal}_k(t)$ trails the MACD and makes it easier to spot MACD turns. A bullish crossover occurs when the MACD turns up and crosses above the signal line, signaling the time to buy. A bearish crossover occurs when the MACD turns down and crosses below the signal line. The length of crossover depends on the strength of the momentum.

The conventional setting for MACD is to use $s = 12$, $l = 26$, and $k = 9$ for daily prices, and $s = 5$, $l = 35$, and $k = 5$ for weekly prices. An MACD with smaller s and larger l is more sensitive to the market movement, creating more centerline crossovers and signal line crossovers.

A word of caution in using the above indicators: since all those measures are defined in terms of exponential smoothing, a crossovers of MACD over either the centerline or the signal line typically lags behind the direction of the market's change. More importantly a crossover itself does not give any information on the length and the magnitude of a trend. For the latter, the *MACD histogram* is defined as

$$\text{Hist}(t) = \text{MACD}(t) - \text{Signal}(t)$$

is also produced along an MACD chart. Large positive values of Hist(t) indicate strong momentum of bullish market. Large negative values of Hist(t) indicate strong momentum of bearish market.

We now present some real data illustration of the indicators introduced above using the *R*-package "*quantmod*" – a quantitative financial modelling and trading framework in *R* with its homepage at *www.quantmod.com*. This package provides functions `MACD`, `EMA` etc. to calculate those technical indicators directly. It also provides powerful graphical functions for technical analysis. To install and upload the package, we call in *R*.

```
> install.packages("quantmod")
> library(quantmod)
```

The *quantmod* package offers a variety of tools for downloading, extracting and displaying daily prices in the OHLC format. (The abbreviation OHLC stands for open, high, low and close prices.) For example, to download the Goldman Sachs share prices from *Yahoo!Finance* into the current *R* session.

```
> getSymbols("GS")
```

To load yahoo prices from the Google Finance,

```
> getSymbols("YHOO", src="google")
```

The option `src` specifies the source from which the data are to be downloaded. The current available sources for downloading include `yahoo` (default site), `google`, `FRED` (the Federal Reserve Bank of St. Louis which provide 11,000 economic series), and `Oanda` (a currency site). To download the S&P 500 index prices from yahoo,

```
> getSymbols("^GSPC")
```

This downloads the currently available daily prices of the S&P 500 index in an *R* object `GSPC`. To extract the prices in 2011 only and to plot the daily close prices together with the MACD charts,

```
> SP500 <- GSPC["2011"]
> chartSeries(SP500,type="line",TA=c(addMACD(),
        addMACD(8,30,9)), theme="white")
```

see Figure 2.21. To change the option `type="line"` to `type="candlesticks"` will produce a candlestick chart which combines the information of OHLC together and is often used in technical analysis by trading professionals. Check out `?chartSeries` for the variety of options in producing various charts.

It is easy to see from Figure 2.21 that the MACD may serve as a capable indicator for the price movement. Looking at the middle panel where $\text{MACD}_{12,26}(t)$ and $\text{Signal}_9(t)$ are plotted, most the time the price increases are signalled by the $\text{MACD}(t) > \text{Signal}(t)$. The large positive values of $\text{Hist}(t)$ indicate that the momentum of the increasing is strong. On the other hand, the times when the price decreases are typically indicated by the fact $\text{MACD}(t) < \text{Signal}(t)$, and the large negative values of $\text{Hist}(t)$ signal the strong momentum of the price decreasing. The sharp drop in prices in the first half of August 2011, due to the S&P's downgrade of the U.S. treasury long-term debt rating, was clearly marked by the large negative values of $\text{Hist}(t)$. The close scrutiny of the first two panels together reveals that delays occur between the price movements and the MACD signals, which is due to the nature of moving averages adopted in defining the indicators.

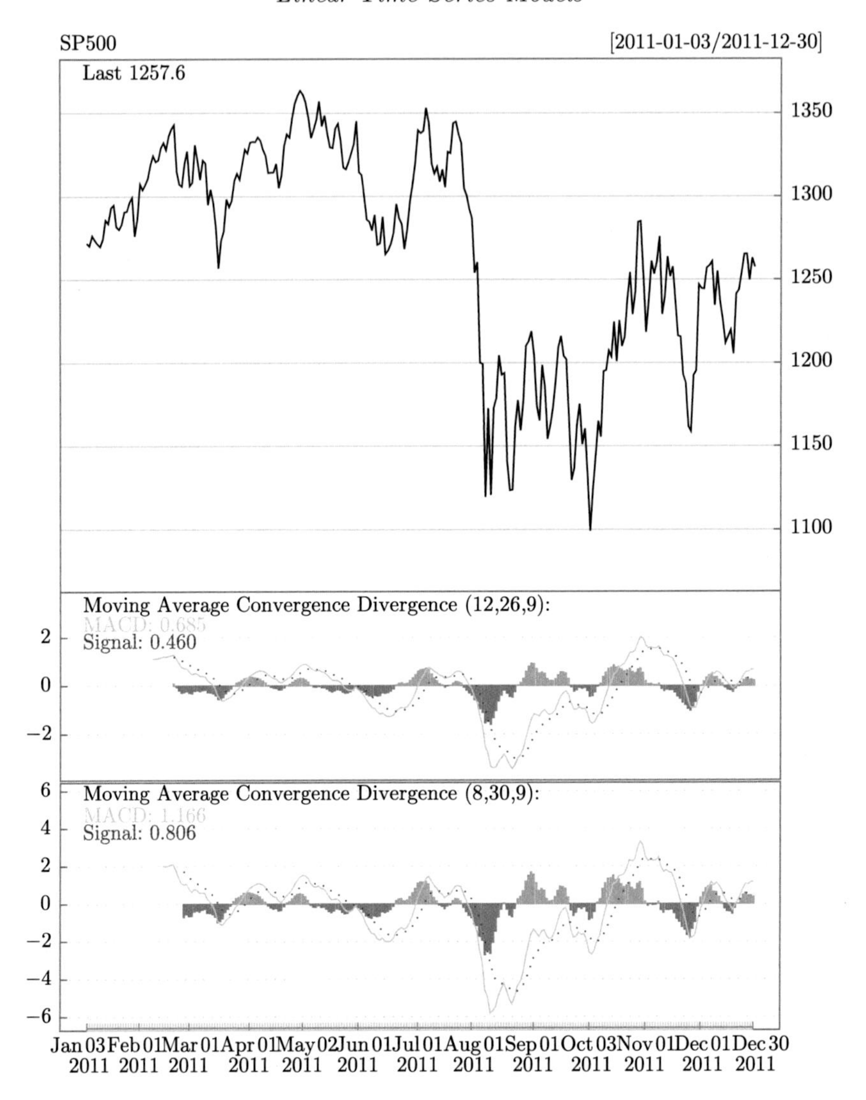

Figure 2.21 The upper panel: the daily close prices of S&P 500 index in 2011. Two lower panels: the curves $\text{MACD}_{12,26}(t)$ (solid lines) with $\text{Signal}_9(t)$ (dotted lines), and $\text{MACD}_{8,30}(t)$ (solid lines) with $\text{Signal}_9(t)$ (dotted lines). The $\text{Hist}(t) = \text{MACD}(t)\text{-Signal}(t)$ are plotted as bar-charts.

Furthermore, the MACD with $s = 12$ and $l = 26$ overlooks the price fluctuations at much higher frequencies, for example, between the August and the October. Note that the MACD line is far below the central line in this period; indicating clearly the bearish market in the same period. Change the period's parameters from (12,26,9) to (8, 30,9) (i.e. the bottom panel in Figure 2.21) mitigates the delay problem of the MACD signals, and it sharpens the signals for this particular data set; compare the two lower panels in Figure 2.21.

RSI

While MACD provides useful indicators for market trends and momentum, it is not bounded. Hence it is not particularly good for identifying some extreme movements of market such as overbought and oversold. The *relative strength index* (RSI), developed by J. Welles Welles, is a momentum oscillator that measures the speed and the magnitude of the price movements. See, e.g. Brown (2012).

Let $I(\cdot)$ be the indicator function. Then, the time series $\{(P_t - P_{t-1})\, I(P_t > P_{t-1})\}$ and $\{(P_t - P_{t-1})\, I(P_t < P_{t-1})\}$ are gains and losses series. The average gain and loss at time t can be defined through exponential smoothing techniques. For a given positive integer k, the average gain and loss at time t over the past k period are defined as

$$\begin{aligned}\text{avGain}_k(t) &= \frac{1}{k}\,(P_t - P_{t-1})\, I(P_t > P_{t-1}) + \frac{k-1}{k}\,\text{avGain}_k(t-1),\\ \text{avLoss}_k(t) &= \frac{1}{k}\,(P_{t-1} - P_t)\, I(P_t < P_{t-1}) + \frac{k-1}{k}\,\text{avLoss}_k(t-1).\end{aligned}$$

Similar to (2.28), we have

$$\text{avGain}_k(t) = \frac{1}{k}\sum_{j=0}^{\infty}\left(\frac{k-1}{k}\right)^j (P_{t-j} - P_{t-j-1}) I(P_{t-j} > P_{t-j-1}),$$

and

$$\text{avLoss}_k(t) = \frac{1}{k}\sum_{j=0}^{\infty}\left(\frac{k-1}{k}\right)^j (P_{t-j-1} - P_{t-j}) I(P_{t-j} < P_{t-j-1}).$$

The k-period RSI is defined as

$$\text{RSI}_k(t) = 100 \times \frac{\text{avGain}_k(t)}{\text{avGain}_k(t) + \text{avLoss}_k(t)}.$$

Since both avGain and avLoss are non-negative, $\text{RSI}_k(t)$ is always between 0 and 100. Furthermore $\text{RSI}_k(t) = 100$ if and only if the prices up to time t are monotonically increasing (or, more precisely, non-decreasing). $\text{RSI}_k(t) = 0$ if and only if the prices up to time t are monotonically decreasing. Those two extreme cases are rare.

The term "overbought" describes a situation in which the price has risen to an overvalued level due to high demand of the underlying asset, and may experience a pullback in the near future. Hence overbought signals the time to sell. In contrast, "oversold" describes a situation in which the price of an underlying asset has fallen sharply, and to a level below its fair value. Thus oversold signals the time to buy. The market is considered overbought when RSI > 70, and oversold when RSI < 30. These traditional levels can be adjusted to better fit the security or analytical requirement. For daily prices, the conventional look-back period is $k = 14$. But this can be lowered to increase sensitivity or raised to decrease sensitivity. For example, a 10-day RSI is more likely to reach overbought or oversold levels than a 14-day RSI. Short-term traders sometimes use 2-period RSI to look for overbought readings above 80 and oversold readings below 20.

Figure 2.22 displays the RSI curves with two periods $k = 14$ and $k = 10$ for the daily closing prices of S&P 500 index in 2011. The figure was produced by calling

```
> chartSeries(SP500, type="line", TA=c(addRSI(), addRSI(10)),
      theme="white")
```

RSI(t; 14) identifies one overbought period in the middle of February and one oversold period in the early of August. Those days can be clearly identified by the *R*-function `RSI`:

```
> rsiSP500 <- RSI(SP500[,4]) # 4th column are the close prices
> rsiSP500[rsiSP500>70] # list the overbought dates
2011-02-11 70.12352
2011-02-14 71.27811
2011-02-16 70.71710
2011-02-17 72.19894
2011-02-18 73.11855
> rsiSP500[rsiSP500<30] # list the oversold dates
2011-08-04 23.99090
2011-08-05 23.90388
2011-08-08 16.45753
2011-08-10 26.58873
```

Looking at the prices at the top panel of Figure 2.22, they may be viewed at the correct moments to sell or to buy if one is flexible on the holding periods. Reducing $k = 14$ to $k = 10$ leads to more oversold and overbought positions for this particular period. It is noticeable that the RSI fails to identify the seemingly obvious "overbought" opportunity immediately before the sharp price drop in early August. However this sharp drop was not due to the S&P 500 being overbought. Instead, it was caused by the exter-

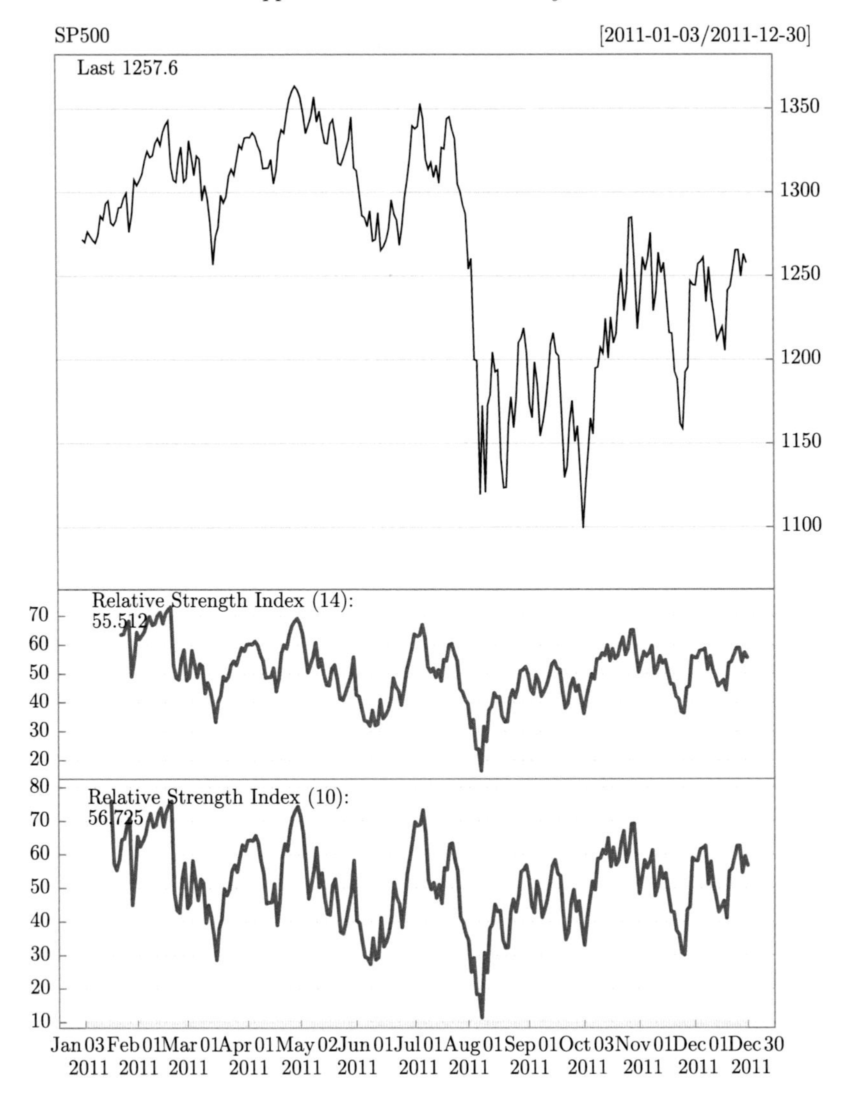

Figure 2.22 The upper panel: the daily close prices of S&P 500 index in 2011. Two lower panels: the curves RSI(t;14) and RSI(t;10).

nal force, namely the downgrading of the U.S. credit rating by Standard & Poor's. Unlike the MACD, there are no delays in the RSI signals.

2.10 Appendix: Time series analysis in R

2.10.1 Start up with R

R is an environment for data analysis and graphics based on *S* language.

It is also a full-featured programming language, freely available for various computer operating systems (including, for example, windows, Mac OS and linux) with complete source codes. Its official homepage is at

`http://www.R-project.org`

R consists of two major parts: the base system and a collection of (over 2K) user contributed add-on packages, all available from the above website. The base system includes some high-priority add-on packages such as graphic systems, linear regression models, time series analysis etc. To install *R*, click on *CRAN* on the left-site of the screen of the above website and choose a mirror site nearest to you geographically, then click on the name matching the operating system of your computer. It should be fairly straightforward to install the base system by following the on-screen instructions.

Before you start *R* for the first time, you should create a folder or directory, say `Rwork`, to hold data files that you will use with *R* for a project. (You should create separate folders/directories for different projects.) After *R* is successfully installed on your computer, you may start it by double-clicking the logo "R" on your desktop if you use *Windows*. An *R*-console will pop up with a prompt character ">". You may enter *R*-commands at this prompt, and they will be carried out when you press the "Enter" key. Some most frequently used commands/tasks are available through the menus in the *R*-console.

Your first task is to change the work directory to the one you have created, say, `Rwork`. To do so, select the `File` menu and then click on the choice `Change dir` $\cdots$. To quit *R*, type at the prompt "`q()`". Every time you quit *R*, you will receive a message asking whether or not R should `save the workspace image`. Click `Yes`, *R* will save your working image in your working directory `Rwork`: all the objects used in your session (including data sets) are saved in the file with the last name `RData`, and all the commands issued in the sessions are saved in the file with the last name `Rhistory`. You may restart your session by clicking on the file with the last name `RData`, then all the objects will be restored.

There are many excellent reference manuals for *R* now. Click on `Manuals` at the *R* official webpage will bring you to some useful documents such as *An Introduction to R*, *The R Language Definition* etc. To access on-screen manual for any *R*-function, e.g. `plot`, enter command

```
> ?plot
```

You may also like to check out, for example, `?rnorm`, `?qqplot` and `?lm`.

2.10.2 R-functions for time series analysis

Within the context of time series analysis, we list some of the most used *R*-functions and commands below.

`arima.sim` : simulate time series data from an ARIMA model specified. For example, the command

```
> arima.sim(n=100, list(ar=c(0.7, -0.4), ma=0.5), sd=2)
```

generates a series of length 100 from the ARMA model

$$X_t = 0.7X_{t-1} - 0.4X_{t-2} + \varepsilon_t + 0.5\varepsilon_{t-1}, \quad \varepsilon_t \sim_{iid} N(0, 2^2).$$

The command

```
> x <- arima.sim(n=150, list(order=c(1,1,0), ar=0.7),
rand.gen=rt, df=3)
```

generates a series of length 150 from ARIMA(1,1,0) model with the AR coefficient 0.7 and the independent innovations from the t distributions with 3 degrees of freedom, and the 150 data are stored in the vector `x`.

Find out more options of this function from

```
> ?arima.sim
```

`diff` : return differenced data. For example, to take differences of `x` at lag 1 and store the differences in `y`:

```
> y <- diff(x, lag=1)
```

`plot.ts` : produce time series plots. For example, try out `plot.ts(x)`, `plot.ts(x, type='o')`, and `plot.ts(diff(x), type='o')`.

`acf` : compute and plot sample ACF or sample ACVF. Note that `acf(y)` plots $\widehat{\rho}(k)$ against k with the constant confidence bounds at $\pm 1.96/\sqrt{T}$, while `acf(y, ci.type='ma')` produces the plot with varying confidence bounds $\pm 1.96\{1 + 2\sum_{j=1}^{k-1} \widehat{\rho}(j)^2\}^{1/2}/\sqrt{T}$ at each time lag k; see (2.10).

`pacf` : compute and plot sample PACF (with the confidence bounds at $\pm 1.96/\sqrt{T}$).

`arima` : fit an ARIMA model.

A simple command

```
> fitx <- arima(x, order=c(p,d,q))
```

fits an ARIMA(p, d, q) model to data `x`, and stores the output of the fitting in `fitx`. To view the essential information on the fitted model,

```
> fitx
```

To list all the elements in the output,

```
> summary(fitx)
```

To view, for example, the estimated coefficients (only) in the fitted model,
`> fitx$coef`

It is often convenient to write an *R* script in a plain text file, and save it in your working folder. All the commands in the script will be executed when you `source` the file into an *R* session. We often add comments to remind us what we are doing. To facilitate this, *R* ignores the text behind the sign `#`. Only users remarks, reminders and explanations should come after the `#` sign. For example, we write a plain text file "fig23.r" as follows:

```
par(mfrow=c(5,1))  # set 5 panels in 1 column in the figure
par(mar=c(2,3,2,1), mgp=c(1.8,0.8,0)) # control margins and
                                       # space between panels
x1 <- rnorm(100,1,sqrt(1.81)) # generate 100 N(1,1.81) data
plot(x1, type='o', pch=19, col='blue', main='Normal white noise',
     xlab='', ylab='') # pch=19 specifies solid cycles as
                       # point charater
abline(1,0) # abline(a,b) superimposes line y=a+bx
x2 <- arima.sim(n=100, list(ma=0.9))+1 #generate 100 AR(1) data
plot(x2, type='o', pch=19, col='blue', main='MA(1) with a=0.9',
      xlab='', ylab='')
abline(1,0)
x3 <- arima.sim(n=100, list(ar=0.669))+1
plot(x3, type='o', pch=19, col='blue', main='AR(1) with
      b=0.669', xlab='', ylab='')
abline(1,0)
x4 <- arima.sim(n=100, list(ar=-0.669))+1
plot(x4, type='o', pch=19, col='blue', main='AR(1) with
      b=-0.669', xlab='', ylab='')
abline(1,0)
x5 <- arima.sim(n=100, list(ar=0.5, ma=0.279))+1
plot(x5, type='o', pch=19, col='blue', main='ARMA(1,1) with
      b=0.5 and a=0.279', xlab='', ylab='')
abline(1,0)
```

We save the file in our working folder `Rwork`. Then Figure 2.23 is produced by the command
`> source('fig21.r')`

2.10.3 TSA – an add-on package

There exist quite a few add-on packages for time series analysis. For example the TSA is a package associated with the textbook Cryer and Chan (2010). It includes some new *R*-functions such as `eacf`, `LB.test`. It also modifies some existing functions in the *R* base system. For example, the function `arima` has been modified to include exogenous variables in the fitting. See Cryer and Chan (2010) pp. 468 for the full list of new functions in the TSA.

To install the TSA, click the Packages menu in an R-console and select `Set CRAN mirror`: choose a mirror site close to you geographically. Then click the Packages menu once more, click `Install packages`, select `TSA` from the list of the packages, and finally click `OK`. Now *R* should install the TSA and all the dependent packages automatically. Once the installation finishes, type

```
> library('TSA') # To load all objects in package TSA to
                 # the current session
> help(package='TSA') # To access the info on objects in TSA
```

Although you only need to install the package once, you have to load it (i.e. `library ('TSA')`) in each session in order to use its functions.

Like all programming languages, *R* may be initially challenging to the beginners (and also to non-frequent users). The best and the most effective way to learn *R* is to use it, as hands-on experience is the most illuminating!

2.11 Exercises

2.1 Let $X_1, \ldots, X_T$ be a stationary time series with the autocovariance function $\gamma(\cdot)$. Denote by $\boldsymbol{X} = (X_1, \ldots, X_T)$.

1. What is the variance–covariance matrix $\text{var}(\boldsymbol{X})$?
2. Show that $\text{var}(\boldsymbol{a}^T\boldsymbol{X}) = \boldsymbol{a}^T\text{var}(\boldsymbol{X})\boldsymbol{a}$ for any constant vector $\boldsymbol{a}$.

2.2 Consider a path dependent payoff function $Y_t = a_1 r_{t+1} + \cdots + a_k r_{t+k}$ where $\{a_i\}_{i=1}^k$ are given weights. If the return time series is weakly stationary in the sense that $\text{cov}(r_t, r_{t+j}) = \gamma(j)$. Show that

$$\text{var}(Y_t) = \sum_{i=1}^k \sum_{j=1}^k a_i a_j \gamma(i-j).$$

A natural estimate of this variance is the following substitution estimator:

$$\widehat{\text{var}}(Y_t) = \sum_{i=1}^k \sum_{j=1}^k a_i a_j \hat{\gamma}(i-j),$$

where $\hat{\gamma}(i-j)$ is defined by (2.4). Show that $\widehat{\text{var}}(Y_t) \geqslant 0$.

2.3 Which of the following models define a stationary and causal time series? Answer the question by finding the roots of their characteristic polynomials.

1. AR(2): $X_t = 0.3X_{t-1} - 0.1X_{t-2} + \varepsilon_t$.
2. MA(2): $X_t = \varepsilon_t + 2\varepsilon_{t-1} - 5\varepsilon_{t-2}$.
3. ARMA(2,2): $X_t = -X_{t-1} + 6X_{t-2} + \varepsilon_t + 2\varepsilon_{t-1} - 5\varepsilon_{t-2}$.

2.4 Suppose that a stock return follows the nonlinear dynamic $X_t = 0.8\varepsilon_{t-1}^2/(1+\varepsilon_{t-1}^2) + \varepsilon_t$, and that $\{\varepsilon_t\} \sim_{i.i.d.} N(0, \sigma^2)$.

1. Simulate the time series of length 500 with $\sigma = 1$, and show the plots of ACF and PACF.
2. Show that the ACF of $\{X_t\}$ is zero except at lag 0;
3. Use (b) to show that the PACF of $\{X_t\}$ is zero.

This example shows that ACF and PACF are useful mainly for linear time series.

2.5 Consider the Yule–Walker equation (2.27). Show that the solution to the difference equation (2.27) admits the form

$$\gamma(k) = c_1\alpha_1^{-k} + \cdots + c_p\alpha_p^{-k}$$

for sufficiently large k, where $\alpha_1, \ldots, \alpha_p$ are the roots of the characteristic function $b(x)$ and are assumed to be distinct. *Hint*: use $b(B)\gamma(k) = (1 - \alpha_1 B)\cdots(1-\alpha_p B)\gamma(k) = 0$ and hence one of $(1-\alpha_j B)\gamma(k) = 0$.

2.6 For an $AR(p)$ process

$$X_t = b_0 + b_1 X_{t-1} + \cdots + b_p X_{t-p} + \varepsilon_t,$$

show that

1. $\pi(k) = 0$ for $k > p$ and
2. it is stationary when $\sum_{j=1}^p |b_j| < 1$.

2.7 After fitting an AR(3) model to the monthly log-return of CRSP data from Jan. 1926 to Dec. 1997, it was obtained that

$$\hat{b}_0 = 0.0103, \hat{b}_1 = 0.104, \hat{b}_2 = -0.010, \hat{b}_3 = -0.120$$

with the estimated covariance matrix as follows:

$$\mathbf{S} = 1000^{-2}\begin{pmatrix} 2^2 & 34 & 0 & 0 \\ 34 & 34^2 & 0 & 0 \\ 0 & 0 & 34^2 & 0 \\ 0 & 0 & 0 & 34^2 \end{pmatrix}$$

1. What is the standard error of $\hat{b}_1$?
2. Test $H_0 : b_1 = 0$ at significance level 1%.
3. The annual return is estimated as $\hat{r} = (1+\hat{\mu})^{12} - 1$, where $\hat{\mu} = \hat{b}_0/(1 - \hat{b}_1 - \hat{b}_2 - \hat{b}_3)$. Construct a 95% confidence interval of the annual return (computing directly from the information given above).
4. Obtain directly the standard error of the annual return $\hat{r}$ if $\mathrm{SE}(\hat{\mu}) = 0.002186$ was already computed.

2.8 Suppose that the log-returns of a monthly bond index follow the MA(2) model:

$$X_t = 0.005 + \varepsilon_t + 0.2\varepsilon_{t-1} - 0.1\varepsilon_{t-2}, \quad \sigma = 0.03$$

1. Compute $\mathrm{var}(X_t)$.
2. Given $\varepsilon_{199} = 0.01$, $\varepsilon_{200} = -0.02$, compute the one-step and two-step ahead prediction at the time $t = 200$.
3. Obtain the associated mean prediction errors in (b).

2.9 Suppose that the daily simple-returns of a stock follow the ARMA model

$$X_t = 0.1X_{t-1} + \varepsilon_t - 0.2\varepsilon_{t-1}, \quad \sigma = 0.1.$$

1. Compute $\text{cov}(X_t, \varepsilon_t)$ and $\text{var}(X_t)$.
2. Express the ARMA model as an MA(∞) model.
3. Given $\varepsilon_{99} = 0.01$, $\varepsilon_{100} = -0.02$ and $X_{99} = 0.2$, compute the one-step and two-step prediction at time $t = 100$.
4. Compute the associated mean prediction errors in (b).

2.10 Suppose that the weekly returns $\{r_t\}$ (in percentage) of the Fancy Investment Inc. follow the following ARMA(2,1) model:

$$r_t = 0.22 + 0.1r_{t-1} + 0.02r_{t-2} + \varepsilon_t - 0.1\varepsilon_{t-1}, \qquad \varepsilon_t \sim N(0, 4^2).$$

1. Is the time series stationary? Show your work.
2. What is the dynamic (model) of the time series $Y_t = (r_t - 0.2)/5$?
3. Express r_t in terms of MA(∞) model. Hint: $\dfrac{1}{(1-az)(1-bz)} = \left(\dfrac{a}{1-az} - \dfrac{b}{1-bz}\right)/(a-b)$.
4. What are the expected return and volatility (SD) per annum? You may truncate the computation of volatility at MA(2).
5. What is the conditional distribution of r_{t+1} given the data observed up to time t for any $t > 2$?
6. Write down the Yale–Walker equation.
7. Suppose that, based on the past 200 weeks data, it was computed that

$$\hat{\rho}(1) = 0.1, \hat{\rho}(2) = -0.1, \hat{\rho}(3) = 0.1, \hat{\rho}(4) = -0.1, \hat{\rho}(5) = 0.1.$$

In absence of knowing the true dynamic of the returns, we test whether the return series is a white noise or not. What do you conclude at significant level $\alpha = 1\%$?

2.11 Let $X_t = b_0 + b_1X_{t-1} + \cdots + b_pX_{t-p} + \varepsilon_t + a_1\varepsilon_{t-1} + \cdots + a_q\varepsilon_{t-q}$ be a stationary ARMA(p, q) process.

1. What is the expected value of the process EX_t?
2. What is the best two-step prediction? You may assume that realized noises $\{\varepsilon_t\}_{t=1}^T$ are known at time T.
3. What is the mean prediction error of the best two-step ahead forecasting?

2.12 Suppose the prices of a stock follow the AR(1) model: $X_t = \mu + \rho X_{t-1} + \varepsilon_t$, where $\{\varepsilon_t\} \sim_{i.i.d.} N(0, \sigma^2)$ and σ is unknown.

1 Derive the conditional maximum likelihood estimator of $\tilde{\mu}$ and $\tilde{\rho}$ for μ and ρ. (Here, the conditional likelihood refers to the density of $(X_2, \ldots, X_T)$ given X_1).

2 Use simulation (1000 times with the initial value $X_0 = 0$) to calculate the 95%-tile and 99%-tile of the null distributions of the Dickey–Fuller tests (without drift and with drift) for $T = 100$ and 400. For concreteness, set $\mu = 0$ and $\sigma = 1$ in your simulation experiment.

2.13 Download the daily closing prices from January 1, 2001 to December 31, 2013 of the Merck company. Analyze the daily data with necessary supporting tables and figures. Answer the following questions.

1. Do the log-prices follow a random walk with a drift?
2. Are the log-returns predictable? Use the Ljung–Box test with lags 5 and 10 at significance level 1%.
3. Fit an ARMA(p,q) ($p + q \leqslant 2$) model to the data with order chosen by AIC.
4. Check if the residual series is white noise.
5. Predict the time series for the first two weeks in 2014.

2.14 Repeat the analysis in Problem 13 (b)–(e) for the squared log-returns.

3

Heteroscedastic Volatility Models

Volatility is a measure for the uncertainty of asset returns. It is usually defined as the conditional standard deviation of an asset return given all the available information up to the present time. The concept of volatility pervades almost every facet of finance. It is related to option pricing (e.g. Black–Scholes formula), risk measures (e.g. value-at-risk), risk-adjusted return (e.g. Sharpe ratio), securities regulations (e.g. capital requirement in Basel III), portfolio allocation and proprietary trading.

The models introduced in Chapter 2 are for modeling the conditional mean of a time series. It facilitates forecasting returns of financial time series or more generally the conditional expectation given the information up to the present time. In fact, an ARMA model defined with white noise innovations such as (2.30) only specifies up to the unconditional first two moments. It says nothing about the conditional variance and higher moments, and hence the uncertainties of forecasting errors. In order to carry out statistical inference for ARMA models, we often assume that the innovations are independent and identically distributed. This additional assumption is often questionable in the context of modeling equity returns which exhibit pronounced time varying volatility; see Sections 1.2 and 1.3.

In this chapter, we introduce the time series models that account for conditional heteroscedasticity (i.e. time-varying volatility). In the context of financial time series, those models are often refereed to as volatility models, as the volatility of a financial market is often, though not always, measured by the conditional standard deviation of the returns given the information up to the present time. With the ever-increasing importance of measuring, forecasting or managing the uncertainty and the risk of financial positions and markets, there has been a strong surge of research in the development of heteroscedasticity models in the last two decades. This chapter introduce several popular heteroscedasticity models such as ARCH, GARCH, and stochastic volatility models. We also give a brief description on a selection of other models designed to catch the various stylized features of financial returns listed in Section 1.2.

3.1 ARCH and GARCH models

Let P_t denote the price and $r_t = \log(P_t/P_{t-1})$ be the log-return at time t. The standard approach for modeling return data is to use the decomposition

$$r_t = \mu_t + X_t, \tag{3.1}$$

where μ_t denotes the (conditional) mean of the return, X_t is a diffusion term which may be modeled as

$$X_t = \sigma_t \varepsilon_t, \qquad \varepsilon_t \sim \mathrm{IID}(0, 1), \tag{3.2}$$

where $\sigma_t > 0$ is determined by the information available *before* time t, hence it is predictable at time t, and ε_t is assumed to be independent of σ_t. The time-varying random quantity σ_t is often referred to as a *volatility function*. For most financial data, the return is dominated by the diffusion term X_t. In fact, we often assume $\mu_t \approx 0$ for high frequency data (i.e. daily or higher frequency returns). Even with weekly or monthly data, due to the weak predictability of asset returns, it is reasonable to take $\mu_t = \mu$, a constant, which can be estimated by the average returns. Therefore, one frequently takes the input time series r_t as X_t for daily returns and demeaned time series for weekly or monthly returns. For these reasons, we focus on the volatility model (3.2) in this chapter. More refined modeling of μ_t is possible. See Section 3.3.

3.1.1 ARCH models

A simple specification for the volatility function σ_t in (3.2) is the *autoregressive conditional heteroscedastic (ARCH) model*:

$$\sigma_t^2 = a_0 + a_1 X_{t-1}^2 + \cdots + a_p X_{t-p}^2, \tag{3.3}$$

where $a_0 > 0$, $a_j \geqslant 0$ $(1 \leqslant j \leqslant p)$ are constants, and p is a positive integer. As ε_t is independent of $X_{t-1}, X_{t-2}, \ldots$ in (3.2), it it easy to see that

$$E_{t-1}X_t = \sigma_t E_{t-1}\varepsilon_t = 0, \quad \text{and} \quad \sigma_t^2 = \mathrm{var}_{t-1}(X_t) = E_{t-1}X_t^2, \tag{3.4}$$

where E_t and var_t are the conditional means and conditional variances given the information up to time t, i.e. $E_t(\cdot) = E(\cdot|X_t, X_{t-1}, \ldots)$ for example. The volatility function σ_t is the conditional standard deviation of X_t given its lagged values. We write $X_t \sim \text{ARCH}(p)$ for the processes defined by (3.2) and (3.3).

The ARCH model was initially introduced by Engle (1982) for modeling the predictive variances for the U.K. inflation rates. Since then it has been widely used for modeling volatilities of financial returns. The basic idea is intuitive: the predictive distribution for X_t based on its own past is a scale-transform of the distribution of ε_t, with the scaling constant σ_t depending on the lagged values of X_t. For example, if $\varepsilon_t \sim N(0,1)$, the predictive distribution is $N(0, \sigma_t^2)$. Hence we only need to predict the volatility σ_t, which boils down to the estimation for the coefficients $a_0, a_1, \ldots, a_p$ in (3.3).

Property (3.4) shows that X_t is a martingale difference and hence it is unpredictable, in lines with the efficiency market hypothesis in Section 1.3. Furthermore, letting $\eta_t = (\varepsilon_t^2 - 1)\sigma_t^2$, then

$$X_t^2 = \sigma_t^2 + \eta_t = a_0 + a_1 X_{t-1}^2 + \cdots + a_p X_{t-p}^2 + \eta_t. \tag{3.5}$$

Using the predictability of σ_t, independence of ε_t with its past, and (3.2), we have

$$E_{t-1}\eta_t = \sigma_t^2(E_{t-1}\varepsilon_t^2 - 1) = \sigma_t^2(E\varepsilon_t^2 - 1) = 0. \tag{3.6}$$

Therefore, $\{\eta_t\}$ is a martingale difference sequence and hence a white noise series. This and (3.5) entail $\{X_t^2\}$ follows an AR-model. Hence, stochastic properties of AR processes apply to the time series $\{X_t^2\}$. We illustrate this with the following simple example.

Example 3.1 To gain further appreciation of the dynamics of ARCH structure, let us look at a simple ARCH(1) model:

$$X_t = \sigma_t \varepsilon_t, \quad \sigma_t^2 = a_0 + a_1 X_{t-1}^2, \quad \varepsilon_t \sim \text{IID}(0,1), \tag{3.7}$$

where $a_0 > 0$ and $a_1 \geqslant 0$.

First we note that if (3.7) admits a strictly stationary solution $\{X_t\}$ with $E(X_t^2) < \infty$, it is necessary that $a_1 \in (0,1)$ as then

$$E(X_t^2) = E(\sigma_t^2) = a_0/(1 - a_1),$$

which is obtained by taking expectation on both sides of second equation in (3.7). In fact this condition is also sufficient; see Theorem 3.1 below. For this specific case,

$$X_t^2 = a_0 + a_1 X_{t-1}^2 + \eta_t.$$

Furthermore, this AR(1) representation entails that

$$\text{Corr}(X_t^2, X_{t+k}^2) = a_1^{|k|}, \qquad k = 0, \pm 1, \pm 2, \ldots, \tag{3.8}$$

provided that X_t is strictly stationary and $E(|X_t|^4) < \infty$ (see Exercise 3.1 for a condition). Hence X_t^2 are always positively auto-correlated, indicating the predictability of the volatility. On the other hand, since the returns $\{X_t\}$ are a martingale difference sequence, it is uncorrelated and unpredictable, conforming with the efficient market hypothesis in Section 1.3.

Prediction follows also from the AR(1) representation. For any $k \geqslant 1$,

$$\text{var}_t(X_{t+k}) = E_t X_{t+k}^2 = a_0 + a_1 E_t X_{t+k-1}^2$$

Then it follows from iterating the above expression that

$$\text{var}_t(X_{t+k}) = \frac{a_0(1 - a_1^k)}{1 - a_1} + a_1^k X_t^2. \tag{3.9}$$

This indicates that a large value of $|X_t|$ will lead to large volatilities for a while in the immediate future. Hence ARCH models can produce the volatility clustering – a stylized feature of financial returns; see §1.2.

To measure the heaviness of tail distributions, let us compare the kurtosis of the return series $\{X_t\}$ with that of the innovation series $\{\varepsilon_t\}$. The sample kurtosis was defined in Section 1.5.2, but we compute the population version here. Under the assumption that $E(\varepsilon_t^4) < \infty$ and $E(X_t^4) < \infty$, using the predictability of σ_t, it holds that

$$EX_t^2 = E(E_{t-1}\sigma_t^2\varepsilon_t^2) = E(\sigma_t^2 E_{t-1}\varepsilon_t^2) = E(\sigma_t^2)E\varepsilon_t^2, \tag{3.10}$$

where $E_{t-1}\varepsilon_t^2 = E\varepsilon_t^2$ follows from the independence of ε_t and $X_{t-1}, X_{t-2}, \ldots$. Similarly,

$$E(X_t^4) = E\{\sigma_t^4\}E(\varepsilon_t^4).$$

Recall that the kurtosis is defined as

$$\kappa_x = \frac{E(X_t^4)}{E(X_t^2)^2} \quad \text{and} \quad \kappa_\varepsilon = \frac{E(\varepsilon_t^4)}{E(\varepsilon_t^2)^2} \tag{3.11}$$

for X_t and ε_t, respectively. It follows from (3.10) and (3.11) that

$$\kappa_x = \kappa_\varepsilon \frac{E\sigma_t^4}{(E\sigma_t^2)^2} \geqslant \kappa_\varepsilon, \tag{3.12}$$

where the last inequality follows from the fact that $E\sigma_t^4 - (E\sigma_t^2)^2 = \text{var}(\sigma_t^2) \geqslant 0$. This implies that the tails of the distribution of X_t is always heavier, in terms of Kurtosis, than that of ε_t regardless of the distribution of ε_t.

Figure 3.1 was produced using a realization of the time series with length 1000 from model (3.7) with $a_0 = 0.1$, $a_1 = 0.9$ and $\varepsilon_t \sim N(0, 1)$. In this figure, panel (c) depicts the conditional volatility as a function of time t,

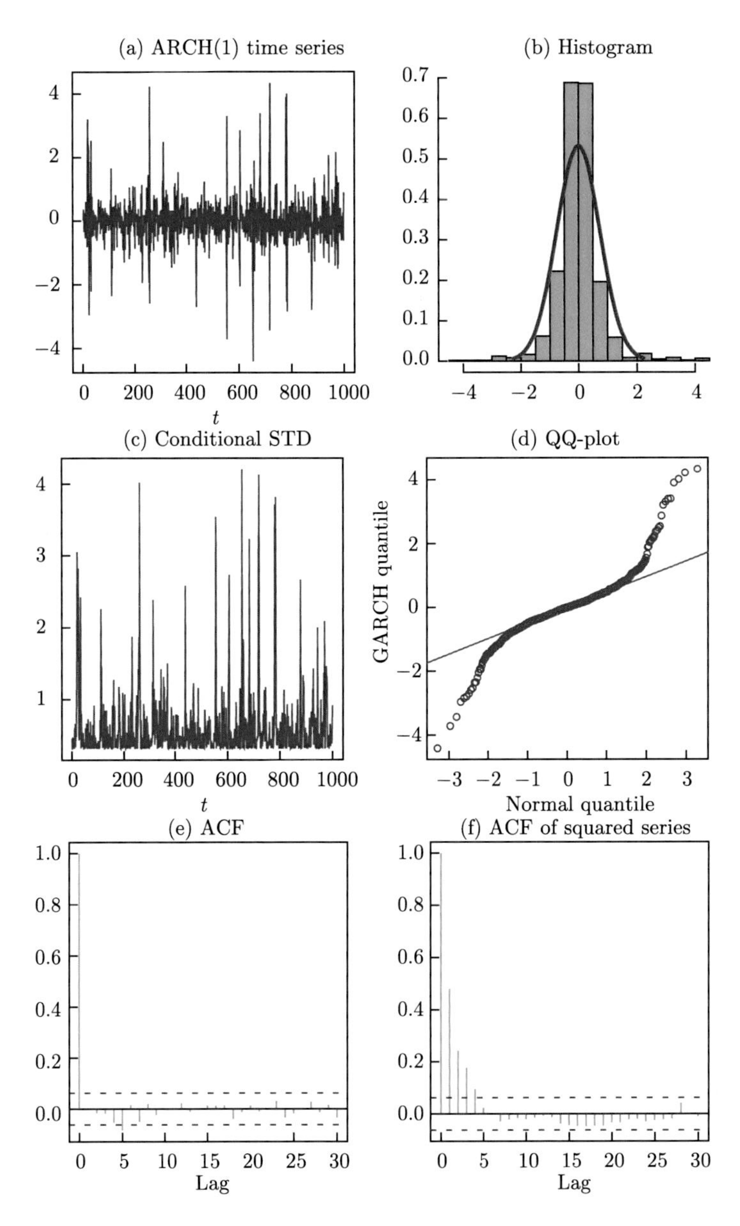

Figure 3.1 A time series of length 1000 was generated from ARCH(1) model with $a_0 = 0.1$, $a_1 = 0.9$ and $\varepsilon_t \sim N(0, 1)$: (a) and (c) time series plots of the first 250 subseries of X_t and σ_t; (b) the normalized histogram of X_t and the normal density function with the same mean and variance; (d) QQ-plot: the sample quantiles of X_t versus the quantiles of $N(0, 1)$; (e) and (f) the sample ACFs of X_t and X_t^2 respectively.

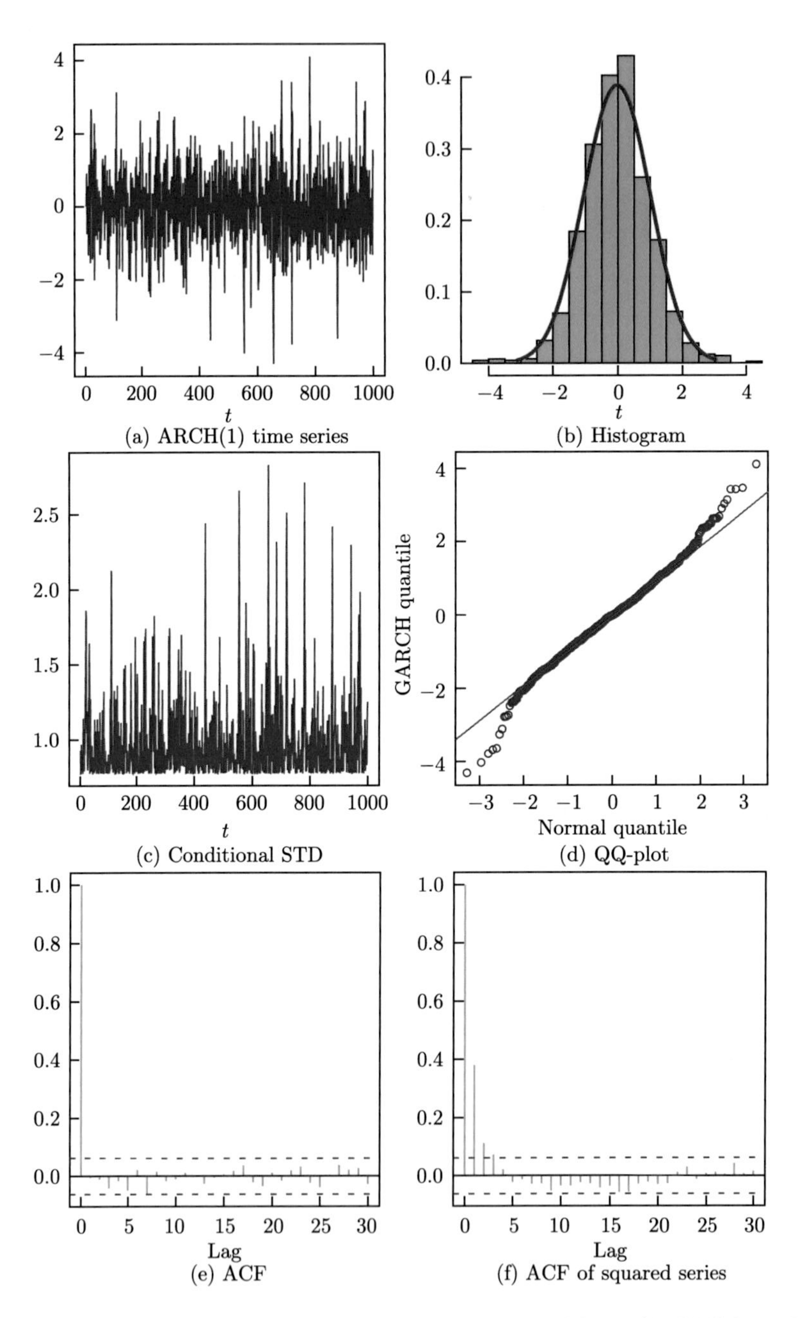

Figure 3.2 A time series of length 1000 was generated from ARCH(1) model with $a_0 = 0.6; a_1 = 0.4$ and everything else the same as that in Figure 3.1.

and panels (a) and (c) indicate that the large values of $|X_t|$ and σ_t tend to associate each other. The unconditional variance of the process $\{X_t\}$ is

$$0.1/(1-0.9) = 1 \quad \text{or} \quad \text{SD} = \sqrt{1} = 1.$$

This matches the standard deviation of the time series depicted in Figure 3.1(a). The heavy-tail behavior of X_t is shown in the histogram plot (b) and, more prominently, the QQ-plot (d). Panels (e) and (f) indicate that there exist significant autocorrelations in the series X_t^2 but not X_t.

From Exercise 3.1, a necessary condition for $EX_t^4 < \infty$ is that $a_1 \leqslant 1/\sqrt{E\varepsilon_t^4} = 1/\sqrt{3} = 0.557$. Therefore, the process with $a_1 = 0.9$ does not have a finite fourth moment. Hence, the ACVF and also ACF of X_t^2 cannot be defined. Nevertheless, the sample ACF of X_t^2 is shown in Figure 3.1(f), which does not resemble equation (3.8). For example, the sample ACF at lag one is around 0.4 rather than 0.9. This is due to the fact that $EX_t^4 = \infty$ when $a_1 = 0.9$.

We repeat the exercise with $a_1 = 0.4$, under which $EX_t^4 < \infty$. The results are shown in Figure 3.2. In this case, the sample ACF in Figure 3.2(f) matches its theoretical counterpart (3.8) with $a_1 = 0.4$. The population (theoretical) kurtosis is (see Exercise 3.1)

$$\kappa_x = \frac{(1-0.4^2)3}{1-3\times 0.4^2} = 4.846.$$

To put this kurtosis into prospective, note that the kurtosis of t-distribution with degree of freedom ν is

$$\kappa_{t_\nu} = \frac{3(\nu-2)}{(\nu-4)}.$$

Thus, $\kappa_x = 4.846$ matches the t-distribution with degree of freedom

$$\frac{3(\nu-2)}{(\nu-4)} = 4.846, \quad \text{or} \quad \nu = 7.25.$$

This explains why the tails of $\{X_t\}$ are much lighter than that those in Figure 3.1.

In contrast to the linear processes presented in Chapter 2, an ARCH process depends on the innovations very nonlinearly and in a complex manner. This can be seen by recursively using the first two equations in (3.7) to obtain

$$\begin{aligned}\sigma_t^2 &= a_0 + a_1 X_{t-1}^2 = a_0 + a_1\varepsilon_{t-1}^2(a_0 + a_1 X_{t-2}^2) \\ &= a_0 + a_0 a_1 \varepsilon_{t-1}^2 + a_1^2 \varepsilon_{t-1}^2 \varepsilon_{t-2}^2 (a_0 + a_1 X_{t-3}^2) \\ &= a_0 + \sum_{j=1}^{k} a_0 a_1^j \prod_{i=1}^{j} \varepsilon_{t-i}^2 + a_1^{k+1} X_{t-k-1}^2 \prod_{i=1}^{k} \varepsilon_{t-i}^2 \\ &= a_0 + \sum_{j=1}^{\infty} a_0 a_1^j \prod_{i=1}^{j} \varepsilon_{t-i}^2.\end{aligned} \tag{3.13}$$

Hence

$$X_t = \varepsilon_t \left(a_0 + \sum_{j=1}^{\infty} a_0\, a_1^j \varepsilon_{t-1}^2 \cdots \varepsilon_{t-j}^2 \right)^{1/2}.$$

This demonstrates the points above.

3.1.2 GARCH models

Although ARCH models catch some stylized features in financial return data, fitting real financial returns with an ARCH model often leads to an ARCH(p) with a large value of p. Similar to the relation between ARMA models and purely AR models, a much more parsimonious representation for volatilities could be obtained by using *generalized autoregressive conditional heteroscedastic (GARCH) models* due to Bollerslev (1986) and Taylor (1986).

A GARCH(p, q) model is defined as

$$X_t = \sigma_t \varepsilon_t, \qquad \sigma_t^2 = a_0 + \sum_{i=1}^{p} a_i X_{t-i}^2 + \sum_{j=1}^{q} b_j \sigma_{t-j}^2, \tag{3.14}$$

where $a_0 > 0$, $a_i \geqslant 0$ $(i > 0)$ and $b_j \geqslant 0$, and $\varepsilon_t \sim \text{IID}(0, 1)$.

The class of GARCH models is among the most important extensions of ARCH models. Effectively a GARCH(p, q) model is an ARCH(∞). In fact GARCH(1,1) remains as a benchmark model among most applications. In the above definition, we require $a_0 > 0$ as otherwise the only stationary solution of (3.14) is $X_t \equiv 0$. When all $a_1 = \cdots = a_p = 0$ and some $b_j > 0$, b_j are unidentifiable. We avoid those trivial cases. Obviously when $q = 0$, GARCH(p, q) reduces to ARCH(p).

The recursive formula (3.14) admits very intuitive interpretation. It is a

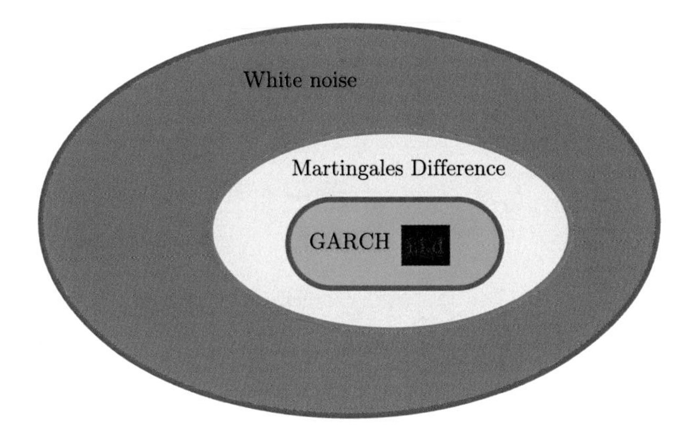

Figure 3.3 Relationship among different white noise processes.

weighted average of squared returns $\{X_{t-i}^2\}$ with weights $\{a_i\}$ and realized volatility $\{\sigma_{t-j}^2\}$ with weights $\{b_j\}$. This interpretation suggests that

$$\sum_{i=1}^{p} a_i + \sum_{i=1}^{q} b_i < 1,$$

and the remaining weights $1 - \sum_{i=1}^p a_i - \sum_{i=1}^q b_i$ is on $a_0/(1 - \sum_{i=1}^p a_i - \sum_{i=1}^q b_i)$, the long-run variance (the unconditional variance of the process). This is indeed the true condition for the stationarity of the GARCH process. See Theorem 3.1. In particular, for GARCH(1,1), it is the weighted average of what happens yesterday, the realized volatility yesterday, and the long-run variance.

Using the same calculation as (3.4), $\{X_t\}$ is a martingale difference sequence. Hence, $EX_t = 0$ and is uncorrelated if $EX_t^2 < \infty$. Figure 3.3 illustrates the relationship among different white noise processes: martingale difference sequences form a large subset of white noise. They are too large to be useful for modeling and simulating financial returns. On the other hand, GARCH is a useful parametric specification of martingale differences. The IID sequences are special cases of GARCH processes with $p = q = 0$.

By definition (3.14), we have

$$\mathrm{var}_t(X_t) = \sigma_t^2 E_{t-1}(\varepsilon_t^2) = \sigma_t^2. \tag{3.15}$$

Hence, σ_t is the conditional volatility given the information up to time t. It is time-varying, determined by the history of the time series and model parameters. Indeed, using the backshift operator B, we have

$$\sigma_t^2 = b(B)^{-1}\left(a_0 + \sum_{i=1}^{p} a_i X_{t-i}^2\right)$$

where $b(z) = 1 - \sum_{j=1}^{q} b_j z^j$. The operator $b(B)$ is invertible if $\sum_{j=1}^{q} b_j < 1$, since

$$|b(z)| \geqslant 1 - \sum_{j=1}^{q} b_j |z|^j > 0, \quad \text{for all } |z| \leqslant 1, \tag{3.16}$$

namely all roots of $b(z)$ lie outside the unit circle. Hence, σ_t^2 is a function of past square returns. Indeed, expanding the above polynomials into the infinite series, we can explicitly obtain

$$\begin{aligned}\sigma_t^2 =& \frac{a_0}{1 - \sum_{j=1}^{q} b_j} + \sum_{i=1}^{p} a_i X_{t-i}^2 \\ &+ \sum_{i=1}^{p} a_i \sum_{k=1}^{\infty} \sum_{j_1=1}^{q} \cdots \sum_{j_k=1}^{q} b_{j_1} \cdots b_{j_k} X_{t-i-j_1-\cdots-j_k}^2,\end{aligned} \tag{3.17}$$

where the multiple sum vanishes if $q = 0$. In the above expansion, we assume that we observe the time series $\{X_t\}_{t=-\infty}^{T}$. In practice, we truncate the expansion, which is equivalent to assume that $X_t = 0$ for $t \leqslant 0$.

Example 3.2 Consider GARCH(1,1) model

$$\sigma_t^2 = a_0 + a_1 X_{t-1}^2 + b_1 \sigma_{t-1}^2. \tag{3.18}$$

A convenient condition for stationarity is $a_1 + b_1 < 1$. Since $b_1 < 1$, we have

$$\begin{aligned}\sigma_t^2 &= (1 - b_1 B)^{-1}(a_0 + a_1 X_{t-1}^2) \\ &= (1 + b_1 B + b_1^2 B^2 + \cdots)(a_0 + a_1 X_{t-1}^2) \\ &= \frac{a_0}{1 - b_1} + a_1(X_{t-1}^2 + b_1 X_{t-2}^2 + b_1^2 X_{t-3}^2 + \cdots),\end{aligned} \tag{3.19}$$

i.e. σ_t^2 is an exponentially weighted average of past squared returns. Hence GARCH(1,1) is ARCH(∞).

3.1.3 Stationarity of GARCH models

While the GARCH model appears unfamiliar, a squared GARCH process is an ARMA process. Hence we may gain some insight on GARCH models from the probabilistic properties of ARMA models.

Let $\eta_t = \sigma_t^2(\varepsilon_t^2 - 1) = X_t^2 - \sigma_t^2$. Then, the same argument leading to (3.6) also implies that $\{\eta_t\}$ is a sequence of martingale differences and hence is a white noise process. It follows from (3.14) that

$$X_t^2 = a_0 + \sum_{i=1}^{p} a_i X_{t-i}^2 + \sum_{j=1}^{q} b_j \sigma_{t-j}^2 + \eta_t.$$

Substituting $\sigma_t^2 = X_t^2 - \eta_t$ into the above equation, we have

$$\begin{aligned} X_t^2 &= \sum_{i=1}^{p} a_i X_{t-i}^2 + \sum_{j=1}^{q} b_j (X_{t-j}^2 - \eta_{t-j}) + \eta_t \\ &= a_0 + \sum_{i=1}^{p\vee q} (a_i + b_i) X_{t-i}^2 + \eta_t - \sum_{j=1}^{q} b_j \eta_{t-j}, \end{aligned} \tag{3.20}$$

where $a_{p+j} = b_{q+j} = 0$ for $j \geqslant 1$, $p \vee q = \max\{p, q\}$. The characteristic polynomial of $\{X_t^2\}$ is

$$a(z) = 1 - \sum_{i=1}^{p\vee q} (a_i + b_i) z^i.$$

Since for all $|z| \leqslant 1$

$$|a(z)| \geqslant 1 - \Big|\sum_{i=1}^{p\vee q} (a_i + b_i) z^i\Big| \geqslant 1 - \sum_{i=1}^{p\vee q} (a_i + b_i),$$

which is positive if

$$\sum_{i=1}^{p} a_i + \sum_{j=1}^{q} b_j < 1. \tag{3.21}$$

Thus, $a(z)$ has all roots outside the unit circle under condition (3.21). Furthermore, the polynomial $b(z) = 1 - \sum_{j=1}^{q} b_j z^j$ also has roots outside the unit circle under condition (3.21); see (3.16). Hence (3.21) implies that X_t^2 is a causal and invertible ARMA$(p \vee q, q)$ process provided $E(X_t^4) < \infty$.

Theorem 3.1 *The necessary and sufficient condition for* (3.14) *defining a unique strictly stationary process* $\{X_t, t = 0, \pm 1, \pm 2, \ldots\}$ *with* $EX_t^2 < \infty$ *is* (3.21). *Furthermore, for this unique stationary solution it holds that* $EX_t = 0$ *and*

$$\mathrm{var}(X_t) = \frac{a_0}{1 - \sum_{i=1}^{p} a_i - \sum_{j=1}^{q} b_j}$$

Proof If $\{X_t\}$ is stationary with $EX_t^2 < \infty$, then from the same argument as (3.10), we have $EX_t^2 = E\sigma_t^2$. Therefore, taking expectation on both sides of (3.14), we obtain

$$EX_t^2 = a_0 + \sum_{i=1}^{p} a_i EX_t^2 + \sum_{i=1}^{p} b_i EX_t^2$$

Solving this gives

$$\text{var}(X_t) = EX_t^2 = \frac{a_0}{1 - \sum_{i=1}^{p} a_i - \sum_{j=1}^{q} b_j}.$$

This implies condition (3.21), since the variance is positive.

On the other hand, if condition (3.21) holds, then $\{X_t^2\}$ is a stationary and invertible ARMA process: σ_t^2 can be represented ARCH(∞) and is hence stationary. Therefore, $X_t = \sigma_t \varepsilon_t$ is also stationary. □

The unconditional variance in Theorem 3.1 is also called *long-run variance*. By the law of large numbers, the sample variance of a long time series of a stationary process approaches to the unconditional variance. Therefore, the unconditional variance is also referred to as the long-run variance. Another way of interpreting the long-run variance is as follows. Standing from the current time t, the variance of X_{t+m} in the remote future (m is large) should be close to the unconditional one, as the history has very little predictive power on X_{t+m}.

Theorem 3.1 presents a necessary and sufficient condition (3.21) for the GARCH model (3.14) defining a strictly stationary process with a finite second moment. The requirement of a finite second moment is important and reasonable (see, e.g. Figure 1.9), which makes the condition as explicit as (3.21). This condition is easy to check in practice. On the other hand, Bougerol and Picard (1992) established a necessary and sufficient condition for the existence of a strictly stationary solution which does not necessarily have a finite second moment; see also Kazakevičius and Leipus (2002). The condition is defined in terms of Lyapunov exponents for some random matrices associated with the model and is difficult to check in practice. Theorem 3.1 combines the results of Chen and An (1998) and Giraitis, Kokoszka and Leipus (2000). See also the proof of Theorem 4.4 of Fan and Yao (2003).

Example 3.2 (continued) Consider GARCH(1,1) model (3.18) again. By (3.20), it has the following ARMA(1,1) representation:

$$X_t^2 = a_0 + (a_1 + b_1) X_{t-1}^2 + \eta_t - b_1 \eta_{t-1}.$$

When $E(X_t^4) < \infty$, the ACVF of X_t^2 can be determined by (2.33) and (2.34). Specifically, let

$$\sigma^2 = E\eta_t^2 = E(\varepsilon_t^2 - 1)^2 E\sigma_t^4$$

be the variance of the above ARMA process. Then, noticing $\gamma(0)$ in (2.33) is the variance of the process, we have

$$\text{var}(X_t^2) = \frac{1 + b_1^2 - 2(a_1 + b_1)b_1}{1 - (a_1 + b_1)^2}\sigma^2,$$

$$\text{cov}(X_t^2, X_{t-1}^2) = \frac{-b_1 + (a_1 + b_1)(1 + b_1^2 - (a_1 + b_1)b_1)}{1 - (a_1 + b_1)^2}\sigma^2,$$

and

$$\text{cov}(X_t^2, X_{t-k}^2) = (a_1 + b_1)^{k-1}\text{cov}(X_t^2, X_{t-1}^2), \quad \text{for } k \geqslant 2.$$

3.1.4 Fourth moments

Not all stationary GARCH processes have finite fourth moments, but this is needed for many statistical problems of GARCH models as we have seen before. Additional conditions are needed, as shown below.

Theorem 3.2 *The GARCH process $\{X_t\}$ has $EX_t^4 < \infty$, provided that*

$$\{E(\varepsilon_t^4)\}^{1/2}\sum_{i=1}^{p} a_i < 1 - \sum_{j=1}^{q} b_j. \tag{3.22}$$

In addition,

$$\kappa_x = \kappa_\varepsilon \frac{E(\sigma_t^4)}{\{E(\sigma_t^2)\}^2} \geqslant \kappa_\varepsilon.$$

Proof Since $E(\varepsilon_t^4) \geqslant \{E(\varepsilon_t^2)\}^2 = 1$, condition (3.22) implies (3.21). Therefore X_t is strictly stationary with $E(X_t^2) < \infty$. Using the same argument leading to (3.10), we have

$$E(X_t^4) = E(\varepsilon_t^4)E(\sigma_t^4) = \kappa_\varepsilon E(\sigma_t^4),$$

where $\kappa_\varepsilon = E(\varepsilon_t^4)$ is the kurtosis parameter as $E\varepsilon_t^2 = 1$. Hence we only need to show that $E(\sigma_t^4) < \infty$. The general arguments are tedious and we use GARCH(1,1) to illustrate the main idea.

For GARCH(1,1) model, by using (3.14), we have

$$\begin{aligned} E(\sigma_t^4) = a_0^2 + a_1^2 E(X_{t-1}^4) + b_1^2 E(\sigma_{t-1}^4) + 2a_0 a_1 E(X_{t-1}^2) \\ + 2a_0 b_1 E(\sigma_{t-1}^2) + 2a_1 b_1 E(X_{t-1}^2 \sigma_{t-1}^2). \end{aligned} \tag{3.23}$$

Note that

$$E(X_{t-1}^2\sigma_{t-1}^2) = E\{\sigma_{t-1}^4\varepsilon_{t-1}^2\} = E(\sigma_{t-1}^4).$$

Substituting this into (3.23), we have

$$E(\sigma_t^4) = a_0^2 + (a_1^2\kappa_\varepsilon + 2a_1b_1 + b_1^2)E(\sigma_t^4) + 2a_0(a_1+b_1)E(\sigma_t^2).$$

Since each term above is nonnegative, it is easy to see that $E(\sigma_t^4) < \infty$ if and only if

$$a_1^2\kappa_\varepsilon + 2a_1b_1 + b_1^2 < 1.$$

Under condition (3.22), we have

$$\begin{aligned} a_1^2\kappa_\varepsilon + 2a_1b_1 + b_1^2 &< (1-b_1)^2 + 2a_1b_1 + b_1^2 \\ &= 1 - 2b_1(1 - b_1 - a_1), \end{aligned}$$

which is less than 1. Hence the fourth moment exists.

The second conclusion follows exactly the same arguments that lead to (3.12). □

Theorem 3.2 shows that a GARCH process is a fat-making factory: the tails of the distribution of X_t are fatter than those of ε_t. It is not unlikely that the GARCH equation makes the tails of X_t too fat. The second moment of X_t^2 (i.e. the fourth moment of X_t) is finite under additional condition (3.22) which is too restrictive and is usually unfulfilled in practice. For example, fitting financial returns with GARCH(1,1) model typically leads to b_1 very close to 1, and the fitted model implies that $E(X_t^4) = \infty$. This explains why it is not recommended to carry out the inference for a GARCH model based on its ARMA representation (3.20), namely, using ARMA techniques.

A case in point is Figure 3.1(f) for an ARCH(1) process with $a_1 = 0.9$. The figure indicates that the estimated ACF for X_t^2 at lag 1 is smaller than 0.5 based on a sample of size 1000. Now (3.20) reduces to

$$X_t^2 = a_0 + a_1X_{t-1}^2 + \eta_t,$$

therefore, formally the 'true' value is $\rho(1) = a_1 = 0.9$. As explained before, the poor estimation is due to the fact that $E(X_t^4) = \infty$ (and, therefore, $\rho(1)$ is not well-defined), as condition (3.22) reduces to $a_1 < 1/\sqrt{3} = 0.577$, which is not fulfilled. In contrast, when $a_1 = 0.4$, the fourth moment exists. Hence, the sample standard deviation and the autocorrelations are close to the theoretical counterpart, as shown in Figure 3.2.

We generated a series of length 1000 from (3.14) with $p = q = 1$, $a_0 = 0.1$, $a_1 = 0.08$, $b_1 = 0.82$ and ε_t independent $N(0,1)$. The time series $\{X_t\}$

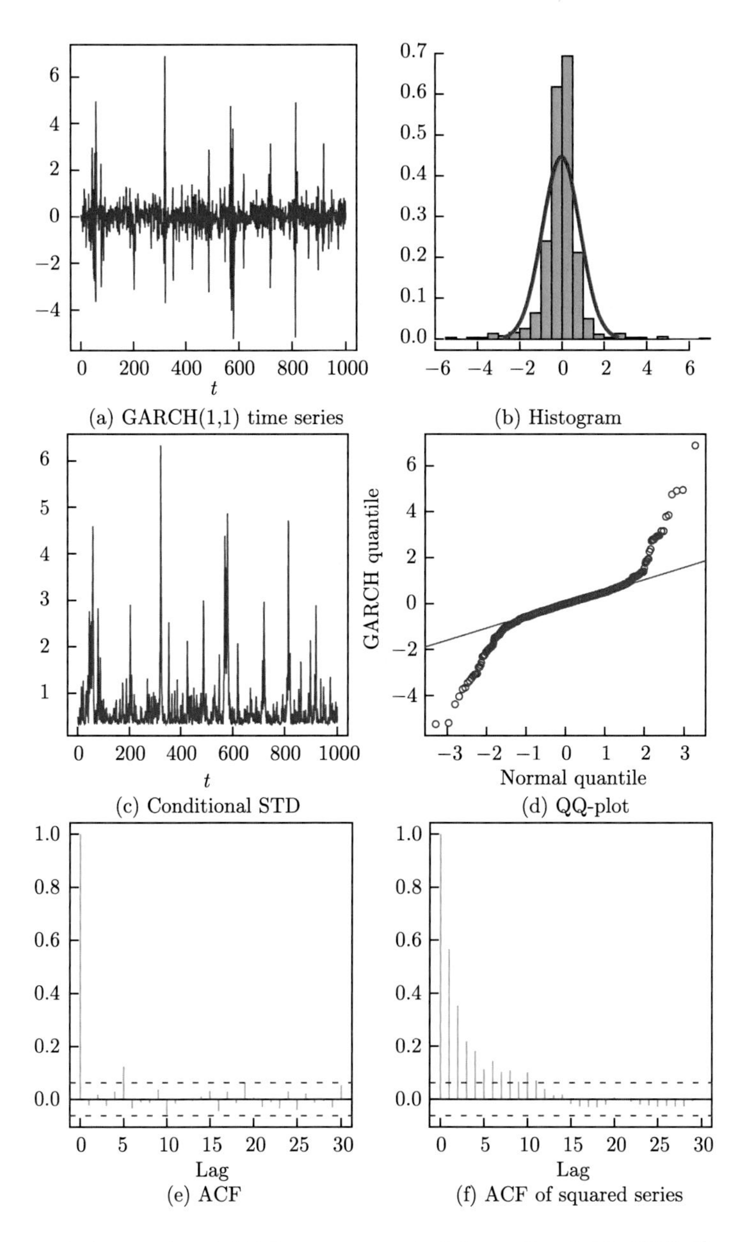

Figure 3.4 A sample of 1000 was generated from the GARCH(1,1) model with $a_0 = 0.1$, $a_1 = 0.08$ and $b_1 = 0.82$: (a) and (c) time series plots of X_t and σ_t; (b) normalized histogram and the normal density function with same mean and variance; (d) QQ-plot: the sample quantiles versus the quantiles of $N(0,1)$; (e) and (f) sample ACFs of $\{X_t\}$ and $\{X_t^2\}$ respectively.

together with their conditional standard deviations $\{\sigma_t\}$ are plotted in Figure 3.4 (a) and (c). Note that the GARCH(1,1) process generated here has the same (unconditional) variance as the ARCH(1) process displayed in Figure 3.1. However the conditional variance of the GARCH process appears more volatile; see Figure 3.1(c) and Figure 3.4(c). Comparing Figure 3.1(d) with Figure 3.4(d), the tails of this GARCH(1,1) process are heavier than those of the ARCH(1) process. Hence we may say that GARCH models are more effective than ARCH models in catching the volatilities. The sample ACF plots in Figure 3.4(e) and (f) indicate some significant autocorrelation in the squared X_t but not X_t itself.

Figures 3.5 presents a simulated time series with $p = q = 1$, $a_0 = 0.6$, $a_1 = 0.2$, $b_1 = 0.2$ and ε_t independent $N(0, 1)$. In contrast with the case presented in Figure 3.4, according to Theorem 3.2, we have $EX_t^4 < \infty$. Therefore, the tails of the distribution are much lighter.

3.1.5 Forecasting volatility

Let $\sigma_T^2(k) = E_T X_{T+k}^2$ be the conditional volatility in k-period from time T. Using the ARMA representation (3.20) and applying (2.82), we have the following recursive formula:

$$\sigma_T^2(k) = a_0 + \sum_{i=1}^{p \vee q} (a_i + b_i)\sigma_T^2(k - i) - \sum_{j=1}^{q} b_j \eta_T(k - j) \tag{3.24}$$

where $\sigma_T^2(m) = X_{T-m}^2$ if $m \leqslant 0$, $\eta_T(m) = 0$ if $m > 0$, and $\eta_T(m) = \eta_{T-m}$ if $m \leqslant 0$.

We now illustrate the probability aspects of GARCH model using the GARCH (1,1) model as follows.

Example 3.2 (continued) Consider GARCH(1,1) model (3.18) again. For any $k \geqslant 1$, we have $\eta_T(k - 1) = 0$, as we are predicting the future noise. Hence, by (3.24), we have for $k \geqslant 1$,

$$\sigma_T^2(k) = a_0 + (a_1 + b_1)\sigma_T^2(k - 1).$$

Iterating this k time, we get

$$\sigma_T^2(k) = \frac{a_0(1 - (a_1 + b_1)^k)}{1 - (a_1 + b_1)} + (a_1 + b_1)^k \sigma_T^2,$$

where σ_T^2 is given by (3.19).

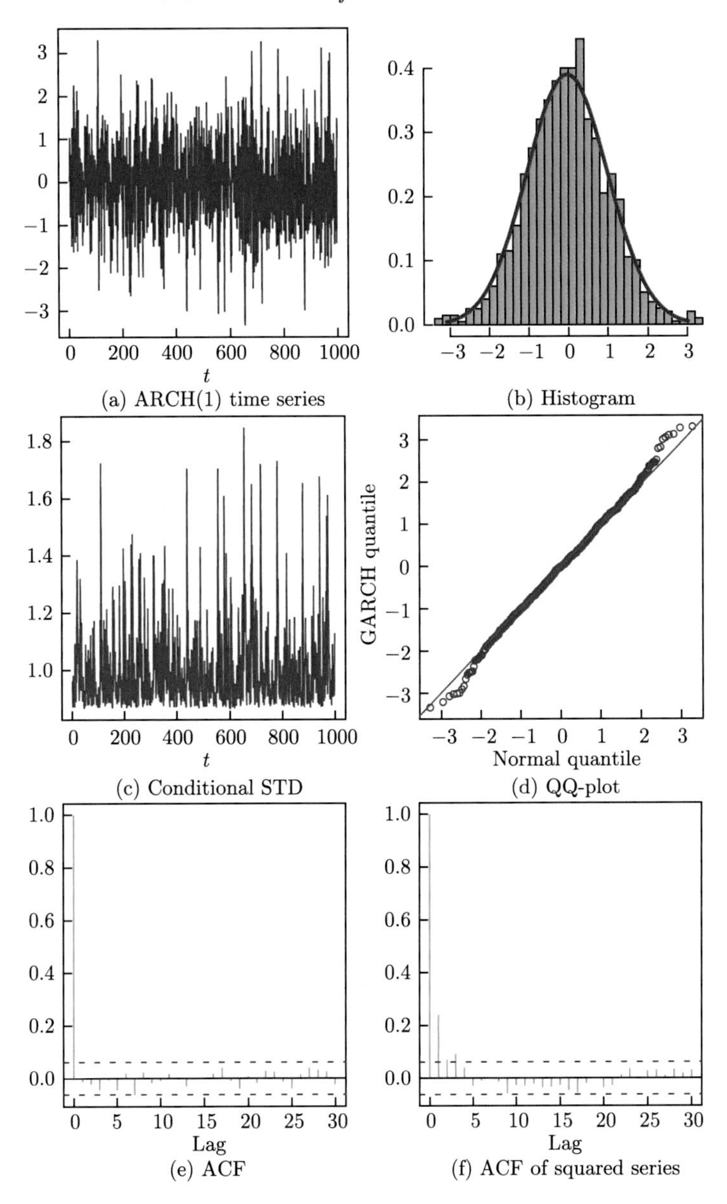

Figure 3.5 A sample of 1000 was generated from GARCH(1, 1) model with $a_0 = 0.6, a_1 = 0.2, b_1 = 0.2$ and everything else the same as that in Figure 3.4.

3.2 Estimation for GARCH models

Suppose that $X_1, \ldots, X_T$ are observations from GARCH model (3.14). We assume that condition (3.21) is always satisfied, and the order (p, q) is known.

Our task is to estimate the $(p+q+1)\times 1$ parameter vector

$$\boldsymbol{\theta} = (a_0, a_1, \ldots, a_p, b_1, \ldots, b_q)'.$$

We introduce two estimation procedures below. Note that the estimation for an ARCH model is a special case of that for a GARCH model with $q = 0$.

Since $\boldsymbol{\theta}$ only occurs in the second moments of X_t, the estimation is more difficult. The likelihood functions, even when correctly specified, tend to be rather flat, leading to large estimation errors. Large sample sizes are often required in order to obtain reliable estimates.

3.2.1 Conditional maximum likelihood estimation

The most often used estimators for ARCH/GARCH models are conditional maximum likelihood estimator. From Section 2.5.4, the likelihood function is the joint density of observed data $(X_1, \ldots, X_T)$

$$f(X_1, \ldots, X_T; \boldsymbol{\theta})$$

as a function of $\boldsymbol{\theta}$. By applying the product rule of probability, the above density can be written as

$$f_\nu(X_1, \ldots, X_\nu; \boldsymbol{\theta}) \prod_{t=\nu+1}^{T} f_c(X_t | X_1, \ldots, X_{t-1}; \boldsymbol{\theta}), \tag{3.25}$$

where $f_\nu(X_1, \ldots, X_\nu; \boldsymbol{\theta})$ is the joint density of $X_1, \ldots, X_\nu$, and $f_c(\cdot)$ is the conditional density of X_t given the history up to time $t-1$. All of them depend on $\boldsymbol{\theta}$. The conditional density is very easy to derive. Given the data up to time $t-1$, σ_t is already known. By (3.14), if ε_t has a density f, $X_t = \sigma_t \varepsilon_t$ has a density

$$f_c(X_t | X_{t-1}, X_{t-2}, \ldots,) = \frac{1}{\sigma_t} f(X_t/\sigma_t)$$

The first factor in (3.25) is usually hard to compute. Take ARCH(p) model, for example. By taking $\nu = p$, all conditional densities can be computed, but not the marginal density in the first factor. For this reason, it is often neglected, as it does not contribute much to the whole summation below. This leads to the conditional log-likelihood function or *conditional likelihood function* for short:

$$\ell(\boldsymbol{\theta}) = \sum_{t=\nu+1}^{T} [-\log \sigma_t + \log f(X_t/\sigma_t)]. \tag{3.26}$$

The *(conditional) maximum likelihood estimator* is defined as

$$\widehat{\boldsymbol{\theta}} = \arg\max_{\boldsymbol{\theta}} \sum_{t=\nu+1}^{T} [-\log \sigma_t + \log f(X_t/\sigma_t)] \tag{3.27}$$

When $\varepsilon_t \sim N(0,1)$, the conditional likelihood function after ignoring constants, reduces to conditional Gaussian likelihood function

$$-\sum_{t=\nu+1}^{T} (\log \sigma_t^2 + X_t^2/\sigma_t^2). \tag{3.28}$$

For example, for an ARCH(p) model, by taking $\nu = p$, one minimizes the criteria function

$$\sum_{t=p+1}^{T} \left[\log(a_0 + a_1 X_{t-1}^2 + \cdots + a_p X_{t-p}^2) + \frac{X_t^2}{a_0 + a_1 X_{t-1}^2 + \cdots + a_p X_{t-p}^2} \right]$$

This is drastically different from the least-squares estimation based on AR(p) representation:

$$\sum_{t=\nu+1}^{T} \left[X_t^2 - a_0 - a_1 X_{t-1}^2 - \cdots - a_p X_{t-p}^2 \right]^2.$$

Such a least-squares method is not recommended for ARCH/GARCH models, as pointed out before, due to non-existence of the fourth moments in many applications.

In addition to the Gaussian distribution in (3.28), two other commonly used families of distributions are

- *t-distribution*: $\varepsilon_t \sim_{i.i.d.} \sqrt{(\nu - 2)/v}\, t_\nu$. The rescaling constant is so taken since we require that $E\varepsilon_t^2 = 1$. Then, it is easy to see from the density of t_ν that ε_t has the density

$$f(x) = \frac{1}{\sqrt{\nu - 2}\, B(1/2, \nu/2)} \left(1 + \frac{x^2}{\nu - 2}\right)^{-(\nu+1)/2}, \quad \nu > 2, \tag{3.29}$$

  where $B(\cdot,\cdot)$ is the Beta function. The degree of freedom ν does not have to be an integer. When it is unknown, it can be estimated jointly from the data by maximizing (3.27) with respect to both ν and $\boldsymbol{\theta}$.
- *Generalized Gaussian distribution*: ε_t has the density function

$$f(x) = c_\nu \exp\left\{ -\frac{1}{2} \left|\frac{x}{\lambda}\right|^\nu \right\}, \quad (0 < \nu \leqslant 2), \tag{3.30}$$

  where $\lambda = \nu\{2^{-2/\nu}\Gamma(\frac{1}{\nu})/\Gamma(3/\nu)\}^{\frac{1}{2}}$ so that $E\varepsilon_t^2 = 1$. The distribution has heavier tail than the normal distribution but lighter tails than any t-distribution.

When ε_t has a density f_ε, which is typically unknown (but can also be known), we nevertheless use the likelihood (3.26), with f being normal density, t-distribution (3.29) or generalized Gaussian distribution (3.30), the resulting estimator is call *quasi-likelihood estimator*. Not all quasi-likelihood estimators are statistically consistent. It was shown by Hall and Yao (2003), Francq, Lepage and Zakoian (2011) and Fan, Qi, and Xiu (2014) that conditional Gaussian quasi-likelihood (3.28) is always consistent, even when ε_t does not follow the normal distribution. See also Theorem 3.3 below. However, it is not the most efficient distribution. Other quasi-likelihood estimators such as (3.26) with a t_{ν_0}-distribution with a given ν_0 are not unbiased in general, but the biases can be corrected to yield more efficient estimators. See Francq, Lepage and Zakoian (2011), Fiorentini and Sentana (2013), and Fan, Qi, and Xiu (2014).

3.2.2 Model diagnostics

GARCH models are a statistical fiction. Data were not necessarily generated from our idealized models. Therefore, the GARCH model might not be a good fit, and model diagnostics are needed. The raw materials are the residuals after GARCH fit:

$$\widehat{\varepsilon}_t = X_t/\widehat{\sigma}_t, \qquad \text{where } \widehat{\sigma}_t = \sigma_t(\widehat{\boldsymbol{\theta}}).$$

Ideally, ε_t should behave like an i.i.d white noise with density the same as the density f used in (3.27). Therefore, we ask the following questions:

- Is $\{\hat{\varepsilon}\}$ an i.i.d. white noise series? This is usually accomplished by the examination of the ACF plot and performing the Ljung–Box test.
- Is $\{\hat{\varepsilon}^2\}$ an i.i.d. white noise series? This can again be checked by using the ACF plot and the Ljung–Box test.
- Does the distribution $\{\hat{\varepsilon}\}$ look like the distribution used to derive the QMLE? For example, if t_{ν_0} is used in (3.26), we would ideally like to have the distribution of $\{\hat{\varepsilon}\}$ look like t_{ν_0}. This can be checked by looking at the Q–Q plot of the residuals against t_{ν_0}-distribution, or by applying the Kolmogorov–Smirnov test. When normal distribution is used in deriving GMLE, we may also test the Gaussianity of $\{\hat{\varepsilon}\}$ using the Jarque–Bera test introduced in Section 1.5.2.

Numerically the conditional maximum likelihood estimator for a given data set may be obtained by using the function `garch` in the *R*-package *TSA*.

In the illustration below, we generate 1000 observations from GARCH(1,1) model with $a_0 = 0.1$, $a_1 = 0.06$, $b_1 = 0.82$, and $\varepsilon_t \sim N(0, 1)$. We then use the function `garch` to fit the data set with a GARCH(1,1) model.

```
> x <- garch.sim(alpha=c(0.1, 0.06), beta=0.82, n=1000)
> garch11 <- garch(x, order=c(1,1))
> summary(garch11)

Residuals:
     Min       1Q   Median       3Q      Max
-3.32041 -0.71230 -0.03774  0.62927  3.03393


Coefficient(s):
    Estimate  Std. Error  t value Pr(>|t|)
a0   0.13989     0.08284    1.689   0.0913 .
a1   0.06833     0.02847    2.400   0.0164 *
b1   0.77494     0.11270    6.876   6.16e-12 ***
---
Signif. codes:  0 *** 0.001 ** 0.01 * 0.05 . 0.1  1

Diagnostic Tests:
        Jarque-Bera test
data:  Residuals
X-squared = 0.2929, df = 2, p-value = 0.8638


        Box--Ljung test
data:  Squared.Residuals
X-squared = 0.0068, df = 1, p-value = 0.9341
```

The estimated parameters are $\widehat{a}_0 = 0.140$, $\widehat{a}_1 = 0.068$ and $\widehat{b}_1 = 0.775$ with the standard errors 0.083, 0.028 and 0.113 respectively. The standard errors are calculated based on the asymptotic variances of the estimators. See Theorem 3.3 below.

The R-function also performs two diagnostic tests based on the residuals. The *Jarque–Bera test* is a goodness-of-fit test for the null hypothesis that the skewness is 0 and the kurtosis is 3. Therefore it is also viewed as a test for the normality. See Section 1.5.2 for additional details. In the above illustration, the p-value of Jarque–Bera test is 0.8638. Hence no significant evidence against the hypothesis that the residuals are normal. Note also that the p-value for the Ljung–Box test (see §1.4.3) for the squared residuals is 0.9341. Hence again no significant evidence against the hypothesis that the squared residuals are white noise.

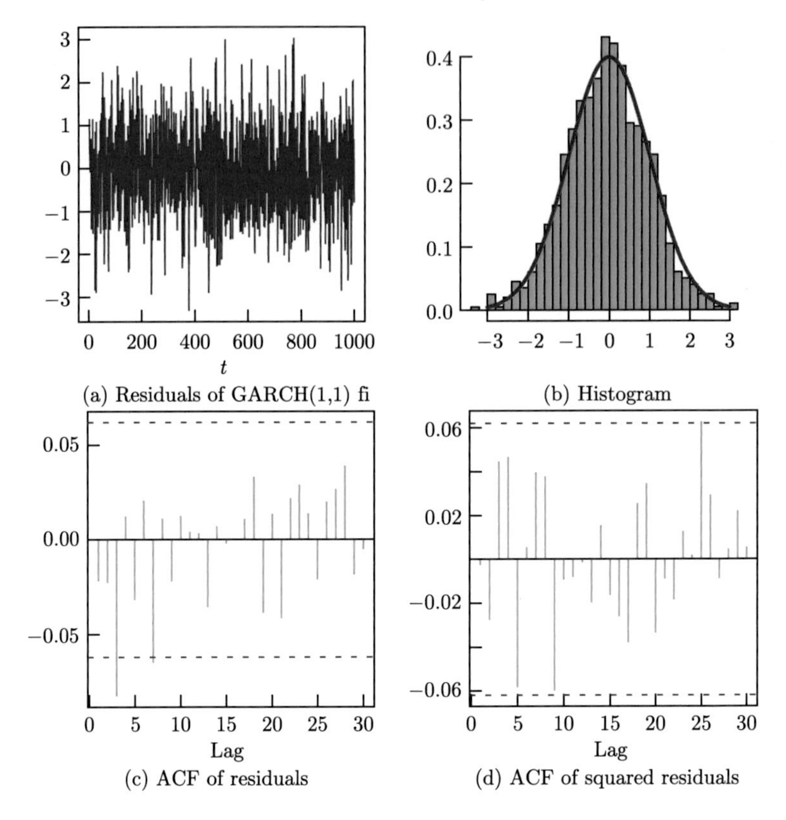

Figure 3.6 Diagnostic plots after GARCH(1,1) fit to a simulated data. (a) Time series plot of residuals, (b) Distribution of residuals, (c) & (d) Autocorrelation functions of residuals and squared residuals.

We also present the diagnostic via graphical plots. Figure 3.6(a) shows that the residuals time series are homoscedastic noises, Figure 3.6(b) confirms that the distribution of residuals are normally distributed, and hence the GARCH(1,1) fit is the maximum likelihood estimate. Figures 3.6(c) and (d) confirms that the residual and squared time series are white noises. All these point to the same direction: the residuals behave like an i.i.d. Gaussian white noise.

3.2.3 Applications of GARCH modeling

We model the daily returns (in percentages) of S&P 500 index, the Goldman Sachs share price, and the Ford share price in January 4, 2010 to August 10, 2012 using GARCH(1,1) model. There were 658 trading days during this period, and hence the sample size is $T = 657$. The adjusted daily close prices are plotted in Figure 3.7. The share prices for both Goldman Sachs and

Table 3.1 *Fitted coefficients and diagnostic statistics for the daily returns of S&P 500 index, Goldman Sachs stock, and Ford stock in January 4, 2010 to August 10, 2012*

	$\widehat{a}_0$	$\widehat{a}_1$	$\widehat{b}_1$	JB test	LB test
S&P 500	0.032 (0.011)	0.115 (0.018)	0.864 (0.022)	0.000	0.003
GS	0.237 (0.043)	0.106 (0.020)	0.845 (0.023)	0.000	0.633
Ford	0.329 (0.100)	0.086 (0.020)	0.856 (0.033)	0.000	0.996

Ford are low and overall bearish since 2011 due to the economic downturn triggered by the Euro debt crisis. However the S&P 500 index maintained an overall positive gain over the period, although it recorded 3.9% drop in a week in July 2011. The estimated coefficients and their standard errors (in parentheses) together with the p-values of the Jarque–Bera (JB) test and the Ljung–Box (LB) test are listed in Table 3.1.

The estimated coefficients satisfy the stationary condition $a_1 + b_1 < 1$ for all the three data sets, but the sum $\widehat{a}_1 + \widehat{b}_1$ are close to 1. The estimate $\widehat{b}_1$ is always close to 1, which is common for financial returns, reflecting the persistence of price volatilities. This unfortunately would imply that the 4th moments of those returns are infinity, contradicting to the stylized features observed in Section 1.2. This is arguably a weakness of GARCH models, making the tails of return distributions too heavy. Thus, when the GARCH fitted models are used to simulate financial returns, they tend to overstate the severity of tail events.

The three fitted GARCH(1,1) models are similar, except the constant term $\widehat{a}_0$ is much smaller for the S&P 500 returns; reflecting the smaller volatilities of a portfolio consisting of 500 stocks than those for individual stocks such as the Goldman Sachs or the Ford. It is easy to see from Figure 3.8 that the returns are in the range of -7% to 5% for the S&P 500 index price, of -15% and 10% for both the Goldman Sachs stock and the Ford stocks. Also plotted in Figure 3.8 are the two curves of $\pm 2\widehat{\sigma}_t$, where $\widehat{\sigma}_t^2$ is the fitted volatilities defined recursively as follows:

$$\widehat{\sigma}_t^2 = \widehat{a}_0 + \widehat{a}_1 X_{t-1}^2 + \widehat{b}_1 \widehat{\sigma}_{t-1}^2. \tag{3.31}$$

See also (3.18). Those two curves seem to form an envelope for the volatility in the sense that

$$X_t \in [-2\,\widehat{\sigma}_t,\ 2\,\widehat{\sigma}_t] \tag{3.32}$$

holds with approximately 95% of the time. Indeed, the relative frequency for the above to hold is 93.91% for the S&P 500 returns, 95.13% for the

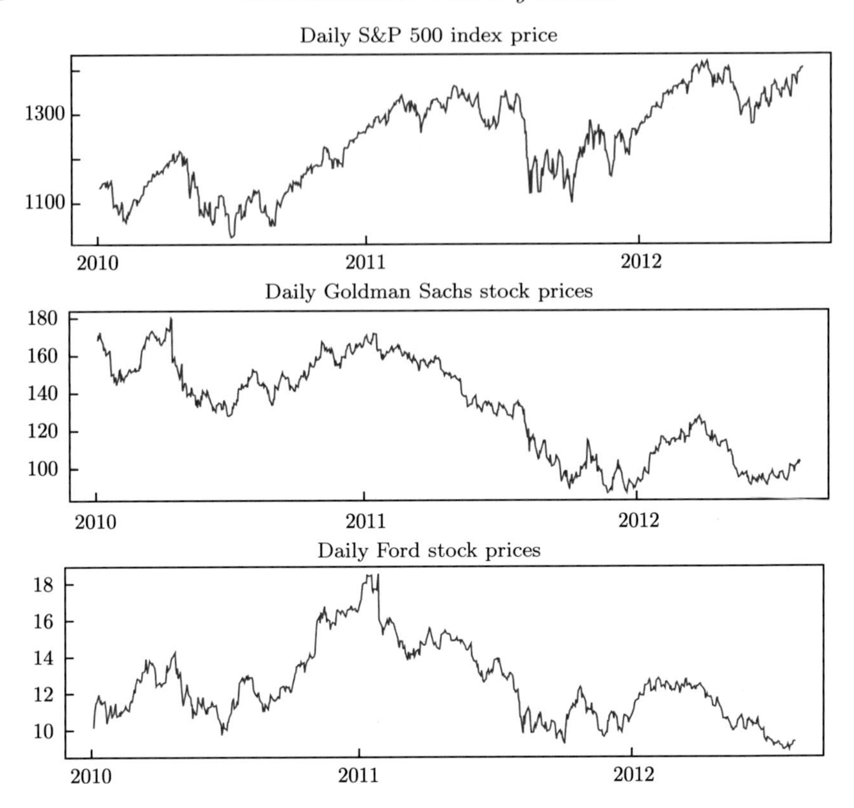

Figure 3.7 Time series plots of the daily prices in January 4, 2010 to August 10, 2012 of the S&P 500 index (the top panel), the Goldman Sachs share (the middle panel) and the Ford share (the bottom panel).

Goldman Sachs returns, and 94.82% for the Ford returns. Those are in line with the nominal confidence level 95%, as the fitted model is obtained by assuming the innovation ε_t in model (3.18) were normal, in spite of the fact that the normality assumption was rejected by the Jarque–Bera test with the p-value equal to 0; see the table above.

The Ljung–Box test for the squared residuals yields the p-value 0.003 for the S&P 500 data, and 0.633 and 0.996 for the other two series. Figure 3.9 displays the sample ACF plots for both the squared and the absolute residuals. There are hardly any significant autocorrelations in both the squared and the absolute residuals for the Ford returns, and also for the Goldman Sachs returns except one leak at lag 10 for the squared residuals. For S&P 500 index, the ACF for both the squared and the absolute returns are significant at lags 1 and 2. This is in line with the result of the Ljung–Box test.

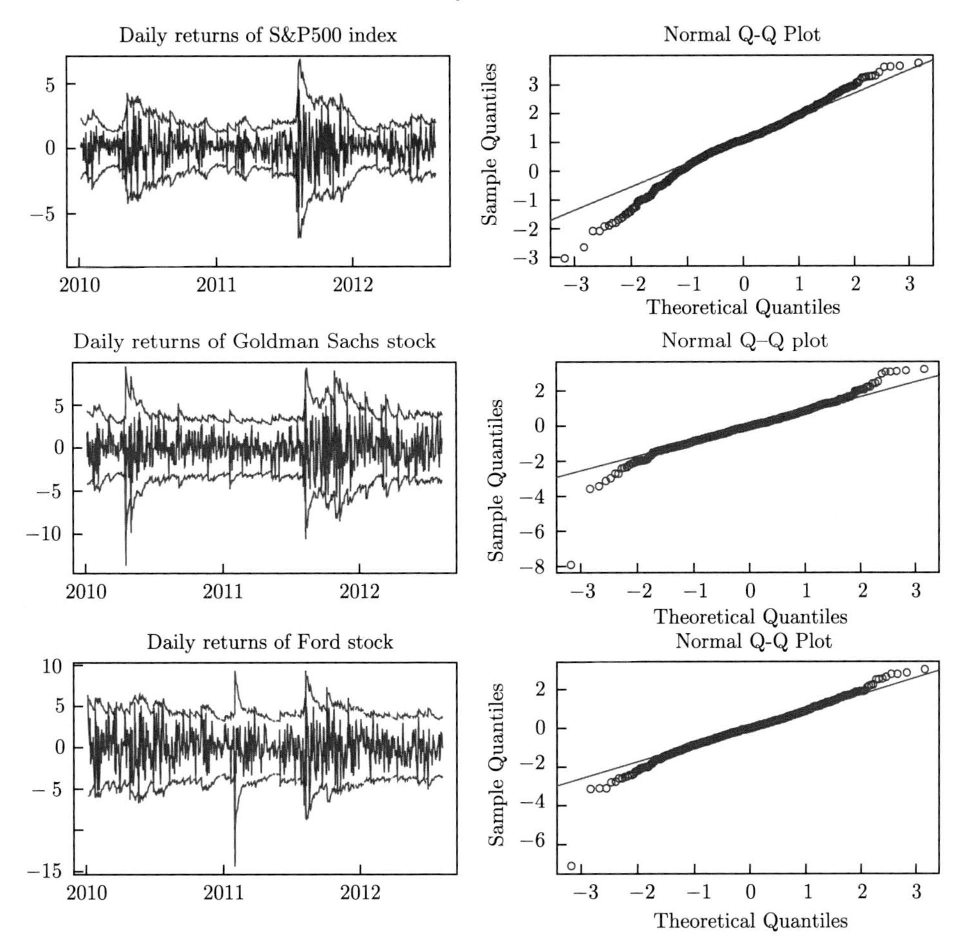

Figure 3.8 Time series plots (left panel) of the daily returns in January 4, 2010 to August 10, 2012 of the S&P 500 index (the top panel), the Goldman Sachs share (the middle panel) and the Ford share (the bottom panel). Two blues lines are $\pm 2\widehat{\sigma}_t$ estimated based on the fitted GARCH(1,1) models. Right panel is the Q–Q plots for normality checking of the residuals on the left panel.

The kurtosis for the residuals from the fitted GARCH(1,1) models are 4.04, 7.34, and 4.96 respectively for the returns of S&P 500, Goldman Sachs, and Ford. They are heavier than the normal distribution, conforming with the Jarque–Bera tests above. Putting them into the prospective of the t-distributions, they correspond to t-distributions with degrees of freedom 9.75, 3.96, and 6.34 respectively. Apparently, the S&P 500 have the lightest tails of the innovation, as expected. While there are deviations from the normal distribution, the deviations are not too serious. For the returns of

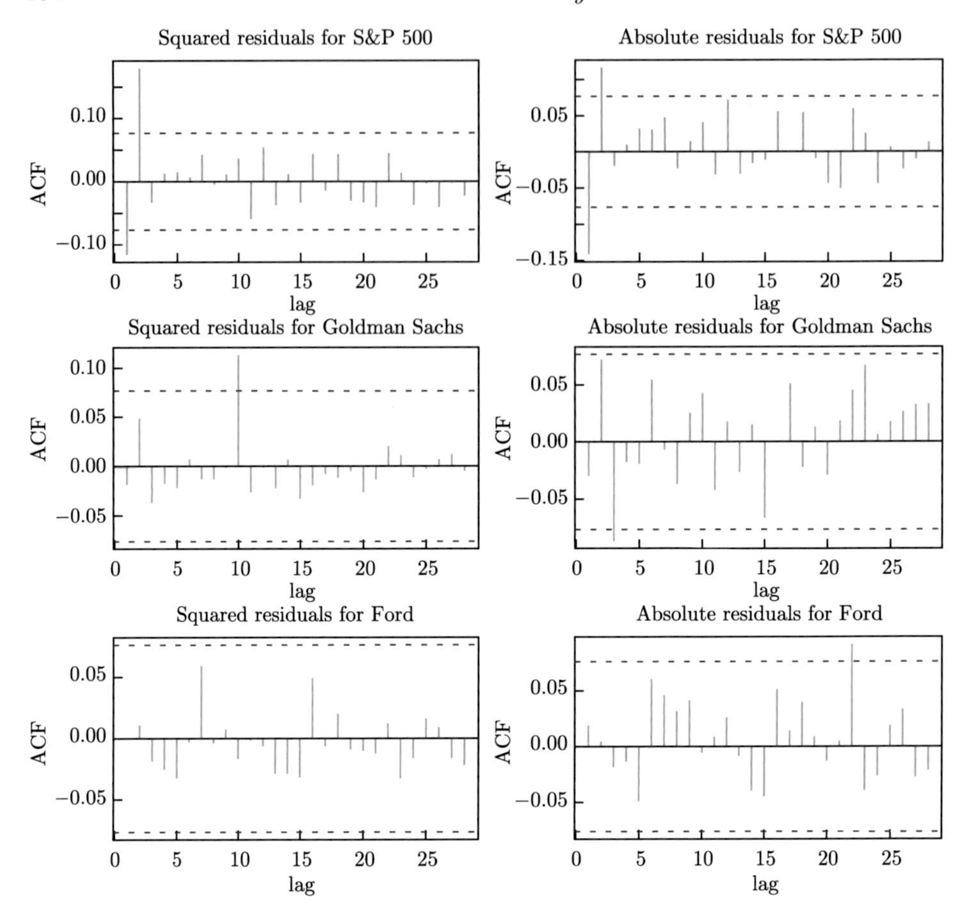

Figure 3.9 Sample ACF plots for the squared and the absolute residuals resulted from the fitted GARCH(1,1) models for the three series.

Goldman Sachs and Ford, the above analysis suggests that one may employ the QMLE (3.27) based on t-distributions with degree of freedom 3.96 and 6.34. See (3.29).

Volatility forecast

GARCH models can also be used to forecast the future volatility. With the MLE $\widehat{\boldsymbol{\theta}}$, the forecast for σ_{T+1}^2 is defined as $\widehat{\sigma}_{T+1}^2 = \sigma_{T+1}^2(\widehat{\boldsymbol{\theta}})$. See for example (3.18). Multiple step forecasting formula was also given in Example 3.2.

Values-at-Risk

An important concept in financial risk management is the so-called *Value at Risk (VaR)*. It is a widely used measure for the risk of loss on a specific asset (or portfolio).

Suppose that the return X_t of an asset follows from GARCH model (3.14). Given the information up to time t, the conditional distribution of X_{t+1} is a scale-transformation of the distribution of ε_{t+1}. More precisely, let $f(\cdot)$ denote the probability density function of ε_{t+1}. Then the conditional distribution of X_{t+1} given $X_t, X_{t-1}, \ldots$ has the density function $\frac{1}{\sigma_t} f(\frac{\cdot}{\sigma_t})$. In terms of risk management, we are interested in the critical value such that a loss exceeding this critical value is of the probability not greater than α, where α is a pre-specified small number such as 5% or 1%. This critical value is called the VaR at the level α. Within the context of GARCH model (3.14), the VaR at the time $t+1$ and the level α is of the form

$$\begin{aligned} \mathrm{VaR}_{\alpha,t+1} &\equiv \max\{L : P(X_{t+1} \leqslant L \,|\, X_t, X_{t-1}, \ldots) \leqslant \alpha\} \\ &= \sigma_{t+1} Q_{\alpha,\varepsilon}, \end{aligned} \tag{3.33}$$

where $Q_{\alpha,\varepsilon}$ denotes the αth quantile of ε_{t+1}. In practice we may estimate $\mathrm{VaR}_{\alpha,t+1}$ by

$$\widehat{\mathrm{VaR}}_{\alpha,t+1} = \widehat{\sigma}_{t+1} \widehat{Q}_{\alpha,\varepsilon}, \tag{3.34}$$

where $\widehat{Q}_{\alpha,\varepsilon}$ is an estimated αth quantile of ε_{t+1}, and the $\widehat{\sigma}_{t+1}$ is the volatility forecast defined as

$$\widehat{\sigma}_{t+1} = \sigma_t(\widehat{\boldsymbol{\theta}}). \tag{3.35}$$

In the above expression σ_t is defined as in (3.17) with convention that $X_t = 0$ when $t \leqslant 0$, and $\widehat{\boldsymbol{\theta}} \equiv \widehat{\boldsymbol{\theta}}_t$ is the MLE of $\boldsymbol{\theta}$ based on data $X_1, \ldots, X_t$ only; see (3.27).

The estimation for $Q_{\alpha,\varepsilon}$ can be carried out either parametrically or nonparametrically. The most convenient approach is to assume that the distribution of ε_t is given, such as $N(0,1)$. Then $Q_{\alpha,\varepsilon}$ is -1.64 for $\alpha = 5\%$, -1.96 for $\alpha = 2.5\%$, and -2.33 for $\alpha = 1\%$. The implication is that the loss beyond $196\sigma_{t+1}\%$ at the time $t+1$ is of the probability 0.025, the loss beyond $164\sigma_{t+1}\%$ is of the probability 0.05, and the loss exceeding $233\sigma_{t+1}\%$ is of the probability 0.01.

However, the above analysis (see Figure 3.8) indicates that the normality assumption is typically inadequate for real financial returns. It often underestimates the risks substantially. One alternative is to estimate $Q_{\alpha,\varepsilon}$ based on the empirical distribution of the residuals $\widehat{\varepsilon}_t = X_t/\widehat{\sigma}_t$. Namely we use the sample αth quantile of the residuals as our estimator for $Q_{\alpha,\varepsilon}$.

We now illustrate the GARCH-based VaR estimation with the three data sets, namely daily returns for the S&P 500 index, the Goldman Sachs stock and the Ford stock in January 4, 2010 to August 10, 2012. For each of 154 trading days in 2012, we use the data from the previous 500 trading days (i.e.

Table 3.2 *Percentage of exceedances of returns over predicted VaR based on GARCH(1,1)*

Distribution of ε_t	Empirical residuals		$N(0,1)$	
Level	5%	2.5%	5%	2.5%
S&P 500	3.90%	1.30%	4.55%	3.25%
Goldman Sachs	2.60%	0.65%	1.30%	1.92%
Ford	3.25%	1.95%	3.25%	2.60%

about two years) to fit a GARCH(1,1) model, and we predict the VaR at the levels of 5% and 2.5% according to (3.34) in which $\widehat{Q}_{\alpha,\varepsilon}$ is estimated either by assuming the normality for ϵ_t or by the empirical distribution of the residuals derived from the fitted GARCH models. For the latter, $\widehat{Q}_{\alpha,\varepsilon}$ in (3.34) varies with respect to t, as the residuals were resulted from fitting a GARCH(1,1) model to the previous 500 data points. The returns in 2012 of the 3 series together with their predicated VaR curves are plotted in Figure 3.10. This figure shows that for more extreme VaR like at the 2.5% level, the estimates based on the normality (the red curves) are less extreme than those based on the empirical residual distribution functions (the green curves); resulting possible underestimating risks based on the normal assumption.

The relative frequencies of the exceedances of the returns over the VaR curves among the 154 trading days in 2012 are listed in Table 3.2. Overall both methods tend to over estimate the VaR. The only two cases for underestimation are for S&P 500 and Ford returns at the level 2.5% based on the normality assumption. However this should not be necessarily taken as the evidence that the methods have the tendency to overestimate the risks. The pattern exhibited in Table 3.2 is more likely caused by the fact that the volatility was particular large in most the second half of 2011 while the returns in 2012 were much less volatile; see Figure 3.8. Therefore prediction was stretched over a period with which the data are remarkably nonstationary. Hence we need to be cautious in interpreting the prediction outcome.

*3.2.4 Asymptotic properties**

To state the asymptotic properties of the Gaussian MLE $\widehat{\boldsymbol{\theta}}$ defined in (3.27) with the standard Gaussian density f, we introduce some notation first. Let $\boldsymbol{U}_t = \partial\sigma_t^2/\partial\boldsymbol{\theta}$ be the gradient vector of σ_t^2 as a function of $\boldsymbol{\theta}$. It may be shown that $\boldsymbol{U}_t/\sigma_t^2$ has all its moments finite. We assume that the matrix

$$\boldsymbol{M} \equiv E(\boldsymbol{U}_t\boldsymbol{U}_t'/\sigma_t^4) > 0 \tag{3.36}$$

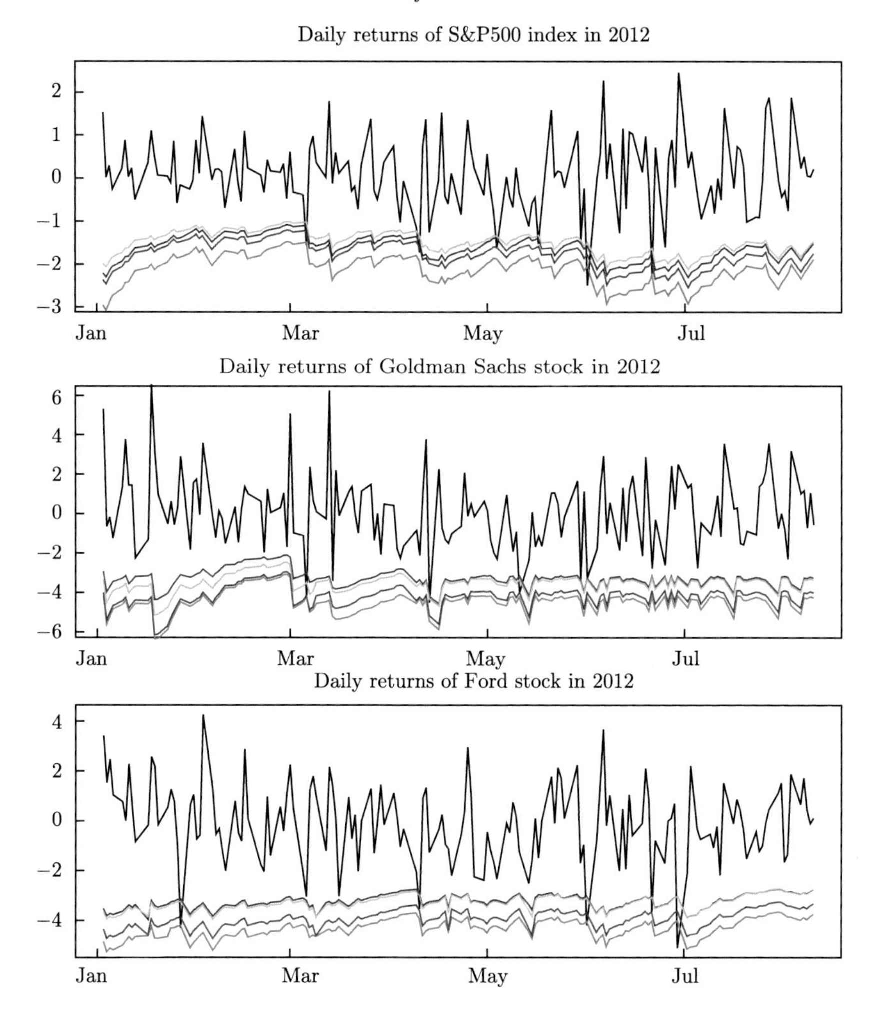

Figure 3.10 Time series plots of the daily returns in January 4 to August 10 2012 together with the predicted VaR from GARCH(1,1) models based on the previous 500 returns at each time: VaR based on normal innovation at the 5% level (yellow lines) and 2.5% level (red lines); VaR based on empirical residual distribution at the 5% level (blue lines) and 2.5% level (green lines).

is positive-definite. Note that Theorem 3.2 below does not require the innovations ε_t in model (3.14) to be normal. In this sense, $\widehat{\boldsymbol{\theta}}$ is a QMLE.

Theorem 3.3 *For GARCH model* (3.14)*, let $a_i > 0$ for all $0 \leqslant i \leqslant p$, and $b_j > 0$ for all $1 \leqslant j \leqslant q$. Let the stationarity condition* (3.21) *hold, and $\boldsymbol{M}$ defined in* (3.36) *be positive-definite. For ν used in* (3.27)*, $\nu \to \infty$ and $\nu/T \to 0$, as the sample size $T \to \infty$.*

(i) *If* $E(\varepsilon_t^4) < \infty$,

$$\frac{T^{1/2}}{\{E(\varepsilon_t^4) - 1\}^{1/2}}(\widehat{\boldsymbol{\theta}} - \boldsymbol{\theta}) \xrightarrow{D} N(0, \boldsymbol{M}^{-1}).$$

(ii) *If* $E(\varepsilon_t^4) = \infty$ *and* $E(\varepsilon_t^{4-\eta}) < \infty$ *for any small positive constant* η,

$$\frac{T^{1/2}}{\lambda_T}(\widehat{\boldsymbol{\theta}} - \boldsymbol{\theta}) \xrightarrow{D} N(0, \boldsymbol{M}^{-1}),$$

where

$$\lambda_T = \inf\left\{\lambda > 0 : E\{\varepsilon_1^4 I(\varepsilon_1^2 \leqslant \lambda)\} \leqslant \lambda^2/T\right\}.$$

The above theorem was established by Hall and Yao (2003). See also Berkes, Horváth and Kokoszka (2003), Mikosch and Straumann (2006) and Straumann and Mikosch (2006). Early attempts dealing with some special models such as GARCH(1,1) include Lee and Hansen (1994) and Lumsdaine (1996). Note that Theorem 3.2 also applies to pure ARCH models for which all ARCH coefficients are required to be positive.

*3.2.5 Least absolute deviations estimation**

Theorem 3.3 indicates that the condition $E(\varepsilon_t^4) < \infty$ is almost necessary for the QMLE $\widehat{\boldsymbol{\theta}}$ defined in (3.36) to be asymptotically normal. Indeed if $E(\varepsilon_t^{4-\eta}) = \infty$ for a constant $\eta > 0$, the convergence rate is slower than the standard rate of $T^{1/2}$. Hence it is difficult to find the standard errors of the estimator and the standard bootstrap methods are not applicable either (see e.g. Hall and Yao 2003). To overcome these drawbacks, Peng and Yao (2003) propose a log-transform-based least absolute deviations estimator (LADE) as an alternative. In fact, the LADE is asymptotically normal with the standard convergence rate $T^{1/2}$ under the condition $E(\varepsilon_t^2) < \infty$. Hence its standard errors can be readily calculated. Alternative approaches to this problem is to apply QMLE with a heavier tailed distribution. See, for example, Francq, Lepage and Zakoian (2011), Fiorentini and Sentana (2013), and Fan, Qi, and Xiu (2014).

In order to introduce the LADE, a different parameterization is required. Let $C_0 > 0$ be the constant such that the median of e_t^2 is equal to 1, where $e_t = C_0^{1/2}\varepsilon_t$. Note that such a C_0 always exists. Then model (3.14) may now be expressed as

$$X_t = s_t e_t, \quad s_t^2 \equiv s_t(\boldsymbol{\beta})^2 = \alpha_0 + \sum_{i=1}^{p} \alpha_i X_{t-i}^2 + \sum_{j=1}^{q} b_j s_{t-j}^2, \tag{3.37}$$

where $s_t^2 = \sigma_t^2/C_0$, $\alpha_i = a_i/C_0$ and

$$\boldsymbol{\beta} = (\alpha_0, \alpha_1, \ldots, \alpha_p, b_1, \ldots, b_q)'.$$

Now (3.37) implies that

$$\log(X_t^2) = \log\{s_t(\boldsymbol{\beta})^2\} + \log(e_t^2), \tag{3.38}$$

and the median of $\log(e_t^2)$ is 0. Thus, the true value of $\boldsymbol{\beta}$ minimizes

$$E\big|\log(X_t^2) - \log\{s_t(\boldsymbol{\beta})^2\}\big|.$$

This leads to our *LADE*

$$\widehat{\boldsymbol{\beta}} = \arg\min_{\boldsymbol{\beta}} \sum_{t=\nu+1}^{T} \big|\log(X_t^2) - \log\{\widetilde{s}_t(\boldsymbol{\beta})^2\}\big|, \tag{3.39}$$

where $\widetilde{s}_t(\boldsymbol{\beta})^2$ is a truncated version of $s_t(\boldsymbol{\beta})^2$ defined as

$$\widetilde{s}_t^2 \equiv \widetilde{s}_t(\boldsymbol{\beta})^2 = \frac{\alpha_0}{1 - \sum_{1\leqslant j\leqslant q} b_j} \sum_{i=1}^{\min(p,t-1)} \alpha_i X_{t-i}^2 \tag{3.40}$$

$$+ \sum_{i=1}^{p} \alpha_i \sum_{k=1}^{\infty} \sum_{j_1=1}^{q} \cdots \sum_{j_k=1}^{q} b_{j_1} \cdots b_{j_k} X_{t-i-j_1-\cdots-j_k} I(t - i - j_1 - \cdots - j_k \geqslant 1),$$

which follows directly, with adaptation of notation, from (3.17).

To state the asymptotic properties of the LADE $\widehat{\boldsymbol{\beta}}$, let

$$\boldsymbol{V}_t = (v_0, v_{t1}, \ldots, v_{t,p+q})'$$

be a $(p+q) \times 1$ vector, where $v_0 = 1/(1 - \sum_{1\leqslant j\leqslant q} b_j)$, and

$$v_{ti} = X_{t-i}^2 + \sum_{k=1}^{\infty} \sum_{j_1=1}^{q} \cdots \sum_{j_k=1}^{q} b_{j_1} \cdots b_{j_k} X_{t-i-j_1-\cdots-j_k}^2, \quad 1 \leqslant i \leqslant p,$$

$$v_{t,p+j} = \frac{\alpha_0}{\left(1 - \sum_{1\leqslant i\leqslant q} b_i\right)^2} + \sum_{i=1}^{p} \alpha_i X_{t-i-j}^2$$

$$+ \sum_{i=1}^{p} \alpha_i \sum_{k=1}^{\infty} (k+1) \sum_{j_1=1}^{q} \cdots \sum_{j_k=1}^{q} b_{j_1} \cdots b_{j_k} X_{t-i-j-j_1-\cdots-j_k}^2, \; 1 \leqslant j \leqslant q.$$

We assume that the matrix

$$\boldsymbol{M}_1 = E(\boldsymbol{V}_t \boldsymbol{V}_t / s_t^4) > 0 \tag{3.41}$$

is positive definite.

Theorem 3.4 *Let all the coefficients in* GARCH *model* (3.14) *be positive. Let the stationarity condition* (3.21) *hold, and the matrix* $\boldsymbol{M}_1$ *defined by* (3.41) *be positive-definite. For* ν *used in* (3.39), *let* $\nu \to \infty$ *and* $\nu/T \to 0$ *as* $T \to \infty$. *Then*

$$T^{1/2}(\widehat{\boldsymbol{\beta}} - \boldsymbol{\beta}) \xrightarrow{D} N\big(0,\ M_1^{-1}/\{4f(0)^2\}\big)$$

provided that the probability density function $f(\cdot)$ *of* $\log(e_t^2)$ *is positive and continuous at 0.*

The above theorem was established by Peng and Yao (2003). Comparing with Theorem 3.3, it does not require the 4th moments of innovation to be finite. It is well expected that the LADE is more robust against heavy tails.

A simulation study was carried out to compare the performance of the Gaussian MLE $\widehat{\boldsymbol{\theta}}$ defined in (3.27) and the LADE $\widehat{\boldsymbol{\beta}}$ defined in (3.39) for GARCH(1,1) model

$$X_t = \sigma_t \varepsilon_t, \qquad \sigma_t^2 = a_0 + a_1 X_t^2 + b_1 \sigma_t^2.$$

We let the true values in the above model be $a_0 = 1.5$, $a_1 = 0.15$ and $b_1 = 0.75$. We consider three different distributions for the innovation ε_t: $N(0,1)$, t_4 and t_3. Setting the sample size $T = 150$ and 300, we drew 500 samples from each setting.

We use $\nu = 20$ in both (3.27) and (3.39). Note that the LADE is based on a different parameterization (3.37), which implies $\alpha_1/\alpha_0 = a_1/a_0$. We measure the estimation error for the MLE $\widehat{\boldsymbol{\theta}} = (\widehat{a}_0, \widehat{a}_1, \widehat{b}_1)'$ by

$$\frac{1}{2}\big(|\widehat{a}_1/\widehat{a}_0 - a_1/a_0| + |\widehat{b}_1 - b_1|\big),$$

and for LADE $\widehat{\boldsymbol{\beta}} = (\widehat{\alpha}_0, \widehat{\alpha}_1, \widetilde{b}_1)'$ by

$$\frac{1}{2}\big(|\widehat{\alpha}_1/\widehat{\alpha}_0 - \alpha_1/\alpha_0| + |\widetilde{b}_1 - b_1|\big).$$

Figure 3.11 shows the boxplots of those absolute errors. Overall both the estimators become more accurate when the sample size T increases from 150 to 300. Furthermore, the heavier the tails of the innovation distribution, the larger the estimation errors are. Note that when $\varepsilon_t \sim t_\nu$, $E(|\varepsilon_t|^\nu) = \infty$ and $E(|\varepsilon_t|^{\nu-\delta}) < \infty$ for any small constant $\eta > 0$. When $\varepsilon_t \sim N(0,1)$, the Gaussian MLE (3.27) is the genuine (conditional) MLE, and it performs better than the LADE. For the cases with t_4 innovations, it follows from Theorem 3.3 (ii) that the MLE is asymptotically normal but with the convergence rate slower than $T^{1/2}$. This is a marginal case for which the performances of the two estimation methods are comparable. When $\varepsilon_t \sim t_3$,

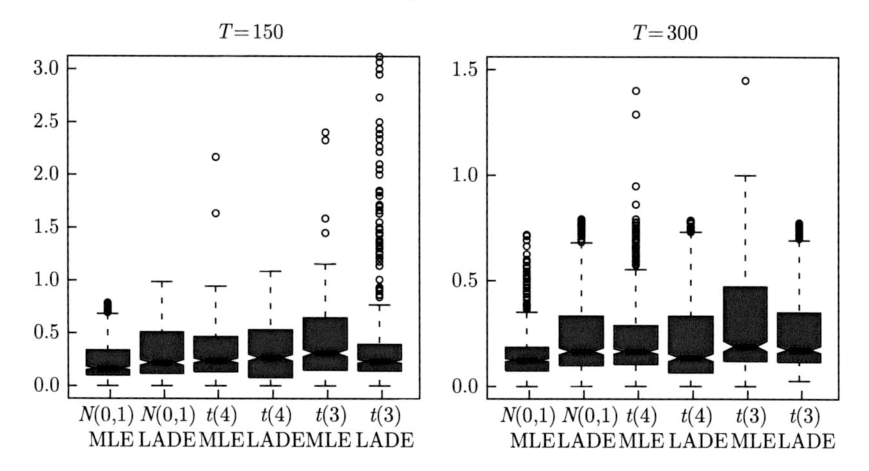

Figure 3.11 Boxplots of the absolute errors of the Gaussian MLE (3.27) and the LADE (3.39) for a GARCH(1,1) model. Labels $N(0,1)$, $t(4)$ and $t(3)$ indicate that ε_t has, respectively, a standard normal, and t-distributions with 3 and 4 degrees of freedom.

the MLE is no longer asymptotically normal but the LADE is. Indeed Figure 3.11 shows clearly that the LADE performs better than the MLE in this heavy tail situation.

When to use what?

The simulation study reported above indicates that the relative performance of the Gaussian MLE $\widehat{\boldsymbol{\theta}}$ and the LADE $\widehat{\boldsymbol{\beta}}$ hinges critically on the tail heaviness of the innovation distribution. Since this distribution is unknown, a practical relevant question is when to use what.

If we knew the distribution of ε_t, the genuine maximum (conditional) likelihood estimator would be used. Note that the LADE would be the maximum likelihood estimator if $\log(e_t^2)$ has a Laplace distribution with density function

$$\frac{\lambda}{2}\exp\{-\lambda|x|\},$$

where $\lambda \in (0,1)$ is a constant. Hence, intuitively, we expect that the Gaussian MLE is a better estimator when the distribution of ε_t is close to $N(0,1)$, and the LADE is better when the distribution of $\log(e_t^2)$ is approximately a Laplace distribution. Hence we may compare the closeness of those two distributions to select a good estimation procedure. We outline such a selection procedure proposed by Huang, Wang and Yao (2008).

Denoted by $\Phi(\cdot)$ the $N(0,1)$ distribution function, and by $G(\cdot)$ the distribution function of the standard Laplacian distribution with the density function $0.25\exp(-|x|/2)$. Let $\widehat{\varepsilon}_t = X_t/\widetilde{\sigma}_t(\widehat{\boldsymbol{\theta}})$ be the residuals derived from the Gaussian MLE. In practice, we standardize $\widehat{\varepsilon}_t$ such that the first two sample moments are 0 and 1. Let $\widehat{e}_t = X_t/\widetilde{s}_t(\widehat{\boldsymbol{\beta}})$ be the residuals derived from the LADE. In practice, we 'standardize' $\widehat{e}_t$ such that the sample median of $\widehat{e}_t^2$ is 1 and the sample mean of $|\log(\widehat{e}_t^2)|$ is 2. This may be achieved by letting

$$\log(\widehat{e}_t^2) = c_1 \log\{c_2 X_t^2/\widetilde{s}_t(\widehat{\boldsymbol{\beta}})^2\}$$

for appropriate positive constants c_1 and c_2. Note that $\Phi(\varepsilon_t) \sim U(0,1)$ when $\varepsilon_t \sim N(0,1)$, and $G\{\log(e_t^2)\} \sim U(0,1)$ when $G(\cdot)$ is the distribution function of $\log(e_t^2)$. Let $\widehat{F}_{n,1}(\cdot)$ be the empirical distribution of $\{\Phi(\widehat{\varepsilon}_t),\ \nu < t \leqslant n\}$, and $\widehat{F}_{n,2}(\cdot)$ the empirical distribution of $[G\{\log(\widehat{e}_t^2)\},\ \nu < t \leqslant n]$. We define the goodness-of-fit statistics below to measure the distances between $\widehat{F}_{n,i}$ and the uniform distribution on $(0,1)$.

$$T_{\text{MLE}} = \int_0^1 |\widehat{F}_{n,1}(x) - x|dx, \quad T_{\text{LADE}} = \int_0^1 |\widehat{F}_{n,2}(x) - x|dx. \tag{3.42}$$

Obviously these statistics are reminiscent of the Cramer–von Mises goodness-of-fit statistics. In practical implementation, we use the Riemann approximations of these integrals:

$$T_{\text{MLE}} = \sum_{t=\nu+1}^{n} \Big|\frac{t-\nu}{n-\nu} - u_t\Big|(u_t - u_{t-1}), \quad T_{\text{LADE}} = \sum_{t=\nu+1}^{n} \Big|\frac{t-\nu}{n-\nu} - v_t\Big|(v_t - v_{t-1}), \tag{3.43}$$

where $u_{\nu+1} \leqslant u_{\nu+2} \leqslant \cdots \leqslant u_n$ are the order statistics of $\{\Phi(\widehat{\varepsilon}_t),\ \nu < t \leqslant n\}$, and $v_{\nu+1} \leqslant v_{\nu+2} \leqslant \cdots \leqslant v_n$ the order statistics of $[G\{\log(\widehat{e}_t^2)\},\ \nu < t \leqslant n]$.

Selection rule We use the LADE if $T_{\text{MLE}} > T_{\text{LADE}}$, and the Gaussian MLE otherwise.

It has been shown that the above selection rule is consistent in the sense that $P(T_{\text{MLE}} > T_{\text{LADE}}) \to 1$ provided

$$\int_0^1 |F_1(x) - x|dx > \int_0^1 |F_2(x) - x|dx,$$

where F_1 and F_2 denote, respectively, the distribution function of $\Phi(\varepsilon_t)$ and $G\{\log(e_t^2)\}$. See Huang, Wang and Yao (2003).

Huang, Wang and Yao (2003) also applied the above selection rule to two daily return series: the Switzerland stock index (SWI) in January 1991 to December 1998, and the B Share of the Shanghai Stock Exchange (SHB) in January 2001 to December 2004. The length of the series are, respectively,

1859 and 946. The P-value of the Jarque–Bera test is 0.000 for both the series, and the kurtosis is 5.72665 for SWI and 5.761476 for SHB. For each of those two series, we fit the first half series with GARCH(1,1) models using both the GMLE and the LADE. The sample size used in the estimation is $n = 930$ for SWI, and 473 SHB. The values of the goodness-of-fit statistics $(T_{\text{MLE}}, T_{\text{LADE}})$ are (0.026, 0.057) for SWI, and (0.044, 0.041) for SHB. Thus our selection rule prefers the GMLE for SWI, and the LADE for SHB.

With the sample size fixed at $n = 930$ for SWI and $n = 473$ for SHB, we also perform one-step ahead prediction of the squared returns for each of the second half series. The prediction is based on the fitted GARCH(1,1) models using both the GMLE and the LADE. With LADE, the predicted squared returns are of the form $\widehat{s}_t^2 S_e$, where S_e is the sample variance of the residuals $\widehat{e}_j \equiv X_j/\widehat{s}_j$ $(j < t)$; see (3.37). The root mean squares error of the prediction based on the GMLE is 1.750 for SWI, and 4.757 for SHB. The root mean squares error based on the LADE is 2.715 for SWI, and 2.894 for SHB. Thus the GMLE provided the more accurate prediction for SWI while the LADE predicted SHB better. This shows that the estimation method preferred by our selection rule also provided better prediction.

3.3 ARMA–GARCH models

In ARMA models, we formally assume that the innovations (i.e. the noise process) is white noise, i.e. we do not impose any explicit conditions beyond their first two moments. In order to reflect time-varying variances observed in financial data, we may assume that the innovations in an ARMA model follow a GARCH structure. This leads to the ARMA(p, q)–GARCH(p_1, q_1) specification as follows:

$$\begin{cases} r_t = \mu + \beta_1 r_{t-1} + \cdots + \beta_p r_{t-p} + X_t + \alpha_1 X_{t-1} + \cdots + \alpha_q X_{t-q}, \\ X_t = \sigma_t \varepsilon_t, \quad \sigma_t^2 = a_0 + \sum_{i=1}^{p} a_i X_{t-i}^2 + \sum_{j=1}^{q} b_j \sigma_{t-j}^2. \end{cases} \tag{3.44}$$

The model allows us to specify simultaneously time-varying conditional mean $\mu_t = E_{t-1} r_t$ and time-varying conditional variance $\sigma_t^2 = \text{var}_{t-1}(r_t)$.

By (3.26), the conditional likelihood function (given $r_1, \ldots, r_\nu$) is

$$L(\mu, \boldsymbol{\alpha}, \boldsymbol{\beta}, \mathbf{a}, \boldsymbol{b}) = \prod_{t=\nu+1}^{T} \frac{1}{\sigma_t} f\Big(\frac{r_t - \mu_t}{\sigma_t}\Big), \tag{3.45}$$

where $\nu > \max(p, p_1)$ is an integer. Maximizing (3.45) yields conditional maximum likelihood estimators for the parameters $(\mu, \boldsymbol{\alpha}, \boldsymbol{\beta}, \mathbf{a}, \boldsymbol{b})$, with their standard errors given in the Hessian matrix of the log-likelihood function at the maximizer. See Section 2.5.4.

Both μ_t and σ_t in (3.45) depend on the parameters nonlinearly. The maximization above are typically solved by numerical iteration using recursive formulas for calculating X_t and σ_t $(1 \leqslant t \leqslant T)$, noting that the available observations are $r_1, \ldots, r_T$ only. When model (3.44) is applied to financial returns, the estimated coefficients $\boldsymbol{\alpha}$ and $\boldsymbol{\beta}$ are typically small. This is reminiscent the efficient market hypothesis. Therefore, it is not unreasonable to replace the first equation in (3.44) by the approximation

$$r_t \approx \mu + X_t, \tag{3.46}$$

where μ is a constant representing the average return in the unit time. For the returns at daily or higher frequency, $\mu \approx 0$ and $r_t \approx X_t$. This explains why we often feed directly the returns into a GARCH model. At weekly and monthly frequencies, one can estimate μ along with GARCH parameters $\boldsymbol{a}$ and $\boldsymbol{b}$ simultaneously by maximizing (3.45) with constraints $\boldsymbol{\alpha} = 0$ and $\boldsymbol{\beta} = 0$, namely maximizing

$$\prod_{t=\nu+1}^{T} \frac{1}{\sigma_t} f\left(\frac{r_t - \mu}{\sigma_t}\right) \quad \text{or} \quad \sum_{t=\nu+1}^{T} \left[-\log(\sigma_t) + \log\left\{ f\left(\frac{r_t - \mu}{\sigma_t}\right)\right\}\right]. \tag{3.47}$$

The latter is the logarithm of the conditional likelihood function.

For other time series such as temperatures for weather derivatives or earnings per share, one can fit more general ARMA–GARCH models to capture time dependent conditional means and conditional variances simultaneously.

3.4 Extended GARCH models

ARCH/GARCH have enjoyed substantial successes in catching some important stylized features of financial returns, such as heavy tails, and volatility clustering. However they cannot reflect other features such as asymmetry, leverage effect and long range dependence. See the discussion in §1.2. Extension of the classic GARCH form to catch various features has received ample attention in both statistical and econometric literature. We list below a few examples. A fuller exposure in this direction can be found in, for example, Chapter 8 of Teräsvirta, Tjostheim and Granger (2011).

3.4.1 EGARCH models

A stylized feature in the financial returns is that investors react more strongly to negative news than to positive news. As the standard GARCH model links the volatility with the squared returns, it is incapable to reflect those directional changes due to the symmetry of the squared returns. To accommodate asymmetric effect of random shocks, Nelson (1991) introduced the following exponential GARCH (EGARCH) model

$$X_t = \varepsilon_t \exp(h_t), \quad h_t = \omega + \beta h_{t-1} + \alpha \varepsilon_{t-1} + \gamma(|\varepsilon_{t-1}| - E|\varepsilon_{t-1}|), \tag{3.48}$$

where $\varepsilon_t \sim \text{IID}(0,1)$, and $\omega, \alpha, \gamma, \beta$ are unknown parameters. Note that those parameters may take negative values, since we are now modeling the logarithm of volatility: $h_t = \log(\sigma_t)$. This can be an advantage in optimization as the non-negativity constraints are now removed, and forms another motivation of the EGARCH model. The dynamics of the volatility is driven by the process h_t, which can be written as

$$h_t = \omega + \beta h_{t-1} + u_t, \quad \text{where} \quad u_t = \alpha \varepsilon_{t-1} + \gamma(|\varepsilon_{t-1}| - E|\varepsilon_{t-1}|). \tag{3.49}$$

Note that u_t are i.i.d. with zero mean and a finite variance. The complicated expression of u_t is to make its mean equal to zero while it reacts to positive and negative ε_{t-1} asymmetrically. For example, the ratio of the volatility caused by a negative shock $\varepsilon_t = -a$ to that by a positive shock $\varepsilon_t = a$ is

$$\frac{\sigma_t(\varepsilon_{t-1} = -a)}{\sigma_t(\varepsilon_{t-1} = a)} = \exp(-2a\alpha), \tag{3.50}$$

and α is typically negative. Note that $\varepsilon_t = a$ represents the shock in the size of a times of the conditional standard deviation of X_t; see the first equation in (3.48). This asymmetry is potentially useful as it allows the volatility to respond more rapidly to the drops of the price than to corresponding rises. This is typically true for financial data; see Table 3.3 below. Of course, when $\alpha = 0$, the effect of the shock ε_{t-1} is symmetric.

Model (3.48) is referred to as EGARCH(1,1). Obviously the second equation of (3.48) can be extended such that it includes the p lagged values of h_t and q lagged values of ε_t on its right hand side; leading to a general EGARCH(p, q) model. We focus on (3.48) only to highlight the essence of EGARCH models. Furthermore EGARCH(1,1) remains as the most frequently used EGARCH models to date.

Although (3.48) looks somehow complicated, it is straightforward to identify its properties. By (3.49), h_t is a stationary and causal AR(1) process if $|\beta| < 1$. Therefore X_t is also a strictly stationary process. Since ε_t is an i.i.d.

sequence, ε_t and h_t are independent with each other. Hence, for any $k \geqslant 1$,

$$E(X_t^k) = E(\varepsilon_t^k)E\{\exp(kh_t)\}.$$

Therefore, the kurtosis of X_t satisfies the following expression:

$$\kappa_x = \frac{E(X_t^4)}{\{E(X_t^2)\}^2} = \frac{E(\varepsilon_t^4)}{\{E(\varepsilon_t^2)\}^2}\frac{E\{\exp(4h_t)\}}{\{E(e^{2h_t})\}^2} = \kappa_\varepsilon \cdot \kappa_\sigma > \kappa_\varepsilon, \tag{3.51}$$

where κ_ε and κ_σ denote, respectively, the kurtosis of ε_t and $\sigma_t = \exp(h_t)$. The inequality in the above expression follows from the fact that $\kappa_\sigma > 1$ as long as σ_t^2 is not a constant, which is guaranteed by Jensen's inequality. (3.51) indicates that the distribution of X_t has heavier tails than those of ε_t. Note that this property also holds for the standard ARCH/GARCH processes.

The statistical inference for EGARCH models may be carried out based on likelihood methods by assuming, for example, normal innovations ε_t. Although the direct expansions such as (3.17) are no longer available, the likelihood functions for EGARCH models may be evaluated in a recursive manner. Denote by $\boldsymbol{\theta} = (\omega^*, \alpha, \beta, \gamma)$ the parameters in the model, where $\omega^* = \omega - \gamma E|\varepsilon_{t-1}|$. Let $\varepsilon_0 = 0$ and $h_0 = \omega^*$. For $t \geqslant 1$, it follows from (3.48) that

$$\begin{aligned} h_t(\boldsymbol{\theta}) &= \omega^* + \alpha\varepsilon_{t-1}(\boldsymbol{\theta}) + \gamma|\varepsilon_{t-1}(\boldsymbol{\theta})| + \beta h_{t-1}(\boldsymbol{\theta}), \\ \varepsilon_t(\boldsymbol{\theta}) &= X_t \exp\{-h_t(\boldsymbol{\theta})\}. \end{aligned} \tag{3.52}$$

Assuming that ε_t follows the standard normal distribution, substituting $h_t = \log \sigma_t$ into (3.27), we obtain the log-likelihood function

$$\ell(\boldsymbol{\theta}) = -\sum_{t=\nu}^{T} h_t(\boldsymbol{\theta}) - \frac{1}{2}\sum_{t=\nu}^{T} X_t^2 e^{-2h_t(\boldsymbol{\theta})},$$

where $\nu > 1$ is an integer. For each given $\boldsymbol{\theta}$, $h_t(\boldsymbol{\theta})$ can be computed recursively according to (3.52).

We choose ν sufficiently large to get rid of the effect of the initial values ε_0 and h_0 in recursion (3.52). A QMLE for $\boldsymbol{\theta}$ may be obtained by maximizing $\ell(\boldsymbol{\theta})$, even though the distribution ε_t may not follow a normal distribution. The asymptotic properties of such an estimator are derived in Straumann and Mikosch (2006). See also §8.3 of Teräsvirta, Tjostheim and Granger (2010) for the statistical tests for EGARCH models.

The *R*-package `rugarch` provides a comprehensive set of functions for statistical inference of univariate GARCH and its extended models, including EGARCH, as well as asymmetric power GARCH and GARCH-in-Mean presented below. The tools provided cover model fitting/estimation, filtering,

Table 3.3 *Estimated coefficients for EGARCH(1,1) models for the daily returns of S&P 500, Goldman Sachs stock, and Fords stock in January 4, 2010 to August 10, 2012*

	$\widehat{\omega}$	$\widehat{\alpha}$	$\widehat{\gamma}$	$\widehat{\beta}$
S&P 500	0.006 (0.006)	−0.210 (0.034)	0.128 (0.031)	0.949 (0.010)
GS	0.113 (0.041)	0.038 (0.032)	0.258 (0.060)	0.930 (0.027)
Ford	0.177 (0.063)	−0.017 (0.028)	0.204 (0.054)	0.894 (0.038)

forecasting, simulation, diagnostic checking using various plots and statistical tests.

Example 3.3 We illustrate the EGARCH(1,1) fitting with three real data sets used in Section 3.2.3 and compare the results with those obtained from the standard GARCH(1,1) model. To fit the data with model (3.48) using `rugarch`, it requires two steps:

```
mspec=ugarchspec(variance.model=list(model="eGARCH",
          garchOrder=c(1,1)), mean.model=list(armaOrder=c(0,0),
          include.mean=F))
egarch11SP=ugarchfit(mspec, spLogR)
```

The first step uses `ugarchspec` to specify model (3.48) with Gaussian ε_t, and then `ugarchfit` fits the percentage log returns of S&P 500, i.e. `spLogR`, with the specified model. The output of the fitting is recorded as `egarch11SP`. The fitted volatilities can be extracted by `sigma(egarch11SP)`. Various plots related fitting and diagnostic checking are obtained by `plot(egarch11SP)`. Note that `residuals(egarch11SP)` return the original return series as the mean model is set to 0 in the above setting. Changing the option `"eGarch"` to `"sGARCH"` leads to a standard GARCH(1,1)fitting. Applying the same fitting to the returns of both Goldman Sachs and Ford stocks, we obtain the estimates and their standard errors (in parentheses) for the coefficients in model (3.48), which are listed in Table 3.3.

Although those coefficients are not comparable with the estimated coefficients for GARCH(1,1) models in Table 3.1, there exist similar patterns. For example, the estimated ω is much larger for individual stock returns than that for the returns of the S&P 500, indicating that the volatility for single stock is much greater than that for a portfolio index. The estimated β is close to 1, implying the volatility persistence or clustering. The estimated sign effect coefficient α is negative for both S&P 500 returns and

Ford returns, indicating that returns and volatilities tend to move in opposite direction, although this asymmetric effect is pronounced for Goldman Sachs returns according to the fitted model.

The asymmetric effect of positive and negative shocks can also been seen. For S&P 500 index, the coefficient $\widehat{\alpha}$ is statistically significantly negative. By (3.50), the ratios of volatilities caused by negative and positive shocks are

$$\frac{\sigma_t(\varepsilon_{t-1} = -1)}{\sigma_t(\varepsilon_{t-1} = 1)} = \exp(2 \times 0.21) = 1.52,$$

and

$$\frac{\sigma_t(\varepsilon_{t-1} = -2)}{\sigma_t(\varepsilon_{t-1} = 2)} = \exp(4 \times 0.21) = 2.32,$$

i.e. the volatility caused by the shock of one negative standard deviation is 1.52 times of that caused by the shock of one positive standard deviation. The multiple increases to 2.32 when the sizes of the shocks are twice that of standard deviation. The asymmetry effect is indeed very strong.

Figure 3.12 plots the returns of the three series together with the bounds $\pm 2\widehat{\sigma}_t = \pm 2\exp(\widehat{h}_t)$; see (3.48). Although the profiles of those plots are similar to those in Figure 3.8, a close examination shows that there are more local variations in the volatility curves fitted by EGARCH(1,1) models. The relative frequency for the two bounds $\pm 2\widehat{\sigma}_t$ to sandwich the returns is 93.91% for S&P 500, 95.13% for Goldman Sachs, and 94.82% for Ford; similar to or the same as those provided by the GARCH(1,1) fitting in Section 3.2.1.

Figure 3.13 displays the quantile–quantile plots for the residuals $X_t/\widehat{\sigma}_t$ from both GARCH(1,1) and EGARCH(1,1) against normal distributions. It is clear that the residuals resulted from both fittings are still heavy-tailed, though the tails from EGARCH(1,1) seem slightly less heavy than those from GARCH(1,1).

3.4.2 Asymmetric power GARCH

Another way to model the asymetric effects is to use *asymmetric power GARCH* (APGARCH) model introduced by Ding, Engle and Granger (1993):

$$X_t = \varepsilon_t \sigma_t, \quad \sigma_t^\delta = a_0 + a_1(|X_{t-1}| - d\,X_{t-1})^\delta + b\sigma_{t-1}^\delta, \tag{3.53}$$

where the parameters a_0, a_1, b are non-negative, $d \in (-1, 1)$, and $\delta \in (0, 2]$, and $\varepsilon_t \sim \text{IID}(0, 1)$. In this setting, the asymmetric effect is accommodated by the asymmetric parameter d. A power index $\delta \in (0, 2]$ is also introduced in the model in an attempt to exploit stronger autocorrelation of power

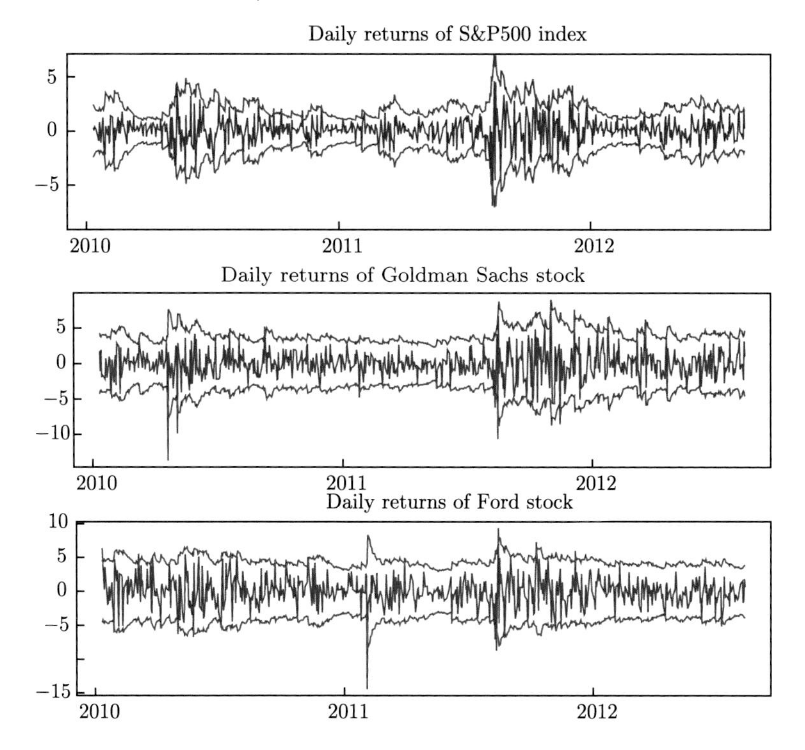

Figure 3.12 Time series plots of the daily returns in January 4, 2010 to August 10, 2012 of the S&P 500 index (the top panel), the Goldman Sachs share (the middle panel), and the Ford share (the bottom panel). Two blues lines are $\pm 2\widehat{\sigma}_t$ estimated based on the fitted EGARCH(1,1) models with $\widehat{\sigma}_t = \exp(\widehat{h}_t)$.

functions of returns for modeling volatility. It is well-documented that the absolute returns typically exhibit stronger autocorrelations than the most commonly used squared returns. See also Figures 1.7 and 1.8.

Model (3.53) admits a strictly stationary solution with $E(|X_t|^\delta) < \infty$ if $a_1 E\{(|\varepsilon_t| - d\varepsilon_t)^\delta\} + b < 1$. Furthermore, this condition also implies that

$$\sigma_t^\delta = \frac{a_0}{1-b} + a_1 \sum_{k=0}^{\infty} b^k (|X_{t-k-1}| - dX_{t-k-1})^\delta.$$

See (3.19). Hence the conditional likelihood function (3.27) can be calculated in a straightforward manner if, for example, we assume $\varepsilon_t \sim N(0,1)$. See Ding, Engle and Granger (1993) and Penzer, Wang and Yao (2009) for further discussion on the properties and statistical inference for model (3.53).

Example 3.4 We illustrate the APGARCH fitting with again three real

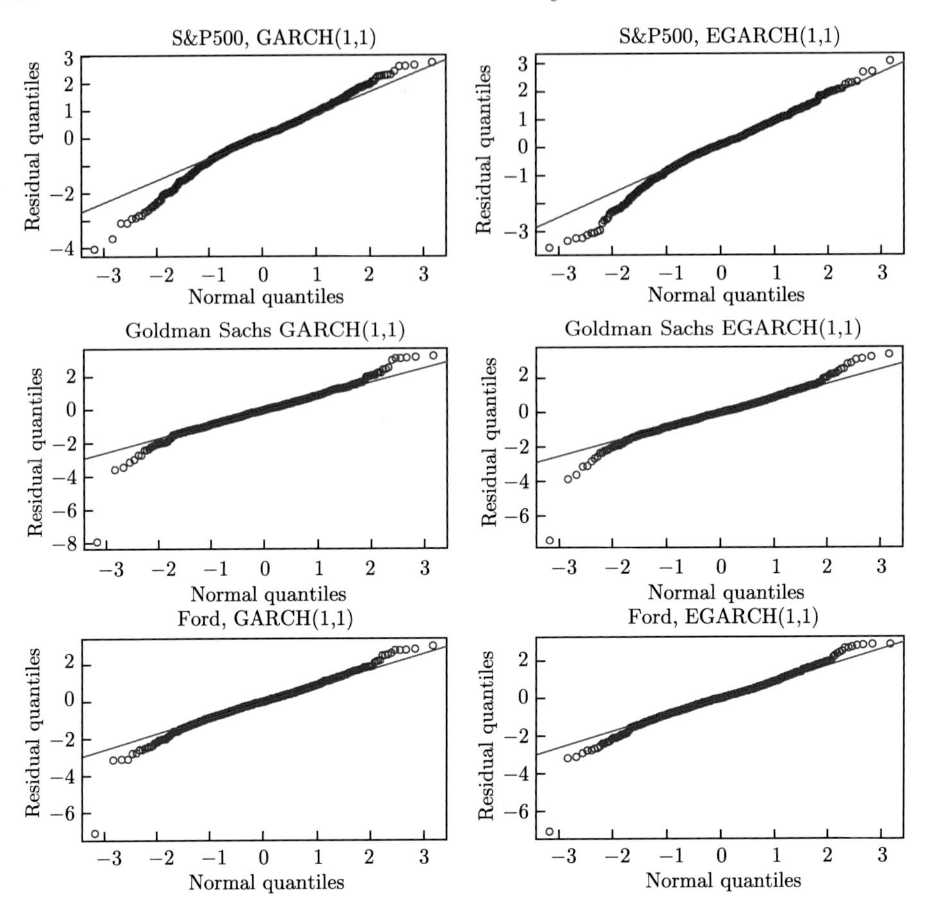

Figure 3.13 The quantile-quantile plots of residuals $X_t/\widehat{\sigma}_t$ against normal distributions for both GARCH(1,1) and EGARCH(1,1) fittings for the returns of S&P 500 index, Goldman Sachs share, and the Ford share in January 4, 2010 to August 10, 2012.

data sets used in Section 3.2.3. To fit APGARCH models, we use the functions in the *R*-package `rugarch` as follows:

```
mspec=ugarchspec(variance.model=list(model="apARCH",
          garchOrder=c(1,1)), fixed.pars=list(delta=1),
          mean.model=list(armaOrder=c(0,0), include.mean=F))
apgarch11SP=ugrarchfit(mspec, spLogR)
```

This fits model (3.53) with the power index pre-set at $\delta = 1$ to the percentage returns of S&P 500 (i.e. `spLogR`). One may also estimate the power index δ by data by removing the option `fixed.pars=list(delta=1)` in the spec `mspec` above. By applying the fitting to both the returns for Goldman Sachs and

Table 3.4 *Estimated coefficients for APGARCH(1,1) models for the daily returns of S&P 500 index, Goldman Sachs stock, and Ford stock in January 4, 2010 to August 10, 2012*

	$\widehat{a}_0$	$\widehat{a}_1$	$\widehat{d}$	$\widehat{b}$
S&P 500	0.051 (0.011)	0.099 (0.018)	1.000 (0.221)	0.882 (0.019)
GS	0.150 (0.054)	0.139 (0.033)	-0.156 (0.132)	0.826 (0.044)
Ford	0.242 (0.098)	0.108 (0.029)	0.110 (0.154)	0.811 (0.057)

Ford, the estimated coefficients with their standard errors (in parentheses) are listed Table 3.4.

The asymmetric effect now is reflected by positive values of $\widehat{d}$ for both S&P 500 and Ford returns. This is consistent with the findings from the fitted EGARCH(1,1) models in Section 3.4.1 in which $\widehat{\alpha} < 0$ for both S&P 500 and Ford returns too. Note for the Goldman Sachs returns, both the fitted EGARCH(1,1) and APGARCH(1,1) reveal no asymmetric effect.

Figure 3.14 plots the returns of the three series together with the bounds $\pm 2\widehat{\sigma}_t$ obtained from APGARCH(1,1) with $\delta = 1$, i.e. $\widehat{\sigma}_t$ were modeled directly instead of deduced from the fitted σ_t^2; see model (3.53). The profiles of the plots are similar to Figures 3.8 and 3.12 in spite of the fact that the volatility models behind those plots are very different from each other. The relative frequency for the two bounds $\pm 2\widehat{\sigma}_t$ to sandwich the returns is 94.07% for S&P 500, 94.98% for Goldman Sachs, and 94.82% for Ford. They are similar to or the same as those with GARCH(1,1) or EGARCH(1,1) fittings.

Now for each trading day in January 4 to August 10, 2012, we use the daily returns in its previous 500 trading days (i.e. about two years) to fit an APGARCH(1,1) model with $\delta = 1$. We then predict the VaR at the levels of 5% and 2.5% according to (3.34) in which $\widehat{Q}_{\alpha,\varepsilon}$ is estimated either by assuming the normality for ε_t or by the empirical distributions of the residuals derived from the fitted APGARCH(1,1) models to the previous 500 data points. The returns in 2012 of the three series together with their predicted VaR curves are plotted in Figure 3.15. Comparing it with Figure 3.10, the VaR curves obtained from different models are different. Nevertheless the differences due to using either the normality assumption or the empirical quantiles of the results are more pronounced at the more extreme level of 2.5% than those at 5% level. This is true with both APGARCH(1,1) fitting and GARCH(1,1) fitting (in Figure 3.10). The relative frequencies that the returns cross over the VaR curves among the 154 trading days in 2012 are

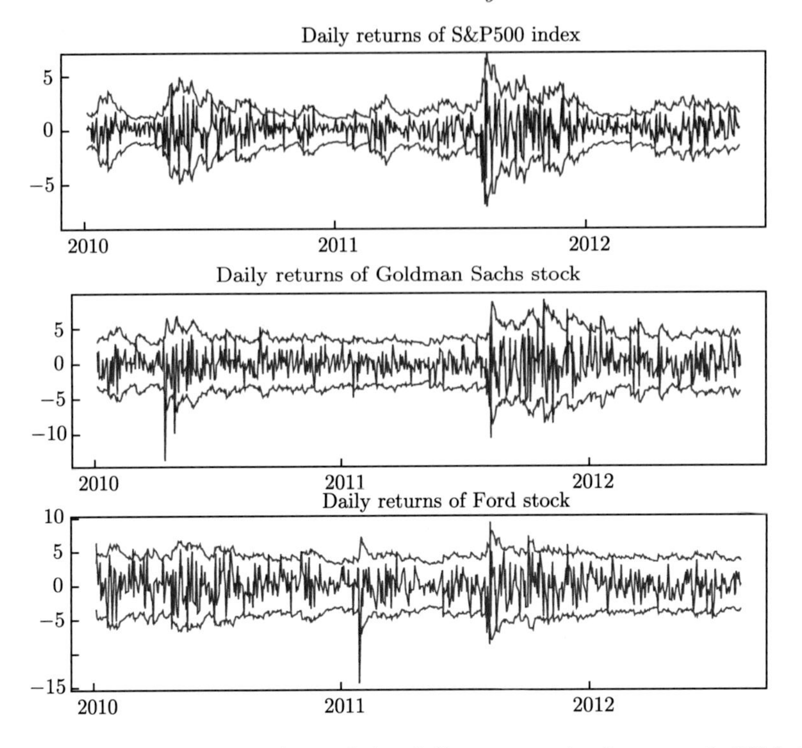

Figure 3.14 Time series plots of the daily returns in January 4, 2010 to August 10, 2012 of the S&P 500 index (the top panel), the Goldman Sachs share (the middle panel) and the Ford share (the bottom panel). Two blues lines are $\pm 2\widehat{\sigma}_t$ estimated based on the fitted APGARCH(1,1) models with $\delta = 1$.

Table 3.5 *Percentage of exceedances of returns over predicted VaR based on* APGARCH*(1,1)*

Distribution of ε_t	Empirical residuals		$N(0,1)$	
Level	5%	2.5%	5%	2.5%
S&P 500	1.95%	0.65%	3.25%	1.95%
Goldman Sachs	2.60%	0.00%	1.94%	0.65%
Ford	3.25%	1.30%	3.25%	1.95%

listed in Table 3.7. Those numbers seem to indicate severe overestimation for risk. However this is more likely caused by the obvious reduction of volatility in 2012 in comparison to late 2011. See the relevant discussion at the end of §3.2.1.

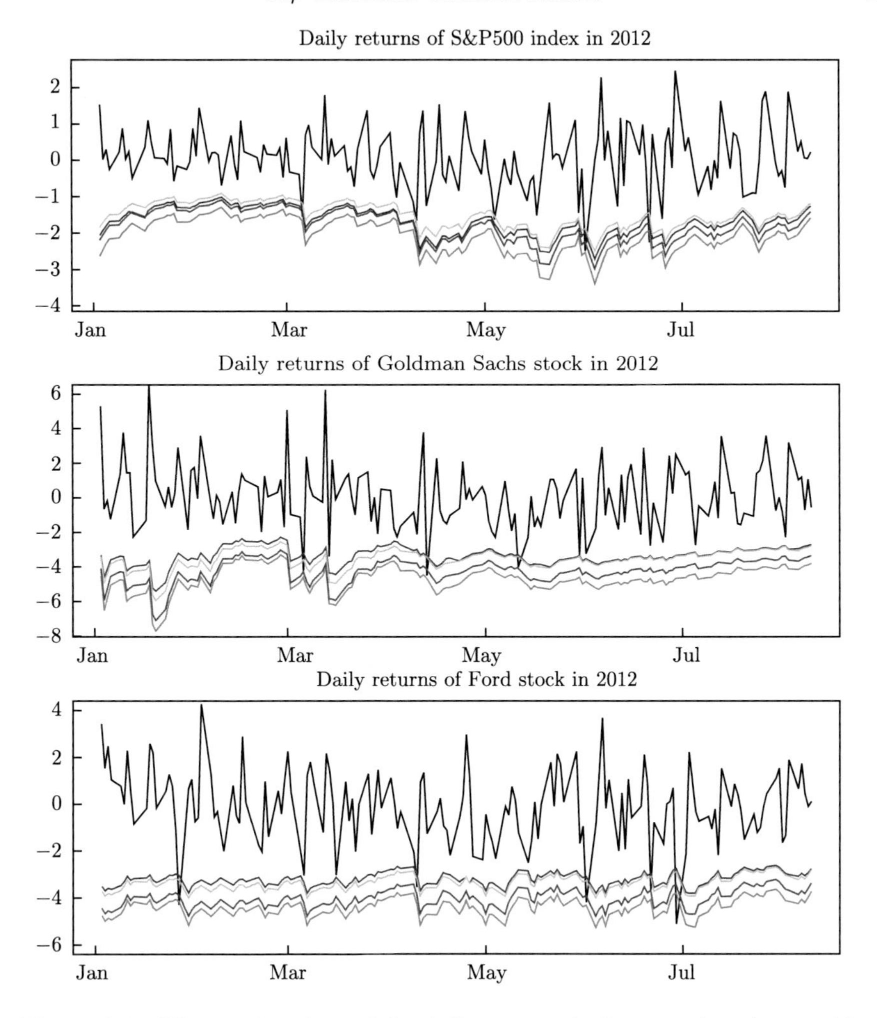

Figure 3.15 Time series plots of the daily returns in January 4 to August 10 2012 together with the predicted VaR from APGARCH(1,1) models based on the previous 500 returns at each given time: VaR based on normal innovation at the 5% level (orange lines) and 2.5% level (red lines); VaR based on empirical residual distribution at the 5% level (blue lines) and 2.5% level (green lines).

3.4.3 Excess returns and GARCH-in-Mean

It is well known from the capital asset pricing model (CAPM; see Chapter 5)that risks are compensated by the returns. To account a risk premium the following GARCH-in-Mean model is proposed:

$$Y_t = g(\sigma_t) + X_t, \qquad X_t = \varepsilon_t \sigma_t$$
$$\sigma_t^2 = a_0 + a_1 X_{t-1}^2 + b_1 \sigma_{t-1}^2, \tag{3.54}$$

and $\varepsilon_t \sim \text{IID}(0,1)$. In the above model, the first term represents the conditional expected return which is related to the underlying risk, and the second term is the volatility around the return. In the original proposal by Engle, Lilien, and Robins (1987), a linear function $g(x) = \theta_0 + \theta_1 x$ is used, in which the risk premium is expressed in terms of the variance. This facilitates parameter estimation. However, from the CAPM in Chapter 5, a more reasonable form is that $g(x) = \alpha_0 + \alpha_1\sqrt{x}$, which now represents the risk premium in terms of the standard deviation.

There is nothing fundamentally new in the probabilistic properties of the GARCH-in-mean process $\{Y_t\}$, as it is the sum of a standard GARCH process and a function of the conditional standard deviation of the GARCH process. Likelihood inference is again straightforward.

GARCH-in-mean models can be fitted using *R*-package `rugarch`. For example to fit model (3.54) to data set `Xdata`:

```
mspec=ugarchspec(variance.model=list(model="sGARCH",
      garchOrder =c(1,1)), mean.model=list(armaOrder=c(0,0),
      include.mean=F, archm=T, archpow=1))
ugarchfit(mspec, Xdata)
```

3.4.4 Integrated GARCH model

As seen in Table 3.1, the sum of the estimated coefficients of for each fitted GARCH model is close to one. A further parsimonious representation is to assume directly

$$a_1 + \cdots + a_p + b_1 + \cdots + b_q = 1$$

In this case, the characteristic polynomial of its ARMA representation (3.20) has a unit root. Therefore, it is called *integrated GARCH* model or *IGARCH* for short. See Engle and Bollerslev (1986). Let us illustrate this with the IGARCH(1,1) model

$$X_t = \sigma_t \varepsilon_t, \quad \sigma_t^2 = a_0 + a_1 X_{t-1}^2 + b_1 \sigma_{t-1}^2, \quad \text{with } b_1 = 1 - a_1. \tag{3.55}$$

By (3.19), we have

$$\sigma_t^2 = \frac{a_0}{1-b_1} + (1-b_1)(X_{t-1}^2 + b_1 X_{t-2}^2 + b_1^2 X_{t-3}^2 + \cdots).$$

The second component is referred to as the exponential smoothing. In the JP Morgan's RiskMetrics, based on daily returns, it is recommended to take $a_0 = 0$ and $b_1 = 0.94$ for computing daily volatility and $b_1 = 0.97$

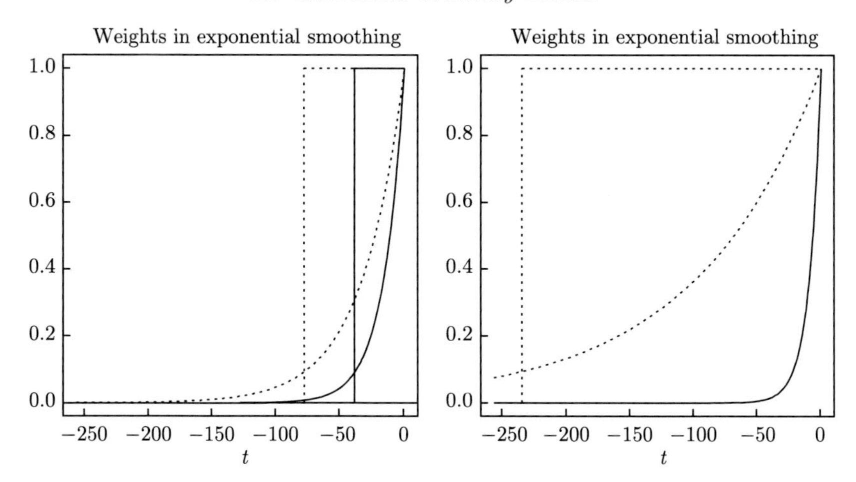

Figure 3.16 Weights in exponential smoothing with $b_1 = 0.94$ (left, solid), 0.97 (left, dashed), 0.90 (right, solid), 0.99 (right, dash). The presented windows is the equivalent amount of smoothing using unweighted average, according to the canonical kernel theory (Marron and Nolan, 2003).

for computing monthly volatility (the final answer should be multiplied by $\sqrt{21}$).

A sufficient condition for GARCH (1,1) to be strong stationary is $E\{\log(a_1 + b_1\varepsilon_t^2)\} < 0$ (Nelson, 1990). By Jensen's inequality

$$E\{\log(a_1 + b_1\varepsilon_t^2)\} < \log(a_1 + b_1) \leqslant 0, \quad \text{if } a_1 + b_1 \leqslant 1.$$

Thus, IGARCH(1,1) is stationary. However, $EX_t^2 = \infty$, since if $X_t^2 < \infty$ the necessary condition for stationarity is $a_1 + b_1 < 1$.

From Example 3.2, k-period ahead forecasting admits the form

$$\sigma_T^2(k) = a_0 + \sigma_T^2(k-1) = (k-1)a_0 + \sigma_T^2(1),$$

where $\sigma_T^2(1) = a_0 + a_1X_T^2 + (1-a_1)\sigma_T^2$. It is typically computed recursively from (3.55) with

$$\sigma_t^2 = a_1X_{t-1}^2 + (1-a_1)\sigma_{t-1}^2$$

with initial value σ_0^2 being the sample average of $\{X_t^2\}_{t=1}^T$.

3.5 Stochastic volatility models

The stochastic volatility model offers an alternative approach to modeling and forecasting heteroscedasticity. Like ARCH/GARCH, it treats volatilities over times as a stochastic process. However it differs from ARCH/GARCH in the sense that the volatility is modeled as a latent state space variable not explicitly linked with observed returns. We provide a gentle introduction on stochastic volatility below, and refer to Shephard and Anderson (2009) and Shephard (1986) for further readings on this topic.

A general form of stochastic volatility models may be written as

$$X_t = \varepsilon_t g(h_t) \qquad \text{and} \qquad h_t = c + \sum_{j=1}^{p} b_j h_{t-j} + e_t, \tag{3.56}$$

where $\varepsilon_t \sim \text{IID}(0,1)$, $e_t \sim \text{IID}(0, \sigma_\varepsilon^2)$, $\{\varepsilon_t\}$ and $\{e_t\}$ are independent, and $g(\cdot) > 0$ is a known function. For most applications, p takes small values such as 1 or 2. The key idea is that the latent process h_t may represent the random and uneven flow of new information that is too complex to be modeled as a function of the lagged values $X_{t-1}, X_{t-2}, \ldots$ only. This is a marked difference from ARCH/GARCH model: a stochastic volatility model is driven by two independent noise processes while an ARCH/GARCH model is driven by one noise process. As a consequence of this modeling strategy, the stochastic volatility model (3.56) has a simpler probabilistic structure. However its statistical inference is more demanding, as the likelihood function does not admit an explicit expression.

3.5.1 Probabilistic properties

In model (3.56), the stochastic heteroscedasticity is driven by h_t which itself is a linear AR(p) process. As $e_t \sim \text{IID}(0, \sigma_e^2)$, h_t is a strictly stationary process with the mean $Eh_t = c/(1 - b_1 - \cdots - b_p)$ and a finite variance, provided that all the roots of the equation $1 - b_1 x - \cdots - b_p x^p = 0$ are outside the unit circle. See Section 2.2.2. Then it is easy to see from the first equation in (3.56) that X_t is also strictly stationary with a finite second moment.

Arguably the most popular form of stochastic volatility model is due to Taylor (1986), which specifies the function $g(\cdot)$ in (3.56) as an exponential function. Hence the model is of the form

$$X_t = \varepsilon_t \exp(h_t/2), \qquad h_t = c + \sum_{j=1}^{p} b_j h_{t-j} + e_t. \tag{3.57}$$

Now it is easy to calculate the moments of X_t. For example, for any $k \geqslant 1$,

$$E(X_t^k) = E(\varepsilon_t^k) E\{\exp(k h_t/2)\}.$$

Hence the kurtosis inequality (3.51) also holds for the stochastic volatility process. Therefore, the distribution of X_t has heavier tails than those of ε_t. This is a property also shared by ARCH/GARCH processes.

By the first equation in (3.57), it holds that

$$\log(X_t^2) = h_t + \log(\varepsilon_t^2).$$

Now we consider a special case of $p = 1$ and $e_t \sim N(0, \sigma_e^2)$ in (3.57). Then $\log(\varepsilon_t^2)$ is a white noise and h_t is a normal AR(1) process. Thus $\log(X_t^2)$ is an ARMA(1,1) process as far as its first two moment properties are concerned; see, e.g. Example 2.7 in Fan and Yao (2003). Furthermore it can be shown that for any $k \neq 0$,

$$\text{Corr}(X_t^2, X_{t-k}^2) = \frac{\exp\{b_1^{|k|}\sigma_e^2/(1-b_1^2)\}}{3\exp\{\sigma_e^2/(1-b_1^2)\} - 1} \approx \frac{\sigma_e^2/(1-b_1^2)}{3\exp\{\sigma_e^2/(1-b_1^2)\} - 1} b_1^{|k|}.$$

The approximation above holds for large $|k|$, which follows from a Taylor expansion. Note that the term on the right-hand side of the above expression is in the form of an ACF for an ARMA(1,1) process.

Further properties of stochastic volatility models can be found in Davis and Mikosch (2009) which shows, among others, the α-mixing properties of stochastic volatility processes.

3.5.2 Parameter estimation

In spite of the simple probabilistic properties stated above, stochastic volatility models, unfortunately, do not facilitate a straightforward statistical estimation and inference. Due to the latent variables h_t, the likelihood function cannot be evaluated directly. Various procedures, both Bayesian (via MCMC) and non-Bayesian such as generalized method of moments (GMM) estimation, approximate quasi-maximum likelihood estimation, and the EM algorithm, have been proposed; see Shephard and Anderson (2009) and the references within. We illustrate below how a simple Taylor's model (3.57) with $p = 1$ can be formulated into a state space model and how the algorithms such as the Kalman filter (§3.6.2 below) and the particle filter (§3.6.4 below) can be used to carry out the (approximate) maximum likelihood estimation.

When $\varepsilon_t \sim N(0, 1)$, $\log(\varepsilon_t^2)$ has mean -1.27 and variance $\pi^2/2$. Furthermore it has the density function

$$f(x) = \frac{1}{\sqrt{2\pi}} \exp\left\{-\frac{1}{2}(e^x - x)\right\}, \quad -\infty < x < \infty. \tag{3.58}$$

Let $Y_t = \log(X_t^2) - E\{\log(X_t^2)\}$, $u_t = \log(\varepsilon_t^2) + 1.27$, and $Z_t = h_t - E(h_t) =$

$h_t - c/(1 - b_1)$. When $p = 1$, it follows from (3.57) that

$$\begin{cases} Z_t = b_1 Z_{t-1} + e_j, \\ Y_t = Z_t + u_t. \end{cases} \tag{3.59}$$

This is a special case of the linear state space model (3.61). In practice, we use $Y_t = \log(X_t^2) - T^{-1}\sum_j \log(X_j^2)$ instead. Harvey, Ruiz and Shephard (1994) applied the Kalman filter (see §3.6.2 below) by assuming u_t being normal. They presented the numerical results indicating that the method could work well when the sample size is large enough.

While it is a common practice to assume that both ε_t and e_t are normal, the distribution of u_t is therefore not normal. In fact its density function (3.58) is skewed with a long tail on the left. Hence the Kalman recursion presented in Section 3.6.2 is no longer relevant, as it was derived under the normality assumption. To overcome this obstacle, various attempts have been made to model the distribution of u_t as a mixture of more than one normal distributions while some versions of Kalman filters can still be applied via, for example, a conditional argument; see Section 6.9 in Shumway and Stoffer (2011).

On the other hand, a particle filter can be easily constructed for model (3.59) for computing its likelihood function; see Section 3.6.4 below. More precisely, the one-step-ahead prediction particles in (3.82) can be easily generated from the AR(1) model (i.e. the first equation) in (3.38). The sampling weights for the filtering particles are defined with $\beta_{tj} = f(Y_t - z_{t|t-1}(j))$ in (3.35), where $f(\cdot)$ is given in (3.58). The implementation for particle filters can be made using two functions `nlf` and `pfilter` in the R-package `pomp` available from the `R-CRAN` server.

Example 3.5 Using a particle filter we fit the following simple stochastic volatility model to each of the three real data sets used in Examples 3.2–3.4:

$$X_t = \varepsilon_t e^{h_t/2}, \qquad h_t = c + b h_{t-1} + e_t, \tag{3.60}$$

where $\varepsilon_t \sim N(0, 1)$ and $e_t \sim N(0, \sigma_\varepsilon^2)$. The estimated coefficients are listed in Table 3.6. Note that $E(h_t) = c/(1 - b)$. Both the estimated c and b are at their smallest for S&P 500 returns. This indicates once again that the returns for S&P 500 are less volatile than the returns for the two stocks.

For each of 154 trading days in 2012, we use the daily returns in its previous 500 trading days to estimate the coefficients in model (3.60) and use the fitted model to predict the VaR at the levels 5% and 2.5% according to (3.34) in which $\widehat{Q}_{\alpha,\varepsilon}$ is estimated either based on the normality assumption on ε_t or by the empirical distributions of the residuals from the fitted mod-

Table 3.6 *Estimated coefficients for the stochastic volatility model (3.60) for the daily returns of S P 500, Goldman Sachs stock, and Ford stock in January 4, 2010 to August 10, 2012*

	$\hat{c}$	$\hat{b}$	$\hat{\sigma}_e^2$
S&P 500	−0.1648	0.2508	2.2275
GS	0.5196	0.2781	1.1365
Ford	0.1337	0.3505	0.0839

Table 3.7 *Percentage of exceedances of returns over predicted VaR based on stochastic volatility model* (3.60)

Distribution of ε_t	Empirical	residuals	$N(0,1)$	
Level	5%	2.5%	5%	2.5%
S&P 500	5.84%	3.89%	4.56%	2.59%
Goldman Sachs	5.19%	2.59%	3.25%	2.59%
Ford	5.03%	2.87%	5.03%	3.59%

els. The returns in 2012 of the three series together with their predicted VaR curves are plotted in Figure 3.17. The relative frequencies that the returns cross over the VaR curves among the 154 trading days are reported in Table 3.6. Comparing this with Tables 3.2 and 3.7, the relative frequencies in Table 3.6 are much closer to their nominal levels for those three data sets.

3.5.3 Leverage effects

The *leverage effect* or statistical leverage effect refers to the empirical evidence of the negative correlation between the return and the volatility of a financial asset; see, for example, Figure 1.10. When a stock price drops, the firm becomes mechanically more leveraged in terms of debt to equity ratio and the volatility of its return typically rises. The changes for the price and the volatility typically go in opposite directions, hence the name of the leverage effect.

Another attractive feature of stochastic volatility models is the ability to model the leverage effect directly in the original formulation. To this end, we let ε_t and e_t in (3.56) be negatively correlated with each other. Then the direction of return, reflected by the sign of ε_t, impacts the future movements of the volatility h_t. For example, negative returns tend to increase the volatilities while positive returns tend to decrease the volatilities. The statistical estimation for such a model involves one more (negative) param-

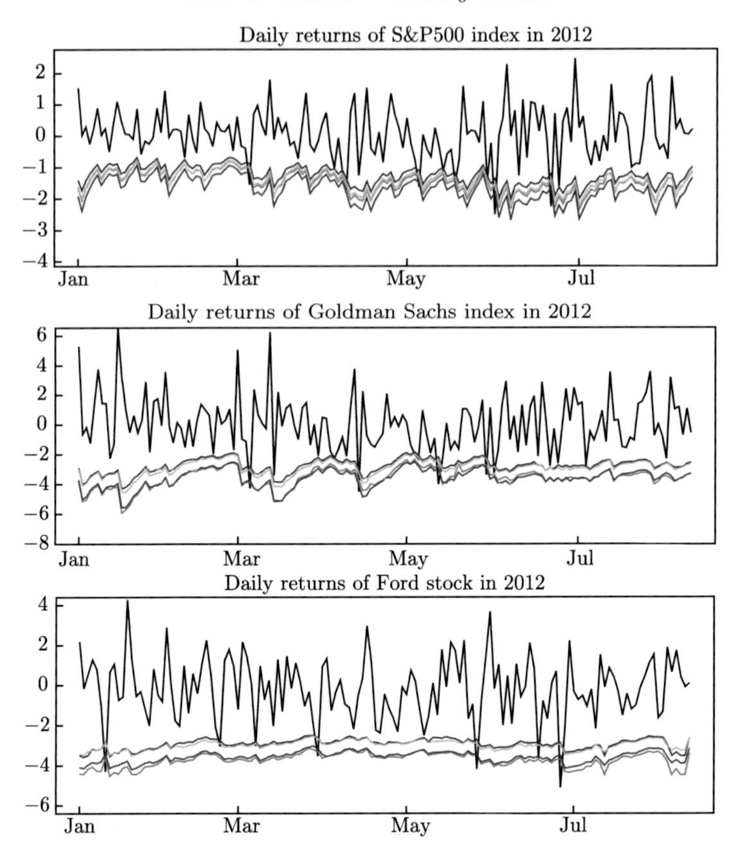

Figure 3.17 Time series plots of the daily returns in January 4 to August 10, 2012 of together with the predicted VaR from stochastic volatility models based on the previous 500 returns at each time:VaR based on normal innovation at the 5% level (orange lines) and 2.5% level (red lines); VaR based on empirical residual distribution at the 5% level (blue lines) and 2.5% level (green lines).

eter representing the correlation of ε_t and e_t. Also note that this negative correlation also causes skewness, i.e. the conditional distribution of X_{t+k} given h_t, for $t \geqslant 2$, may be skewed.

3.6 Appendix: State space models*

The state space methodology originated in the field of engineering, starting with the path-breaking paper of Kalman (1960). Kalman's original contribution is two-fold: (i) linear state space models can encapsulate a wide class of problems, and, perhaps more importantly, (ii) both estimation and forecasting can be carried out by recursive updating due to the *Kalman filter*. The state space methodology is based on a structural analysis: different effects such as trend, seasonal, regression elements, structure changes, interventions are modeled separately and explicitly. It can also handle multiple processes with ease. Furthermore no stationarity conditions are required as far as all the algorithms are concerned. We refer to Durbin and Koopman (2012) for a modern treatment for state space models for time series analysis.

3.6.1 Linear models

A linear *state space model* for time series is of the form:

$$\begin{cases} \boldsymbol{Z}_t = \boldsymbol{A}_t \boldsymbol{Z}_{t-1} + \boldsymbol{e}_t, & \text{(state equation)} \\ \boldsymbol{Y}_t = \boldsymbol{B}_t \boldsymbol{Z}_t + \boldsymbol{\varepsilon}_t, & \text{(observation equation)} \end{cases} \tag{3.61}$$

where $\boldsymbol{Z}_t$ is a q-variate unobservable vector and is referred to as the *state variable*, $\boldsymbol{e}_t$ are independent *state noises* with zero mean and $\text{var}(\boldsymbol{e}_t) = \boldsymbol{\Sigma}_{e,t}$, $\boldsymbol{Y}_t$ is a p observed vector, $\boldsymbol{\varepsilon}_t$ are independent observation noises with zero mean and $\text{var}(\boldsymbol{\varepsilon}_t) = \boldsymbol{\Sigma}_{\varepsilon,t}$, and $\boldsymbol{A}_t, \boldsymbol{B}_t$ are coefficient matrices. Furthermore, $(\boldsymbol{\varepsilon}_t, \boldsymbol{e}_t)$ are independent of $\{(\boldsymbol{Y}_j, \boldsymbol{Z}_j), j = t-1, t-2, \ldots\}$.

Given the observations $\mathcal{Y}_s \equiv (\boldsymbol{Y}'_1, \ldots, \boldsymbol{Y}'_s)'$, a particularly important task in state space modeling is to estimate the state variables $\boldsymbol{Z}_t$, as the inference such as likelihood computation can be carried out using the state variables. The estimation for $\boldsymbol{Z}_t$ is classified into three different scenarios:

$$\begin{aligned} \textit{Prediction}: \quad & \boldsymbol{Z}_{t|s} = E(\boldsymbol{Z}_t|\mathcal{Y}_s) \quad \text{for } t > s, \\ \textit{Filtering}: \quad & \boldsymbol{Z}_{t|t} = E(\boldsymbol{Z}_t|\mathcal{Y}_t), \\ \textit{Smoothing}: \quad & \boldsymbol{Z}_{t|s} = E(\boldsymbol{Z}_t|\mathcal{Y}_s) \quad \text{for } t < s. \end{aligned} \tag{3.62}$$

We also use the notation

$$\boldsymbol{V}_{t_1,t_2|s} = E\{(\boldsymbol{Z}_{t_1} - \boldsymbol{Z}_{t_1|s})'(\boldsymbol{Z}_{t_2} - \boldsymbol{Z}_{t_2|s})\}, \quad \text{and} \quad \boldsymbol{V}_{t|s} = \boldsymbol{V}_{t,t|s}. \tag{3.63}$$

3.6.2 Kalman recursions for Gaussian models

When $\boldsymbol{e}_t \sim N(\boldsymbol{0}, \boldsymbol{\Sigma}_{e,t})$ and $\boldsymbol{\varepsilon}_t \sim N(\boldsymbol{0}, \boldsymbol{\Sigma}_{\varepsilon,t})$, all the variables in model (3.61) are jointly normal. In this case,

$$\begin{aligned} \boldsymbol{Z}_{t|s} &= E\boldsymbol{Z}_t + \text{cov}(\boldsymbol{Z}_t, \mathcal{Y}_s)\{\text{var}(\mathcal{Y}_s)\}^{-1}(\mathcal{Y}_s - E\mathcal{Y}_s), \\ \boldsymbol{V}_{t|s} &= \text{var}(\boldsymbol{Z}_t|\mathcal{Y}_s) = \text{var}(\boldsymbol{Z}_t) - \text{cov}(\boldsymbol{Z}_t, \mathcal{Y}_s)\{\text{var}(\mathcal{Y}_s)\}^{-1}\text{cov}(\mathcal{Y}_s, \boldsymbol{Z}_t). \end{aligned} \tag{3.64}$$

Note that $\boldsymbol{Z}_{t|s}$ is a linear function of $\mathcal{Y}_s$, a ps-dimensional observed vector, regardless the relative sizes of t and s. This is a nice property of the Gaussianity. As an inverse of a $ps \times ps$ matrix is involved in the above expressions, we may not be able to compute $\boldsymbol{Z}_{t|s}$ and $\boldsymbol{V}_{t|s}$ based on those formulas directly. The *Kalman filter*, or more precisely, the *Kalman recursion* provides an efficient alternative which requires only to calculate inverses of $p \times p$ matrices instead. The recursion is based on the three sets of formulas: forecasting, filtering and smoothing, which we list below and can be derived based on (3.64). We refer to Appendix C in Kitagawa (2010) for the further details.

Prediction

$$\begin{aligned}
\textit{One-step-ahead}: \boldsymbol{Z}_{t+1|t} &= \boldsymbol{A}_{t+1}\boldsymbol{Z}_{t|t}, \qquad (3.65)\\
\boldsymbol{V}_{t+1|t} &= \boldsymbol{A}_{t+1}\boldsymbol{V}_{t|t}\boldsymbol{A}'_{t+1} + \boldsymbol{\Sigma}_{e,t+1}.\\
\textit{Multi-step-ahead}: \boldsymbol{Z}_{t+k|t} &= \boldsymbol{A}_{t+k}\boldsymbol{Z}_{t+k-1|t}, \qquad (3.66)\\
\boldsymbol{V}_{t+k|t} &= \boldsymbol{A}_{t+k}\boldsymbol{V}_{t+k-1|t}\boldsymbol{A}'_{t+k} + \boldsymbol{\Sigma}_{e,t+k}, \quad k \geqslant 2.
\end{aligned}$$

Filtering: Put $\boldsymbol{H}_t = \boldsymbol{B}_t\boldsymbol{V}_{t|t-1}\boldsymbol{B}'_t + \boldsymbol{\Sigma}_{\varepsilon,t}$.

$$\begin{aligned}
\boldsymbol{Z}_{t|t} =& \boldsymbol{Z}_{t|t-1} + \boldsymbol{V}_{t|t-1}\boldsymbol{B}'_t\boldsymbol{H}_t^{-1}(\boldsymbol{Y}_t - \boldsymbol{B}_t\boldsymbol{Z}_{t|t-1}) \qquad (3.67)\\
\boldsymbol{V}_{t|t} =& \boldsymbol{V}_{t|t-1} - \boldsymbol{V}_{t|t-1}\boldsymbol{B}'_t\boldsymbol{H}_t^{-1}\boldsymbol{B}_t\boldsymbol{V}_{t|t-1}.
\end{aligned}$$

Smoothing: Put $\boldsymbol{G}_t = \boldsymbol{V}_{t|t}\boldsymbol{A}'_{t+1}\boldsymbol{V}_{t+1|t}^{-1}$. For $t < T$,

$$\begin{aligned}
\boldsymbol{Z}_{t|T} =& \boldsymbol{Z}_{t|t} + \boldsymbol{G}_t(\boldsymbol{Z}_{t+1|T} - \boldsymbol{Z}_{t+1|t}), \qquad (3.68)\\
\boldsymbol{V}_{t|T} =& \boldsymbol{V}_{t|t} + \boldsymbol{G}_t(\boldsymbol{V}_{t+1|T} - \boldsymbol{V}_{t+1|t})\boldsymbol{G}'_t.
\end{aligned}$$

We may start the recursion with the initial condition $\boldsymbol{Z}_{1|0} = \boldsymbol{0}$ and $\boldsymbol{V}_{1|0} = \boldsymbol{0}$. At $t = 1$, compute $\boldsymbol{Z}_{1|1}$ and $\boldsymbol{V}_{1|1}$ using (3.67) in the filtering step. After that, we can compute multi-forecasting using (3.65) and (3.66), and smoothing using (3.68). The iteration now moves to the time $t = 2$.The diagram below, also taken from Kitagawa (2010), illustrates how Kalman recursion works: it consists of the steps of one-step-ahead prediction ($\Rightarrow$), filtering ($\Downarrow$), smoothing ($\leftarrow$), and multi-step-ahead prediction ($\rightarrow$) in the particular orders. The one-step-ahead predictions and filtering are the key steps, while multi-step-prediction are usually performed only for $\boldsymbol{Z}_{T+k|T}$ for $k \geqslant 1$, where T is the sample size.

$$
\begin{array}{ccccccccccc}
\boldsymbol{Z}_{1|1} & \Rightarrow & \boldsymbol{Z}_{2|1} & \rightarrow & \boldsymbol{Z}_{3|1} & \rightarrow & \boldsymbol{Z}_{4|1} & \rightarrow & \boldsymbol{Z}_{5|1} & \rightarrow \\
 & & \Downarrow & & & & & & & \\
\boldsymbol{Z}_{1|2} & \leftarrow & \boldsymbol{Z}_{2|2} & \Rightarrow & \boldsymbol{Z}_{3|2} & \rightarrow & \boldsymbol{Z}_{4|2} & \rightarrow & \boldsymbol{Z}_{5|2} & \rightarrow \\
 & & & & \Downarrow & & & & & \\
\boldsymbol{Z}_{1|3} & \leftarrow & \boldsymbol{Z}_{2|3} & \leftarrow & \boldsymbol{Z}_{3|3} & \Rightarrow & \boldsymbol{Z}_{4|3} & \rightarrow & \boldsymbol{Z}_{5|3} & \rightarrow \\
 & & & & & & \Downarrow & & & \\
\boldsymbol{Z}_{1|4} & \leftarrow & \boldsymbol{Z}_{2|4} & \leftarrow & \boldsymbol{Z}_{3|4} & \leftarrow & \boldsymbol{Z}_{4|4} & \Rightarrow & \boldsymbol{Z}_{5|4} & \rightarrow \\
 & & & & & & & & \Downarrow &
\end{array}
$$

Note that the red and blue parts are not essential in the iterations. They are just outputs of filtering and multi-step forecasting.

Let $\boldsymbol{\theta}$ denote collectively the unknown parameters in model (3.61). Now we illustrate how to use the Kalman recursion to compute the likelihood function of $\boldsymbol{\theta}$ based on observations $\boldsymbol{Y}_T, \ldots, \boldsymbol{Y}_1$. The joint density function of $\boldsymbol{Y}_1, \ldots, \boldsymbol{Y}_T$ can be factorized as

$$f(\boldsymbol{Y}_1) \prod_{t=2}^{T} f(\boldsymbol{Y}_t | \mathcal{Y}_{t-1}), \tag{3.69}$$

where $f(\boldsymbol{Y}_t|\mathcal{Y}_{t-1})$ is the conditional density function of $\boldsymbol{Y}_t$ given $\mathcal{Y}_{t-1}$. Hence it is a normal density function with mean $\boldsymbol{Y}_{t|t-1}$ and variance $\boldsymbol{W}_{t|t-1}$, where

$$\boldsymbol{Y}_{t+k|t} \equiv E(\boldsymbol{Y}_{t+k}|\mathcal{Y}_t) = \boldsymbol{B}_{t+k}\boldsymbol{Z}_{t+k|t}, \tag{3.70}$$

and

$$\boldsymbol{W}_{t+k|t} \equiv \text{var}(\boldsymbol{Y}_{t+k} - \boldsymbol{Y}_{t+1|t}) = \boldsymbol{B}_{t+k}\boldsymbol{V}_{t+k|t}\boldsymbol{B}'_{t+k} + \boldsymbol{\Sigma}_{\varepsilon,t+k}. \tag{3.71}$$

The equalities above are implied by the observation equation in (3.61). Hence, log-likelihood (conditionally on $\boldsymbol{Y}_1$) $l(\boldsymbol{\theta})$ can be expressed as

$$-2\,l(\boldsymbol{\theta}) = \sum_{t=2}^{T} \big\{ \log|\boldsymbol{W}_{t|t-1}| + (\boldsymbol{Y}_t - \boldsymbol{Y}_{t|t-1})'\boldsymbol{W}_{t|t-1}^{-1}(\boldsymbol{Y}_t - \boldsymbol{Y}_{t|t-1}) \big\}. \tag{3.72}$$

This represents a unified algorithm for computing Gaussian likelihood functions for a large class of processes, since the state space model (3.61) encapsulates stationary ARMA models as well as nonstationary models such as trend and seasonal models as special cases; see, e.g. Harvey (1989) and Durbin and Koopman (2012). Further, the observation equation in (3.61) may contain an observable exogenous variable $\boldsymbol{\xi}_t$:

$$\boldsymbol{Y}_t = \boldsymbol{B}_t\boldsymbol{Z}_t + \boldsymbol{C}_t\boldsymbol{\xi}_t + \boldsymbol{\varepsilon}_t. \tag{3.73}$$

In this case the Kalman recursion algorithm still holds with only one modification: replace $\boldsymbol{Y}_t$ on the RHS of (3.67) by $\boldsymbol{Y}_t - \boldsymbol{C}_t\boldsymbol{\xi}_t$. Of course now (3.70) has one more term $\boldsymbol{C}_{t+k}\boldsymbol{\xi}_{t+k}$ on its RHS.

Note the computation of the likelihood function (3.72) only involves the steps of the one-step-ahead predictions ($\Rightarrow$), and the filtering ($\Downarrow$) in a Kalman recursion, i.e. the steps in black in the diagram above.

For an illustration, we consider an ARMA(p, q) model

$$X_t = b_1 X_{t-1} + \cdots + b_p X_{t-p} + \varepsilon_t + a_1\varepsilon_{t-1} + \cdots + a_q\varepsilon_{t-q}, \tag{3.74}$$

where ε_t are independent and $N(0, \sigma^2)$. Let $r = \max(p, q+1)$. Then the above model can be written as

$$X_t = \sum_{i=1}^{r} b_i X_{t-1} + \varepsilon_t + \sum_{j=1}^{r} a_j \varepsilon_{t-j}$$

with some coefficient being 0. Now define an $r \times 1$ state variable

$$\boldsymbol{Z}_t = \begin{pmatrix} X_t \\ b_2 X_{t-1} + \cdots + b_r X_{t-r+1} + a_1\varepsilon_t + \cdots + a_{r-1}\varepsilon_{t-r+2} \\ b_3 X_{t-1} + \cdots + b_r X_{t-r+2} + a_2\varepsilon_t + \cdots + a_{r-1}\varepsilon_{t-r+3} \\ \vdots \\ b_r X_{t-1} + a_{r-1}\varepsilon_t \end{pmatrix}.$$

Then the ARMA model (3.74) can be written as a linear state space model as follows:

$$\begin{cases} X_t = \boldsymbol{A}\boldsymbol{Z}_t, \\ \boldsymbol{Z}_t = \boldsymbol{B}\boldsymbol{Z}_{t-1} + \boldsymbol{\varepsilon}_t, \end{cases} \tag{3.75}$$

where $\boldsymbol{A} = (1, 0, \ldots, 0)$, and

$$\boldsymbol{B}_t = \begin{pmatrix} b_1 & 1 & & 0 \\ \vdots & & \ddots & \\ b_{r-1} & 0 & & 1 \\ b_r & 0 & \cdots & 0 \end{pmatrix}, \qquad \boldsymbol{\varepsilon}_t = \begin{pmatrix} 1 \\ a_1 \\ \vdots \\ a_{r-1} \end{pmatrix} \varepsilon_t.$$

Hence the MLE (2.48) for ARMA models can be calculated using a Kalman filter via (3.72). In fact this is the algorithm adopted by many statistical packages.

There are quite a few *R* functions for calculating Kalman filters. For example, there are several functions for computing Kalman filters in the package `astsa` associated with Schumway and Stoffer (2011). See also `http://www.stat.pitt.edu/stoffer/tsa3/`.

3.6.3 Nonlinear models

When $\boldsymbol{e}_t$ in model (3.61) is not normal, $\boldsymbol{Z}_t$ and $\boldsymbol{Y}_t$ are not normal. Then the Kalman recursion presented in the section above is no longer relevant, as the conditional expectation $\boldsymbol{Z}_{t|s}$ and the conditional variance $\text{var}(\boldsymbol{Z}_t|\mathcal{Y}_s)$ do not admit the simple expressions (3.64). Furthermore, the conditional distribution of $\boldsymbol{Z}_t$ given $\mathcal{Y}_s$ may depend on $\mathcal{Y}_s$ more intimately than merely via its mean and variance functions. Below we state a general nonlinear state space model and present a set of recursion formulas for the associated conditional distributions. Those recursion formulas also apply to linear model (3.61) with non-Gaussian $\boldsymbol{e}_t$ and/or $\boldsymbol{\varepsilon}_t$.

Consider the following nonlinear state-space model

$$\begin{cases} \boldsymbol{Z}_t = \boldsymbol{G}_t(\boldsymbol{Z}_{t-1},\ \boldsymbol{e}_t), & \text{(state equation)} \\ \boldsymbol{Y}_t = \boldsymbol{F}_t(\boldsymbol{Z}_t,\ \boldsymbol{\varepsilon}_t), & \text{(observation equation)} \end{cases} \tag{3.76}$$

where $\boldsymbol{G}_t(\cdot)$ and $\boldsymbol{F}_t(\cdot)$ are known vector functions up to some unknown parameters, and $\{\boldsymbol{e}_t\}$ and $\{\boldsymbol{\varepsilon}_t\}$ are two sequences of independent random vectors.

With the abuse of notation, we denote by $p(\boldsymbol{X}|\boldsymbol{Y})$ the conditional density function of $\boldsymbol{X}$ given $\boldsymbol{Y}$. Then it follows from (3.76) that

$$\begin{aligned} p(\boldsymbol{Z}_t|\boldsymbol{Z}_{t-1}, \mathcal{Y}_{t-1}) &= p(\boldsymbol{Z}_t|\boldsymbol{Z}_{t-1}), \\ p(\boldsymbol{Y}_t|\boldsymbol{Z}_t, \mathcal{Y}_{t-1}) &= p(\boldsymbol{Y}_t|\boldsymbol{Z}_t). \end{aligned} \tag{3.77}$$

For the state-space model with additive noise, i.e.

$$\boldsymbol{Z}_t = \boldsymbol{G}_t(\boldsymbol{Z}_{t-1}) + \boldsymbol{e}_t, \qquad \boldsymbol{Y}_t = \boldsymbol{F}_t(\boldsymbol{Z}_t) + \boldsymbol{\varepsilon}_t,$$

it is easy to see that

$$p(\boldsymbol{Z}_t|\boldsymbol{Z}_{t-1}, \mathcal{Y}_{t-1}) = g_t(\boldsymbol{Z}_t - \boldsymbol{G}_t(\boldsymbol{Z}_{t-1}))$$

and

$$p(\boldsymbol{Y}_t|\boldsymbol{Z}_t, \mathcal{Y}_{t-1}) = f_t(\boldsymbol{Y}_t - \boldsymbol{F}_t(\boldsymbol{Z}_t)),$$

where g_t and f_t are, respectively, the density functions of $\boldsymbol{e}_t$ and $\boldsymbol{\varepsilon}_t$.

Based on (3.77), we can derive the following distributional recursion formulas.

One-step-ahead prediction

$$\begin{aligned} p(\boldsymbol{Z}_t|\mathcal{Y}_{t-1}) &= \int p(\boldsymbol{Z}_t, \boldsymbol{Z}_{t-1}|\mathcal{Y}_{t-1}) d\boldsymbol{Z}_{t-1} \\ &= \int p(\boldsymbol{Z}_t|\boldsymbol{Z}_{t-1}, \mathcal{Y}_{t-1}) p(\boldsymbol{Z}_{t-1}|\mathcal{Y}_{t-1}) d\boldsymbol{Z}_{t-1} \\ &= \int p(\boldsymbol{Z}_t|\boldsymbol{Z}_{t-1}) p(\boldsymbol{Z}_{t-1}|\mathcal{Y}_{t-1}) d\boldsymbol{Z}_{t-1}. \end{aligned} \tag{3.78}$$

Filtering

$$\begin{aligned} p(\boldsymbol{Z}_t|\mathcal{Y}_t) &= p(\boldsymbol{Z}_t|\boldsymbol{Y}_t, \mathcal{Y}_{t-1}) \\ &= p(\boldsymbol{Y}_t|\boldsymbol{Z}_t, \mathcal{Y}_{t-1}) p(\boldsymbol{Z}_t|\mathcal{Y}_{t-1}) / p(\boldsymbol{Y}_t|\mathcal{Y}_{t-1}) \\ &= p(\boldsymbol{Y}_t|\boldsymbol{Z}_t) p(\boldsymbol{Z}_t|\mathcal{Y}_{t-1}) / p(\boldsymbol{Y}_t|\mathcal{Y}_{t-1}), \end{aligned} \tag{3.79}$$

where $p(\boldsymbol{Y}_t|\mathcal{Y}_{t-1}) = \int p(\boldsymbol{Y}_t|\boldsymbol{Z}_t) p(\boldsymbol{Z}_t|\mathcal{Y}_{t-1}) d\boldsymbol{Z}_t$ as in (3.78). With an initial value for $\boldsymbol{Z}_0$ and an initial density for $p(\boldsymbol{Z}_0) \equiv p(\boldsymbol{Z}_0|\mathcal{Y}_0)$, we may evaluate $p(\boldsymbol{Z}_1|\mathcal{Y}_0)$, $p(\boldsymbol{Z}_1|\mathcal{Y}_1), p(\boldsymbol{Z}_2|\mathcal{Y}_1)$, $p(\boldsymbol{Z}_2|\mathcal{Y}_2), \ldots$ recursively using (3.78) and (3.79).

Denote by $\boldsymbol{\theta}$ collectively all the unknown parameters in model (3.76). With the available observations $\boldsymbol{Y}_1, \ldots, \boldsymbol{Y}_T$, the log likelihood function (conditional on an initial value $\mathcal{Y}_0$) can be written as

$$\begin{aligned} l(\boldsymbol{\theta}) &= \sum_{t=1}^{T} \log p(\boldsymbol{Y}_t|\mathcal{Y}_{t-1}) \\ &= \sum_{t=1}^{T} \log \int p(\boldsymbol{Y}_t|\boldsymbol{Z}_t, \mathcal{Y}_{t-1}) p(\boldsymbol{Z}_t|\mathcal{Y}_{t-1}) d\boldsymbol{Z}_t \\ &= \sum_{t=1}^{T} \log \int p(\boldsymbol{Y}_t|\boldsymbol{Z}_t) p(\boldsymbol{Z}_t|\mathcal{Y}_{t-1}) d\boldsymbol{Z}_t. \end{aligned} \tag{3.80}$$

The second equality follows the second identity in (3.77).

Although the likelihood function (3.80) can be calculated in principle using the recursion (3.78)–(3.79), the computation involves many integrations which cause various technical difficulties. Kitagawa (1987) proposed a grid approximation method which updates the conditional density functions on a finite grid of points and replaces the integrals by their Riemann sums over the grids. However the method may suffer from numerical instabilities, especially when the dimension of $\boldsymbol{Y}_t$ and/or $\boldsymbol{X}_t$ is high. With the modern computing power, it becomes a common practice to compute integrals by Monte Carlo simulation. We introduce one such method in the section below.

3.6.4 Particle filters

The particle filter, also called the sequential Monte Carlo method or bootstrap filter, is an algorithm for executing the recursion (3.78)–(3.79) by Monte Carlo simulation. The key is to generate the samples (i.e. particles) alternatively from the prediction distribution $p(\boldsymbol{Z}_t|\mathcal{Y}_{t-1})$ and the filter distribution $p(\boldsymbol{Z}_t|\mathcal{Y}_t)$ for $t = 1, 2, \ldots, T$. We use the notation:

$\{\boldsymbol{z}_{t|t-1}(1), \ldots, \boldsymbol{z}_{t|t-1}(m)\}$ is a sample from $p(\boldsymbol{Z}_t|\mathcal{Y}_{t-1})$, and
$\{\boldsymbol{z}_{t|t}(1), \ldots, \boldsymbol{z}_{t|t}(m)\}$ is a sample from $p(\boldsymbol{Z}_t|\mathcal{Y}_t)$.

Before we spell out the algorithm, let us illustrate how those particles can be used to calculate, for example, the likelihood function (3.80). Let $h_t(\cdot|\boldsymbol{z})$ be the conditional density function of $\boldsymbol{Y}_t$ given $\boldsymbol{Z}_t = \boldsymbol{z}$. Then by the law of large numbers, as $m \to \infty$

$$\frac{1}{m}\sum_{j=1}^{m} h_t\{\boldsymbol{Y}_t|\boldsymbol{z}_{t|t-1}(j)\} \xrightarrow{a.s.} \int h_t(\boldsymbol{Y}_t|\boldsymbol{Z}_t)p(\boldsymbol{Z}_t|\mathcal{Y}_{t-1})d\boldsymbol{Z}_t$$

Thus the log-likelihood function $l(\boldsymbol{\theta})$ in (3.80) can be approximated by

$$\widetilde{l}(\boldsymbol{\theta}) = \sum_{t=1}^{T} \log\left(\frac{1}{m}\sum_{j=1}^{m} h_t\{\boldsymbol{Y}_t|\boldsymbol{z}_{t|t-1}(j)\}\right). \tag{3.81}$$

Note that $h_t(\cdot|\boldsymbol{z})$ is determined by the observation equation in model (3.76). When it has additive noise, i.e.

$$\boldsymbol{Y}_t = \boldsymbol{F}_t(\boldsymbol{Z}_t) + \boldsymbol{\varepsilon}_t, \qquad h_t(\boldsymbol{y}|\boldsymbol{z}) = f_t(\boldsymbol{y} - \boldsymbol{F}_t(\boldsymbol{z})),$$

where f_t is the density function $\boldsymbol{\varepsilon}_t$.

We now state the *particle filter* for generating the samples $\{\boldsymbol{z}_{t|t-1}(j)\}$ and $\{\boldsymbol{z}_{t|t}(j)\}$ in the order of $t = 1, 2, \ldots, T$. Let $\boldsymbol{z}_{0|0}(1), \ldots, \boldsymbol{z}_{0|0}(m)$ be the initial values.

One-step-ahead prediction particles: For $j = 1, \ldots, m$, let

$$\boldsymbol{z}_{t|t-1}(j) = \boldsymbol{G}_t\big(\boldsymbol{z}_{t-1|t-1}(j), \boldsymbol{e}_t(j)\big), \tag{3.82}$$

where $\boldsymbol{e}_t(1), \ldots, \boldsymbol{e}_t(m)$ are drawn independently from the distribution of $\boldsymbol{e}_t$.

Filtering particles: $\boldsymbol{z}_{t|t}(1), \ldots, \boldsymbol{z}_{t|t}(m)$ are drawn independently from the following discrete distribution.

Value	$z_{t\|t-1}(1)$	$z_{t\|t-1}(2)$	$\cdots$	$z_{t\|t-1}(m)$
Probability	π_{t1}	π_{t2}	$\cdots$	π_{tm}

where $\pi_{tj} = \beta_{tj} / \sum_{1 \leqslant i \leqslant m} \beta_{ti}$, and

$$\beta_{tj} = h_t\{\boldsymbol{Y}_t | \boldsymbol{z}_{t|t-1}(j)\}, \quad j = 1, \ldots, m, \tag{3.83}$$

where $h_t(\cdot|\boldsymbol{z})$ is the conditional density of $\boldsymbol{Y}_t$ given $\boldsymbol{Z}_t = \boldsymbol{z}$.

The proof of the above algorithm is simple. First, since $\boldsymbol{z}_{t-1|t-1}$ is drawn from $p(\boldsymbol{Z}_{t-1}|\mathcal{Y}_{t-1})$, it follows from (3.78) that $\boldsymbol{z}_{t|t-1}$ can be drawn from the conditional distribution of $\boldsymbol{Z}_t$ given $\boldsymbol{Z}_{t-1} = \boldsymbol{z}_{t-1|t-1}$. Now the state equation in model (3.76) implies (3.82). To justify the algorithm for the filtering particles, it follows from (3.79) that $p(\boldsymbol{Z}_t|\mathcal{Y}_t)$ as a density for $\boldsymbol{Z}_t$ is proportional to $p(\boldsymbol{Y}_t|\boldsymbol{Z}_t)p(\boldsymbol{Z}_t|\mathcal{Y}_{t-1})$, i.e.

$$p(\boldsymbol{Z}_t|\mathcal{Y}_t) \propto p(\boldsymbol{Y}_t|\boldsymbol{Z}_t)p(\boldsymbol{Z}_t|\mathcal{Y}_{t-1}). \tag{3.84}$$

Hence a sample from $p(\boldsymbol{Z}_t|\mathcal{Y}_t)$ can be obtained by resampling from a sample drawn from $p(\boldsymbol{Z}_t|\mathcal{Y}_{t-1})$ according to the importance weights – this is an application of the standard *importance sampling* procedure. Now $\{\boldsymbol{z}_{t|t-1}(1), \ldots, \boldsymbol{z}_{t|t-1}(m)\}$ is a sample from $p(\boldsymbol{Z}_t|\mathcal{Y}_{t-1})$. According to (3.84), the importance weight for $\boldsymbol{z}_{t|t-1}(j)$ is proportional to

$$\beta_j = p(\boldsymbol{Y}_t|\boldsymbol{Z}_t = \boldsymbol{z}_{t|t-1}(j)) = h_t\{\boldsymbol{Y}_t|\boldsymbol{z}_{t|t-1}(j)\}.$$

Normalizing $\{\beta_j\}$ leads to the resampling probability weights $\{\pi_j\}$. This completes the proof.

The particle filter is a general framework which can be applied to various time series models. Variants of the method can be found in, among others, Gordon, et al.(1993), Kitagawa (1996), Berzuini et al.(1997), Liu and Chen (1998), and Lin et al.(2005). We also refer to Chapter 15 of Kitagawa (2010) for a more detailed account of the method.

3.7 Exercises

3.1 For ARCH(1) model, if X_t is strong stationary and $EX_t^4 < \infty$, show that

$$EX_t^4 = \frac{a_0^2(1 + a_1)E\varepsilon_t^4}{(1 - a_1)(1 - a_1^2 E\varepsilon_t^4)}.$$

and the kurtosis

$$\kappa_x = \frac{(1 - a_1^2)\kappa_\varepsilon}{1 - a_1^2 E\varepsilon_t^4}.$$

Therefore a necessary condition for the existence of the fourth moment is that $a_1 < 1/\sqrt{E\varepsilon_t^4}$.

3.2 For the GARCH(1,2) model

$$X_t = \sigma_t \varepsilon_t, \quad \sigma_t^2 = a_0 + a_1 X_{t-1}^2 + b_1 \sigma_{t-1}^2 + b_2 \sigma_{t-2}^2$$

1. Express X_t^2 as an ARMA(2,2) process.
2. Express the GARCH(1,2) as an ARCH(∞) model. Here for simplicity, you may assume that $1 - b_1 B - b_2 B^2 = (1 - c_1 B)(1 - c_2 B)$ for two real constants $c_1 \neq c_2$.
3. Derive the recursive formula for the multi-step ahead forecasts of volatility.

3.3 Suppose that the daily log-returns $\{r_t\}$ of General Electric Inc. from January 1995 to December 2012 follows the AR(1)-GARCH(1,1) model:

$$r_t = \alpha_0 + \alpha_1 r_{t-1} + \eta_t, \quad \eta_t = \sigma_t \varepsilon_t, \quad \sigma_t^2 = a_0 + a_1 \eta_{t-1}^2 + b_1 \sigma_{t-1}^2.$$

1. Write down the conditional likelihood function when ε_t follows the standard normal distribution.
2. Write down the conditional likelihood function when ε_t follows a standardized t-distribution with degree of freedom ν.
3. Simulate the series of length 2520 with $\alpha_0 = 0.002, \alpha_1 = -0.12$, $a_0 = 0.000015$, $a_1 = 0.0414$, $b_1 = 0.921$, and the Gaussian innovation, and show the series.
4. What is the variance of this AR(1)-GARCH(1,1) model? Compare it with the sample variance? Compare the sample kurtosis of the time series with that of the innovative time series. **Hint**: Use the model to write down $\mathrm{var}(r_t)$ and use the fact r_{t-k} and η_t are uncorrelated for $k > 0$.

3.4 Suppose that the volatilities of the daily log-returns of the Coco-Cola company follow the GARCH(1,1) model

$$X_t = \sigma_t \varepsilon_t, \quad \sigma_t^2 = a_0 + a_1 X_{t-1}^2 + b_1 \sigma_{t-1}^2$$

with $a_1 + b_1 < 1$ and $\varepsilon_t \sim N(0,1)$.

1. If $a_0 = 0.006$, $a_1 = 0.05$ and $b_1 = 0.55$, is the tail of the distribution lighter than that of t_4 in terms of kurtosis?
2. What is the auto-correlation function of the series $\{X_t^2\}$?
3. If a_0, a_1 and b_1 are estimated as 0.006, 0.1 and 0.4 respectively with associated covariance matrix

$$10^{-4} \begin{pmatrix} 15 & 5 & 0 \\ 5 & 4 & 0 \\ 0 & 0 & 30 \end{pmatrix},$$

   what is the estimated long-run variance (unconditional variance)? What is the associated standard error?
4. With the parameters in (a), if $X_T^2 = 0.02$ and $\sigma_T^2 = 0.03$, give the one-step and two-step forecast of the volatility.
5. Suppose that we have observed the data and wish to fit the GARCH(p, q) with $p + q \leqslant 2$. Outline the key steps (including diagnostics) for fitting the data.

3.5 After fitting a GARCH(1,1) model with the Gaussian innovation, what kind of diagnostic steps should we follow? What else would you try if the residuals do not follow the normal distribution?

3.6 What are two key motivations for introducing the family of EGARCH(p,q) in comparison with original ARCH(p) models?

3.7 Suppose that the monthly return of a stock follows EGARCH(1,0)

$$h_t = -0.873 + 0.735 h_{t-1} + g(\varepsilon_{t-1}),$$
$$g(\varepsilon_{t-1}) = 0.087\varepsilon_{t-1} + 0.327(|\varepsilon_{t-1}| - \sqrt{2/\pi}),$$

where $h_t = \log \sigma_t^2$ and $\varepsilon_t \sim i.i.d. N(0,1)$. What is the volatility σ_{t+1} with a shock of magnitude $\varepsilon_t = -1$ and what is volatility σ_{t+1} with a shock of magnitude $\varepsilon_t = 1$.

3.8 Suppose that the log-return $\{r_t\}$ of a portfolio follow an GARCH (p,q) model: $r_t = \sigma_t \varepsilon_t$, with

$$\sigma_t^2 = a_0 + \sum_{i=1}^{p} a_i X_{t-i}^2 + \sum_{j=1}^{q} b_j \sigma_{t-j}^2.$$

1. Show that $\text{cov}(r_t^2, \sigma_t^2) > 0$.
2. What is the unconditional volatility of the τ-period log-returns: $R_{t,\tau} = r_{t+1} + \cdots + r_{t+\tau}$?
3. What is the conditional volatility of $R_{T,\tau}$. Express it in terms of $\sigma_T^2(m) = E_T r_{T+m}^2$.

3.9 Consider the log-monthly returns of Intel from January, 1990 to December 2013.

1. Are the returns predictable?
2. Use the PACF plot of the series to determine the order of the fit of the autoregressive model for the return. Plot also the PACF for the squared return series.
3. Fit a GARCH(1,1) model to the return series using the Gaussian innovation.
4. Compute the mean return and long-run volatility (unconditional standard deviation).
5. Use the Delta-method to get the SE of the mean return and long-run volatility.
6. Provide necessary model diagnostics using graphs and test statistics.

3.10 Consider the log-weekly returns of the Ford company from January 1990 to December 2013.

1. Fit an ARMA(1,1)-GARCH(1,1) model to the return series using a t-innovation with unknown degree of freedom as innovation.
2. Fit an EGARCH(1,1) model to the return series using the normal distribution.
3. Compare empirically the fitted mean and volatility of returns from two parametric models. Compare them also with the sample mean and sample variance of the original time series.
4. Which residual series from the above fits more closely resembles the white noise? You may compare the Ljung–Box statistics for the residual and square-residual series, in addition to their ACF plots.

4

Multivariate Time Series Analysis

In many practical situations, time series data over multiple subjects are recorded together. Often data are correlated across different subjects and over different times. A simple example is the S&P 500 index and the future of the S&P 500 index. Therefore it is legitimate to analyze multiple time series together. We argue that the techniques for handling multiple time series are ever more important in this information age marked by the acceleration of economic globalization and the explosion of data collection. Price fluctuations in one market may cause or be caused by the movements in other markets. An infamous example is the "Asian Contagion", which occurred in 1997 and started in Thailand, spread quickly to southeast Asian countries, and then spilled over to the rest of the world. Hence it is important to take into account the dependence or correlation across different series. Furthermore a time delay of a certain movement in one series in comparison with others is often of practical importance. Finally each individual time series may be nonstationary, but nonstationary features in different series may cancel each other. Therefore analyzing multiple series together, we may be able to identify some latent stationary properties. In this chapter, we introduce some basic linear models and methods for analyzing multiple time series data. We also cover the topics such as the Granger causality, impulse response functions, cointegration, which have direct relevance to modelling economic and financial data.

4.1 Stationarity and autocorrelation matrices

4.1.1 Stationary vector processes

Let $\boldsymbol{X}_t = (X_{t1}, \dots, X_{td})'$ be d time series, where X_{tj} is the jth component series. The series $\boldsymbol{X}_t$ is *weakly stationary*, or simply *stationary*, if all its first and second moments are time-invariant, i.e.

$$\boldsymbol{\mu} \equiv E\boldsymbol{X}_t, \qquad \boldsymbol{\Gamma}(k) \equiv E\{(\boldsymbol{X}_{t+k} - \boldsymbol{\mu})(\boldsymbol{X}_t - \boldsymbol{\mu})'\} \tag{4.1}$$

are independent of t. The matrix valued function $\boldsymbol{\Gamma}(\cdot)$ is called the autocovariance matrix function, or *cross covariance* function. For the latter, the emphasis is on the covariance of difference component series. Let μ_j be the jth element of $\boldsymbol{\mu}$, and $\gamma_{ij}(k)$ be the (i, j)th element of $\boldsymbol{\Gamma}(k)$. Then

$$\gamma_{ij}(k) = \text{cov}(X_{t+k,i}, X_{tj}) = \text{cov}\{(X_{t+k,i} - \mu_i)(X_{tj} - \mu_j)\} \tag{4.2}$$

is called the cross covariance between the ith and jth component series at time lag k for $i \neq j$. It is the autocovariance for the jth component series when $i = j$, Note that $\boldsymbol{\Gamma}(k)$ is not a symmetric matrix unless $k = 0$. In general, $\boldsymbol{\Gamma}(-k) = \boldsymbol{\Gamma}(k)'$.

The autocorrelation matrix, which is also called the *cross correlation matrix*, of $\boldsymbol{X}_t$ is defined as

$$\boldsymbol{R}(k) \equiv \big(\rho_{ij}(k)\big) = \boldsymbol{D}^{-1/2}\boldsymbol{\Gamma}(k)\boldsymbol{D}^{-1/2}, \tag{4.3}$$

where $\boldsymbol{D}^{-1/2}$ is the diagonal matrix with $\gamma_{jj}(0)^{-1/2}$, the reciprocal of the standard deviation of the jth series, as its jth main diagonal element. Then

$$\rho_{ij}(k) = \text{Corr}(X_{t+k,i}, X_{tj}), \tag{4.4}$$

is the *cross correlation coefficient* between the two component series at lag k. Note that an autocorrelation coefficient $\rho_{jj}(k)$ is symmetric in the sense $\rho_{jj}(k) = \rho_{jj}(-k)$ for all integer k, but a cross correlation is not necessarily symmetric, namely for any $i \neq j$, typically, $\rho_{ij}(k) \neq \rho_{ij}(-k), k = 1, 2, \ldots$. On the other hand,

$$\rho_{ij}(k) = \rho_{ji}(-k), \qquad \text{i.e.} \quad \boldsymbol{R}(-k) = \boldsymbol{R}(k)'.$$

If $\boldsymbol{X}_t$ is a weakly stationary vector time series, all its component series are weakly stationary univariate time series. However the converse is not necessarily true, which requires an additional condition that the cross correlations (4.4) is also time-invariant. Similarly to the univariate case, we call $\{\boldsymbol{X}_t\}$ *strictly stationary* if the joint distribution of $\boldsymbol{X}_1, \ldots, \boldsymbol{X}_k$ is the same as that of $\boldsymbol{X}_{t+1}, \ldots, \boldsymbol{X}_{t+k}$ for any $k \geqslant 1$ and integer t. Obviously if $\boldsymbol{X}_t$ is strictly stationary and $E(\|\boldsymbol{X}_t\|^2) < \infty$, $\boldsymbol{X}_t$ is also weakly stationary, where $\|\cdot\|$ denotes the Euclidean norm of a vector. Linear dependence (i.e. correlation) is entirely determined by the first two moments. Therefore weak stationarity is often adopted in the context of linear time series models.

The simplest type of weakly stationary vector processes is *vector white noise* denoted by $\text{WN}(\boldsymbol{a}, \boldsymbol{\Sigma}_\varepsilon)$. We say $\boldsymbol{\varepsilon}_t \sim \text{WN}(\boldsymbol{a}, \boldsymbol{\Sigma}_\varepsilon)$ if $E\boldsymbol{\varepsilon}_t = \boldsymbol{a}$, $\text{var}(\boldsymbol{\varepsilon}_t) =$

$\boldsymbol{\Sigma}_\varepsilon$, and $\text{cov}(\boldsymbol{\varepsilon}_t, \boldsymbol{\varepsilon}_s) = \mathbf{0}$ for any $t \neq s$. Hence there exists no serial correlation across all the components of $\boldsymbol{\varepsilon}_t$. However different components of $\boldsymbol{\varepsilon}_t$ may be correlated with each other contemporaneously as $\boldsymbol{\Sigma}_\varepsilon$ is not necessarily a diagonal matrix. Most frequently used white noise is the one with mean $\mathbf{0}$, i.e. $\text{WN}(\mathbf{0}, \boldsymbol{\Sigma}_\varepsilon)$.

Vector white noise processes serve as building blocks for constructing vector stationary processes. For example, a *vector moving average* process is defined as

$$\boldsymbol{X}_t = \boldsymbol{\mu} + \boldsymbol{\varepsilon}_t + \boldsymbol{B}_1\boldsymbol{\varepsilon}_{t-1} + \cdots + \boldsymbol{B}_q\boldsymbol{\varepsilon}_{t-q}, \tag{4.5}$$

where $\boldsymbol{B}_1, \ldots, \boldsymbol{B}_q$ are coefficient matrices, and $\boldsymbol{\varepsilon} \sim \text{WN}(\mathbf{0}, \boldsymbol{\Sigma}_\varepsilon)$. We call such a process as a moving average process with order q, denoted by $\boldsymbol{X}_t \sim \text{MA}(q)$. It is easy to see that any vector MA(q) process is weakly stationary. Furthermore, $E\boldsymbol{X}_t = \boldsymbol{\mu}$, and the cross covariance is equal to

$$\boldsymbol{\Gamma}(k) = \begin{cases} \boldsymbol{B}_k\boldsymbol{\Sigma}_\varepsilon + \boldsymbol{B}_{k+1}\boldsymbol{\Sigma}_\varepsilon\boldsymbol{B}_1' + \cdots + \boldsymbol{B}_q\boldsymbol{\Sigma}_\varepsilon\boldsymbol{B}_{q-k}' & 0 \leqslant k \leqslant q, \\ 0 & k > q, \end{cases} \tag{4.6}$$

and $\boldsymbol{\Gamma}(-k) = \boldsymbol{\Gamma}(k)'$. Similar to univariate MA(q), the cross-covariance and the cross-correlation matrices cut off at q. In the above expression, $\boldsymbol{B}_0 = \boldsymbol{I}_d$ is the $d \times d$ identity matrix.

Example 4.1 Consider a simple bivariate MA(4) process defined by

$$X_{t1} = \varepsilon_t + 0.5\varepsilon_{t-4}, \qquad X_{t2} = \varepsilon_t,$$

where ε_t is a univariate WN(0, 1). It is easy to see that

$$\boldsymbol{\mu} = E\begin{pmatrix} X_{t1} \\ X_{t2} \end{pmatrix} = \mathbf{0}, \quad \boldsymbol{\Gamma}(\mathbf{0}) = \text{var}\begin{pmatrix} X_{t1} \\ X_{t2} \end{pmatrix} = \begin{pmatrix} 1.25 & 1 \\ 1 & 1 \end{pmatrix},$$

$$\boldsymbol{\Gamma}(4) = E\left\{\begin{pmatrix} \varepsilon_{t+4} + 0.5\varepsilon_t \\ \varepsilon_{t+4} \end{pmatrix}(\varepsilon_t + 0.5\varepsilon_{t-4}, \varepsilon_t)\right\} = \begin{pmatrix} 0.5 & 0.5 \\ 0 & 0 \end{pmatrix} = \boldsymbol{\Gamma}(-4)',$$

and $\boldsymbol{\Gamma}(k) = 0$ for all $k \neq 0, \pm 4$. As $\boldsymbol{D} = \text{diag}(1/\sqrt{1.25}, 1) = \text{diag}(0.894, 1)$, the cross correlation function is

$$\boldsymbol{R}(0) = \begin{pmatrix} 1 & 0.894 \\ 0.894 & 1 \end{pmatrix}, \quad \boldsymbol{R}(4) = \begin{pmatrix} 0.4 & 0.447 \\ 0 & 0 \end{pmatrix} = \boldsymbol{R}(-4)'$$

and $\boldsymbol{R}(k) = \mathbf{0}$ otherwise.

Although vector MA models are among the simplest multivariate time series model, it exhibits an innate difficulty, i.e. overparametrization. For example, for a simple d-vector MA(1) model, i.e. $q = 1$ in (4.5), the number of unknown parameters in coefficient matrix $\boldsymbol{B}_1$ is d^2. Hence even for moderately large d, some regularization and/or dimension-reduction techniques are often employed to reduce the number of parameters in vector time series models.

4.1.2 Sample cross-covariance/correlation matrices

With available observations $\boldsymbol{X}_1, \ldots, \boldsymbol{X}_T$ from a weakly stationary process, a natural estimator of the cross-covariance matrix is the *sample cross-covariance matrix*

$$\widehat{\boldsymbol{\Gamma}}(k) \equiv (\widehat{\gamma}_{ij}(k)) = \frac{1}{T}\sum_{t=1}^{T-k}(\boldsymbol{X}_{t+k} - \widehat{\boldsymbol{\mu}})(\boldsymbol{X}_t - \widehat{\boldsymbol{\mu}})', \quad \widehat{\boldsymbol{\mu}} = \frac{1}{T}\sum_{t=1}^{T}\boldsymbol{X}_t. \tag{4.7}$$

Furthermore the cross-correlation matrix can be estimated by the *sample cross-correlation matrix*

$$\widehat{\boldsymbol{R}}(k) \equiv \big(\widehat{\rho}_{ij}(k)\big) = \widehat{\boldsymbol{D}}^{-\frac{1}{2}}\widehat{\boldsymbol{\Gamma}}(k)\,\widehat{\boldsymbol{D}}^{-\frac{1}{2}}, \tag{4.8}$$

where $\widehat{\boldsymbol{D}} = \mathrm{diag}\big(\widehat{\gamma}_{11}(0), \ldots, \widehat{\gamma}_{dd}(0)\big)$. Sample cross-correlation matrix at each time lag k is a $d \times d$ matrix. For small d, it can be displayed as a cross-correlogram produced by, for example, the `R` function `acf`.

Figure 4.1 displays the daily close log prices of the FTSE 100 index, FTSE MidCap index and FTSE SmallCap index in 2011. The FTSE index series are designed to represent the performance of UK companies on equity markets. FTSE 100 comprises the 100 most highly capitalized blue chip companies listed on London Stock Exchange, representing over 80% of the entire market capitalization. It is used extensively as a basis for investment products, such as derivatives and exchange-traded funds. FTSE MidCap (also called FTSE 250) includes the next 250 largest companies after the FTSE 100. It represents approximately 15% of UK market capitalization. FTSE SmallCap index consists of the 351st to the 619th largest listed companies on the London Stock Exchange market. In spite of the substantial difference in the percentages of the market capitalization, there is clearly a certain degree of synchrony in the fluctuation in the three indices. The sharp drop in August of the year was due to fears of contagion of the European sovereign debt crisis started in Italy, Spain and then France. While those price series are clearly nonstationary, the co-movements suggest that some linear combinations of the three series (such as a difference of any two of them) may resemble some stationary features. Furthermore are there any contagious phenomena in the sense that one index leads to the others? For now

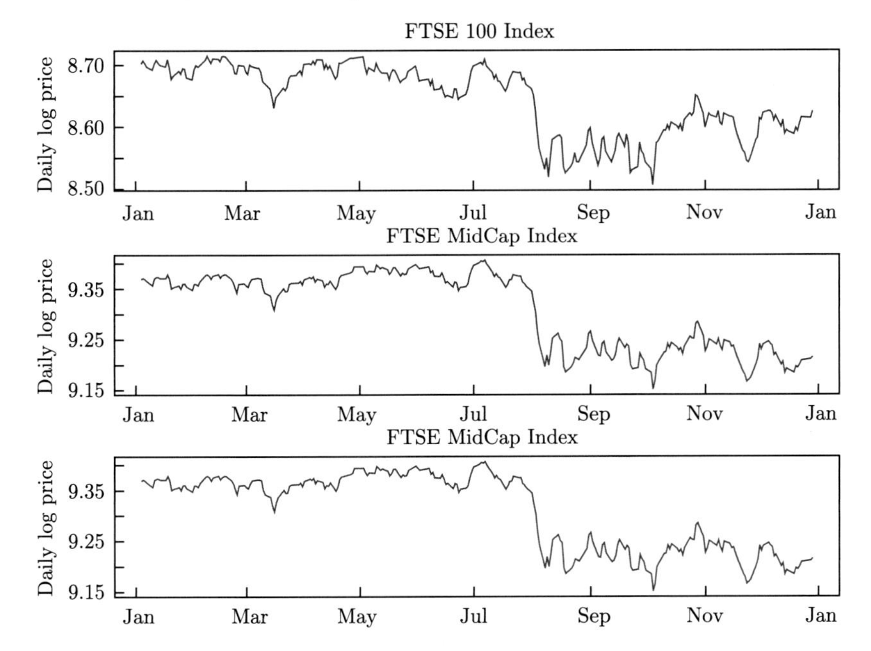

Figure 4.1 Daily log close prices of FTSE 100 index, FTSE MidCap index, and FTSE SmallCap index in 2011.

we only calculate the sample cross-correlation matrices for the log returns of the three daily price series. The sample cross-correlogram produced by `R` is presented in Figure 4.2. The three panels on the main diagonal are the auto-correlograms for the three return series: there exist hardly any significant autocorrelations for each of the three series. This is consistent with the martingale hypothesis in Section 1.3. For $i < j$, the cross-correlation coefficients $\widehat{\rho}_{ij}(k)$ are plotted against k in the (i, j)th panel for $k = 0, 1, 2, \ldots$, and in the (j, i)th panel for $k = 0, -1, -2, \ldots$. (Note $\widehat{\rho}_{ji}(k) = \widehat{\rho}_{ij}(-k)$.) All the panels off the main diagonal indicate clearly the very significant positive contemporaneous correlations among the three return series, as $\widehat{\rho}_{ij}(0)$ are greater than 0.8. Perhaps more interestingly, $\widehat{\rho}_{12}(-1)$, $\widehat{\rho}_{13}(-1)$, $\widehat{\rho}_{23}(-1)$ are all significantly positive, as they are above the significance bound $1.96/\sqrt{T}$, while $\widehat{\rho}_{12}(1)$, $\widehat{\rho}_{13}(1)$ are not significant. This indicates that, for example, the changes in FTSE 100 have 1-day delay effect in FTSE MidCap and FTSE SmallCap, but not vise versa. This may suggest using the changes in FTSE 100 to predict the changes next day in FTSE MidCap and FTSE SmallCap.

A word of caution: all statistical packages (including `R`) plot $\pm 1.96/\sqrt{T}$ (or $\pm 2/\sqrt{T}$) as significance bounds in ACF plots by default. They are the valid asymptotic bounds for sample autocorrelation coefficients $\widehat{\rho}_{jj}(k)$ when

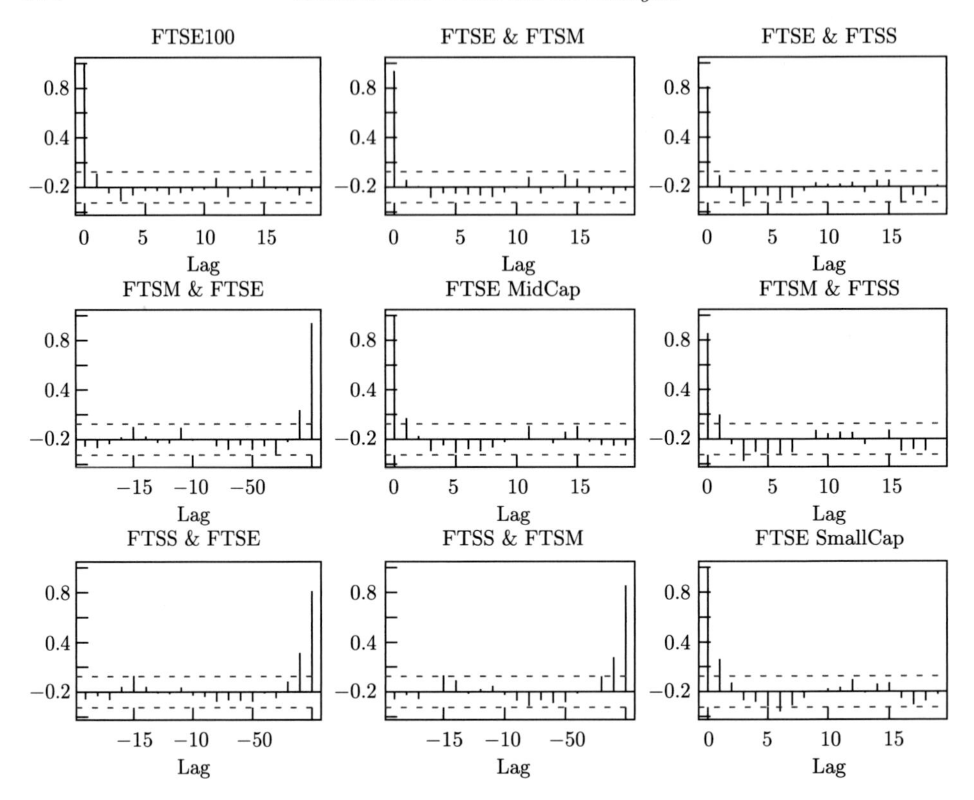

Figure 4.2 Sample cross-correlations of the log returns for the daily prices of FTSE 100 index, FTSE MidCap index, and FTSE SmallCap index in 2011.

the jth component series is white noise (see (2.32)). They are also valid asymptotic bounds for sample cross-correlation coefficients $\widehat{\rho}_{ij}(k)$ if (and only if) at least one of the ith and jth component series is white noise (Theorem 7.3.1 in Brockwell and Davis 1996). Hence when some component series of $\boldsymbol{X}_t$ are not white noise, the confidence bounds presented in ACF plots should be taken with extra care.

However for Figure 4.2, there are no evidences suggesting that any one of the three return series is not white noise. Therefore the bounds $\pm 1.96/\sqrt{T}$ can be taken as the thresholds for the significance of both sample autocorrelation coefficients and sample cross-correlation coefficients.

4.2 Vector autoregressive models

There exist no conceptual difficulties to extend univariate ARMA models to multivariate cases. However vector ARMA models without further restrictions in parameters are of little practical use, as they are often not identifiable, in addition to the issue of overparametrization mentioned earlier. See a simple example with $d = 2$ below.

Example 4.2 Consider a bivariate MA(1) process

$$\begin{pmatrix} X_{t1} \\ X_{t2} \end{pmatrix} = \begin{pmatrix} \varepsilon_{t1} \\ \varepsilon_{t2} \end{pmatrix} + \begin{pmatrix} 0 & 1 \\ 0 & 0 \end{pmatrix} \begin{pmatrix} \varepsilon_{t-1,1} \\ \varepsilon_{t-1,2} \end{pmatrix},$$

where $(\varepsilon_{t1}, \varepsilon_{t2})' \sim \text{WN}(\mathbf{0}, \boldsymbol{I}_2)$. Since $X_{t2} = \varepsilon_{t2}$, the above equation can be written as

$$\begin{pmatrix} X_{t1} \\ X_{t2} \end{pmatrix} = \begin{pmatrix} \varepsilon_{t1} \\ \varepsilon_{t2} \end{pmatrix} + \begin{pmatrix} 0 & 1 \\ 0 & 0 \end{pmatrix} \begin{pmatrix} X_{t-1,1} \\ X_{t-1,2} \end{pmatrix}.$$

Hence it is also a bivariate AR(1) process.

One way to avoid the identification issue is to use autoregressive models only. On the other hand, the inference for multivariate time series models is remarkably similar to that in univariate case. We illustrate in the context of vector autoregressive models below how the ideas from univariate time series can be taken for handling multivariate cases with matrix operations.

4.2.1 Stationarity

A d-vector autoregressive model with order p is of the form

$$\boldsymbol{X}_t = \boldsymbol{c} + \boldsymbol{A}_1 \boldsymbol{X}_{t-1} + \cdots + \boldsymbol{A}_p \boldsymbol{X}_{t-p} + \boldsymbol{\varepsilon}_t, \tag{4.9}$$

where $\boldsymbol{\varepsilon}_t \sim \text{WN}(\mathbf{0}, \boldsymbol{\Sigma}_\varepsilon)$, $\boldsymbol{c}$ is a $d \times 1$ constant vector, and $\boldsymbol{A}_1, \ldots, \boldsymbol{A}_p$ are $d \times d$ autoregressive coefficient matrices. We write $\boldsymbol{X}_t \sim \text{AR}(p)$ or $\text{VAR}(p)$ to emphasize the vector part of the AR model. In this model, each component of $\boldsymbol{X}_t$ is a linear combination of its lagged values and the lagged values of other components. Given $\boldsymbol{X}_{t-1}, \ldots, \boldsymbol{X}_{t-p}$, the concurrent linear relationship among the $X_{t1}, \ldots, X_{tp}$ is reflected by the non-zero off-diagonal elements in $\boldsymbol{\Sigma}_\varepsilon$.

Example 4.3 Let us look at a simple VAR(1) model first:

$$\boldsymbol{X}_t = \boldsymbol{A} \boldsymbol{X}_{t-1} + \boldsymbol{\varepsilon}_t. \tag{4.10}$$

We did not include a constant term in the model for the sake of simplicity. Replacing $\boldsymbol{X}_{t-k}$ by $\boldsymbol{\varepsilon}_{t-k} + \boldsymbol{A} \boldsymbol{X}_{t-k-1}$ in the above equation recursively for

$k = 1, 2, \ldots$, we obtain that for any integer $\ell > 0$,

$$\begin{aligned}
\boldsymbol{X}_t &= \boldsymbol{\varepsilon}_t + \boldsymbol{A}\boldsymbol{X}_{t-1} = \boldsymbol{\varepsilon}_t + \boldsymbol{A}\boldsymbol{\varepsilon}_{t-1} + \boldsymbol{A}^2\boldsymbol{X}_{t-2} \\
&= \boldsymbol{\varepsilon}_t + \boldsymbol{A}\boldsymbol{\varepsilon}_{t-1} + \cdots + \boldsymbol{A}^\ell \boldsymbol{\varepsilon}_{t-\ell} + \boldsymbol{A}^{\ell+1}\boldsymbol{X}_{t-\ell-1} \\
&= \boldsymbol{\varepsilon}_t + \sum_{j=1}^{\infty} \boldsymbol{A}^j \boldsymbol{\varepsilon}_{t-j},
\end{aligned}$$

provided $\boldsymbol{A}^\ell \to 0$ as $\ell \to \infty$. Then $\boldsymbol{X}_t$ is a vector MA(∞) model and, therefore, is weakly stationary. The necessary and sufficient condition for $\boldsymbol{A}^\ell \to 0$ (as $\ell \to \infty$) is that all eigenvalues of $\boldsymbol{A}$ are smaller than 1 in module, under which (4.10) defines a weakly stationary solution. If, in addition, $\boldsymbol{\varepsilon}_t$ are strictly stationary, such a weakly stationary process is also strictly stationary.

For weakly stationary $\boldsymbol{X}_t$ defined by (4.10), it is easy to see $E(\boldsymbol{X}_t) = 0$. By multiplying $\boldsymbol{X}'_{t-k}$, for $k = 0, 1, 2, \ldots$, from behind on both sides of (4.10) and then taking the expectation, we obtain

$$\boldsymbol{\Gamma}(0) = \boldsymbol{A}\boldsymbol{\Gamma}(1)' + \boldsymbol{\Sigma}_\varepsilon, \qquad \boldsymbol{\Gamma}(k) = \boldsymbol{A}\boldsymbol{\Gamma}(k-1), \;\; k = 1, 2, \ldots. \tag{4.11}$$

The second equation of the above is the Yule–Walker equation for vector AR(1). Hence the model parameters $(\boldsymbol{A}, \boldsymbol{\Sigma}_\varepsilon)$ and the cross-covariance function $\boldsymbol{\Gamma}(\cdot)$ are uniquely determined by each other as in (4.11).

Now let us consider a simple case with $d = 2$. The implied model for the first component series under (4.10) is

$$X_{t1} = a_{11}X_{t-1,1} + a_{12}X_{t-1,2} + \varepsilon_{t1},$$

where a_{ij} is the (i,j)th element of $\boldsymbol{A}$. By substituting $X_{t-1,2}$ using the second component model in (4.10), X_{t1} is then a linear function of its two lagged values plus other terms. Hence the PACF for X_{t1} will not cut off at lag 1, in contrast to the univariate cases (see §2.2.2). On the other hand, if we set $Z_t \equiv X_{t-1,2}$ as an input and $Y_t \equiv X_{t1}$ as an output, this is a simple input-output system (with feedback when $a_{11} \neq 0$), provided that $a_{21} = 0$ and $\boldsymbol{\Sigma}_\varepsilon$ is diagonal.

In general the capacity of vector autoregressive model is large. By properly regularizing the autoregressive coefficient matrices, it may include, for example, transfer function models for dynamic input-and-output systems.

For VAR(p) model (4.9), if

$$\left|\boldsymbol{I}_d - \boldsymbol{A}_1 x - \cdots - \boldsymbol{A}_p x^p\right| \neq 0 \;\; \text{for all complex } x \text{ with } |x| \leqslant 1, \tag{4.12}$$

there exists a weakly stationary solution $\{\boldsymbol{X}_t\}$ satisfying (4.9), for which

$$E\boldsymbol{X}_t = (\boldsymbol{I}_d - \boldsymbol{A}_1 - \cdots - \boldsymbol{A}_p)^{-1}\boldsymbol{c}, \tag{4.13}$$

and the Yule–Walker equation admits the form

$$\boldsymbol{\Gamma}(k) = \boldsymbol{A}_1\boldsymbol{\Gamma}(k-1) + \cdots + \boldsymbol{A}_p\boldsymbol{\Gamma}(k-p), \qquad k = 1, 2, \ldots. \tag{4.14}$$

In addition,

$$\boldsymbol{\Gamma}(0) = \boldsymbol{A}_1\boldsymbol{\Gamma}(1)' + \cdots + \boldsymbol{A}_p\boldsymbol{\Gamma}(p)' + \boldsymbol{\Sigma}_\varepsilon.$$

4.2.2 Parameter estimation

Estimation for vector autoregressive models is similar to that for univariate models: it can be done by one of the three methods: least squares method, Yule–Walker estimation (i.e. method of moments estimation), and quasi maximum likelihood estimation based on, for example, Gaussian innovation distribution. The estimators obtained by all the three methods are asymptotically normal and also asymptotically equivalent under some mild conditions. The likelihood functions can be calculated via either prewhitenning or a Kalman filter. To avoid cumbersome notation, we illustrate below those methods for a vector AR(2) model.

Let $\boldsymbol{X}_1, \ldots, \boldsymbol{X}_T$ be observations from

$$\boldsymbol{X}_t = \boldsymbol{c} + \boldsymbol{A}_1\boldsymbol{X}_{t-1} + \boldsymbol{A}_2\boldsymbol{X}_{t-2} + \boldsymbol{\varepsilon}_t, \tag{4.15}$$

where $\boldsymbol{\varepsilon}_t \sim \mathrm{WN}(\boldsymbol{0}, \boldsymbol{\Sigma}_\varepsilon)$. The goal is to estimate the parameters $\boldsymbol{c}, \boldsymbol{A}_1, \boldsymbol{A}_2, \boldsymbol{\Sigma}_\varepsilon$.

Least squares estimation

Perhaps the least squares estimation (LSE) is the easiest way in this context. Note that the parameters in $\boldsymbol{c}, \boldsymbol{A}_1, \boldsymbol{A}_2$ can be decoupled row by row. For example, the equation for the first components under model (4.15) admits the form

$$X_{t1} = c_1 + \boldsymbol{X}_{t-1}'\boldsymbol{a}_1^{(1)} + \boldsymbol{X}_{t-2}'\boldsymbol{a}_1^{(2)} + \varepsilon_{t1},$$

where c_1 is the first component of $\boldsymbol{c}$, and $\boldsymbol{a}_i^{(j)}$ is the ith row vector of the matrix $\boldsymbol{A}_j$. Note that vectors are always in column in our notation, i.e. a vector is a matrix with only one column. Thus, the parameter associated with first time series can be estimated by the least-squares method, which minimizes the squared fitting errors with respect to $c_1, \boldsymbol{a}_1^{(1)}$ and $\boldsymbol{a}_1^{(2)}$:

$$\sum_{t=3}^{T}(X_{t1} - c_1 - \boldsymbol{X}_{t-1}'\boldsymbol{a}_1^{(1)} - \boldsymbol{X}_{t-2}'\boldsymbol{a}_1^{(2)})^2.$$

Applying the above estimation method to each component of (4.15), we obtain the estimators $\widehat{\boldsymbol{c}}, \widehat{\boldsymbol{A}}_1, \widehat{\boldsymbol{A}}_2$. Then the estimator for $\boldsymbol{\Sigma}_\varepsilon$ can be defined as

$$\widehat{\boldsymbol{\Sigma}}_\varepsilon = \frac{1}{T-2}\sum_{t=3}^{T}\widehat{\boldsymbol{\varepsilon}}_t\widehat{\boldsymbol{\varepsilon}}_t', \tag{4.16}$$

where

$$\widehat{\boldsymbol{\varepsilon}}_t = \boldsymbol{X}_t - \widehat{\boldsymbol{c}} - \widehat{\boldsymbol{A}}_1\boldsymbol{X}_{t-1} - \widehat{\boldsymbol{A}}_2\boldsymbol{X}_{t-2}.$$

Yule–Walker estimation

The Yule–Walker estimation (YWE) can be obtained by solving the equations

$$\widehat{\boldsymbol{\Gamma}}(1) = \boldsymbol{A}_1\widehat{\boldsymbol{\Gamma}}(0) + \boldsymbol{A}_2\widehat{\boldsymbol{\Gamma}}(1)', \qquad \widehat{\boldsymbol{\Gamma}}(2) = \boldsymbol{A}_1\widehat{\boldsymbol{\Gamma}}(1) + \boldsymbol{A}_2\widehat{\boldsymbol{\Gamma}}(0), \tag{4.17}$$

which replaces the cross covariance matrices in (4.14) by their sample version. The equations (4.17) can be written as

$$\left(\boldsymbol{A}_1, \boldsymbol{A}_2\right)\begin{pmatrix}\widehat{\boldsymbol{\Gamma}}(0) & \widehat{\boldsymbol{\Gamma}}(1)\\ \widehat{\boldsymbol{\Gamma}}(1)' & \widehat{\boldsymbol{\Gamma}}(0)\end{pmatrix} = \left(\widehat{\boldsymbol{\Gamma}}(1), \widehat{\boldsymbol{\Gamma}}(2)\right).$$

The solution can be found explicitly and is given by

$$\left(\widehat{\boldsymbol{A}}_1, \widehat{\boldsymbol{A}}_2\right) = \left(\widehat{\boldsymbol{\Gamma}}(1), \widehat{\boldsymbol{\Gamma}}(2)\right)\begin{pmatrix}\widehat{\boldsymbol{\Gamma}}(0) & \widehat{\boldsymbol{\Gamma}}(1)\\ \widehat{\boldsymbol{\Gamma}}(1)' & \widehat{\boldsymbol{\Gamma}}(0)\end{pmatrix}^{-1}. \tag{4.18}$$

The YWE (and also the LSE) for $\boldsymbol{c}$ is

$$\widehat{\boldsymbol{c}} = \bar{\boldsymbol{X}}_{3,T} - \widehat{\boldsymbol{A}}_1\bar{\boldsymbol{X}}_{2,T-1} - \widehat{\boldsymbol{A}}_2\bar{\boldsymbol{X}}_{1,T-2}, \tag{4.19}$$

where $\bar{\boldsymbol{X}}_{i,j} = \frac{1}{j-i+1}\sum_{i\leqslant t\leqslant j}\boldsymbol{X}_t$. The negligible difference between the LSE and the YWE is caused by using $\bar{\boldsymbol{X}} = \bar{\boldsymbol{X}}_{1,n}$ instead of $\bar{\boldsymbol{X}}_{i,j}$ in calculating the sample cross covariance functions in (4.7). In practice and almost all statistical packages, all $\bar{\boldsymbol{X}}_{3,T}, \bar{\boldsymbol{X}}_{2,T-1}, \bar{\boldsymbol{X}}_{1,T-2}$ are replaced by $\bar{\boldsymbol{X}}$. The resulting LSE and YWE are then indeed the same.

Maximum likelihood estimation

Assuming $\boldsymbol{\varepsilon}_t \sim N(0, \boldsymbol{\Sigma}_\varepsilon)$ in (4.15), then the conditional distribution of $\boldsymbol{X}_t$ given $\boldsymbol{X}_{t-1}$ and $\boldsymbol{X}_{t-2}$ is

$$N(\boldsymbol{X}_t - \boldsymbol{c} - \boldsymbol{A}_1\boldsymbol{X}_{t-1} - \boldsymbol{A}_2\boldsymbol{X}_{t-2}, \boldsymbol{\Sigma}_\varepsilon).$$

The density of such a multivariate normal distribution is given by

$$\frac{1}{(2\pi)^{d/2}|\boldsymbol{\Sigma}_\varepsilon|^{1/2}} \exp\big(-\boldsymbol{\varepsilon}_t^{\mathrm{T}}\boldsymbol{\Sigma}_\varepsilon^{-1}\boldsymbol{\varepsilon}_t\big),$$

where $\boldsymbol{\varepsilon}_t = \boldsymbol{X}_t - \boldsymbol{c} - \boldsymbol{A}_1\boldsymbol{X}_{t-1} - \boldsymbol{A}_2\boldsymbol{X}_{t-2}$. Now, by the chain rule, the joint density is

$$\begin{aligned} f(\boldsymbol{X}_1, \ldots, \boldsymbol{X}_T) &= f_1(\boldsymbol{X}_1, \boldsymbol{X}_2) f_2(\boldsymbol{X}_3, \ldots, \boldsymbol{X}_T|\boldsymbol{X}_1, \boldsymbol{X}_2) \\ &= f_1(\boldsymbol{X}_1, \boldsymbol{X}_2) \prod_{t=3}^{T} f(\boldsymbol{X}_t|\boldsymbol{X}_1, \ldots, \boldsymbol{X}_{t-2}). \end{aligned}$$

Using the multivariate density formula above, the product can be written as

$$\left\{\frac{1}{(2\pi)^{d/2}|\boldsymbol{\Sigma}_\varepsilon|^{1/2}}\right\}^{T-2} \exp\left(-\sum_{t=3}^{T} \boldsymbol{\varepsilon}_t^{\mathrm{T}}\boldsymbol{\Sigma}_\varepsilon^{-1}\boldsymbol{\varepsilon}_t\right).$$

The term $f_1(\boldsymbol{X}_1, \boldsymbol{X}_2)$ in the likelihood function has a negligible impact on the likelihood function. Ignoring it yields the conditional density function of $(\boldsymbol{X}_3, \ldots, \boldsymbol{X}_T)$ given $\boldsymbol{X}_1, \boldsymbol{X}_2$. It becomes the conditional likelihood function, when it is regarded as a function of the parameter (see Section 2.5.4). Let $\boldsymbol{\theta}$ denote the parameters $\boldsymbol{c}$, $\boldsymbol{A}_1, \boldsymbol{A}_2$ amd $\boldsymbol{\Sigma}_\varepsilon$. Taking the logarithm of the above conditional likelihood function and noticing that

$$\sum_{t=3}^{T} \boldsymbol{\varepsilon}_t^{\mathrm{T}}\boldsymbol{\Sigma}_\varepsilon^{-1}\boldsymbol{\varepsilon}_t = \sum_{t=3}^{T} \mathrm{tr}(\boldsymbol{\Sigma}_\varepsilon^{-1}\boldsymbol{\varepsilon}_t\boldsymbol{\varepsilon}_t^{\mathrm{T}}) = \mathrm{tr}\left(\boldsymbol{\Sigma}_\varepsilon^{-1}\sum_{t=3}^{T}\boldsymbol{\varepsilon}_t\boldsymbol{\varepsilon}_t^{\mathrm{T}}\right),$$

the log conditional likelihood function (given $\boldsymbol{X}_1, \boldsymbol{X}_2$) is of the form

$$\ell(\boldsymbol{\theta}) = -\frac{T-2}{2}\log|\boldsymbol{\Sigma}_\varepsilon| - \frac{1}{2}\mathrm{tr}\{\boldsymbol{\Sigma}_\varepsilon^{-1}\boldsymbol{M}(\boldsymbol{c}, \boldsymbol{A}_1, \boldsymbol{A}_2)\}, \tag{4.20}$$

where we ignore the constant term and substitute the definition of $\boldsymbol{\varepsilon}_t$ to obtain

$$\begin{aligned} \boldsymbol{M}(\boldsymbol{c}, \boldsymbol{A}_1, \boldsymbol{A}_2) = \sum_{t=3}^{T}\boldsymbol{\varepsilon}_t\boldsymbol{\varepsilon}_t^{\mathrm{T}} = \sum_{t=3}^{T}&(\boldsymbol{X}_t - \boldsymbol{c} - \boldsymbol{A}_1\boldsymbol{X}_{t-1} - \boldsymbol{A}_2\boldsymbol{X}_{t-2}) \\ &(\boldsymbol{X}_t - \boldsymbol{c} - \boldsymbol{A}_1\boldsymbol{X}_{t-1} - \boldsymbol{A}_2\boldsymbol{X}_{t-2})'. \end{aligned}$$

It can be shown that for $\widehat{\boldsymbol{A}}_1, \widehat{\boldsymbol{A}}_2$ and $\widehat{\boldsymbol{c}}$ given by (4.18) and (4.19),

$$\boldsymbol{M}(\boldsymbol{c}, \boldsymbol{A}_1, \boldsymbol{A}_2) - \boldsymbol{M}(\widehat{\boldsymbol{c}}, \widehat{\boldsymbol{A}}_1, \widehat{\boldsymbol{A}}_2)$$

is a non-negative definite matrix, if we replace all the $\bar{\boldsymbol{X}}_{i,j}$ by $\bar{\boldsymbol{X}}$, as we did before. Hence the conditional MLE for $(\boldsymbol{c}, \boldsymbol{A}_1, \boldsymbol{A}_2)$ are the same as the LSE and the YWE; see Exercise 4(ii). By substituting those estimators into the likelihood function (4.20), it follows from the standard multivariate normal theory (Anderson, 2003) that the conditional MLE for $\boldsymbol{\Sigma}_\varepsilon$ is the same as $\widehat{\boldsymbol{\Sigma}}_\varepsilon$ given in (4.16). The actual derivation is somewhat complicated and omitted. The resulting maximum log-likelihood value is the log-likelihood function (4.20), evaluated at its MLE, and is given by

$$\max_{\boldsymbol{\theta}} \ell(\boldsymbol{\theta}) = \ell(\hat{\boldsymbol{\theta}}) = -\frac{T-2}{2} \log |\hat{\boldsymbol{\Sigma}}_\varepsilon| - \frac{1}{2} d. \tag{4.21}$$

For vector AR models, the LSE, the YWE and the conditional MLE are all the same. When T is not large and/or d is large, a full (quasi-)MLE often makes more efficient use of available data. Under the assumption that $\boldsymbol{\varepsilon}_t \sim N(\mathbf{0}, \boldsymbol{\Sigma}_\varepsilon)$, $\boldsymbol{X}_1, \ldots, \boldsymbol{X}_T$ are jointly normal, and its joint density function can be calculated by, for example, either the innovation algorithm via prewhitenning (§3.2 of Fan and Yao 2003), or a Kalman filter (§3.5.2 before). The resulting MLE is different from the three estimators stated above, although they are all asymptotically equivalent.

4.2.3 Model selection and diagnostics

In fitting an AR model, we need to determine the order p. This can be done either via hypothesis testing or in terms of some information criteria.

Determining p by hypothesis tests

To determine if a model with order $p > 1$ is sufficiently large, we may test the hypotheses:

$$\begin{aligned} H_0 &: \boldsymbol{X}_t = \boldsymbol{c} + \boldsymbol{A}_1 \boldsymbol{X}_{t-1} + \cdots + \boldsymbol{A}_p \boldsymbol{X}_{t-p} + \boldsymbol{\varepsilon}_t, \quad \text{against} \\ H_1 &: \boldsymbol{X}_t = \boldsymbol{c} + \boldsymbol{A}_1 \boldsymbol{X}_{t-1} + \cdots + \boldsymbol{A}_{p+1} \boldsymbol{X}_{t-p-1} + \boldsymbol{\varepsilon}_t, \end{aligned}$$

where $\boldsymbol{\varepsilon}_t \sim \mathrm{WN}(0, \boldsymbol{\Sigma}_\varepsilon)$. Under the AR($p$) model above, as a generalization of (4.16), define an estimator for $\boldsymbol{\Sigma}_\varepsilon$ as

$$\hat{\boldsymbol{\Sigma}}_\varepsilon(p) = \frac{1}{T-2p-1} \sum_{t=p+1}^{T} \left(\boldsymbol{X}_t - \widehat{\boldsymbol{c}} - \sum_{j=1}^{p} \widehat{\boldsymbol{A}}_j \boldsymbol{X}_{t-j} \right) \left(\boldsymbol{X}_t - \widehat{\boldsymbol{c}} - \sum_{j=1}^{p} \widehat{\boldsymbol{A}}_j \boldsymbol{X}_{t-j} \right)',$$

where $\widehat{\boldsymbol{c}}$ and $\widehat{\boldsymbol{A}}_j$ are the LSE for $\boldsymbol{c}$ and $\boldsymbol{A}_j$, respectively. Using the maximum conditional likelihood function (4.21), it is easy to see that twice the conditional likelihood ratio test is $(T-2)\log\left(|\hat{\boldsymbol{\Sigma}}_\varepsilon(p)|/|\hat{\boldsymbol{\Sigma}}_\varepsilon(p+1)|\right)$ (See also section 2.5.4). Adjusting the constant multiplier to make the null distribution more accurate at finite sample, the test statistic for the hypotheses above is defined as

$$S = (T-d-p-1/2)\log\left(|\hat{\boldsymbol{\Sigma}}_\varepsilon(p)|/|\hat{\boldsymbol{\Sigma}}_\varepsilon(p+1)|\right).$$

We reject H_0 (i.e. AR(p)) in favor of H_1 (i.e. AR($p+1$)) for large values of S. Tiao and Box (1981) show that S is asymptotically χ^2-distributed with d^2 degrees of freedom under H_0.

Information criteria

Alternatively we may use an information criterion to select the order p. In the context of vector AR fitting, The model parameters are pd^2 for autoregressive matrices and d for vector c. The AIC, BIC (also called the Schwartz criterion) and HQIC(Hannan and Quinn 1979) are defined respectively as

$$\begin{aligned}
\mathrm{AIC}(p) &= \log(|\widehat{\boldsymbol{\Sigma}}_\varepsilon(p)|) + 2d^2p/T,\\
\mathrm{BIC}(p) &= \log(|\widehat{\boldsymbol{\Sigma}}_\varepsilon(p)|) + d^2p\log(T)/T,\\
\mathrm{HQIC}(p) &= \log(|\widehat{\boldsymbol{\Sigma}}_\varepsilon(p)|) + 2d^2p\log(\log T)/T,
\end{aligned}$$

where we drop a term like $2d/T$ in AIC(p), as it does not depend on p. We may choose p such that one of those criteria is minimized. Let $\widehat{p}$(AIC), $\widehat{p}$(BIC), $\widehat{p}$(HQIC) denote, respectively, the value of p selected by AIC, BIC, HQIC. Then as long as $T \geqslant 16$, it holds that with high probability

$$\widehat{p}(\mathrm{AIC}) \;\geqslant\; \widehat{p}(\mathrm{HQIC}) \;\geqslant\; \widehat{p}(\mathrm{BIC}).$$

It can be proved that both BIC and HQIC provide consistent order determination, while AIC overestimates p with a positive probability, under the assumption that the true model is indeed AR(p) with a finite p. See §4.3 of Lütkepohl (2006). In practice it is alway advisable to examine several models selected by different criteria, or also the second best or the third best models selected by one criterion, and use the one which makes practical sense for the task in hand.

Portmanteau tests

The most frequently used method for model diagnostic is to examine if the residuals

$$\widehat{\boldsymbol{\varepsilon}}_t = \boldsymbol{X}_t - \widehat{\boldsymbol{c}} - \sum_{j=1}^{p} \widehat{\boldsymbol{A}}_j \boldsymbol{X}_{t-j}, \quad t = p+1, \ldots, T$$

behave like vector white noise. Let $\widehat{\boldsymbol{\Gamma}}(k)$ denote the sample cross-covariance matrix of the residuals at lag k. Then the portmanteau test statistic is defined as

$$Q_m = T^2 \sum_{j=1}^{m} \frac{1}{T-j} \mathrm{tr}\{\widehat{\boldsymbol{\Gamma}}(k)' \widehat{\boldsymbol{\Gamma}}(0)^{-1} \widehat{\boldsymbol{\Gamma}}(k) \widehat{\boldsymbol{\Gamma}}(0)^{-1}\}, \tag{4.22}$$

where $m > p$ is an integer. If the true model is indeed AR(p) with $\boldsymbol{\varepsilon}_t \sim \mathrm{IID}(0, \boldsymbol{\Sigma}_\varepsilon)$, Q_m is asymptotically χ^2-distributed with degrees of freedom $d^2(m-p)$. This is an extension of the Ljung–Box test in (1.15) and in Section 2.6.2. We compute the P-value by computing the left tail probability of $\chi^2_{d^2(m-p)}$-distribution at the observed test statistic Q_m. This is equivalent to rejecting the hypothesis of AR(p) at the significance level $\alpha \in (0, 1)$ if Q_m is greater than the top-α quantile of the χ^2-distribution with $d^2(m-p)$ degrees of freedom.

Unfortunately the above portmanteau test may be sensitive to the choice of m. In practice, we often carry out the test with different values of m. In addition, it is not a very powerful test when d is large.

4.2.4 Illustration with real data

For the time being, R-function `arima` only works with univariate time series, the function `ar` is too simple to produce satisfactory analysis for vector time series. The illustration below uses the functions contained in an R-package `vars` which is available from the R-CRAN project.

Let `X` be the matrix with 3 columns containing the log daily returns in 2011 of FTSE 100, FTSE MidCap, FTSE SmallCap. To fit a vector AR(p) model with p determined by BIC (i.e. the `Schwartz Criterion`), we issue the following commands and obtain the summary of fitted results:

```
> FTSEvar = VAR(X, lag.max=3, ic="SC"); summary(FTSEvar)

Estimation results for equation FTSE100:
========================================
FTSE100 = FTSE100.l1 + FTSE.MidCap.l1 + FTSE.SmallCap.l1
          + const
```

```
                  Estimate Std. Error t value Pr(>|t|)
FTSE100.l1         0.4623839  0.1822584   2.537   0.0118 *
FTSE.MidCap.l1    -0.5277461  0.2154000  -2.450   0.0150 *
FTSE.SmallCap.l1   0.2438484  0.1981300   1.231   0.2196
const             -0.0003319  0.0008497  -0.391   0.6964
---
Signif. codes:  0 *** 0.001 ** 0.01 * 0.05 . 0.1   1

Residual standard error: 0.01329 on 244 degrees of freedom
Multiple R-Squared: 0.03624,    Adjusted R-squared: 0.02439
F-statistic: 3.058 on 3 and 244 DF,  p-value: 0.02896

Estimation results for equation FTSE.MidCap:
============================================
FTSE.MidCap = FTSE100.l1 + FTSE.MidCap.l1 + FTSE.SmallCap.l1
              + const
                  Estimate Std. Error t value Pr(>|t|)
FTSE100.l1         0.6041161  0.1689675   3.575 0.000422 ***
FTSE.MidCap.l1    -0.5709383  0.1996923  -2.859 0.004616 **
FTSE.SmallCap.l1   0.2651595  0.1836817   1.444 0.150139
const             -0.0005873  0.0007877  -0.746 0.456663
---
Residual standard error: 0.01232 on 244 degrees of freedom
Multiple R-Squared: 0.08665,    Adjusted R-squared: 0.07542
F-statistic: 7.716 on 3 and 244 DF,  p-value: 6.044e-05


Estimation results for equation FTSE.SmallCap:
==============================================
FTSE.SmallCap = FTSE100.l1 + FTSE.MidCap.l1 + FTSE.SmallCap.l1
                + const
                  Estimate Std. Error t value Pr(>|t|)
FTSE100.l1         0.2692036  0.1076958   2.500   0.0131 *
FTSE.MidCap.l1    -0.1276323  0.1272791  -1.003   0.3170
FTSE.SmallCap.l1   0.0778511  0.1170743   0.665   0.5067
const             -0.0006338  0.0005021  -1.262   0.2080
---
Residual standard error: 0.007855 on 244 degrees of freedom
Multiple R-Squared: 0.1037, Adjusted R-squared: 0.09271
F-statistic: 9.414 on 3 and 244 DF,  p-value: 6.557e-06

Covariance matrix of residuals:
                FTSE100 FTSE.MidCap FTSE.SmallCap
FTSE100       1.767e-04   1.545e-04     8.617e-05
FTSE.MidCap   1.545e-04   1.519e-04     8.237e-05
FTSE.SmallCap 8.617e-05   8.237e-05     6.170e-05

Correlation matrix of residuals:
              FTSE100 FTSE.MidCap FTSE.SmallCap
FTSE100        1.0000      0.9430        0.8252
FTSE.MidCap    0.9430      1.0000        0.8508
FTSE.SmallCap  0.8252      0.8508        1.0000
```

The selected order by BIC (also by HQIC) is $p = 1$. All three fitted models are statistically significant with the P-value 0.029 for FTSE 100, and practically 0 for the other two models. The predictive power is relatively low, as the multiple squared correlation is 3.62% for FTSE 100, 8.67% for FTSE MidCap and 10.37% for FTSE SmallCap. These multiple R^2's, while small, are not too small to be useful in predicting financial returns. The larger percentages for FTSE MidCap and FTSE SmallCap models are due to the contribution of one-day lagged value of FTSE 100 which is the most significant term in the models. This indicates the lagged effect from FTSE 100 on the other two indices. To refit the model by leaving out insignificant terms, run

```
FTSEvarR = restrict(FTSEvar)
```

which leads to

$$\begin{cases} X_{t1} = 0.469X_{t-1,1} - 0.398X_{t-1,2} + \varepsilon_{t1}, \\ X_{t2} = 0.608X_{t-1,1} - 0.427X_{t-1,2} + \varepsilon_{t2}, \\ X_{t3} = 0.195X_{t-1,1} + \varepsilon_{t3}. \end{cases} \tag{4.23}$$

where X_{t1}, X_{t2}, X_{t3} denote the log returns on day t of, respectively, FTSE 100, FTSE MidCap, and FTSE SmallCap. The significance levels and the multiple squared correlation coefficients of these reduced models hardly changed from the full models reported. Figure 4.3 presents the cross-correlations of the residuals from the above fitted vector AR(1) model, produced by `acf(residuals(FTSEvarR))`. Comparing with Figure 4.2, there are hardly any significant correlations at non-zero lags. More diagnostic plots can be produced by calling the following R-functions:

```
FTSEdiag = serial.test(FTSEvarR); plot(FTSEdiag)
```

To perform the portmanteau test (4.22) for the residual with, for example, $m = 6$, run

```
serial.test(FTSEvar, lags.pt=6, type ="PT.adjusted").
```

Vector AR models can be used to forecast future values in the same manner as univariate AR models. For example, we may forecast the next 15 returns of those three FTSE indices by using the R function `predict` as follow, where we specify the predictive boundaries with the coverage probability 0.95.

```
FTSEpred = predict(FTSEvarR, n.ahead=15, ci=0.95)
plot(FTSEpred)
```

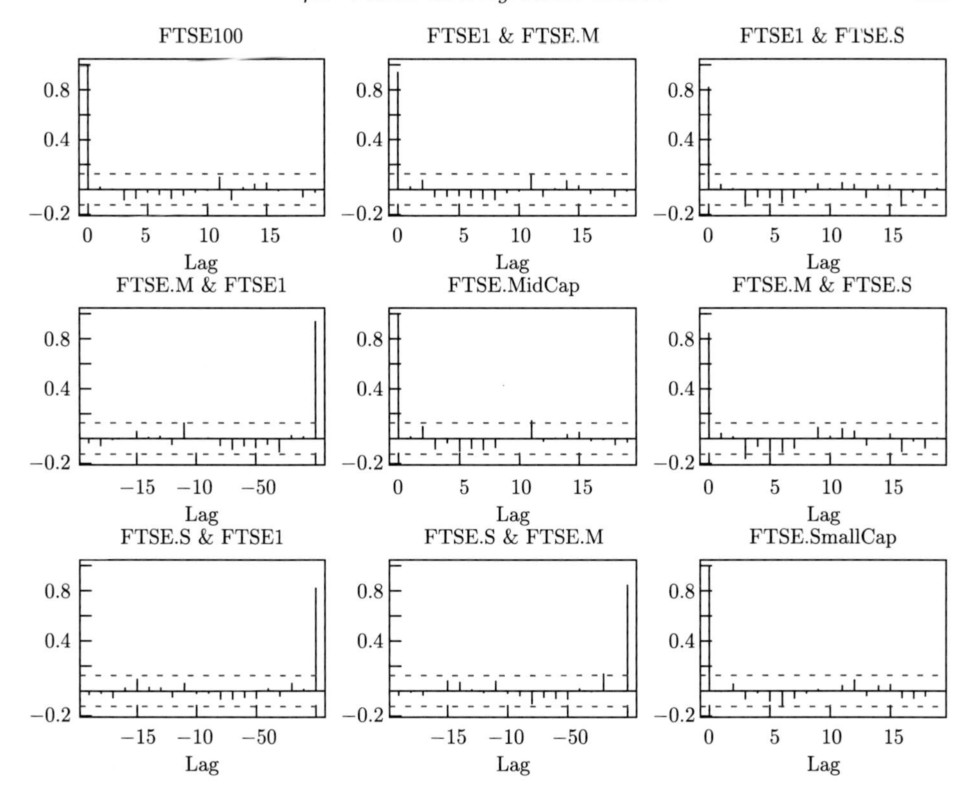

Figure 4.3 Sample cross-correlations of the residuals resulting from the fitted vector AR(1) model (4.23) for the log returns of the daily prices of FTSE 100 index, FTSE MidCap index, and FTSE SmallCap index in 2011-2012.

The produced Figure 4.4 plots the original return series together with the predicted values and the pointwise predictive intervals with the 0.95 coverage probability.

4.2.5 Granger causality

The *Granger causality* (Granger 1969) is an important concept in econometrics. Let Z_t and Y_t be two univariate time series. Let $\mathcal{L}(U|V)$ denote the conditional distribution of U given V.

Time series Z_t is said to Granger cause time series Y_t if

$$\mathcal{L}(Y_t|Y_{t-1}, Z_{t-1}, Y_{t-2}, Z_{t-2}, \ldots) \neq \mathcal{L}(Y_t|Y_{t-1}, Y_{t-2}, \ldots). \tag{4.24}$$

Obviously the Granger causality is present if the changes in the lagged values

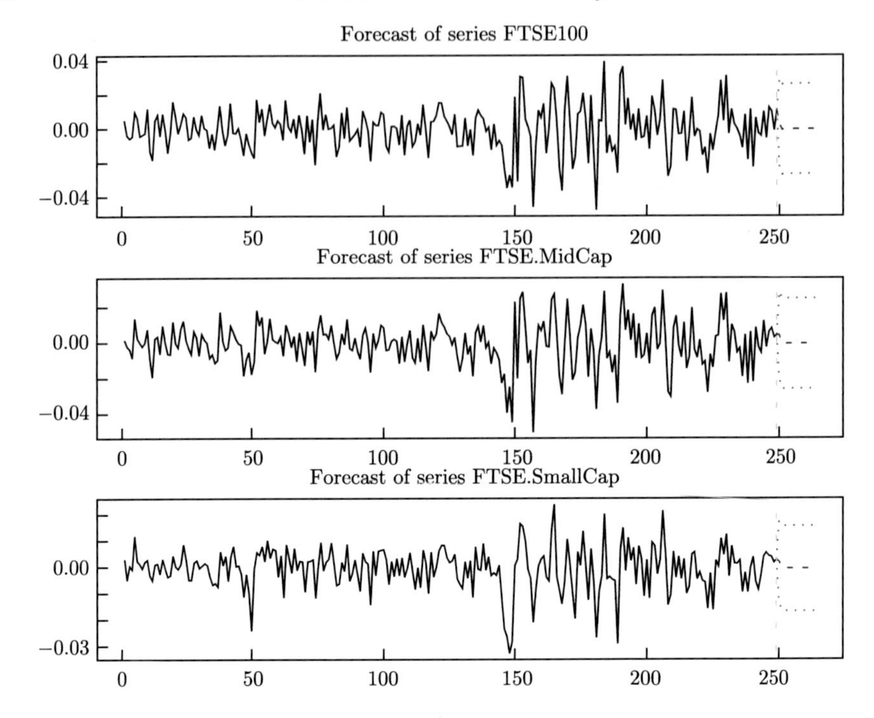

Figure 4.4 Daily returns of FTSE 100, FTSE MidCap and FTSE Small Cap in 2011, together with predicted returns for next 15 trading days and their predictive boundaries with the 0.95 coverage probability.

of Z_t indeed constitute some exogenous causes for the changes in Y_t. However the converse is not always true, as (4.24) only implies the dependence between Y_t and Z_{t-k} (for some $k \geqslant 1$), conditioning on $Y_{t-1}, Y_{t-2}, \ldots$. It does not say which causes which. In fact the changes in both Y_t and Z_{t-k} may be caused by a common latent process which is not present in the system.

While definition (4.24) is general, practical applications often narrow down to the so-called the Granger causality in mean, for which condition (4.24) is replaced by

$$E(Y_t|Y_{t-1}, Z_{t-1}, Y_{t-2}, Z_{t-2}, \ldots) \neq E(Y_t|Y_{t-1}, Y_{t-2}, \ldots). \tag{4.25}$$

Obviously the Granger causality in variance may be formulated in a similar manner.

The causality (4.25) can easily be verified for the VAR models. Let $\boldsymbol{X}_t = (Z_t, Y_t)'$, and assume $\boldsymbol{X}_t \sim \text{AR}(p)$, i.e.

$$\boldsymbol{X}_t = \boldsymbol{c} + \sum_{j=1}^{p} \boldsymbol{A}_j \boldsymbol{X}_{t-j} + \boldsymbol{\varepsilon}_t, \tag{4.26}$$

where $\boldsymbol{\varepsilon}_t \sim \mathrm{WN}(0, \boldsymbol{\Sigma}_\varepsilon)$. Under the additional assumption that $\boldsymbol{X}_t$ is a Gaussian process, the conditional expectations in (4.25) are linear in the variables that are conditioned upon. The first conditional expectation is determined by the second component of $\boldsymbol{c} + \sum_{j=1}^p \boldsymbol{A}_j \boldsymbol{X}_{t-j}$, which depends on some lags of $Z's$ unless $a_{21}^{(1)} = \cdots = a_{21}^{(p)} = 0$, where $a_{21}^{(k)}$ denotes the $(2,1)$-element of the coefficient matrix $\boldsymbol{A}_k$ in (4.26). Therefore, (4.25) is equivalent to the condition that at least one of $a_{21}^{(k)} \neq 0$ for $1 \leqslant k \leqslant p$. The so-called Granger causality test is to test the hypothesis

$$H_0 : a_{21}^{(1)} = \cdots = a_{21}^{(p)} = 0$$

for model (4.26). When H_0 is rejected, Z_t is regarded as Granger causing Y_t.

To test hypothesis H_0, we compare the goodness of fit with and without using the time series $\{Z_t\}$. Let

$$\mathrm{RSS} = \sum_{t=p+1}^{T} \left\| \boldsymbol{X}_t - \widehat{c} - \sum_{j=1}^{p} \widehat{\boldsymbol{A}}_j \boldsymbol{X}_{t-j} \right\|^2$$

be the residual sum squares under the full model (4.26), with $\widehat{c}, \widehat{\boldsymbol{A}}_1, \ldots, \widehat{\boldsymbol{A}}_p$ being the LSE for the parameters. This measures the quality of fit of the VAR model (4.26), with use of the time series $\{Z_t\}$. Similarly, let RSS_r be the residual sum squares under the restricted model, defined in a similar manner but with the estimated parameters subject to the constraints specified in H_0. In other words, it is the sum of RSS for fitting AR(p) model to Y_t without using Z_t and that of fitting AR(p) model to Z_t using both lagged values of Y_t and Z_t. The contribution of Z_t to Y_t is then measured by the standard F-test statistic:

$$F = \frac{(\mathrm{RSS}_r - \mathrm{RSS})/p}{\mathrm{RSS}/(2T - 4p - 2)}.$$

Under H_0, pF is asymptotically χ^2-distributed with p degrees of freedom. Alternatively F may be approximately regarded as obeying F-distribution with $(p, 2T - 4p - 2)$ degrees of freedom. The latter typically provides a more accurate approximation at the finite sample. See §11.2 of Hamilton (1994).

The maximum likelihood ratio test (2.56) can also be employed to this testing problem. The log-likelihood function of the VAR(p) model is given by (4.20). As in Section 4.2.3, let $\hat{\boldsymbol{\Sigma}}_\varepsilon$ be the estimated covariance matrix of $\boldsymbol{\varepsilon}_t$ under the full model, using both the time series $\{Y_t\}$ and $\{Z_t\}$. This measures the quality of fit of the VAR model (4.26), with use of the time series $\{Z_t\}$. Similarly, let $\hat{\boldsymbol{\Sigma}}_{\varepsilon,r}$ be the estimated covariance matrix with the restriction $a_{21}^{(1)} = \cdots = a_{21}^{(p)} = 0$. Then, by 4.21, the maximum likelihood ratio test is given by

$$(T-p)\log\left(|\hat{\boldsymbol{\Sigma}}_{\varepsilon,r}|/|\hat{\boldsymbol{\Sigma}}_{\varepsilon}|\right).$$

The test statistic follows the asymptotically χ^2-distribution with degree p, the number of restrictions under the null hypothesis.

Instantaneous causality

For the bivariate AR model (4.26), if the off-diagonal element $\boldsymbol{\Sigma}_\varepsilon$ is not zero, then $\mathrm{Corr}(Z_t, Y_t|\boldsymbol{X}_{t-1},\ldots,\boldsymbol{X}_{t-p}) \neq 0$. In this case, Z_t and Y_t have *instantaneous Granger causality*. Again, this definition does not necessarily imply any causality between the two time series.

The instantaneous Granger causality can be tested with the null hypothesis $H_0 : \sigma_{21} = 0$, where σ_{21} denotes the off-diagonal element of $\boldsymbol{\Sigma}_\varepsilon$. Under H_0,

$$\mathrm{cov}(Z_t, Y_t|\boldsymbol{X}_{t-1},\ldots,\boldsymbol{X}_{t-p}) = 0.$$

A Wald test statistic can be constructed based on the estimate for σ_{21} and its asymptotic normality. See, for example, §3.6.3 of Lütkepohl (2006).

The function `causality` in the package `vars` implements the F-test for Granger causality and the Wald test for instantaneous causality. It also offers the option to evaluate the P-values of the tests by a wild bootstrap method instead of the asymptotic distributions. Let `X12` be a data matrix containing the daily log returns of FTSE 100 and FTSE Midcap as its two columns. We run

```
> m12 = VAR(X12, ic="SC")
> causality(m12)

H0: FTSE100 do not Granger-cause FTSE.MidCap
F-Test = 13.3076, df1 = 1, df2 = 490, p-value = 0.0002927

H0: No instantaneous causality between FTSE100 and FTSE.MidCap
Chi-squared = 116.7708, df = 1, p-value < 2.2e-16
```

Since the null hypothesis of no Granger causality is rejected with the P-value 0.0003, there is significant evidence indicating that FTSE 100 Granger causes FTSE MidCap. On the other hand, the null hypothesis of no instantaneous causality is rejected with the P-value 0. Hence, there exists instantaneous causality between FTSE 100 and FTSE MidCap. Applying the above test to different data sets, we also find that FTSE 100 Granger causes FTSE SmallCap with the P-value 0.004, and that there exists no significant evidence indicating that FTSE SmallCap Granger causes FTSE 100 as the P-value is 0.914.

4.2.6 Impulse response functions

In addition to the Granger causality, another way to investigate the effect of a change in one component series on the other components is via the so-called *impulse response functions*, which measure the resulting changes in other components at different time lags due to a unit change in one component series.

The impulse response functions for a stationary vector AR model can be easily derived from its MA(∞) expression. For vector AR(p) model (4.9) satisfying condition (4.12), it admits $MA(\infty)$ representation:

$$\boldsymbol{X}_t = \boldsymbol{c} + \boldsymbol{\varepsilon}_t + \sum_{k=1}^{\infty} \boldsymbol{B}_k \boldsymbol{\varepsilon}_{t-k}, \tag{4.27}$$

with the elements of $\boldsymbol{B}_j$ decaying to 0 exponentially as $j \to \infty$. So in practice, the sum on the right hand side of the above is truncated at a finite integer. An increase in the first component of $\boldsymbol{X}_t$ by a unit at time t can only be caused by the same increase in the first component of ε_t, as the values of ε_{t-k}, for $k \geqslant 1$, were already determined before times t. The impact of such an increase on the ith component at time $t+k$, i.e. the change in $X_{t+k,i}$, is $b_{i1}^{(k)}$ – the $(i,1)$th element of $\boldsymbol{B}_k$. In general,

---

the (i,j)th element of $\boldsymbol{B}_k$ is the impulse response of $X_{t+k,i}$, the ith component at the k units of time ahead, from one unit of extra shock in the jth component of $\boldsymbol{X}_t$.

The above analysis is implicitly based on the assumption that the components of $\boldsymbol{\varepsilon}_t$ are independent with each other, which implies no Granger instantaneous causality among the components of $\boldsymbol{X}_t$. However when the components of $\boldsymbol{\varepsilon}_t$ are not independent, such as when $\boldsymbol{\Sigma}_\varepsilon$ is not diagonal, a change in one component of $\boldsymbol{X}_t$ is typically associated with some changes in the other components. Therefore it is impossible to define the responses with respect to an impulse on a single component of $\boldsymbol{X}_t$. To alleviate the correlations among the components of $\boldsymbol{\varepsilon}_t$, the following transformation is applied to model (4.27):

$$\boldsymbol{\varepsilon}_t = \boldsymbol{\Psi}_0 \boldsymbol{e}_t, \quad \text{for } \boldsymbol{\Psi}_0 \boldsymbol{\Psi}_0' = \boldsymbol{\Sigma}_\varepsilon. \tag{4.28}$$

Then $\text{var}(\boldsymbol{e}_t) = \boldsymbol{I}_d$, i.e. the components of $\boldsymbol{e}_t$ are uncorrelated with each other, and (4.27) can be written as

$$\boldsymbol{X}_t = \boldsymbol{c} + \boldsymbol{\Psi}_0 \boldsymbol{e}_t + \sum_{k=1}^{\infty} \boldsymbol{\Psi}_k \boldsymbol{e}_{t-k}, \tag{4.29}$$

where $\boldsymbol{\Psi}_k = \boldsymbol{B}_k \boldsymbol{\Psi}_0$ for $k \geqslant 1$. The matrices $\boldsymbol{\Psi}_0, \boldsymbol{\Psi}_1, \ldots$ are called the *impulse response functions*. More precisely the plot of $\psi_{ij}^{(k)}$ against k, for $k = 0, 1, \ldots$, is called the response function of X_{ti} to the impulse on the jth component of $\boldsymbol{e}_t$, where $\psi_{ij}^{(k)}$ denotes the (i, j)th element of $\boldsymbol{\Psi}_k$. The computation of $\boldsymbol{\Psi}_k$ can be carried out using R-function `Psi` in the package `vars`.

Unfortunately the above definition is far from ideal. First $\psi_{ij}^{(k)}$ measures the responses to the change in the component of $\boldsymbol{e}_t$ instead of $\boldsymbol{X}_t$. This is arguably an innate problem when the components of $\boldsymbol{X}_t$ are dependent with each other, as then it may be impossible to isolate the change of one component from those of the others. Furthermore, the definition of $\boldsymbol{e}_t$ is not unique, as $\boldsymbol{\Psi}_0$ in (4.28) can be replaced by $\boldsymbol{\Psi}_0 \boldsymbol{H}$ for any $d \times d$ orthogonal matrix $\boldsymbol{H}$. Most packages, including `vars`, adopt the Cholesky decomposition of $\boldsymbol{\Sigma}_\varepsilon$, leading to a lower triangular $\boldsymbol{\Psi}_0$ in (4.28).

Example 4.4 Figure 4.5 plots the daily log prices of S&P 500 index and JP Morgan stock in 2013. To fit vector AR models to the returns of these two price series,

```
> fitX=VAR(X, ic="AIC"); summary(fitX)

Estimation results for equation S.P500:
=========================================
S.P500 = S.P500.l1 + JP.Morgan.l1 + const

              Estimate Std. Error t value Pr(>|t|)
S.P500.l1     -0.07625    0.06887  -1.107    0.269
JP.Morgan.l1 -0.02099    0.03720  -0.564    0.573
const          0.10375    0.04358   2.381    0.018 *
---
Signif. codes:  0 *** 0.001 ** 0.01 * 0.05 . 0.1  1

Residual standard error: 0.6825 on 247 degrees of freedom
Multiple R-Squared: 0.009656, Adjusted R-squared: 0.001637
F-statistic: 1.204 on 2 and 247 DF,  p-value: 0.3017
```

```
Estimation results for equation JP.Morgan:
==========================================
JP.Morgan = S.P500.l1 + JP.Morgan.l1 + const


             Estimate Std. Error t value Pr(>|t|)
S.P500.l1     0.52797    0.12227   4.318 2.28e-05 ***
JP.Morgan.l1 -0.17623    0.06604  -2.669  0.00812 **
const         0.08523    0.07737   1.102  0.27172
---
Signif. codes:  0 *** 0.001 ** 0.01 * 0.05 . 0.1   1

Residual standard error: 1.212 on 247 degrees of freedom
Multiple R-Squared: 0.07407, Adjusted R-squared: 0.06657
F-statistic: 9.879 on 2 and 247 DF,  p-value: 7.456e-05
```

The fitted model for S&P 500 returns is not significant; indicating that it is almost impossible to predict future returns for S&P 500. Nevertheless, the fitted model for JP Morgan returns is highly significant with P value smaller than 0.0001. Furthermore, the estimated coefficient for the lagged S&P 500 return is 0.52797 which is statistically highly significant. The estimated covariance matrix for the innovation is

$$\widehat{\boldsymbol{\Sigma}}_{\varepsilon} = \begin{pmatrix} 0.4658 & 0.3515 \\ 0.3515 & 1.4681 \end{pmatrix}.$$

The estimated correlation coefficient between the two innovation components is 0.425. Figure 4.6 plots the estimated impulse response functions for this fitted AR(1) model, which were calculated by calling the R-function

```
> Psi(fitX).
```

Since $\widehat{\boldsymbol{\Sigma}}_{\varepsilon}$ is not a diagonal matrix, the estimated responses were measured with respect to the impulses in the two components of the standardized innovation $\boldsymbol{e}_t$; see representation (4.29). Nevertheless there exist hardly any responses from S&P 500 return to both components of $\boldsymbol{e}_t$ at non-zero lags. (Note that the response at zero lag has no predictive values.) On the other hand, the response from JP Morgan to the first component of $\boldsymbol{e}_t$ is greater than 0.5 at lag 1. Since there exists no significant autocorrelations at non-zero lags for this JP Morgan return series, this must be the response to the impulse in the return of S&P 500 on the previous day. This analysis indicates that S&P 500 index carries a lead effect on JP Morgan stock at lead 1, but not vice versa.

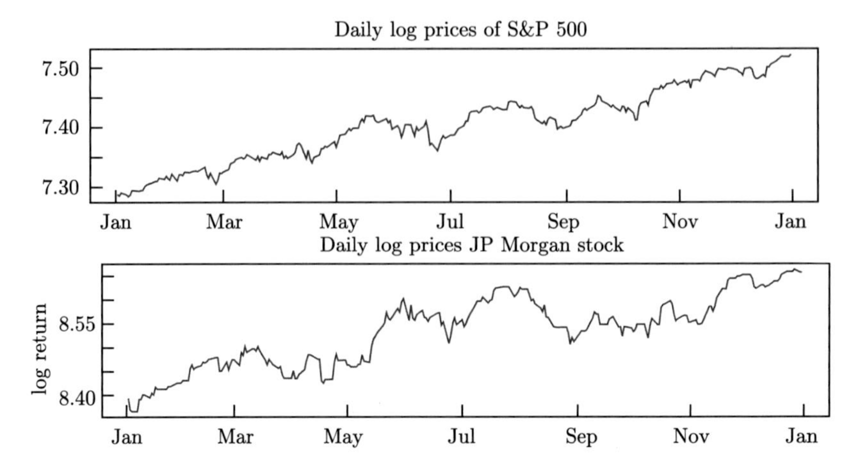

Figure 4.5 Daily log prices of S&P 500 index and JP Morgan stocks in 2013.

Figure 4.6, with additional confidence intervals calculated by a bootstrap method can also be produced by

```
> irfX=irf(fitX); plot(irfX)
```

We omit the plot.

4.3 Cointegration

Asset price time series are typically not stationary, as they often exhibit unit-root property. In practice this means that the current price is usually the best predictor for the next price; conforming with the efficient market hypothesis. On the other hand, there exist many co-movements, in either the same or opposite directions, among the prices of different assets. Therefore instead of differencing each price series individually, we may take differences, or more generally, linear combinations of different price series such that the resulting new series are stationary. This phenomenon, which is the essence for the concept of cointegration, was documented in Box and Tiao (1977), though the term *cointegration* was introduced later by Granger (1981). Cointegration has become an important concept in analyzing economic and financial time series. It is supported by some well-known principles in economics such as the purchasing power parity, the supply and demand model, factor asset pricing models. It paves the way to forecast some nonstationary dynamics based on stationary models.

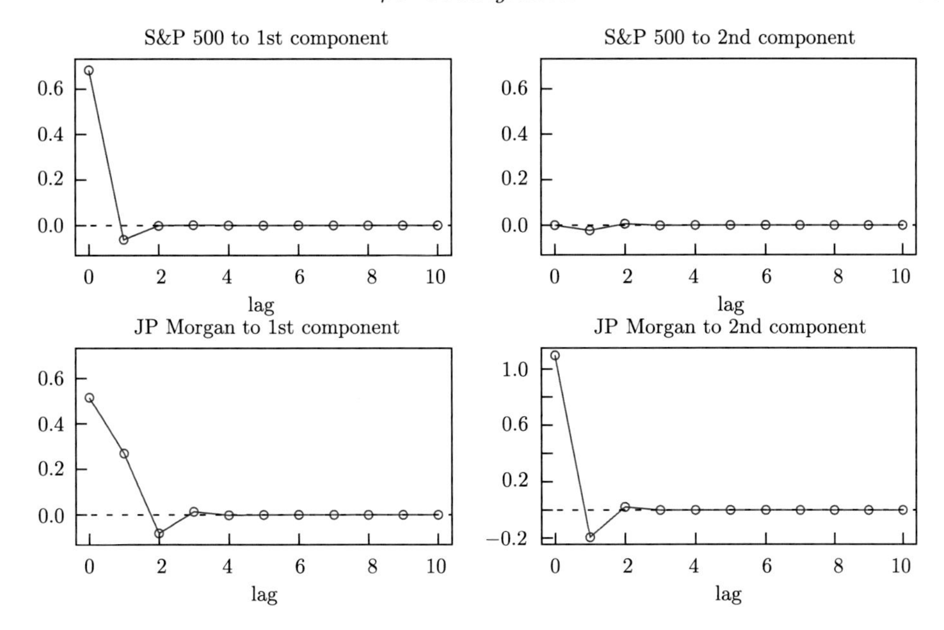

Figure 4.6 Impulse response functions (the 2×2 matrices as a function of k) of the fitted AR(1) model for the daily log returns (in percentage) of S&P 500 index and JP Morgan stocks in 2013.

4.3.1 Unit roots and cointegration

Before we define the cointegration for vector time series more formally, let us recall the concept of unit root for univariate time series. Let $X_t \sim \text{ARIMA}(p, 1, 0)$, where the integration order is supposed to be 1 and the MA-order is set at 0 for simplicity. Then X_t itself is not stationary but differenced X_t, i.e. $\nabla X_t = X_t - X_{t-1}$ is. In econometrics literature, X_t is said to have a unit root and is denoted as $X_t \sim \text{I}(1)$. Similarly for $X_t \sim \text{ARIMA}(p, k, 0)$, X_t is said to have k unit roots and is denoted as $X_t \sim \text{I}(k)$. Note that $X_t \sim \text{I}(0)$ denotes that X_t is a stationary process. The testing for unit roots is often carried out using the augmented Dickey–Fuller (ADF) test (see §2.8.2). For example the ADF test for model (2.66) is to test if $X_t \sim \text{ARIMA}(p, 1, 0)$.

Let Y_t and X_t be two I(1) time series. Those two series are said to be *cointegrated* if there exists non-zero constant β such that $Y_t - \beta X_t \sim \text{I}(0)$, namely stationary. In this definition, we require that both univariate time series have unit root and, therefore, β is non-zero, to avoid trivial cases. A simple example is the time series of the S&P 500 index and its future. Their difference should be stationary. Similarly, stock prices of the same stock (e.g. American depositary receipt) traded simultaneously at different markets should also be co-integrated. One way to understand cointegration is that the deviations from stationarity in both Y_t and X_t can cancel each other. Obviously the use of $Y_t - \beta X_t$ is not essential, as its multiple by any non-zero constant is also a stationary process.

Extension of the definition to d-vector series has been done in several slightly different manners. We adopt the following definition which avoids trivial cases.

A vector time series $\boldsymbol{X}_t$ is said to be *cointegrated* with order (k, h) $(k \geqslant h \geqslant 1)$, denoted as $\boldsymbol{X}_t \sim \mathrm{CI}(k, h)$, if

(i) all component series of $\boldsymbol{X}_t$ are $\mathrm{I}(k)$, and,
(ii) there exists a non-zero vector $\boldsymbol{\beta}$ such that $\boldsymbol{\beta}'\boldsymbol{X}_t \sim \mathrm{I}(k-h)$.

The most frequently used cointegration model is $\mathrm{CI}(1,1)$, i.e. all the components of $\boldsymbol{X}_t$ have one unit root and $\boldsymbol{\beta}'\boldsymbol{X}_t$ is stationary.

Cointegration shows an advantage in analyzing several time series together: although each individual series is nonstationary, the stationarity can be restored by taking appropriate linear combinations of them. The stationarity reflected by cointegration is a long term property within the system. It can be regarded as an equilibrium property within a nonstationary process. It is also worth mentioning that the cointegration transformation from $\boldsymbol{X}_t$ to $\boldsymbol{\beta}'\boldsymbol{X}_t$ is concurrent, involving no lagged values of $\boldsymbol{X}_t$. This is a marked difference from the different operation $\nabla \boldsymbol{X}_t$. While both $\nabla \boldsymbol{X}_t$ and $\boldsymbol{\beta}'\boldsymbol{X}_t$ are stationary when $\boldsymbol{X}_t \sim \mathrm{CI}(1,1)$, $\nabla \boldsymbol{X}_t$ tends to have more complex autocorrelation structure than $\boldsymbol{\beta}'\boldsymbol{X}_t$. Also note the fact that there may exist multiple constant vectors $\boldsymbol{\beta}_1, \ldots, \boldsymbol{\beta}_r$, which are linearly independent (i.e. the rank of matrix $(\boldsymbol{\beta}_1, \ldots, \boldsymbol{\beta}_r)$ is r), such that $\boldsymbol{\beta}_j'\boldsymbol{X}_t \sim \mathrm{I}(0)$. It must be true that $r < d$, as, otherwise, $\boldsymbol{X}_t$ is stationary.

4.3.2 Engle–Granger method and error correction models

Various estimation and testing methods have been proposed for the inference for cointegration processes, including the Engle–Granger two-step method (Engle and Granger 1987), Johansen's likelihood approach (Johansen 1995), and the method developed by Phillips (1991). Below we illustrate the Engle–Granger two-step method with a bivariate $\mathrm{CI}(1,1)$ process. The next section contains a brief introduction of Johansen's likelihood method. We refer to Hatanaka (1996) for a more systematic treatment of cointegration processes, and to Pfaff (2006) for using `R` for cointegration analysis.

Let X_t and Y_t be two $\mathrm{I}(1)$ processes. Engle and Granger (1987) suggests first to run regression

$$Y_t = \alpha + \beta X_t + Z_t, \tag{4.30}$$

where Z_t is the error term. As both X_t and Y_t are non-stationary I(1) processes, the least squares estimators $(\widehat{\alpha}, \widehat{\beta})$ for (α, β) are not asymptotically normal. An important consequence is that the conventional t- or F-tests for regression models do not apply here. Engle and Granger (1987) suggest testing for a unit root in the residual series

$$\widehat{Z}_t \equiv Y_t - \widehat{\alpha} - \widehat{\beta} X_t, \tag{4.31}$$

by the ADF test of §2.8.2. When a unit-root null hypothesis is rejected for $\widehat{Z}_t$, we take the view that the cointegration between Y_t and X_t has been identified. Since the test is based on the data $\widehat{Z}_t$ involving the estimated parameters $\widehat{\alpha}$ and $\widehat{\beta}$, slightly different critical values should be used; see Table B.9 in Hamilton (1994) which also contains the critical values for the cases with more than one regressors on the RHS of (4.30). This test is called the *cointegration augmented Dickey–Fuller (CADF) test*.

Though the CADF test stands well in the comparison with quite a few new tests developed after Engle and Granger (1987), precaution should be taken in applying such an approach. First one should test for unit root for each of Y_t and X_t separately, as otherwise the CADF test does not make sense. Secondly judicious judgement may be required in choosing appropriate null-hypothesis model among (2.66) and (2.68) in applying the CADF test; see the relevant discussion in §2.8.2.

Once the cointegration between Y_t and X_t has been established, the second step of the Engle–Granger method is to fit the so-called *error-correction-model* (ECM) as follows:

$$\nabla Y_t = a_0 + a_1 \widehat{Z}_{t-1} + a_2 \nabla X_{t-1} + a_3 \nabla Y_{t-1} + \varepsilon_{t1}, \tag{4.32}$$

$$\nabla X_t = b_0 + b_1 \widehat{Z}_{t-1} + b_2 \nabla Y_{t-1} + b_3 \nabla X_{t-1} + \varepsilon_{t2}, \tag{4.33}$$

where $\widehat{Z}_{t-1}$ is defined as in (4.31), ∇ denotes the difference operator, and $(\varepsilon_{t1}, \varepsilon_{t2})' \sim \text{WN}(0, \boldsymbol{\Sigma}_\varepsilon)$. In the above equations, the terms with ∇X_{t-1} and ∇Y_{t-1} reflect the short term effects of the changes in the two time series. Obviously further lagged terms of ∇Y_t and ∇X_t may be included in the RHS of the above equations. On the other hand, the terms with $\widehat{Z}_{t-1}$ reflect the long term effect of the equilibrium (4.30). Note that Z_{t-1} is called the error-correction term in the sense that it is the error from the equilibrium (4.30). The coefficients a_1, b_1 determine the speeds at which Y_t and X_t adjust to the equilibrium state. All terms in equations (4.32) and (4.33) are stationary.

Hence the techniques for linear regression can be adopted to carry out the statistical inference for those two models.

Example 4.5 We illustrate error-correction-model using a simple toy model – *a drunk and a dog*. Let a drunk's position at time 0 be 0. Suppose that the drunk moves a step of size ε_t at time t, where ε_t are independent and $N(0,1)$ for different t. Let X_t be the position of the drunk at time t. Then $\{X_t\}$ is a random walk with $X_0 = 0$, and

$$X_t = X_{t-1} + \varepsilon_t, \qquad t \geqslant 1,$$

Let Y_t denote the position of the drunk's dog at time t. Since the dog always follows its master though always wanders a bit, it is reasonable to assume that

$$Y_t = X_t + Z_t, \tag{4.34}$$

where $\{Z_t\}$ are independent and $N(0,\sigma^2)$. Then both X_t and Y_t are I(1) processes, though Y_t is not a random walk. Nevertheless (4.34) represents the cointegration between the two paths, which reflects the fact that the dog always stays with its master, which is a long term property. Suppose we are interested in predicting the change of dog's position

$$\nabla Y_t = \nabla X_t + \nabla Z_t,$$

based on the information available up to time $t-1$. As X_t is a random walk, ∇X_t is unpredictable. While ∇Z_t is an MA(1) process, $\nabla Z_{t-1} = \nabla Y_{t-1} - \nabla X_{t-1}$ carries the information on ∇Y_t as

$$\text{Corr}(\nabla Y_t, \nabla Y_{t-1}) = \frac{\text{cov}(\nabla Z_t, \nabla Z_{t-1})}{\text{var}(\nabla Y_t)} = \frac{-\sigma^2}{1+2\sigma^2}.$$

On the other hand,

$$\begin{aligned}\text{Corr}(\nabla Y_t, Z_{t-1}) &= \frac{\text{cov}(\nabla Z_t, Z_{t-1})}{\{\text{var}(\nabla Y_t)\text{var}(Z_{t-1})\}^{1/2}} = \frac{-\sigma}{\sqrt{1+2\sigma^2}} \\ &\approx \sqrt{2 + \frac{1}{\sigma^2}}\text{Corr}(\nabla Y_t, \nabla Z_{t-1}).\end{aligned}$$

This shows clearly that using the error-correction term Z_{t-1} leads to a significance improvement in the predictive power for ∇Y_t. We include below a numerical example performed in R.

To generate the paths for the drunk and his dog,

```
> n=50; e=rnorm(n)
> X=cumsum(e) # Drunk's random walk path
Y=X+rnorm(n, 0, 1/2) # Dog's path
```

The paths are plotted in Figure 4.7. To estimate the cointegration and to extract the error-correction terms $\widehat{Z}_t$,

```
> t=lm(Y~X); Z1=residuals(t)
> hatZ = Z1[-c(1,n)] # leave out the 1st and last elements
```

The fitted cointegration is

$$\widehat{Z}_t = Y_t - 0.0583 - 0.9733X_t. \tag{4.35}$$

To line up the differenced data together,

```
> dX=diff(X)
> dY=diff(Y)
> YXZ=data.frame(embed(cbind(dY,dX),2), hatZ)
> colnames(YXZ)=c("dY0","dX0","dY1", "dX1", "hatZ")
> attach(YXZ)
```

To fit an ECM for ∇Y_t,

```
> ecm=lm(dY0~hatZ+dX1); summary(ecm)
              Estimate Std. Error t value Pr(>|t|)
(Intercept)     0.2066     0.1678   1.231 0.224598
Z              -1.1516     0.3184  -3.616 0.000752 ***
dX1            -0.2892     0.1601  -1.806 0.077609 .
---
Signif. codes:  0 *** 0.001 ** 0.01 * 0.05 . 0.1   1
Residual standard error: 1.147 on 45 degrees of freedom
Multiple R-squared: 0.2742, Adjusted R-squared: 0.2419
F-statistic:    8.5 on 2 and 45 DF,  p-value: 0.0007389
```

The fitted ECM is

$$\widehat{\nabla Y_t} = 0.2066 - 1.156\widehat{Z}_{t-1} - 0.2892\nabla X_{t-1}$$

with the P-value 0.0007 and the squared regression correlation (the multiple R^2) 27.42%. Note that having included $\widehat{Z}_{t-1}$, defined in (4.35), and ∇X_{t-1} in the model, ∇Y_{t-1} is no longer significant. Adding it to the model, the squared regression correlation is 28.73%. On the other hand, using $\widehat{Z}_{t-1}$ alone leads to the squared regression correlation 22.16% while using ∇X_{t-1} and ∇Y_{t-1} (without $\widehat{Z}_{t-1}$) leads to the squared regression correlation merely 15.66%. Hence among $\widehat{Z}_{t-1}, \nabla X_{t-1}, \nabla Y_{t-1}$, the error-correction term $\widehat{Z}_{t-1}$ possesses the most predictive power for ∇Y_t.

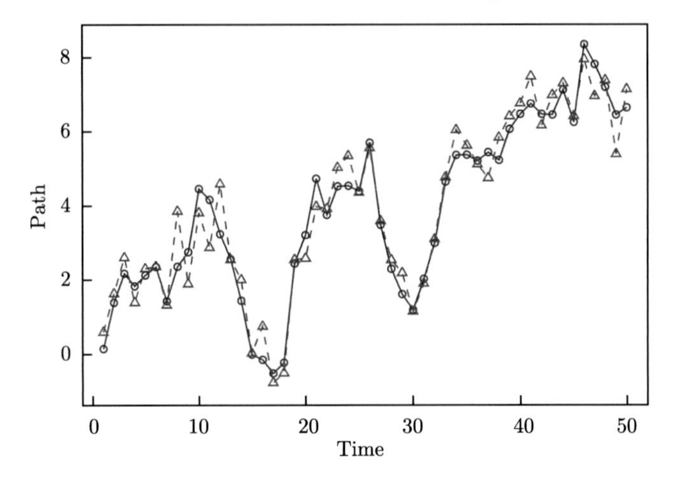

Figure 4.7 A random walk path of a drunk (cycles connected with solid lines) and an associated path of his dog (triangles connected with dashed lines) defined as in (4.34) with $\sigma = 1/2$.

Spurious linear regression

Before we close this section, we make some comments on the so-called *spurious regression* phenomenon related to fitting the models such as (4.30), caused by the nonstationary nature of $\{X_t\}$ and $\{Y_t\}$. If Y_t and X_t are two independent random walks, it holds in (4.30) that $\beta = 0$. However the LSE $\widehat{\beta}$ converges in distribution to a function of a Brownian motion, which is not zero and causes spurious correlation. More worryingly the usual t-statistic for testing the hypothesis $\beta = 0$ diverge at rate $T^{1/2}$, and hence the hypothesis $\beta = 0$ is rejected when the sample size T is sufficiently large. This phenomenon, coined as 'spurious regression', was first observed in simulation by Granger and Newbold (1974), and was then proved by Phillips (1986). In practice spurious regression phenomenon is encountered when we run regression for two I(1) variables Y_t and X_t as in (4.30) while Z_t is also I(1), i.e. the cointegration occurs among more than two I(1) processes and (4.30) is then a misspecified cointegration model. See, e.g. §15.5 of Hatanaka (1996). Thus it is important in practice to apply a unit-root test, such as the CADF test, to the residuals $\widehat{Z}_t$ in (4.31). One should only proceed to fitting an ECM when the unit-root hypothesis for $\widehat{Z}_t$ is rejected.

When Z_t is I(0) and both Y_t and X_t are I(1) in (4.30), namely, (4.30) correctly specifies an integration, the LSE $\widehat{\beta}$ is consistent or super consistent as the convergence rate is T instead of the standard $T^{1/2}$ (Stock 1987). The asymptotic distribution of the t-statistic for testing hypothesis $\beta = 0$ is often not normal and may be asymmetric. One alternative is to estimate β by the difference data $\nabla Y_t = \beta \nabla X_t + \nabla Z_t$. Since all the variables involved are stationary, the standard regression techniques such as t-test apply. However there may be some efficiency loss due to the more complex noise structure of ∇Z_t in comparison to that of Z_t.

4.3.3 Johansen's likelihood method*

Now let us consider a d-variate IC(1,1) process $\boldsymbol{X}_t$. Then there exists a $d \times r$ matrix $\boldsymbol{V} = (\boldsymbol{\beta}_1, \ldots, \boldsymbol{\beta}_r)$ with $\text{rank}(\boldsymbol{V}) = r < d$ such that $\boldsymbol{\beta}_j' \boldsymbol{X}_t \sim I(0)$ for $j = 1, \ldots, r$. By the Granger representation theorem (Engle and Granger 1987), $\boldsymbol{X}_t$ admits the following error-correction-model:

$$\nabla \boldsymbol{X}_t = \boldsymbol{W}\boldsymbol{V}'\boldsymbol{X}_{t-1} + \sum_{i=1}^{p} \boldsymbol{A}_i \nabla \boldsymbol{X}_{t-i} + \boldsymbol{\varepsilon}_t, \tag{4.36}$$

where both $\boldsymbol{W}$ and $\boldsymbol{V}$ are $d \times r$ matrices with ranks equal to r, $\boldsymbol{A}_1, \ldots, \boldsymbol{A}_p$ are $d \times d$ coefficient matrices, and $\boldsymbol{\varepsilon}_t \sim \text{WN}(0, \boldsymbol{\Sigma}_\varepsilon)$. Since $r < d$, (4.36) is a reduced-rank regression. It is a multivariate extension of (2.66). Furthermore if $\boldsymbol{X}_t$ contains some deterministic trends, a constant or linear terms can be added to the RHS of (4.36), as in (2.67) and (2.68). But we present below the simple form (4.36). The inference methodology based on a likelihood approach initiated by Soren Johansen; see Johansen (1995) for a thorough and concise presentation of this approach.

Let

$$\boldsymbol{Z}_{t0} = \nabla \boldsymbol{X}_t, \quad \boldsymbol{Z}_{t1} = \boldsymbol{X}_t, \quad \boldsymbol{Z}_{t2} = (\nabla \boldsymbol{X}_{t-1}', \ldots, \nabla \boldsymbol{X}_{t-p}')'.$$

Then (4.36) can be written as

$$\boldsymbol{Z}_{t0} = \boldsymbol{W}\boldsymbol{V}'\boldsymbol{Z}_{t1} + \boldsymbol{A}\boldsymbol{Z}_{t2} + \boldsymbol{\varepsilon}_t, \tag{4.37}$$

where $\boldsymbol{A} = (\boldsymbol{A}_1, \ldots, \boldsymbol{A}_p)$. One of the key ideas in Johansen's approach is to concentrate on the inference for the cointegration coefficient matrix $\boldsymbol{V}$ and the adjustment matrix $\boldsymbol{W}$ first, including their rank which is the number of latent cointegration components. Once $\boldsymbol{V}$ and $\boldsymbol{W}$ are specified, the estimation for $\boldsymbol{A}$ is a standard linear regression problem. We present the Johansen method in four parts below.

(i) Auxiliary regression

Based on Lemma 4.1 below, we may first run two regressions:

$$\boldsymbol{Z}_{t0} = \widehat{\boldsymbol{H}}_0 \boldsymbol{Z}_{t2} + \boldsymbol{R}_{t0}, \qquad \boldsymbol{Z}_{t1} = \widehat{\boldsymbol{H}}_1 \boldsymbol{Z}_{t2} + \boldsymbol{R}_{t1},$$

where the regression coefficients by the least-squares method are given by

$$\widehat{\boldsymbol{H}}_i = \boldsymbol{W}_{i2}\boldsymbol{W}_{22}^{-1}, \quad \boldsymbol{W}_{ij} = \frac{1}{T}\sum_{t=1}^{T} \boldsymbol{Z}_{ti}\boldsymbol{Z}'_{tj}.$$

Then it holds that

$$\boldsymbol{R}_{t0} = \boldsymbol{W}\boldsymbol{V}'\boldsymbol{R}_{t1} + \boldsymbol{e}_t. \tag{4.38}$$

Comparing with model (4.37), only the cointegration coefficients $\boldsymbol{V}$ and $\boldsymbol{W}$ are present in this model.

Lemma 4.1 ((The Frisch–Waugh theorem)) *For the linear regression* $y_t = b_1 x_{t1} + b_2 x_{t2} + \varepsilon_t$, *the least squares estimator* $\widehat{b}_1$ *for* b_1 *can be obtained in two steps:*

1. *Regress* y_t *on* x_{t2}, *leading to the residual* $r_{t1} = y_t - \widehat{\beta}_1 x_{t2}$. *Regress* x_{t1} *on* x_{t2}, *leading to the residual* $r_{t2} = x_{t1} - \widehat{\beta}_2 x_{t2}$.
2. *Obtain* $\widehat{b}_1$ *from regressing* r_{t1} *on* r_{t2}, *i.e.* $r_{t1} = b_1 r_{t2} + e_t$.

(ii)Profiling likelihood

By assuming $\boldsymbol{e}_t$ independent and $N(0, \boldsymbol{\Sigma}_e)$ in (4.38), the likelihood function is

$$L(\boldsymbol{V}, \boldsymbol{W}, \boldsymbol{\Sigma}_e) \propto |\boldsymbol{\Sigma}_e|^{-T/2} \exp\left\{-\frac{T}{2}\operatorname{tr}(\boldsymbol{\Sigma}_e^{-1}\boldsymbol{M})\right\},$$

where

$$\boldsymbol{M} \equiv \boldsymbol{M}(\boldsymbol{V}, \boldsymbol{W}) = \frac{1}{T}\sum_{t=1}^{T}(\boldsymbol{R}_{t0} - \boldsymbol{W}\boldsymbol{V}'\boldsymbol{R}_{t1})(\boldsymbol{R}_{t0} - \boldsymbol{W}\boldsymbol{V}'\boldsymbol{R}_{t1})'.$$

Let us maximize the likelihood function arguments by arguments.

For given $\boldsymbol{V}$ and $\boldsymbol{W}$, it follows from the standard multivariate normal theory (Anderson, 2003) that maximizing the likelihood with respect $\boldsymbol{\Sigma}_e$ gives the solution $\hat{\boldsymbol{\Sigma}}_e = \boldsymbol{M}$. Substituting this into the likelihood function yields a profile likelihood

$$L(\boldsymbol{V}, \boldsymbol{W}) = \max_{\boldsymbol{\Sigma}_e \geqslant 0} L(\boldsymbol{V}, \boldsymbol{W}, \boldsymbol{\Sigma}_e) \propto |\boldsymbol{M}(\boldsymbol{V}, \boldsymbol{W})|^{-T/2}.$$

We now further maximize the likelihood $L(\boldsymbol{V}, \boldsymbol{W})$ with respect to $\boldsymbol{W}$. This is equivalent to minimizing $|\boldsymbol{M}(\boldsymbol{V}, \boldsymbol{W})|$. An obvious candidate is the least-squares estimator $\widetilde{\boldsymbol{W}}(\boldsymbol{V})$ from equation (4.38), treating $\boldsymbol{V}$ as known. In the matrix form, it is given by

$$\widetilde{\boldsymbol{W}}(\boldsymbol{V}) = \sum_{t=1}^{T} \boldsymbol{R}_{t0}(\boldsymbol{V}'\boldsymbol{R}_{t1})' \left[\sum_{t=1}^{T}(\boldsymbol{V}'\boldsymbol{R}_{t1})(\boldsymbol{V}'\boldsymbol{R}_{t1})'\right]^{-1}.$$

A simple simplification yields

$$\widetilde{\boldsymbol{W}}(\boldsymbol{V}) = \boldsymbol{S}_{01}\boldsymbol{V}(\boldsymbol{V}'\boldsymbol{S}_{11}\boldsymbol{V})^{-1}, \quad \boldsymbol{S}_{ij} = \frac{1}{T}\sum_{t=1}^{T}\boldsymbol{R}_{ti}\boldsymbol{R}'_{tj}. \tag{4.39}$$

Then it can be shown that

$$\boldsymbol{M}(\boldsymbol{V},\boldsymbol{W}) \geqslant \boldsymbol{M}(\boldsymbol{V},\widetilde{\boldsymbol{W}}(\boldsymbol{V})) = \boldsymbol{S}_{00} - \boldsymbol{S}_{01}\boldsymbol{V}(\boldsymbol{V}'\boldsymbol{S}_{11}\boldsymbol{V})^{-1}\boldsymbol{V}'\boldsymbol{S}_{10}.$$

By Lemma 4.2,

$$\begin{aligned} L(\boldsymbol{V}) &\equiv \max_{\boldsymbol{W}} L(\boldsymbol{V},\boldsymbol{W}) = L\big(\boldsymbol{V},\widetilde{\boldsymbol{W}}(\boldsymbol{V})\big) \\ &= |\boldsymbol{S}_{00} - \boldsymbol{S}_{01}\boldsymbol{V}(\boldsymbol{V}'\boldsymbol{S}_{11}\boldsymbol{V})^{-1}\boldsymbol{V}'\boldsymbol{S}_{10}|^{-T/2}, \end{aligned} \tag{4.40}$$

Lemma 4.2 *For any two matrices $\boldsymbol{M}_1$ and $\boldsymbol{M}_2$, if $\boldsymbol{M}_1 \geqslant \boldsymbol{M}_2$, then $|\boldsymbol{M}_1| \geqslant |\boldsymbol{M}_2|$.*

(iii) Maximum likelihood estimation for $\boldsymbol{V}$ and $\boldsymbol{W}$

By (4.40), the MLE for $\boldsymbol{V}$ is the minimizer of

$$|\boldsymbol{S}_{00} - \boldsymbol{S}_{01}\boldsymbol{V}(\boldsymbol{V}'\boldsymbol{S}_{11}\boldsymbol{V})^{-1}\boldsymbol{V}'\boldsymbol{S}_{10}| = |\boldsymbol{S}_{00}|\frac{|\boldsymbol{V}'(\boldsymbol{S}_{11} - \boldsymbol{S}_{10}\boldsymbol{S}_{00}^{-1}\boldsymbol{S}_{01})\boldsymbol{V}|}{|\boldsymbol{V}'\boldsymbol{S}_{11}\boldsymbol{V}|}.$$

The equality above is due to the fact that

$$\begin{aligned} \begin{vmatrix} \boldsymbol{S}_{00} & \boldsymbol{S}_{01}\boldsymbol{V} \\ \boldsymbol{V}'\boldsymbol{S}_{10} & \boldsymbol{V}'\boldsymbol{S}_{11}\boldsymbol{V} \end{vmatrix} &= |\boldsymbol{S}_{00}| \cdot |\boldsymbol{V}'\boldsymbol{S}_{11}\boldsymbol{V} - \boldsymbol{V}'\boldsymbol{S}_{10}\boldsymbol{S}_{00}^{-1}\boldsymbol{S}_{01}\boldsymbol{V}| \\ &= |\boldsymbol{V}'\boldsymbol{S}_{11}\boldsymbol{V}| \cdot |\boldsymbol{S}_{00} - \boldsymbol{S}_{01}\boldsymbol{V}(\boldsymbol{V}'\boldsymbol{S}_{11}\boldsymbol{V})^{-1}\boldsymbol{V}'\boldsymbol{S}_{10}|. \end{aligned}$$

Let $\boldsymbol{C} = \boldsymbol{S}_{11}^{1/2}\boldsymbol{V}$. As $\boldsymbol{V}$ is a $d \times r$ matrix with $r < d$, it holds that

$$\begin{aligned} &\min_{\boldsymbol{V}} \frac{|\boldsymbol{V}'(\boldsymbol{S}_{11} - \boldsymbol{S}_{10}\boldsymbol{S}_{00}^{-1}\boldsymbol{S}_{01})\boldsymbol{V}|}{|\boldsymbol{V}'\boldsymbol{S}_{11}\boldsymbol{V}|} \\ &= \min_{\boldsymbol{C}} \frac{|\boldsymbol{C}'(\boldsymbol{I}_d - \boldsymbol{S}_{11}^{-1/2}\boldsymbol{S}_{10}\boldsymbol{S}_{00}^{-1}\boldsymbol{S}_{01}\boldsymbol{S}_{11}^{-1/2})\boldsymbol{C}|}{|\boldsymbol{C}'\boldsymbol{C}|} \\ &= \prod_{j=1}^{r}(1 - \widehat{\lambda}_j), \end{aligned} \tag{4.41}$$

where $\lambda_1 \geqslant \cdots \geqslant \lambda_r$ are the r largest eigenvalues of the matrix

$$\boldsymbol{S}_{11}^{-1/2}\boldsymbol{S}_{10}\boldsymbol{S}_{00}^{-1}\boldsymbol{S}_{01}\boldsymbol{S}_{11}^{-1/2}.$$

Those eigenvalues are always between 0 and 1. Let $\widehat{\boldsymbol{\gamma}}_1, \ldots, \widehat{\boldsymbol{\gamma}}_r$ be the corresponding orthonormal eigenvectors. Then the extreme value in (4.41) is attained when

$$\boldsymbol{C} = \boldsymbol{S}_{11}^{1/2}\boldsymbol{V} = (\widehat{\boldsymbol{\gamma}}_1, \ldots, \widehat{\boldsymbol{\gamma}}_r). \tag{4.42}$$

Note that $\widehat{\lambda}_1 \geqslant \cdots \geqslant \widehat{\lambda}_r$ are also the r largest eigenvalues of matrix

$$\boldsymbol{S}_{11}^{-1}\boldsymbol{S}_{10}\boldsymbol{S}_{00}^{-1}\boldsymbol{S}_{01} \tag{4.43}$$

with the corresponding eigenvectors satisfying the relation

$$\widehat{\boldsymbol{b}}_j = \boldsymbol{S}_{11}^{-1/2}\widehat{\boldsymbol{\gamma}}_j, \quad j = 1, \ldots, r.$$

By (4.42), the MLE for $\boldsymbol{V}$ can be taken as

$$\widehat{\boldsymbol{V}} = (\widehat{\boldsymbol{b}}_1, \ldots, \widehat{\boldsymbol{b}}_r). \tag{4.44}$$

Note the cointegration coefficient matrix $\boldsymbol{V}$ is only identifiable upto the linear space spanned by the columns of $\boldsymbol{V}$, as we can replace $(\boldsymbol{V}, \boldsymbol{W})$ in (4.36) by $(\boldsymbol{V}\boldsymbol{H}', \boldsymbol{W}\boldsymbol{H}^{-1})$ for any invertible $r \times r$ matrix $\boldsymbol{H}$. This flexibility allows one to choose $\widehat{\boldsymbol{V}}$ (by multiplying such an $\boldsymbol{H}$) which facilitates plausible interpretability for the resulting cointegrated components.

In summary, the MLE for the cointegration coefficient matrix $\boldsymbol{V}$ can be any $d \times r$ matrix such that the linear space spanned by its columns is the same as that spanned by all the eigenvectors corresponding to the r largest eigenvalues of matrix (4.43). Once $\widehat{\boldsymbol{V}}$ is specified, it follows from (4.39) that the MLE for the adjustment matrix $\boldsymbol{W}$ is

$$\widehat{\boldsymbol{W}} = \widetilde{\boldsymbol{W}}(\widehat{\boldsymbol{V}}) = \boldsymbol{S}_{01}\widehat{\boldsymbol{V}}(\widehat{\boldsymbol{V}}'\boldsymbol{S}_{11}\widehat{\boldsymbol{V}})^{-1}.$$

(iv) Testing for the number of cointegration components

In practice the number of columns r of the cointegration matrix $\boldsymbol{V}$ is unknown, which is the number cointegrated components in the system. Knowing r is of immense importance in applications. When $r = d$, i.e. the dimension of process, $\boldsymbol{X}_t$ is stationary in the sense that no components of $\boldsymbol{X}_t$ is I(1). On the other hand, when $r = 0$, there exists no integration at all among all the components of $\boldsymbol{X}_t$.

Johansen proposed two tests for the value of r based on the eigenvalues of matrix (4.43), denoted as $\widehat{\lambda}_1 \geqslant \cdots \geqslant \widehat{\lambda}_d$. To test the hypotheses

$$H_0 : \text{rank}(\boldsymbol{V}) \leqslant r \qquad \text{against} \qquad H_1 : \text{rank}(\boldsymbol{V}) > r,$$

Johansen proposes the trace statistic

$$\tau_1 = -T \sum_{j=r+1}^{d} \log(1 - \widehat{\lambda}_j), \tag{4.45}$$

and H_0 above is rejected for large values of τ_1. Note that $\tau_1 \geqslant 0$. Furthermore, when $\text{rank}(\boldsymbol{V}) \leqslant r$, $\lambda_{r+1} = \cdots = \lambda_d = 0$, and hence τ_1 will take values close to 0 under H_0.

An alternative approach is to test the hypotheses

$$H_0 : \text{rank}(\boldsymbol{V}) = r \qquad \text{against} \qquad H_1 : \text{rank}(\boldsymbol{V}) = r + 1.$$

We reject H_0 for large values of the test statistic

$$\tau_2 = -T \log(1 - \widehat{\lambda}_{r+1}). \tag{4.46}$$

Unfortunately the asymptotic distributions of both τ_1 and τ_2 cannot be evaluated explicitly. The critical values of those tests have been tabulated; see Table B.10 of Hamilton (1994).

The above estimation and testing methods are implemented in the R-package `urca` as `co.jo`. See the illustration in the section below and also Pfaff (2006).

4.3.4 Illustration with real data

We illustrate Johansen's likelihood method with the U.S. Treasury real yield curve (daily) rates at fixed maturities 5, 7, 10, 20, and 30 years in January 2, 2013 to February 11, 2014, displayed in Figure 4.8. Those real market yields are calculated from composites of secondary market quotations obtained by the Federal Reserve Bank of New York. There are in total 278×5 observations.

To fit the data with ECM (4.36), we first check if all five component series are I(1) processes. To this end, suppose the 278×5 data matrix is named as `tbill` in R. We apply the Augmented Dickey–Fuller (ADF) test (see Section 2.8.2) for each component series and its differenced series, which can be carried out using `ur.df` in the R-package `urca`. For example,

```
> t=ur.df(tbill[,1], type="none", lags=4, selectlags="AIC")
> summary(t)
```

performs the test with the first subseries. Since the differenced series exhibit no trends, we apply the ADF test based on model (2.66), specified by the option `type="none"` above with the AR order p determined by AIC. The option `lags=4` sets the upper bound for p. The tests for all five component series were not significant even at the 10% level; indicating no significant evidence against the hypothesis that there was at least one unit root in each of the five interest rates. Applying the same test to the differenced rates, the

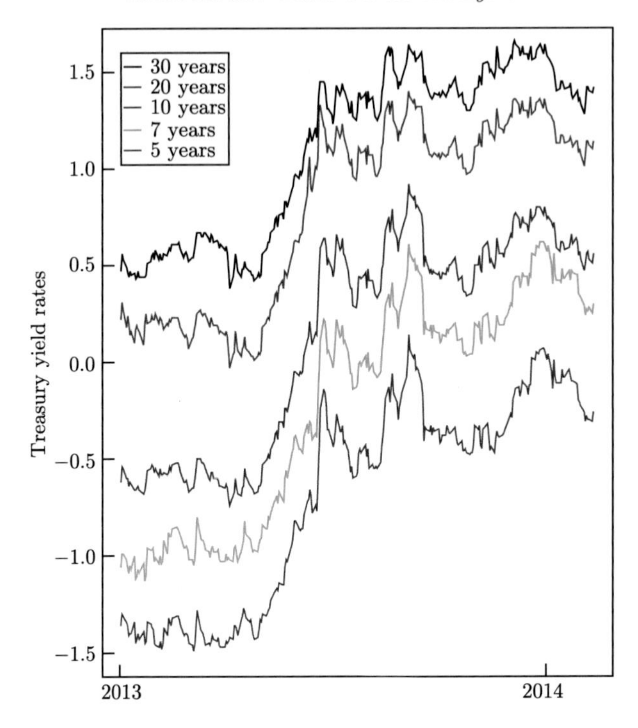

Figure 4.8 The daily U.S. Treasury real yield curve rates at fixed maturities of five, seven, ten, twenty and thirty years in the period of January 2, 2013 to February 11, 2014.

unit-root hypothesis is rejected at 1% significance level for all the five series. Hence it is reasonable to assume that all the five yields are I(1) series.

To fit the data with model (4.37) and to apply the trace test (4.45),

```
> m1=ca.jo(tbill, type="trace", ecdet="none", K=2,
          spec="transitory")
> summary(m1)
Eigenvalues (lambda):
[1] 0.124620 0.116170 0.056899 0.025242 0.004652

Values of test statistic and critical values of test:
          test 10pct  5pct  1pct
r <= 4 |  1.29  6.50  8.18 11.65
r <= 3 |  8.34 15.66 17.95 23.52
r <= 2 | 24.51 28.71 31.52 37.22
r <= 1 | 58.60 45.23 48.28 55.43
r = 0  | 95.33 66.49 70.60 78.87
```

The test rejects H_0: rank($\boldsymbol{V}$) $\geqslant r$ for $r = 1$ at the 1% significance level (the observed test statistic is 58.60, which is larger than 55.43), but cannot reject the hypothesis for $r = 2$ even at the 10% level (observed test statistic 24.51 is smaller than the critical value 28.71). This indicates that there exist two cointegration relations among the five yields series. To apply the test statistic (4.46), run `ca.jo(tbill, type="eigen", ecdet="none", K=2, spec="transitory")`, leading to

```
Values of test statistic and critical values of test:
          test 10pct  5pct  1pct
r <= 4 |  1.29  6.50  8.18 11.65
r <= 3 |  7.06 12.91 14.90 19.19
r <= 2 | 16.17 18.90 21.07 25.75
r <= 1 | 34.08 24.78 27.14 32.14
r = 0  | 36.73 30.84 33.32 38.78
```

i.e. the null hypothesis H_0: rank($\boldsymbol{V}$) $= r$ is rejected at the 1% significance level for $r = 1$, but cannot be rejected at the 10% level for $r = 2$. Once again this test also indicates two cointegration relations. The estimated matrices $\boldsymbol{V}, \boldsymbol{W}$ are stored at `m1@V` and `m1@W` respectively. The five candidate cointegrated variables $\widehat{\boldsymbol{V}}'\boldsymbol{X}_t$, can be extracted by

```
> y = as.matrix(tbill)%*%as.matrix(m1@V)
```

We plot those five transformed series (now in `y`) together with their ACF in Figure 4.9. The first series look stationary, and to a less extent also the 2nd series, as its ACF decays fast. The last three series are clearly not stationary as their ACF do not decay fast enough. Those visual observations lend further support to the assertion that there exist 2 cointegrated relations in the system. The first two columns of `m1@V`, normalized respectively such that each is a unit vector (the sum of squared elements is one), are

```
                ect1         ect2
5Year.l1     0.5808369    0.1115934
7Year.l1    -0.4399423    0.1382447
10Year.l1   -0.1716606   -0.7205056
20Year.l1   -0.3933011    0.6585049
30Year.l1    0.5337847   -0.1252131
```

Hence the first cointegrated variable could be viewed tentatively as the sum of the contrast between the 5-year rate and the 7-year rate and the contrast between the 30-year rate and 20-year rate. The second cointegrated variable is dominated by the contrast between the 20-year rate and the 10-year rate.

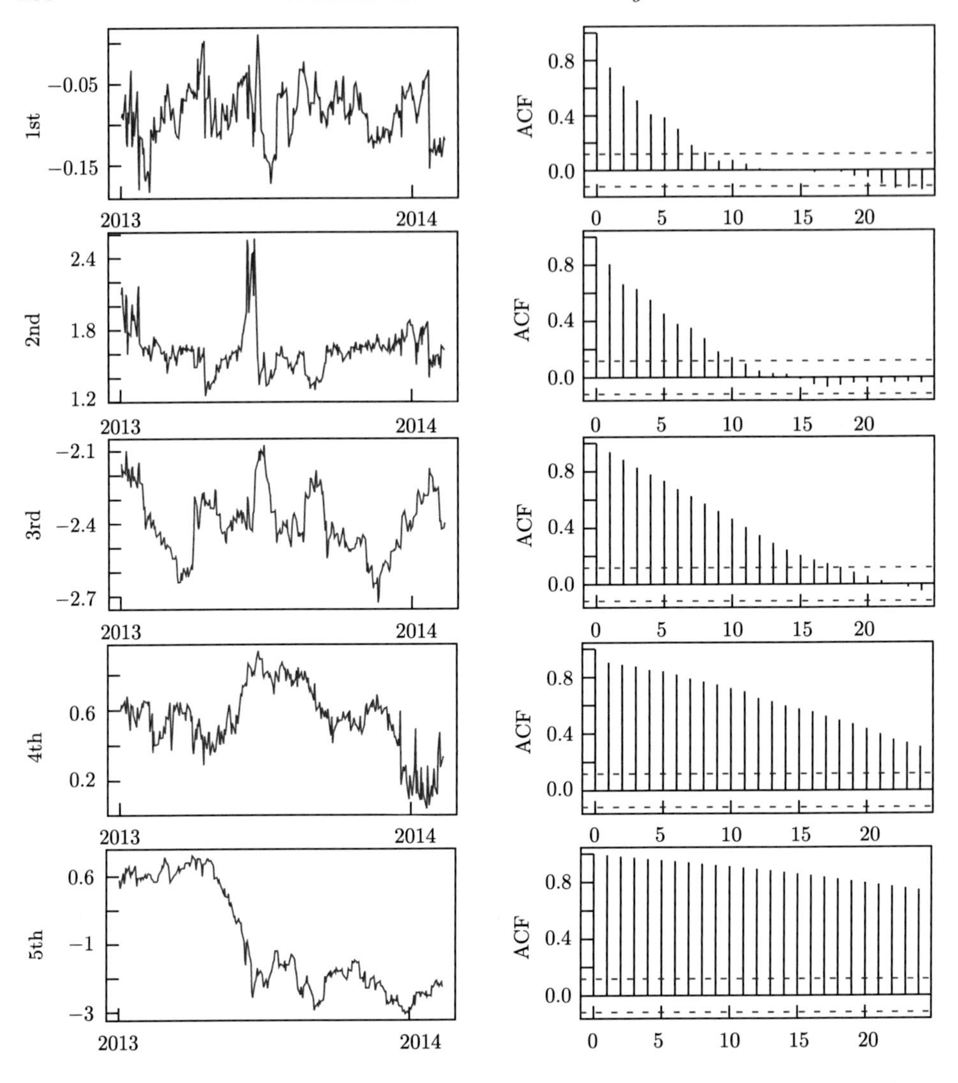

Figure 4.9 Time series and ACF plots for the five candidate cointegrated variables resulting from fitting model (4.36) to the daily U.S. Treasury yield rates displayed in Figure 4.8.

To refit the ECM with $r = 2$ fixed, run R-function `cajorls(m1, r=2)`, which returns the estimated coefficients in the fitted ECM:

```
            5Year.d  7Year.d  10Year.d  20Year.d  30Year.d
ect1        -0.0875  -0.0693   0.0952    0.1381   -0.0727
ect2         0.1377   0.1261  -0.0056   -0.1166    0.0737
constant    -0.0653  -0.0645  -0.0452    0.0251   -0.0192
5Year.dl1    0.3726   0.5228   0.3455    0.2816    0.2697
```

```
7Year.dl1    -0.5103  -0.7809  -0.1995   -0.1299   -0.1751
10Year.dl1    0.4397   0.6145  -0.1263   -0.3357   -0.1019
20Year.dl1    0.0153  -0.0309   0.1355    0.0828    0.1294
30Year.dl1   -0.1897  -0.2339  -0.0130    0.2431   -0.0250
```

In the above R-outputs, `d` indicates difference, `l1` indicates lag 1, and `dl1` indicates a differenced variable at lag 1. For example, the fitted ECM model for the differenced 5-year rate is

$$\begin{aligned}\nabla X_{t1} &= -0.0875U_{t-1,1} + 0.1377U_{t-1,2} - 0.0653 \\ &\quad + 0.3726\nabla X_{t-1,1} - 0.5103\nabla X_{t-1,2} + 0.4397\nabla X_{t-1,3} \\ &\quad + 0.0153\nabla X_{t-1,4} - 0.1897\nabla X_{t-1,5},\end{aligned}$$

where $X_{t1}, \ldots, X_{t5}$ denote, respectively, the 5-year, 7-year, 10-year, 20-year and 30-year rates at time t, U_{t1} and U_{t2} are the two normalized cointegrated variables. The normalization makes the top $r \times r$ submatrix of $\widehat{\boldsymbol{V}}$ is $\boldsymbol{I}_r$. Note that this normalization does not change the identified cointegrated space. The standard errors of the estimated coefficients in the above model can be calculated using the function `abStdErr` available at the book site

`http://orfe.princeton.edu/~jqfan/fan/FinEcon.html`

4.4 Exercises

4.1 Let Z_t be a univariate stationary time series. Let $X_{t1} = Z_t$ and $X_{t2} = Z_{t-6}$. Show that $\boldsymbol{X}_t \equiv (X_{t1}, X_{t2})'$ is a bivariate stationary process. Express the cross correlation function of $\boldsymbol{X}_t$ in terms of the ACF of Z_t.

4.2 Does the ACF (or the cross correlation function) of the MA(1) process $\boldsymbol{X}_t$ defined in Example 4.2 cut off at lag 1?

4.3 Determine the cross covariance function of the ARAM(1,1) process defined by

$$\boldsymbol{X}_t - \boldsymbol{A}\boldsymbol{X}_{t-1} = \boldsymbol{\varepsilon}_t + \boldsymbol{A}\boldsymbol{\varepsilon}_{t-1}, \quad \boldsymbol{\varepsilon}_t \sim \mathrm{WN}(\boldsymbol{0}, \boldsymbol{I_2}), \quad \boldsymbol{A} = \begin{pmatrix} 0.5 & 0 \\ 0.5 & 0.5 \end{pmatrix}.$$

4.4 (i) Show that for any non-negative definite $\boldsymbol{M}$, there exist a non-negative definite matrix denoted as $\boldsymbol{M}^{1/2}$ such that $\boldsymbol{M}^{1/2}\boldsymbol{M}^{1/2} = \boldsymbol{M}$.
(ii) Let $\boldsymbol{A}$ and $\boldsymbol{B}$ be two matrices such that $\boldsymbol{A} - \boldsymbol{B}$ is non-negative definite. Show that $\mathrm{tr}(\boldsymbol{MA}) \geqslant \mathrm{tr}(\boldsymbol{MB})$ for any non-negative definite matrix $\boldsymbol{M}$.

4.5 Download from *Yahoo!Finance* the daily prices in 2003 for the S&P 500 index (^GSPC) and the S&P 600 Smallcap Index (SLYG).

1. The data downloaded from *Yahoo* are in the format of 7 columns: *Date, Open, High, Low, Close, Volume, AdjClose*. You use the adjusted close prices in the last column. Two indices are not necessarily traded on the same day. To extract the prices from the days when both the indices were traded and line them up together with the dates, you may use the following R commends:

```
> D=which(SP500[,1]%in%SP600[,1])
> d=which(SP600[,1]%in%SP500[,1])
> X=data.frame(SP500[D,1], SP500[D,7], SP600[d,7])
```

This forms a matrix `X` with 3 columns: *Date, S&P* 500 *price, S&P* 600 *price.*

2. Plot the log prices of the two indices, and plot their cross correlation function.
3. Plot the returns of the two indices, and plot their cross correlation function. Compare your findings with (b) above.
4. Fit a bivariate AR(p) model to the two return series with the order p determined by BIC. Perform the diagnostic checking on the fitted model.
5. Identify any Granger causality in mean and instantaneous Granger causality.
6. Plot and comment on the impulse response functions.

4.6 On December 1, 2012 *The Economist* published an article entitled "Another game of thrones" which claims the technology giants *Google, Apple, Facebook* and *Amazon* are at each others throats in all sorts of ways.

1. Download the daily prices in 2011–2012 of the shares of those four companies.
2. Analyze those data to identity any cointegration among the log prices of those shares.
3. Fit an ECM for the log prices of those four shares. Comment on your findings.

5

Efficient Portfolios and Capital Asset Pricing Model

So far in this book, especially in the first three chapters, we have focused on the longitudinal aspects of financial returns and risks. The next five chapters concentrate on the cross-sectional aspects. Given a set of risky financial assets, how do we construct an efficient portfolio? How do we quantify the relation between the risk and return of a portfolio?

Markowitz's (1952, 1959) groundbreaking work on portfolio optimization and asset pricing has had a profound impact on financial economics and is a milestone of modern finance. It led to the celebrated Capital Asset Pricing Model (CAPM), developed by Sharpe (1964), under which risk and reward relation is beautifully quantified. It revolutionized the theory of portfolio choice and asset pricing and led Harry M. Markowitz, Merton R.C. Miller and William F. Sharpe to the Nobel Prize in Economics in 1990.

Is CAPM consistent with financial data? Is a portfolio efficient within a given set of risky assets? This chapter will also introduce econometric methods to address these questions.

5.1 Efficient portfolios

Suppose that we have p risky securities at time t with returns $\{R_{i,t+1}\}_{i=1}^{p}$ in the next period, which includes dividend payment. Suppose that there exists a riskless bond earning interest $R_{0,t}$, which is known at time t. How do we construct an efficient portfolio?

5.1.1 Returns and risks of portfolios

A portfolio is characterized by an allocation vector $(\alpha_0, \alpha_1, \ldots, \alpha_p)^{\mathrm{T}}$ with proportion α_i invested in security i. Denote by $\boldsymbol{\alpha} = (\alpha_1, \ldots, \alpha_p)^{\mathrm{T}}$ and $\boldsymbol{R}_t =$

$(R_{1,t}, \ldots, R_{p,t})^{\mathrm{T}}$. Since all money is invested (with cash invested on the riskless bond), the proportion satisfies

$$\alpha_0 + \alpha_1 + \cdots + \alpha_p = \alpha_0 + \mathbf{1}^{\mathrm{T}}\boldsymbol{\alpha} = 1, \tag{5.1}$$

where $\mathbf{1} = (1, \ldots, 1)^{\mathrm{T}}$ is a p-dimensional vector. In the formulation above, some α_i's can be negative, which corresponds to a short position (bet against the i asset). The return of the portfolio is

$$r_{t+1} = \alpha_0 R_{0,t} + \boldsymbol{\alpha}^{\mathrm{T}}\boldsymbol{R}_{t+1},$$

which is the sum of the returns from the risk-free and risky assets. It depends on how the portfolio is allocated.

The expected return and variance of the portfolio are given respectively by

$$\mu_t(\boldsymbol{\alpha}) = \boldsymbol{E}_t r_{t+1} = \alpha_0 R_{0,t} + \boldsymbol{\alpha}^{\mathrm{T}}\boldsymbol{E}_t\boldsymbol{R}_{t+1}$$

and

$$\sigma_t^2(\boldsymbol{\alpha}) = \mathrm{var}_t(r_{t+1}) = \boldsymbol{\alpha}^{\mathrm{T}}\mathrm{var}_t(\boldsymbol{R}_{t+1})\boldsymbol{\alpha}.$$

In the presence of the riskless asset, we introduce the excess returns $\boldsymbol{Y}_{t+1} = \boldsymbol{R}_{t+1} - R_{0,t}\mathbf{1}$ to eliminate the constraint (5.1). The return of the portfolio can now be written as

$$r_{t+1} = R_{0,t} + \boldsymbol{\alpha}^{\mathrm{T}}\boldsymbol{Y}_{t+1}, \tag{5.2}$$

whose expected return and volatility are given respectively by

$$\mu_t(\boldsymbol{\alpha}) = R_{0,t} + \boldsymbol{\alpha}^{\mathrm{T}}\boldsymbol{\xi}_t, \quad \text{and} \quad \sigma_t^2(\boldsymbol{\alpha}) = \boldsymbol{\alpha}^{\mathrm{T}}\boldsymbol{\Sigma}_t\boldsymbol{\alpha}, \tag{5.3}$$

where $\boldsymbol{\xi}_t = E_t\boldsymbol{Y}_{t+1}$ is the expected excess return, and $\boldsymbol{\Sigma}_t = \mathrm{var}_t(\boldsymbol{Y}_{t+1})$. Note that $\boldsymbol{\Sigma}_t$ is the same as $\mathrm{var}(\boldsymbol{R}_{t+1})$ since the risk-free rate $R_{0,t}$ is a known constant at time t.

The advantage of the excess return formulation is now clear. The expected return and variance in (5.3) depend only on the allocation vector $\boldsymbol{\alpha}$ of the risky assets, which is unconstrained in the p-dimensional space. Once the allocation on the risky assets is determined, the riskless free portion α_0 is computed from (5.1).

5.1.2 Portfolio optimization

The mean–variance approach of Markowitz (1952, 1959) is to select the portfolio that maximizes the tradeoff of the expected return and volatility.

Specifically, he advocated to find $\boldsymbol{\alpha}$ that maximizes

$$\mu_t(\boldsymbol{\alpha}) - \frac{A}{2}\sigma_t^2(\boldsymbol{\alpha}) = \boldsymbol{\alpha}^{\mathrm{T}}\boldsymbol{\xi}_t - \frac{A}{2}\boldsymbol{\alpha}^{\mathrm{T}}\boldsymbol{\Sigma}_t\boldsymbol{\alpha} + R_{0,t}, \tag{5.4}$$

where $A > 0$ is a parameter chosen by investors. It will be clear soon that A measures the investor's *risk aversion*. It varies from investor to investor.

Remark 5.1 Through a Lagrangian argument, the portfolio optimization problem (5.4) is equivalent to maximizing the expected returns with a given risk tolerance:

$$\max_{\boldsymbol{\alpha}} \mu_t(\boldsymbol{\alpha}), \qquad \text{subject to} \quad \sigma_t^2(\boldsymbol{\alpha}) \leqslant B,$$

for a given B. In this case, the risk aversion parameter A is indeed the Lagrange multiplier, chosen to satisfy $\sigma_t^2(\boldsymbol{\alpha}) = B$. In other words, the optimization problem (5.4) is the same as maximizing the portfolio expected return subject to the constraint on the portfolio risk bounded by $\sqrt{B}$. It is also equivalent to

$$\min_{\boldsymbol{\alpha}} \sigma_t^2(\boldsymbol{\alpha}), \qquad \text{subject to} \quad \mu_t(\boldsymbol{\alpha}) \geqslant C,$$

for a given C. That is, with the expected return at least C, we wish to find the portfolio whose risk is as small as possible. This equivalence can also be derived from the Lagrange multiplier method.

The unconstrained optimization problem (5.4) is particularly easy to solve. The gradient vector, which consists of the partial derivatives of the objective function with respect to each variable α_j, is given by

$$\boldsymbol{\xi}_t - A\boldsymbol{\Sigma}_t\boldsymbol{\alpha}.$$

This can easily be obtained by expressing (5.4) in the form of summation of each component and then taking partial derivatives. Setting the gradient vector to zero, we obtain the optimal allocation of the portfolio:

$$\boldsymbol{\alpha}_t^* = \frac{1}{A}\boldsymbol{\Sigma}_t^{-1}\boldsymbol{\xi}_t \quad \text{and} \quad \alpha_{0,t}^* = 1 - \mathbf{1}^{\mathrm{T}}\boldsymbol{\alpha}_t^*, \tag{5.5}$$

in which the second equality follows from (5.1). It is easily seen that the larger the value of A, the smaller the proportion allocated on the risky assets in the optimal portfolio. Hence, A is a risk aversion parameter.

The above portfolio allocation problem focuses on one period. For each given time t, the portfolio is optimized and allocated according to (5.5). Therefore, from now on, we will drop the dependence on time t to simplify the notation. In practice, the holding period is usually taken to be one month or one week to reduce transaction costs. At the end of each month or week, the portfolio is rebalanced according to (5.5). More details on implementation of this strategy will be given in Chapter 7.

Example 5.1 Suppose that the riskless asset earns 5% interest and that the returns of 3 risky financial assets are respectively 10%, 25%, and 55% per year with volatility (standard deviation) 12%, 40% and 110% respectively. In addition, suppose that the correlation matrix of the three risky assets is given by

$$\boldsymbol{\Gamma} = \begin{pmatrix} 1 & 0.7 & 0.4 \\ 0.7 & 1 & 0.5 \\ 0.4 & 0.5 & 1 \end{pmatrix}.$$

Then, the covariance matrix of the three risky assets is given by

$$\begin{aligned} \boldsymbol{\Sigma} = \mathrm{var}(\boldsymbol{R}) &= \begin{pmatrix} 0.12 & & \\ & 0.4 & \\ & & 1.1 \end{pmatrix} \boldsymbol{\Gamma} \begin{pmatrix} 0.12 & & \\ & 0.4 & \\ & & 1.1 \end{pmatrix} \\ &= \begin{pmatrix} 0.0144 & 0.0336 & 0.0528 \\ 0.0336 & 0.1600 & 0.2200 \\ 0.0528 & 0.2200 & 1.2100 \end{pmatrix}. \end{aligned}$$

According to (5.5), the optimal portfolio allocation vector for the risky portfolio is

$$\boldsymbol{\alpha}^* = A^{-1}\boldsymbol{\Sigma}^{-1} \begin{pmatrix} 0.05 \\ 0.20 \\ 0.50 \end{pmatrix} = A^{-1} \begin{pmatrix} 0.8722 \\ 0.7346 \\ 0.2416 \end{pmatrix}.$$

The above expression suggests that the larger the choice of A, the smaller the portion of the risky assets is allocated. To complete the problem, we need to specify the risk aversion parameter A.

Suppose that an investor is willing to invest 20% in the riskless asset. Then, we have

$$\alpha_1^* + \alpha_2^* + \alpha_3^* = (0.8722 + 0.7346 + 0.2416)/A = 1.848/A = 0.8.$$

This yields $A = 1.848/0.8$ and

$$\boldsymbol{\alpha}^* = \frac{0.8}{1.848} \begin{pmatrix} 0.8722 \\ 0.7346 \\ 0.2416 \end{pmatrix} = \begin{pmatrix} 0.3775 \\ 0.3180 \\ 0.1046 \end{pmatrix}.$$

In other words, he should invest 37.5%, 31.8%, 10.46% in Assets 1, 2, and 3, respectively, and 20% on the riskless asset. With this allocation, by (5.3), the expected return of the portfolio is

$$5\% + 37.75\% \times 5\% + 31.80\% \times 20\% + 10.46\% \times 50\% = 18.48\%.$$

This portfolio has variance

$$\boldsymbol{\alpha}^{*\mathrm{T}} \boldsymbol{\Sigma} \boldsymbol{\alpha}^* = 0.0583$$

and standard deviation $\sqrt{0.0583} = 24.15\%$. According to Remark 5.1, no portfolios with expected returns of at least 18.48% have risk (standard deviation) smaller than 24.15%, and no portfolios with risk smaller than 24.15% can have an expected return larger than 18.48%.

The risk aversion parameter can also be determined by setting the target expected return or the target risk, solving the problems posted in Remark 5.1.

5.1.3 Efficient portfolios and Sharpe ratios

We now characterize the efficient portfolio. Since only one period returns are concerned, we drop the subscript t to simplify the notation. Let $P = \boldsymbol{\xi}^{\mathrm{T}} \boldsymbol{\Sigma}^{-1} \boldsymbol{\xi}$. With the optimal portfolio allocation, substituting (5.5) into (5.3), the expected return is

$$\mu^* = R_0 + \boldsymbol{\alpha}^{*\mathrm{T}} \boldsymbol{\xi} = R_0 + P/A, \tag{5.6}$$

and the variance is given by

$$\sigma^{*2} = \boldsymbol{\alpha}^{*\mathrm{T}} \boldsymbol{\Sigma} \boldsymbol{\alpha}^* = P/A^2. \tag{5.7}$$

It is clear from (5.7) that A controls the portfolio risk: the larger A is the smaller the proportion allocated to risky assets.

From (5.7), it follows that $A = P^{1/2}/\sigma^*$. Substituting it into (5.6), we obtain

$$\mu^* = R_0 + P^{\frac{1}{2}} \sigma^*. \tag{5.8}$$

This leads to the definition of the *Sharpe ratio* of the efficient portfolio as

$$\text{Sharpe ratio} = \frac{\mu^* - R_0}{\sigma^*} = P^{1/2}. \tag{5.9}$$

This definition extends to any portfolio: For a given portfolio with allocation vector $\boldsymbol{\alpha}$ on the risky assets and $(1 - \mathbf{1}^{\mathrm{T}} \boldsymbol{\alpha})$ on the risk-free asset, its Sharpe ratio is defined as

$$S(\boldsymbol{\alpha}) = \frac{\boldsymbol{\mu}(\boldsymbol{\alpha}) - R_0}{\sigma(\boldsymbol{\alpha})} = \frac{\boldsymbol{\alpha}^{\mathrm{T}} \boldsymbol{\xi}}{(\boldsymbol{\alpha}^{\mathrm{T}} \boldsymbol{\Sigma} \boldsymbol{\alpha})^{1/2}}, \tag{5.10}$$

Table 5.1 *Return, risk and Sharpe ratios of five financial assets given in Example* 5.1

Asset	0	1	2	3	Optimal
return	5%	10%	25%	55%	18.48%
Excessive return	0%	5%	20%	50%	13.48%
risk	0%	12%	40%	110%	24.15%
Sharpe Ratio	–	0.417	0.500	0.455	0.558

where the last equality utilizes (5.3). The Sharpe ratio gives excess return per unit risk. It measures the efficiency of a portfolio and allows us to compare the efficiency of two portfolios with different returns and risks. It is related to *risk-adjusted return*. Note that for the optimal portfolio, the Sharpe ratio does not depend on the risk aversion parameter A, as shown in (5.9). This also follows easily by substituting (5.6) and (5.7) into (5.10).

Example 5.2 The Sharpe ratios or risk-adjusted returns for the financial assets given in Example 5.1 are summarized in Table 5.1, in which Asset 0 represents the riskless bond.

From the table, it is clear that Asset 1 is least efficient, followed by Asset 3, and then Asset 2. The Sharpe ratio is maximized at the efficient portfolio. It can be computed that $P = \boldsymbol{\xi}^{\mathrm{T}}\boldsymbol{\Sigma}^{-1}\boldsymbol{\xi} = 0.3113$. In view of (5.9), the Sharpe ratio of the optimal portfolio is $P^{1/2} = 0.558$. This provides an alternative method to the one presented in Table 5.1 for computing the Sharpe ratio of the optimal portfolio.

To appreciate why Sharpe ratio measures the efficiency of financial assets, let us compare Asset 1 and Asset 2, which have very different expected returns and volatilities. Suppose that Jack invested \$100 in Asset 1 whose expected return is 10% per annum and that Jill has a capital \$100 and chooses to invest only in Asset 2. Jack knows that his expected return will be inferior to Jill's, as his portfolio has a smaller risk. In order to have a comparable risk with Jill's portfolio, he needs to invest $40/12 = 3.3333$ times as much in Asset 1 so that his portfolio risk is also 40%. Thus, he needs to finance \$233.33 from the riskless bond (i.e. go short on Asset 0). With the investment in place, his annual return is expected to be $333.33 * 10\% - 233.33 * 5\% = \$21.67$ whereas Jill's annual expected return is \$25.00, even though their portfolios have the same risk. Therefore, Asset 2 is more efficient than Asset 1, even when their risks are adjusted. This is reflected in the Sharpe ratio.

The Sharpe ratio appears in the above calculation as follows. The excess return of Jill's portfolio is

$$\underbrace{10\% * 40/12}_{\text{leveraged return}} - \underbrace{5\% * (40/12 - 1)}_{\text{cost of loan}} = \frac{10\% - 5\%}{12} \times 40 + 5\% = 21.67\%,$$

which is greater than the expected return of Asset 1 but smaller than the expected return of Asset 2. For the more general result, see Exercise 5. 2.

5.1.4 Efficient frontiers

The above example shows that the optimal mean-variance portfolio has the highest Sharpe ratio. This holds with generality. For any other portfolio with allocation vector $\boldsymbol{\alpha}$ that has the same risk as the optimal portfolio: $\boldsymbol{\alpha}^{\mathrm{T}}\boldsymbol{\Sigma}\boldsymbol{\alpha} = (\sigma^*)^2$, its expected excess return is bounded by

$$\begin{aligned}\boldsymbol{\alpha}^{\mathrm{T}}\boldsymbol{\xi} &= \boldsymbol{\alpha}^{\mathrm{T}}\boldsymbol{\xi} - \frac{A}{2}\boldsymbol{\alpha}^{\mathrm{T}}\boldsymbol{\Sigma}\boldsymbol{\alpha} + \frac{A}{2}\sigma^{*2} \\ &\leqslant \boldsymbol{\alpha}^{*\mathrm{T}}\boldsymbol{\xi} - \frac{A}{2}\boldsymbol{\alpha}^{*\mathrm{T}}\boldsymbol{\Sigma}\boldsymbol{\alpha}^* + \frac{A}{2}\sigma^{*2} \\ &= \boldsymbol{\alpha}^{*\mathrm{T}}\boldsymbol{\xi}. \end{aligned} \tag{5.11}$$

The last inequality follows from the fact that α^* optimizes (5.4). Hence, the expected return of any portfolio $\boldsymbol{\alpha}$ lies below that of the optimal portfolio with the same risk. In other words, the optimal portfolios form efficient frontiers: For each given portfolio with risk σ^*, the expected return is no larger than that of the optimal portfolio. Figure 5.1 depicts the efficient frontiers in the risk-return space. The point $(0, R_0)$ represents the risk-free bond.

By (5.8), in the risk-return space, the efficient frontier is a line, depicted in Figure 5.1. The intercept is the riskless rate, corresponding to the optimal portfolio with risk aversion parameter $A = \infty$. The slope is the Sharpe ratio $P^{1/2}$, independent of risk aversion parameters. Thus, any portfolios on the efficient frontiers have the same Sharpe ratio and hence the same efficiency. They differ only in the risk attitude. As the percentage of risky assets increases, so does the expected return.

For any given portfolio, by choosing an appropriate risk aversion parameter A, there always exists an efficient portfolio with the same risk. Hence, by (5.11), its expected return must lie below the efficient frontier. Therefore, any other portfolios must be in the shaded area in the risk-return space. Its Sharpe ratio (5.10) is the slope of the line passing through the riskless portfolio $(0, R_0)$ and can not be larger than that of the efficient portfolio (see Figure 5.1). This can also be proved mathematically as follows. Note that

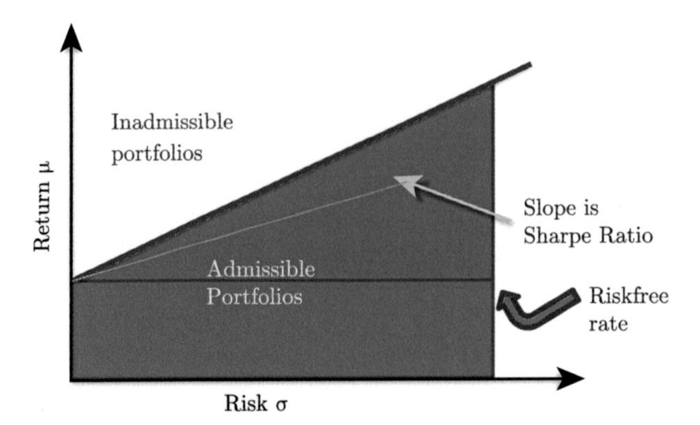

Figure 5.1 The efficient portfolio frontier is shown by red thick line in the risk and return plane. The slope of the line is the Sharpe ratio of optimal portfolios with different risk aversion parameters. The horizontal line is the risk-free interest rate. For any portfolio, the slope of the line passing through the point $(0, R_0)$ is its Sharpe ratio.

$S(\boldsymbol{\alpha})$ in (5.10) is independent of the scale of $\boldsymbol{\alpha}$. Therefore, we can normalize the portfolio by fixing the risk at a given level σ^*. By (5.11), we have

$$\max_{\boldsymbol{\alpha}} S(\boldsymbol{\alpha}) = \max_{\boldsymbol{\alpha}^T \boldsymbol{\Sigma} \boldsymbol{\alpha} = \sigma^{*2}} \frac{\boldsymbol{\alpha}^T \boldsymbol{\xi}}{\sigma^*} = \frac{\mu^* - R_{t,0}}{\sigma^*}.$$

5.1.5 Challenges of implementation

Practical implementation of the optimal portfolio faces a number of challenges, particularly when the number of assets p is large. First of all, the *gross exposure* $c = \sum_{i=0}^{p} |\alpha_i^*|$ of the optimal portfolio can be large. The proportion of short positions is $(c-1)/2$. This follows from the fact that $w^+ + w^- = c$ and $w^+ - w^- = 1$, in which w^+ and w^- are respectively the proportion of long and short positions. If $c = 4$, for example, there are 150% short positions and 250% long positions. Such a portfolio is too risky for most applications. Secondly, each investor has a limited credit to borrow. Thirdly, we need to estimate the parameters such as $\boldsymbol{\mu}$ and $\boldsymbol{\Sigma}$ based on limited amount of data (e.g. one year). For $p = 500$, there are $p(p+1)/2 = 125,250$ parameters. Thus, $\boldsymbol{\Sigma}$ and $\boldsymbol{\mu}$ can not be estimated with good accuracy. As a result, the allocation vector $\hat{\boldsymbol{\alpha}}$ based on the data can be very different from the optimal allocation vector $\boldsymbol{\alpha}^*$. We will address this problem in Chapter 7.

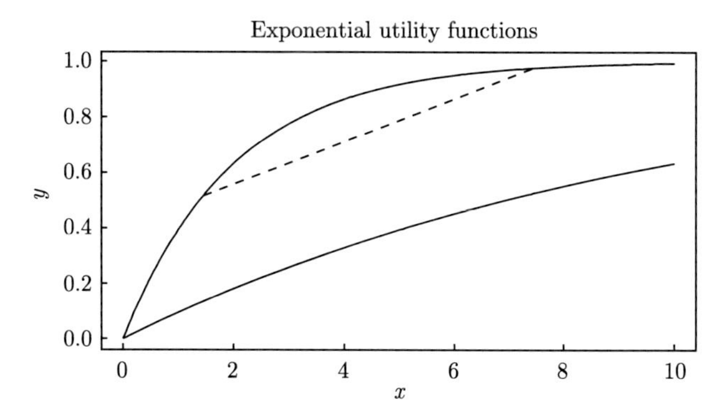

Figure 5.2 Exponential utility functions with $A = 0.5$ (top curve) and $A = 0.10$ (bottom curve).

5.2 Optimizing expected utility function

One drawback of the Markowitz problem is that it is formulated from a mathematical rather than an economic point of view. Economists believe that the investment objective is to optimize the personal utility of wealth. The *utility* $U(w)$ is an increasing and concave function of wealth w. The concavity makes the marginal utility $U'(w)$ decrease as w increases, as dictated by the law diminishing marginal utility. This is a reasonable assumption, as the increase of \$10,000 in wealth brings less utility to a wealthy man than to a poor man.

Utility functions have no units or scale. It is a relative measure. A commonly used utility is the *exponential utility* $U(w) = 1-\exp(-Aw)$ for $A > 0$ or *power utility* $U(w) = w^a$ for $0 < a \leqslant 1$. Figure 5.2 depicts the exponential utility function of two different choices of A. As to be seen soon, A is indeed the risk aversion parameter.

Example 5.3 To understand better the utility function and its applications in making decision under uncertainty, consider the following example with two possible actions:

Action 1 Win \$100 with certainty.

Action 2 Win \$10,000 with probability a or lose \$ 1,000 with probability $(1 - a)$.

The expected outcome of Action 2 is

$$a * 10000 + (1 - a) * 1000 = 11000a - 1000.$$

Since the utility function has no scale, we set $U(-1000) = 0$ and $U(10000) = 1$ in this example. Then the value of $U(100)$ will reflect the risk aversion of an investor.

If you think that actions 1 and 2 are about same with $a = 0.1$, your utility function at 100 should be

$$U(100) = 0.1U(10000) + 0.9U(-1000) = 0.1.$$

In this case, you really behave like a gambler. The expected value of the second action is only \$100, the same as Action 1, yet you do not demand any risk premium. If another investor feels that for $a = 0.2$, he is indifferent between Action 1 and Action 2, since he feels strongly against losing \$1,000 and demands compensation for bearing the risk. Then, his utility function is

$$U(100) = 0.2U(10000) + 0.8U(-1000) = 0.2.$$

Apparently, the second investor is more conservative as he puts more utility at the wealth of \$100.

We now turn to utility optimization. Suppose that initial wealth of an investor is 1 unit. Then, his wealth in the next period is

$$w_1 = 1 + R_0 + \boldsymbol{\alpha}^{\mathrm{T}}\boldsymbol{Y}.$$

If his utility is the exponential function $U(w) = 1 - \exp(-Aw)$, then he wishes to allocate his portfolio to maximize his utility in the next period:

$$\max_{\boldsymbol{\alpha}} \boldsymbol{E}U(w_1) = 1 - \min_{\boldsymbol{\alpha}} \boldsymbol{E}\exp\{-A(1 + R_0) - A\boldsymbol{\alpha}^{\mathrm{T}}\boldsymbol{Y}\}.$$

This is the same as minimizing the function

$$\boldsymbol{E}\exp(-A\boldsymbol{\alpha}^{\mathrm{T}}\boldsymbol{Y}).$$

If the distribution of the excess return is normal, then

$$\boldsymbol{Y} \sim \mathcal{N}(\boldsymbol{\xi}, \boldsymbol{\Sigma}), \tag{5.12}$$

and it follows from a multivariate normal result that

$$\boldsymbol{\alpha}^{\mathrm{T}}\boldsymbol{Y} \sim \mathcal{N}(\boldsymbol{\alpha}^{\mathrm{T}}\boldsymbol{\xi}, \boldsymbol{\alpha}^{\mathrm{T}}\boldsymbol{\Sigma}\boldsymbol{\alpha}). \tag{5.13}$$

Recall that the moment generating function of $X \sim N(\mu, \sigma^2)$ is given by

$$\boldsymbol{E}\exp(sX) = \exp(s\mu + s^2\sigma^2/2), \qquad \text{for any } s, \tag{5.14}$$

which leads to

$$\boldsymbol{E}\exp(\underbrace{-A}_{s}\underbrace{\boldsymbol{\alpha}^{\mathrm{T}}\boldsymbol{Y}}_{X}) = \exp\{\underbrace{-A}_{s}\underbrace{\boldsymbol{\alpha}^{\mathrm{T}}\boldsymbol{\xi}}_{\mu} + \frac{A^2}{2}\underbrace{\boldsymbol{\alpha}^{\mathrm{T}}\boldsymbol{\Sigma}\boldsymbol{\alpha}}_{\sigma^2}\}.$$

Minimizing the above is the same as maximizing

$$\boldsymbol{\alpha}^{\mathrm{T}}\boldsymbol{\xi} - \frac{A}{2}\boldsymbol{\alpha}^{\mathrm{T}}\boldsymbol{\Sigma}\boldsymbol{\alpha},$$

with respect to $\boldsymbol{\alpha}$, the same as the Markowitz problem (5.4). In other words, the Markowitz portfolio optimization is the same as the utility optimization with the exponential utility when the return follows a multivariate normal distribution.

5.3 Capital asset pricing model

The *capital asset pricing model* (CAPM), derived by Sharpe (1964) and Lintner (1965), is a celebrated model in asset pricing. It allows us to quantify exactly how risks are rewarded. While the original derivations differ from what we use here, the financial economics approach below provides a better understanding on the concept of the *market portfolio.*

5.3.1 Market portolio

Suppose that each investor trades at the mean-variance optimal portfolio, but different investors have different risk preference. Let A_i be the risk aversion coefficient for the ith investor with amount of wealth w_i to be invested. Then, his demand on the risky and riskless assets are $\boldsymbol{\alpha}_i^* = w_i\boldsymbol{\Sigma}^{-1}\boldsymbol{\xi}/A_i$ and $w_i - \mathbf{1}^{\mathrm{T}}\boldsymbol{\alpha}_i^*$ respectively. The total demands on the risky assets are the sum of all investors, which are given by

$$\boldsymbol{\alpha}^D = \sum_i \boldsymbol{\alpha}_i^* = \sum_i \frac{w_i}{A_i}\boldsymbol{\Sigma}^{-1}\boldsymbol{\xi}.$$

Without loss of generality, assume that each stock is normalized to have \$1 per share. Then the jth component of $\boldsymbol{\alpha}^D$ is the total demand for the j-asset in terms of number of shares by all market participants.

Suppose that the total supply on the numbers of shares for all of those risky assets is $\boldsymbol{a}$. The equilibrium condition must be met, i.e. $\boldsymbol{\alpha}^D = \boldsymbol{a}$. In other words,

$$\boldsymbol{a} = \sum_i \frac{w_i}{A_i}\boldsymbol{\Sigma}^{-1}\boldsymbol{\xi}.$$

Now, define the *market portfolio* to be the giant portfolio, which consists of all shares of all tradable financial assets, namely it has a_j shares on the jth risky asset for all j. Recall that each asset has been normalized to have a market price \$1 per share. Translating this portfolio in terms of proportions of holdings, rather than the number of shares of holdings, the market portfolio has the allocation vector

$$\boldsymbol{b} = \boldsymbol{a} / \left(\sum_i w_i \right) = \frac{1}{A} \boldsymbol{\Sigma}^{-1} \boldsymbol{\xi} \quad \text{or} \quad \boldsymbol{\xi} = A \boldsymbol{\Sigma} \boldsymbol{b}, \tag{5.15}$$

where $A^{-1} = (\sum_i w_i / A_i)/(\sum_i w_i)$ is the weighted harmonic average of individual risk aversion parameters. The excess return of the market portfolio is $Y^m = \boldsymbol{b}^T \boldsymbol{Y}$, extra return from the risky part of the portfolio. By (5.15), it is mean-variance efficient, since it is an optimal portfolio with risk aversion parameter A.

As a_j is also the market value of the jth asset, the market portfolio is indeed the value-weighted index of all tradable assets at a given time period. In practice, researchers often use the portfolio of S&P 500 or CRSP indices as its proxies. Of course, this focuses mainly on the equity market of the United States. With financial globalization, one would argue that the market portfolio should include international assets and bonds. It should also include real estate and precious metals and other commodities. These are clear from the above supply and demand equilibrium argument.

At the equilibrium, the ith investor invests $\boldsymbol{\alpha}_i^* = (w_i/A_i)\boldsymbol{\Sigma}^{-1}\boldsymbol{\xi}$ on risky assets and the rest $w_i - \mathbf{1}^T \boldsymbol{\alpha}_i^*$ on the riskless asset. This results in the *two fund separation theorem*: each investor should invest only on the market portfolio and risk-free asset. The weights on these two funds (market portfolio and risk-free asset) are used to achieve the risk preference of an investor. Such a portfolio is efficient, regardless of the weights. This can be seen as follows. Suppose that an investor holds proportion w on the market portfolio with return R_m and the rest on the riskless bond with rate r_f. Then, his return is $wR_m + (1-w)r_f$ with expected return $wr_m + (1-w)r_f$ and standard deviation $w\sigma_m$, where $r_m = ER_m$ and $\sigma_m = \text{SD}(R_m)$. The Sharpe ratio of his portfolio is

$$\frac{wr_m + (1-w)r_f - r_f}{w\sigma_m} = \frac{r_m - r_f}{\sigma_m},$$

the same as the market portfolio.

5.3.2 Capital asset pricing model

Note that for any random variables X and Y with bounded first two moments, we can always find α and β that minimize

$$E(Y - a - bX)^2$$

with respect to a and b. Taking derivatives with respect to a and b and setting them to zero, we obtain

$$E(Y - \alpha - \beta X) = 0, \qquad E(Y - \alpha - \beta X)X = 0.$$

Letting $\varepsilon = Y - \alpha - \beta X$, we have $E\varepsilon = 0$ and $\operatorname{cov}(X, \varepsilon) = 0$. In other words, we can always decompose the random variable Y as

$$Y = \alpha + \beta X + \varepsilon, \qquad E\varepsilon = 0 \text{ and } \operatorname{cov}(X, \varepsilon) = 0. \tag{5.16}$$

Applying the above decomposition to the excess return of each risky asset in the market portfolio, we have

$$\boldsymbol{Y} = \boldsymbol{\alpha} + \boldsymbol{\beta} Y^m + \boldsymbol{\epsilon} \tag{5.17}$$

with $\boldsymbol{E\epsilon} = 0$ and $\operatorname{cov}(\boldsymbol{\epsilon}, Y^m) = 0$, where Y^m is the excess return of the market portfolio and $\operatorname{cov}(\boldsymbol{\epsilon}, Y^m) = EY^m\boldsymbol{\epsilon}$ is the vector of covariances. Here $\boldsymbol{Y}$ is the excess return vector of all risky assets. The parameters $\boldsymbol{\alpha}$ and $\boldsymbol{\beta}$ are called respectively the *market* α and *market* β. The market α represents extra returns over the risk that is undertaken.

Theorem 5.1 ((Capital Asset Pricing Model)) *In the decomposition* (5.17), $\boldsymbol{\alpha} = 0$. *Namely, the excess asset returns can be decomposed as*

$$\boldsymbol{Y} = \boldsymbol{\beta} Y^m + \boldsymbol{\epsilon}, \qquad \boldsymbol{E\epsilon} = 0 \text{ and } \operatorname{cov}(\boldsymbol{\epsilon}, Y^m) = 0 \tag{5.18}$$

with the market beta's given by

$$\boldsymbol{\beta} = \operatorname{cov}(\boldsymbol{Y}, Y^m)/\operatorname{var}(Y^m). \tag{5.19}$$

Proof Note that from the decomposition (5.17)

$$\operatorname{cov}(\boldsymbol{Y}, Y^m) = \operatorname{cov}(\boldsymbol{\alpha} + \boldsymbol{\beta} Y^m + \boldsymbol{\epsilon}, Y^m) = \boldsymbol{\beta}\operatorname{cov}(Y^m, Y^m).$$

This results in (5.19). By the definition of the market portfolio, we have its excess return $Y^m = \boldsymbol{b}^{\mathrm{T}}\boldsymbol{Y}$. It follows from (5.19) that

$$\boldsymbol{\beta} = \frac{\boldsymbol{\Sigma b}}{\boldsymbol{b}^{\mathrm{T}}\boldsymbol{\Sigma b}}.$$

Now from the decomposition (5.17), we have

$$\boldsymbol{\alpha} = \boldsymbol{EY} - \boldsymbol{\beta E}Y^m = \boldsymbol{\xi} - \frac{\boldsymbol{\Sigma b}}{\boldsymbol{b}^{\mathrm{T}}\boldsymbol{\Sigma b}}\, \boldsymbol{b}^{\mathrm{T}}\boldsymbol{\xi},$$

where $\boldsymbol{\xi} = \boldsymbol{EY}$. Substituting (5.15) into the above expression, we obtain

$$\boldsymbol{\alpha} = A\boldsymbol{\Sigma b} - \frac{\boldsymbol{\Sigma b}}{\boldsymbol{b}^{\mathrm{T}}\boldsymbol{\Sigma b}} \cdot \boldsymbol{b}^{T} A\boldsymbol{\Sigma b} = 0.$$

This completes the proof. □

Following the same steps of the proof, the reverse of Theorem 5.1 is also true. We leave this proof as a homework (Exercise 5.10), but state the result below.

Theorem 5.2 *Given a portfolio vector* $\boldsymbol{a}$ *whose excess return* $\mathbf{Y}^{\boldsymbol{a}} = \boldsymbol{a}^T\boldsymbol{Y}$*, in the decomposition*

$$\boldsymbol{Y} = \boldsymbol{\alpha_a} + \boldsymbol{\beta_a} Y^{\boldsymbol{a}} + \boldsymbol{\epsilon}^{\boldsymbol{a}}, \qquad E\boldsymbol{\epsilon}^{\boldsymbol{a}} = 0 \text{ and } \operatorname{cov}(\boldsymbol{\epsilon}^{\boldsymbol{a}}, Y^{\boldsymbol{a}}) = 0,$$

the intercepts $\boldsymbol{\alpha}_a = \mathbf{0}$ *if and only if* $\boldsymbol{\alpha}$ *is proportional to* $\boldsymbol{b}$*, i.e. the portfolio with allocation vector* $\boldsymbol{\alpha}$ *is proportional to the market portfolio.*

Theorem 5.1 is the *capital asset pricing model*derived by Sharpe (1964) and Lintner (1965) with existence of the risk-free asset. It quantifies the relations between the risks and returns. First of all, the excess return of any asset can be decomposed into two parts. The first part is the return due to its dependence with the market and the second part is the idiosyncratic noise related only to individual asset but uncorrelated with the market. More precisely, the expected excess return of ith security is given by

$$\boldsymbol{E}Y_i = \beta_i \boldsymbol{E}Y^m. \tag{5.20}$$

It quantifies exactly the relationship between risk and return. The expected return of the asset depends on its degree of dependence on the market, as measured by β_i. The larger the market beta, the higher the expected return. The cross-sectional risk is measured by

$$\beta_i = \operatorname{Cov}(Y_i, Y^m)/\operatorname{var}(Y^m),$$

which is called *market beta*. The larger the riskier since

$$\operatorname{var}(Y_i) = \beta_i^2 \operatorname{var}(Y^m) + \operatorname{var}(\varepsilon_i).$$

The above argument is valid only when the *market risk premium* $EY^m > 0$. This is the premium paid to investors who bear the risks.

5.3.3 Market β and its applications

The market β, which is the sensitivity of an asset to the market portfolio, is often used to measure the cross-sectional risk. In practice, the market β is determined with the following three specifications:

- Use S&P 500 index or CRSP index as a proxy of the market portfolio;
- Take the US Treasury bill rates as proxies of the riskless returns;

- Compute the last 5 years monthly excess returns and run the simple linear regression

$$Y_{it} = \alpha_i + \beta_i Y_t^m + \varepsilon_{it}, \quad t = 1, \ldots, T \tag{5.21}$$

with $T = 60$. The resulting least-squares estimate is the market β for asset i.

Example 5.4 As an illustration, we now compute the market betas for three stocks Goldman Sachs Group(GS), International Business Machines(IBM), and General Electric Company(GE) on January 31, 2011. The excess returns of the stocks are plotted against the excess return of the S&P 500 index for the last 60 months. The least squares fits are presented and the slopes are the market beta. The market β's for GS, IBM and GE are respectively 1.37, 0.74 and 1.66 on January 31, 2011. The market beta is time-dependent. To illustrate this, we plot the market beta for GE from January 2006 to January 2011. The time varying feature is evident. The market beta for GE shoots up during the financial crisis in 2008 and remains high. This is due in part to the division of GE capital, which lost a lot of money during the crisis (recalling the leverage effect in Chapter 2), making the stock behave like a financial stock.

Suppose that a hedge fund holds \$10 million of the Goldman Sachs Group stock, \$20 million of the IBM stock and \$30 million of the GE stock on January 31, 2011. Then, its market equivalent exposure is

$$\$10 \times 1.37 + \$20 \times 0.74 + \$30 \times 1.66 = \$78.3 \text{ million.}$$

To hedge 'completely' the market risk, the fund needs to short \$78.3 million S&P 500 index. This can also be accomplished by holding put options of S&P 500 stocks. With this hedge, the exposure is "market neutral".

There are several important applications of the capital asset pricing models. First of all, it can be used to estimate the covariance matrix of asset returns. By (5.18), we have

$$\operatorname{var}(\boldsymbol{Y}) = \boldsymbol{\beta}\boldsymbol{\beta}^{\mathrm{T}} \operatorname{var}(Y^m) + \operatorname{var}(\boldsymbol{\epsilon}). \tag{5.22}$$

If the market portfolio captures all cross-sectional risks, the idiosyncratic noise should depend only on the individual firm itself, unrelated to other firms. In this case, we may assume that $\operatorname{var}(\boldsymbol{\epsilon})$ is diagonal. When $\operatorname{var}(\boldsymbol{\epsilon})$ is diagonal, the right hand side of (5.22) has only $2p+1$ unknown parameters: p parameters for the market beta, 1 parameter for $\operatorname{var}(Y^m)$ and p parameters for the diagonal elements of $\operatorname{var}(\boldsymbol{\epsilon})$. This parametric based covariance matrix has far less elements than the nonparametric covariance matrix, which has

$p(p+1)/2$ elements. For example, if $p = 1000$, the former has 2001 free parameters while the latter has 500,500. When $\text{var}(\boldsymbol{\epsilon})$ is not diagonal but sparse, the number of parameters is still much less than that of the full nonparametric covariance matrix. See Chapter 7 for additional details.

The CAPM can also be applied to capital budgeting decisions in corporate finance. According to (5.20), the expected return of a firm is given by

$$ER = r_f + \beta(r_m - r_f), \tag{5.23}$$

where r_f and r_m are the risk-free interest rate and expected return of the market portfolio, and β is the market beta of the firm. The line (5.23) is called the firm's market *capital line*. With this estimate, decision makers can decide whether or not to carry out an investment project. See Copeland, Weston, and Shastri (2005) for details.

The CAPM can also be used for portfolio performance evaluation. For example, to investigate whether a fund and a strategy beats the market, one can run the regression (5.21) over a period of time and see whether the fund generates positive α or no α or even negative α. This can also be applied to pick analysts who recommend stocks periodically.

5.4 Validating CAPM

The capital asset pricing model was derived based on several assumptions. They might not be consistent with financial data. Like any theory, it needs to be verified by empirical data. In this section, we introduce econometrics methods to validate the Sharpe–Linter version of CAPM (5.18). For a comprehensive survey, see the overview article by Sentana (2009).

5.4.1 Econometric formulation

Let $\{\boldsymbol{Y}_t\}_{i=1}^T$ be a vector of excess returns of N portfolios. These portfolios are usually constructed from the p assets that form the market portfolio. Correspondingly, let $\{Y_t^m\}_{t=1}^T$ be the excess returns of the proxy of the market portfolio. This is usually the excess returns of S&P 500 or CRSP index over the US Treasury bills. Inspired by the decomposition (5.17), we assume the linear model

$$\boldsymbol{Y}_t = \boldsymbol{\alpha} + \boldsymbol{\beta} Y_t^m + \boldsymbol{\epsilon}_t, \qquad t = 1, \ldots, T \tag{5.24}$$

where

$$\boldsymbol{E}\boldsymbol{\epsilon}_t = 0, \quad \text{var}(\boldsymbol{\epsilon}_t) = \boldsymbol{\Sigma}, \quad \text{cov}(Y_t^m, \boldsymbol{\epsilon}_t) = 0,$$

for $t = 1, \ldots, T$. We assume in addition that $\boldsymbol{\epsilon}_t \sim_{i.i.d.} \mathcal{N}(0, \boldsymbol{\Sigma})$. We do not assume that $\boldsymbol{\Sigma}$ is diagonal for the reasons to be elucidated at the end of Section 5.5. When the CAPM holds, no portfolio should have a nonvanishing α so that the null hypothesis should be $\boldsymbol{\alpha} = \mathbf{0}$. In other words, our statistical testing problem becomes

$$H_0 : \boldsymbol{\alpha} = \mathbf{0} \qquad \text{vs.} \quad H_1 : \boldsymbol{\alpha} \neq \mathbf{0}. \tag{5.25}$$

While the above statistical problem is designed to validate CAPM, it can also be applied to check whether a constructed portfolio is efficient among the assets used to construct the portfolio. If it is efficient, then there should be no α for any assets that are used to construct the portfolio (see Theorem 5.2 or Exercise 5.10). We can then apply the above model with Y_t^m being the excess return of the constructed portfolio. The technique also allows us to check whether an analyst has stock picking ability. Among stocks that an analyst recommends, we can test whether there is any alpha in his/her recommendations.

5.4.2 Maximum likelihood estimation

Under model (5.24), given Y_t^m, the vector $\boldsymbol{Y}_t$ has a distribution $\mathcal{N}(\boldsymbol{\alpha} + \boldsymbol{\beta} Y_t^m, \boldsymbol{\Sigma})$. In addition, the vectors $\{\boldsymbol{Y}_t\}_{t=1}^T$ are independent given $\{Y_t^m\}_{t=1}^T$ since the random variables $\{\boldsymbol{\epsilon}_t\}_{t=1}^T$ are independent. Therefore, the conditional joint density of $\{\boldsymbol{Y}_t\}_{t=1}^T$ given $\{Y_t^m\}_{t=1}^T$ is the product of their marginal normal densities. By the formula for multivariate density (Anderson, 2003), we have

$$\begin{aligned}
& f(\boldsymbol{Y}_1, \ldots, \boldsymbol{Y}_T | Y_1^m, \ldots, Y_T^m) \\
=& \prod_{t=1}^T (2\pi)^{-N/2} |\boldsymbol{\Sigma}|^{-1/2} \exp\left[-\frac{1}{2} (\boldsymbol{Y}_t - \boldsymbol{\alpha} - \boldsymbol{\beta} Y_t^m)^T \boldsymbol{\Sigma}^{-1} (\boldsymbol{Y}_t - \boldsymbol{\alpha} - \boldsymbol{\beta} Y_t^m) \right] \\
=& (2\pi)^{-NT/2} |\boldsymbol{\Sigma}|^{-T/2} \exp\left[-\frac{1}{2} \sum_{t=1}^T (\boldsymbol{Y}_t - \boldsymbol{\alpha} - \boldsymbol{\beta} Y_t^m)^T \boldsymbol{\Sigma}^{-1} (\boldsymbol{Y}_t - \boldsymbol{\alpha} - \boldsymbol{\beta} Y_t^m) \right].
\end{aligned}$$

The conditional log-likelihood function, which is the logarithm of the conditional density as a function of parameters, is given by

$$\begin{aligned}
\ell(\boldsymbol{\alpha}, \boldsymbol{\beta}, \boldsymbol{\Sigma}) =& -\frac{NT}{2} \log(2\pi) - \frac{T}{2} \log |\boldsymbol{\Sigma}| \\
& -\frac{1}{2} \times \sum_{t=1}^T (\boldsymbol{Y}_t - \boldsymbol{\alpha} - \boldsymbol{\beta} Y_t^m)^{\mathrm{T}} \boldsymbol{\Sigma}^{-1} (\boldsymbol{Y}_t - \boldsymbol{\alpha} - \boldsymbol{\beta} Y_t^m). \quad (5.26)
\end{aligned}$$

The *likelihood function* indicates how likely the observed data is generated for a given set of values of parameters. It is the reverse engineering of the probability theory. See Section 2.5.4.

Different parameter values give different likelihood of generating the observed data $\{\boldsymbol{Y}_t\}_{t=1}^T$. The most probable one is the one that maximizes the likelihood function. The resulting estimate is the *maximum likelihood estimate (MLE)*, which is defined as

$$(\hat{\boldsymbol{\alpha}}, \hat{\boldsymbol{\beta}}, \hat{\boldsymbol{\Sigma}}) = \operatorname{argmax} \ell(\boldsymbol{\alpha}, \boldsymbol{\beta}, \boldsymbol{\Sigma}).$$

It can be derived that the maximum likelihood estimate is given by

$$\begin{aligned} \hat{\boldsymbol{\alpha}} &= \bar{\boldsymbol{Y}} - \hat{\boldsymbol{\beta}}\bar{Y}_m, \\ \hat{\boldsymbol{\beta}} &= \sum_{t=1}^{T} (\boldsymbol{Y}_t - \bar{\boldsymbol{Y}})(Y_t^m - \bar{Y}_m) / \sum_{i=1}^{T} (Y_t^m - \bar{Y}_m)^2, \\ \hat{\boldsymbol{\Sigma}} &= T^{-1} \sum_{t=1}^{T} \hat{\boldsymbol{\epsilon}}_t \hat{\boldsymbol{\epsilon}}_t^T, \end{aligned} \qquad (5.27)$$

where $\bar{\boldsymbol{Y}} = T^{-1}\sum_{t=1}^T \boldsymbol{Y}_t$ is the sample average of the excess return of the assets, $\bar{Y}_m = T^{-1}\sum_{t=1}^T Y_t^m$ is the mean of the excess return of the market, and $\hat{\boldsymbol{\epsilon}}_t = \boldsymbol{Y}_t - \hat{\boldsymbol{\alpha}} - \hat{\boldsymbol{\beta}}Y_t^m$ is the residual of the linear regression fit.

The formulas in (5.27) look complicated, but they are in fact very simple to comprehend. The estimators of $\boldsymbol{\alpha}$ and $\boldsymbol{\beta}$ are the same as those obtained by fitting (5.21) separately by the ordinary least squares. Let $(\hat{\alpha}_i, \hat{\beta}_i)$ be the ordinary least-squares fit to (5.21). Then, the least-squares estimates (5.27) are the same as

$$\hat{\boldsymbol{\alpha}} = (\hat{\alpha}_1, \ldots, \hat{\alpha}_N)^{\mathrm{T}}, \qquad \hat{\boldsymbol{\beta}} = (\hat{\beta}_1, \ldots, \hat{\beta}_N)^{\mathrm{T}},$$

whose components are obtained from the separate least-squares fit. The residual vector $\hat{\boldsymbol{\epsilon}}_t$ at time t is also the same as the vector of the residuals obtained from the individual least-squares fit. The cross-sectional covariance estimator $\hat{\boldsymbol{\Sigma}}$ is the sample covariance matrix of residual vectors [divided by T instead of $(T-2)$].

Note that the maximum likelihood estimate for $\boldsymbol{\alpha}$ and $\boldsymbol{\beta}$ are linear in $\{\boldsymbol{Y}_t\}_{t=1}^T$, whose conditional distribution is normal. Therefore, the conditional distributions of $\hat{\boldsymbol{\alpha}}$ and $\hat{\boldsymbol{\beta}}$ should also be normal, whose mean and variance can easily be derived. Since our inference is always conditioned upon $\{Y_t^m\}_{t=1}^T$, we will drop the word "conditioning on $\{Y_t^m\}_{t=1}^T$". It is well known that the least squares estimate is unbiased:

$$E\hat{\boldsymbol{\alpha}} = \boldsymbol{\alpha}, \quad E\hat{\boldsymbol{\beta}} = \boldsymbol{\beta},$$

where the expectation is really the conditional expectation given $\{Y_t^m\}_{t=1}^T$. The conditional variance can also be derived. In particular, it can be shown that

$$\mathrm{var}(\hat{\boldsymbol{\alpha}}) = T^{-1}(1 + \bar{Y}_m^2/\hat{\sigma}_m^2)\boldsymbol{\Sigma},$$

where $\hat{\sigma}_m$ = SD of $\{Y_t^m\}$. In summary,

$$\hat{\boldsymbol{\alpha}} \sim \mathcal{N}(\boldsymbol{\alpha}, T^{-1}(1 + \bar{Y}_m^2/\hat{\sigma}_m^2)\boldsymbol{\Sigma}). \tag{5.28}$$

This facilitates the inference about $\boldsymbol{\alpha}$, which centers our discussion.

5.4.3 Testing statistics

There are three commonly used techniques for testing statistical hypothesis. One of them is the *Wald test*, which is given by

$$\hat{\boldsymbol{\alpha}}^{\mathrm{T}}[\mathrm{var}(\hat{\boldsymbol{\alpha}})]^{-1}\hat{\boldsymbol{\alpha}} = T(1 + \bar{Y}_m^2/\hat{\sigma}_m^2)^{-1}\hat{\boldsymbol{\alpha}}^{\mathrm{T}}\boldsymbol{\Sigma}^{-1}\hat{\boldsymbol{\alpha}}.$$

Using (5.28), under the null hypothesis H_0, the test statistic follows the χ^2-distribution with degree of freedom N, denoted it by χ^2_N. However, this test statistic is infeasible, involving unknown covariance matrix $\boldsymbol{\Sigma}$. Using the estimated covariance matrix (5.27), we have the feasible Wald test defined by

$$T_0 = T(1 + \bar{Y}_m^2/\hat{\sigma}_m^2)^{-1}\hat{\boldsymbol{\alpha}}^{\mathrm{T}}\hat{\boldsymbol{\Sigma}}^{-1}\hat{\boldsymbol{\alpha}} \overset{a}{\sim}_{H_0} \chi^2_N, \tag{5.29}$$

where the notation $\overset{a}{\sim}_{H_0}$ means "distributed approximately under the null hypothesis H_0". It reminds us that distribution holds only approximately under the null hypothesis $H_0 : \boldsymbol{\alpha} = 0$. The last part of (5.29) holds only approximately because of the substitution of $\boldsymbol{\Sigma}$ by $\hat{\boldsymbol{\Sigma}}$. We will soon see that the approximation is not good for our applications.

Account for the substitution error of using $\hat{\boldsymbol{\Sigma}}$ instead of the true $\boldsymbol{\Sigma}$ in (5.29) is possible. Under the normal model, it can be shown from multivariate analysis that under H_0

$$T_1 = \frac{T - N - 1}{NT} T_0 \sim F_{N,T-N-1}. \tag{5.30}$$

This provides a way to determine the exact distribution of T_0 since it is linearly related to T_1.

One of the most commonly used techniques for testing statistical hypotheses is the *maximum likelihood ratio test* or the *Wilks test*. It compares the likelihood of generating the observed data under the whole parameter space with that under the null hypothesis which puts restrictions such as $\boldsymbol{\alpha} = 0$ on the parameters. When the whole parameter space is far more likely to produce the given data than the null parameter space, it provides evidence against the null hypothesis. Therefore, the likelihood ratio, or equivalently, the log-likelihood difference, provides an intuitively appealing approach to test against the null hypothesis. See Section 2.5.4.

The maximum of the log-likelihood under the full parameter space is given by

$$\max \ell(\boldsymbol{\alpha}, \boldsymbol{\beta}, \boldsymbol{\Sigma}) = \ell(\hat{\boldsymbol{\alpha}}, \hat{\boldsymbol{\beta}}, \hat{\boldsymbol{\Sigma}}),$$

which is the logarithm of the likelihood for generating the given data $\{Y_t\}$. Substituting (5.27) into (5.26), we obtain

$$\ell(\hat{\boldsymbol{\alpha}}, \hat{\boldsymbol{\beta}}, \hat{\boldsymbol{\Sigma}}) = -\frac{NT}{2} \log(2\pi) - \frac{T}{2} \log |\hat{\boldsymbol{\Sigma}}| - \frac{T}{2}. \tag{5.31}$$

To see this, let $\operatorname{tr}(\boldsymbol{A})$ be the trace of a square matrix $\boldsymbol{A}$, which is just the sum of the diagonal elements. After substitution of (5.27) into (5.26), the last term in (5.26) is simply $\sum_{t=1}^{T} \hat{\boldsymbol{\epsilon}}_t^{\mathrm{T}} \hat{\boldsymbol{\Sigma}}^{-1} \hat{\boldsymbol{\epsilon}}_t$. Each of the summand is a number, which equals to its trace. Using $\operatorname{tr}(\boldsymbol{AB}) = \operatorname{tr}(\boldsymbol{BA})$, the last term of (5.26) is given by

$$\sum_{t=1}^{T} \operatorname{tr}\{\hat{\boldsymbol{\epsilon}}_t^{\mathrm{T}} \hat{\boldsymbol{\Sigma}}^{-1} \hat{\boldsymbol{\epsilon}}_t\} = \sum_{t=1}^{T} \operatorname{tr}\{\hat{\boldsymbol{\Sigma}}^{-1} \hat{\boldsymbol{\epsilon}}_t \hat{\boldsymbol{\epsilon}}_t^{\mathrm{T}}\} = \operatorname{tr}\left\{\hat{\boldsymbol{\Sigma}}^{-1} \sum_{t=1}^{T} \hat{\boldsymbol{\epsilon}}_t \hat{\boldsymbol{\epsilon}}_t^{\mathrm{T}}\right\},$$

which is T by using the definition of $\hat{\boldsymbol{\Sigma}}$.

We can derive the maximum likelihood analogously under the null hypothesis $\boldsymbol{\alpha} = 0$. In this case, the estimate admits the same form as (5.27), except the intercepts are now known to be 0. More specifically, the maximum likelihood estimate under the null hypothesis is

$$\hat{\boldsymbol{\beta}}_0 = \sum_{t=1}^{T} \boldsymbol{Y}_t Y_t^m / \sum_{i=1}^{T} (Y_t^m)^2, \quad \hat{\boldsymbol{\Sigma}}_0 = T^{-1} \sum_{t=1}^{T} \hat{\boldsymbol{\epsilon}}_t^0 (\hat{\boldsymbol{\epsilon}}_t^0)^{\mathrm{T}}, \tag{5.32}$$

where $\hat{\boldsymbol{\epsilon}}_t^0 = \boldsymbol{Y}_t - \hat{\boldsymbol{\beta}}_0 Y_t^m$ is the residual under the null hypothesis. Again, these quantities can be obtained by running the marginal regression (5.21) separately without using the intercept. For example, $\hat{\boldsymbol{\epsilon}}_t^0$ is simply the vector of residuals at time t from running (5.21) separately without using the intercept. The maximum likelihood is analogously given by

$$\ell(\mathbf{0}, \hat{\boldsymbol{\beta}}_0, \hat{\boldsymbol{\Sigma}}_0) = -\frac{NT}{2} \log(2\pi) - \frac{T}{2} \log |\hat{\boldsymbol{\Sigma}}_0| - \frac{T}{2}. \tag{5.33}$$

By using (5.31) and (5.33), we obtain the maximum likelihood ratio test

$$
\begin{aligned}
T_2 &= 2\{\max \ell(\boldsymbol{\alpha}, \boldsymbol{\beta}, \boldsymbol{\Sigma}) - \max_{H_0:\boldsymbol{\alpha}=0} \ell(\boldsymbol{\alpha}, \boldsymbol{\beta}, \boldsymbol{\Sigma})\} \\
&= 2\{\ell(\hat{\boldsymbol{\alpha}}, \hat{\boldsymbol{\beta}}, \hat{\boldsymbol{\Sigma}}) - \max \ell(\mathbf{0}, \hat{\boldsymbol{\beta}}_0, \hat{\boldsymbol{\Sigma}}_0)\} \\
&= T(\log|\hat{\boldsymbol{\Sigma}}_0| - \log|\hat{\boldsymbol{\Sigma}}|). \qquad (5.34)
\end{aligned}
$$

The general likelihood ratio theory, also referred to as the *Wilks theorem*, shows that the test statistic is asymptotically χ^2-distributed, with the degree of freedom the same as the number of restrictions under the null hypothesis, which is N in this application. Therefore,

$$T_2 \overset{a}{\sim}_{H_0} \chi_N^2.$$

A better approximation (see Table 5.2) can be obtained by the following adjustment

$$T_3 = \frac{T - N/2 - 2}{T} T_2 \overset{a}{\sim}_{H_0} \chi_N^2. \qquad (5.35)$$

It can be shown that

$$T_2 = T \log\left(1 + \frac{NT_1}{T - N - 1}\right). \qquad (5.36)$$

Hence, T_0, T_1, T_2 and T_3 are equivalent, since they are monotonic functions of each other, but they result in different critical values. We summarize the above result in the following theorem.

Theorem 5.3 *The maximum likelihood estimator for* $\boldsymbol{\alpha}$ *and* $\boldsymbol{\beta}$ *in the multi-period model* (5.24) *is given by* (5.27). *The maximum likelihood ratio test statistic for testing* $H_0 : \boldsymbol{\alpha} = 0$ *is given by* (5.34) *whose asymptotic null distribution is* χ_N^2. *Its corrected version* (5.35) *has the same asymptotic null distribution.*

The exact distributions of T_0, T_2 and T_3 can be found via the exact distribution of T_1. Why do we introduce so many versions of the test? When the normality assumption holds, we would use T_1. However, the normality assumption might not hold. In this case, we would employ the test statistic T_3, since it does not explicitly use the normality assumption. Therefore, for practical purpose, considering T_1 and T_3 suffices. The former is more accurate when the normality assumption holds whereas the latter is expected to be more accurate when it does not hold. By the aggregational Gaussianity feature of financial returns, when the monthly returns are used, the Gaussianity assumption is not that unreasonable.

To see how good the asymptotic approximations are, let us assume that

Table 5.2 *The exact size of tests when the approximate null distribution* χ^2_N *is used with the target size of tests 5% for different sample size* T *and number of portfolios* N

	$N=10$			$N=20$			$N=40$		
sample size	T_0	T_2	T_3	T_0	T_2	T_3	T_0	T_2	T_3
60	0.170	0.096	0.051	0.462	0.211	0.057	0.985	0.805	0.141
120	0.099	0.070	0.050	0.200	0.105	0.051	0.610	0.275	0.059
180	0.080	0.062	0.050	0.136	0.082	0.051	0.368	0.164	0.053
240	0.072	0.059	0.050	0.109	0.073	0.050	0.257	0.124	0.052

the data are normal so that T_1 has an exact F-distribution. Using this as the golden standard, we can gauge the accuracy of the approximation of the null distribution of the test statistics T_0, T_2, T_3 by using χ^2_N. For example, when $N=10$ and $T=60$, the upper 5 percentile of χ^2_{10} is 18.31, namely

$$P(T_0 \geqslant 18.31) \approx 5\%,$$

by using the approximated distribution χ^2_{10}. On the other hand, the exact size of the test is

$$P(T_0 \geqslant 18.31) = P\left(T_1 \geqslant \frac{T-N-1}{NT} 18.31\right) = P(T_1 \geqslant 1.495).$$

Since $T_1 \sim F_{10,49}$, the exact probability is 17.0%, which is reported on the first cell of Table 5.2. Thus, the approximation (5.29) is indeed very poor. Similarly, for T_2 statistic, using (5.36), we have

$$\begin{aligned} P(T_2 \geqslant 18.31) &= P\{T_1 \geqslant 49/10[\exp(18.31/60) - 1]\} \\ &= P(T_1 \geqslant 1.749) = 9.6\%, \end{aligned}$$

reported also in Table 5.2. This is certainly a closer approximation, but is still very different from the target level of 5%. On the other hand, for test statistic T_3,

$$P(T_3 \geqslant 18.31) = P(T_2 \geqslant 60/53 \cdot 18.31) = P(T_1 \geqslant 2.022) = 5.1\%,$$

depicted also in Table 5.2. This is a much closer approximation. Table 5.2 summarizes the approximation results, in which the target size of the tests is 5%.

From Table 5.2, it is clear that the approximation of the distribution of T_0 by χ^2_N is very poor, even when the normality assumption holds. While the target size of the test is 5%, the actual size is far from 5%. The approximation gets better when the sample size increases. The null distribution of

Table 5.3 *Summary of individual stocks* $\boldsymbol{\alpha}$ *and* $\boldsymbol{\beta}$ *and expected return*

	Exxon	GE	JNJ
α	0.6848	−0.4155	0.2588
P-value for $H_0 : \alpha = 0$	0.1076	0.4445	0.4585
β	0.5416	1.3848	0.4927
Expected return (monthly)	0.3265%	0.4457%	0.3196%

T_3 approximated by χ^2_N is very accurate, except in the case where $T = 60$ and $N = 40$. This is why we prefer to use T_3 over T_0 or T_2.

Example 5.5 To test the Sharpe–Lintner version of CAPM, we took the S&P 500 index as a proxy for market portfolio, and the 3-month US treasury bill as a proxy for the risk-free asset. The study period is from February 2001 to January 2011, which spans a 10-year period. The monthly excess returns of the three companies Exxon–Mobil, Johnson and Johnson, General Electric are computed. For each individual stock, we run the simple linear regression against the excess returns of the market portfolio. The estimated market α and β as well as the P-values for testing individual $\alpha = 0$ are summarized in Table 5.3

From Table 5.3, the $\alpha's$ for individual stocks are not very statistically significant, as their corresponding P-values are larger than 10%. The test statistic T_3, for example, combines three individual tests into a joint test against CAPM. By forming the vectors of residuals from three individual fits, it can easily be computed that

$$T_0 = 3.90, \qquad d.f. = 3, \qquad \text{p-value} = 27.2\%,$$
$$T_1 = 1.26, \qquad d.f. = (3, 116), \qquad \text{p-value} = 29.1\%.$$

Hence the modified maximum likelihood ratio test can be obtained via

$$T_3 = (T - N/2 - 2)\log\left(1 + \frac{NT_1}{T - N - 1}\right) = 3.727,$$

with degree of freedom 3, giving a P-value of 29.2%. For returns of individual stocks, the normality assumption might not hold due to fat tails. The P-value produced by T_3 has more fidelity. In short, this small study does not provide strong evidence against CAPM, even though we use 10-year data rather than the 5-year data normally recommended for validating CAPM.

The individual regression yields the market β for Exxon to be $\beta = 0.5416$. The average monthly log-return for S&P 500 over last 15 years (Feb. 1996 to Jan. 2011) is $r_m = 0.391\%$ and the average risk-free rate over the same 15-year period is $r_f = 0.250\%$. According to CAPM (5.23), we have that the expected monthly log-return for Exxon is 0.3265%. The same computation applies to the expected returns of GE and Johnson and Johnson. They are also reported in Table 5.3.

Remark 5.2 For the linear model, $y_t = a + bx_t + \varepsilon_t$, the least-squares estimator satisfies [see the first equation of (5.27)]

$$\bar{y} = \hat{a} + \hat{b}\bar{x}.$$

Now, letting x and y be respectively the excess returns of the market portfolio and an asset, we have $\bar{R} - \bar{r}_f = \hat{a} + \hat{b}(\bar{r}_m - \bar{r}_f)$, or

$$\bar{R} = \hat{a} + \bar{r}_f + \hat{b}(\bar{r}_m - \bar{r}_f), \tag{5.37}$$

where $\bar{R}$ is the average return of the asset. Thus, if we use CAPM with $\bar{r}_f$ and $\bar{r}_m$ computed in the same period (2001-2011) in the Example 5.3, the CAPM predicted monthly return is $\bar{r}_f + \hat{b}(\bar{r}_m - \bar{r}_f)$, which differs from the actual average only by an amount of $\hat{a}$, according to (5.37). Thus, CAPM prediction merely replaces $\hat{a}$ by its theoretical value 0 in (5.37). If we compute $\bar{r}_m$ and $\bar{r}_f$ using a different period of data (e.g. 15 years data), the difference is hard to quantify.

5.5 Empirical studies

5.5.1 *An overview*

Empirical evidence on CAPM was largely positive in the early 1970's. There was not enough statistical evidence against the null hypothesis that CAPM holds. See, for example, Jensen, Black, and Scholes (1972) and Fama, and MacBeth (1973). However, the *anomaly* literature started to emerge in the late 1970's. Anomalies can be thought of as portfolios created by grouping characteristics of firms so that they can have higher *Sharpe ratio* than that of the proxy of the market portfolio. For example, Basu (1977) observed the PE effect: Firms with low price-earning ratios have higher sample returns than those predicted by CAPM; Banz (1981) reported the size effect: Low market capitalization firms have higher sample mean returns. Value effect and size were also found (Fama and French, 1992, 1993): Firms with high book-to-market ratio or low market capitalization can have higher average returns than those predicted by the CAPM. Style effects were also postulated: Buying losers and selling winners has higher average return than what the CAPM predicts (De Bondt, and Thaler, 1985, Jegadeesh and Titman, 1993)). See Campbell, Lo and MacKinlay (1997) for additional details.

There are also counter arguments to the above findings. First of all, the proxy of market portfolio is not broad enough. The market portfolio should include broader financial assets such as bonds, real estate, foreign assets, among others. There are issues of *data-snooping* and *bias sampling* in the data collection and measurement stage. Examples of data snooping include choosing study periods to make CAPM less favorable, and constructing portfolios with different parameters and choosing to report only unfavorable results (e.g. long bottom f percent but short top f percent of stocks in terms of market capitalization and choose f to make CAPM fail). The first situation can also be regarded as bias sampling. This is not to say that data snooping has actually been done in the above empirical studies, but to give readers the concept of data snooping and bias sampling. Thirdly, the CAPM was a one-period model, yet multi-period data are used to test the model. When multi-period data are used, the model (5.24) should have time-varying coefficients $\boldsymbol{\alpha}_t$ and $\boldsymbol{\beta}_t$ (see e.g. Figure 5.3) instead of a constant market $\boldsymbol{\alpha}$ and $\boldsymbol{\beta}$.

We outline the above empirical results and counter arguments to help readers understand better CAPM and potential challenges in the empirical studies. We do not take a side in the debates. However, it is widely believed that one market factor alone is not adequate to capture all cross-sectional risks. Multiple factors will work better. This is the subject of the next chapter.

5.5.2 Fama–French portfolios

To test the Sharpe–Lintner version of CAPM by test statistics introduced in §5.4, we use the CRSP value-weighted index as a proxy of the market portfolio. The returns of one-month Treasury Bill are used as the proxy of risk-free rates. Six valued weighted portfolios ($N = 6$) are formed by using two factors: market equity (classified as small and big) and the ratio of book equity to market equity (labeled as value, neutral and growth). The first factor measures the size effect and the second factor represents the value effect. They are constructed according to the intersections of 2 portfolios formed by the size (small and big) and three portfolios based on the book-to-market ratio (value, neutral and growth):

Size	Book-to-Market Ratio		
	Value	Neutral	Growth
Small	small value	small neutral	small growth
Big	big value	big neutral	big growth

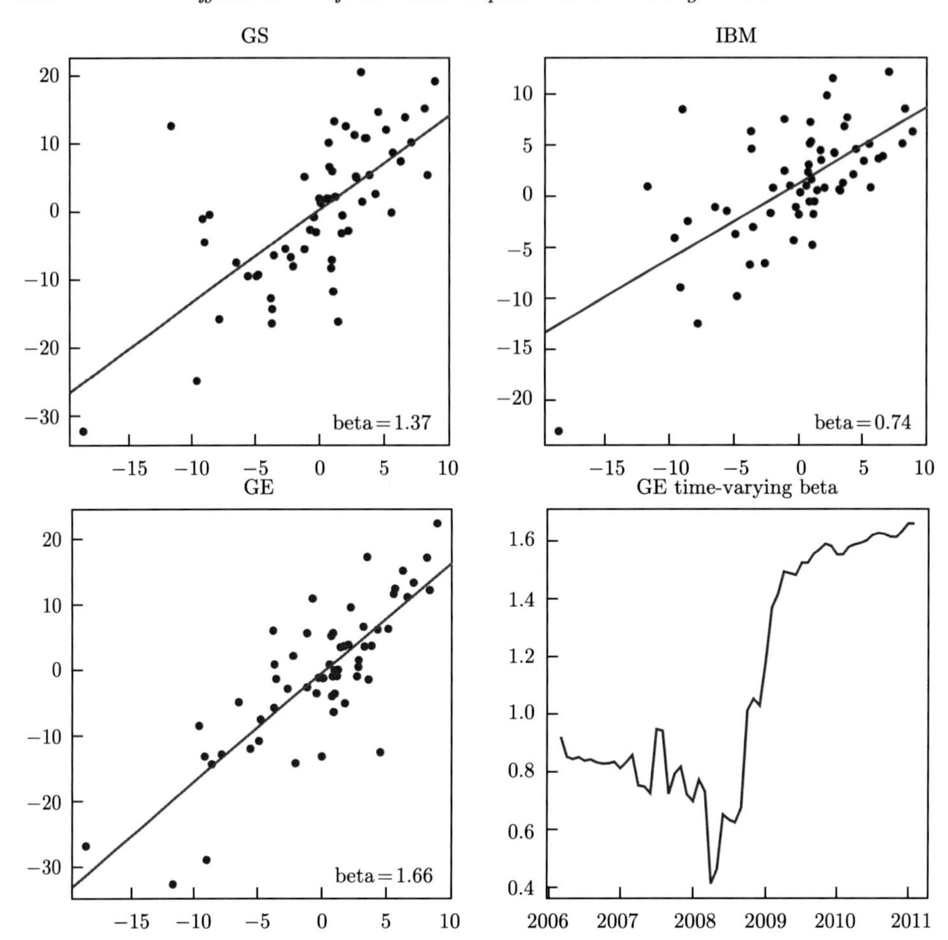

Figure 5.3 The market beta's for Goldman Sachs Group, International Business Machines, General Electric Company on January 31, 2011 and the market beta of GE from January 2006 to January 2011. The horizontal axis is the monthly returns in percent of S&P 500 index and the vertical axis is the monthly returns of stocks (in percent). The slope shows the market β. The bottom right panel shows the time-market beta of GE from 2006 to 2011.

The data were downloaded from the Data Library of Kenneth French. More details can be found on that website. The study focuses on the monthly returns in post-war periods, starting from January 1951 to December 2010. The aggregational Gaussian property makes the normality assumption more reasonable with monthly returns. The results are summarized in Table 5.4 below. The testing periods are grouped by using 12 five-year periods.

The test results are mixed. Among the three test statistics, the χ^2-approximation to the null distributions is most accurate for T_3 statistic. While T_1

Table 5.4 *Empirical results for tests of the Sharpe–Lintner version of the CAPM*

Time	T_0	p-value	T_1	p-value	T_3	p-value
1/51-12/55	6.97	32.36	1.03	41.87	6.04	41.82
1/56-12/60	11.89	6.44	1.75	12.74	9.95	12.70
1/61-12/65	33.72	0.00	4.96	0.04	24.53	0.04
1/66-12/70	10.01	12.42	1.47	20.51	8.49	20.46
1/71-12/75	23.46	0.07	3.45	0.59	18.15	0.59
1/76-12/80	21.09	0.18	3.11	1.11	16.57	1.10
1/81-12/85	69.14	0.00	10.18	0.00	42.16	0.00
1/86-12/90	26.96	0.01	3.97	0.24	20.41	0.23
1/91-12/95	32.06	0.00	4.72	0.06	23.54	0.06
1/96-12/00	17.18	0.86	2.53	3.16	13.85	3.14
1/01-12/05	17.52	0.75	2.58	2.88	14.09	2.86
1/06-12/10	10.32	11.18	1.52	18.98	8.73	18.93

gives the exact P-value, we are uncertain about the validity of the normal assumption. Therefore, we place more faith in the P-values produced by T_3 statistic. The results produced by T_1 and T_3 are approximately the same.

There are many periods that the P-values are less than 5%, which suggests many testing results are statistically significant. However, statistically significant results prove only that the evidence against the null hypothesis $H_0 : \boldsymbol{\alpha} = 0$ is strong, but it does not imply that the difference from zero is important. The estimated values of $\boldsymbol{\alpha}$, in comparison to the level of idiosyncratic noises $\boldsymbol{\Sigma}$, indicate the importance of the deviations. For each of the six portfolios, we present, in Figure 5.4, its estimated value $\hat{\alpha}_i$, its standardized multiple R-square, the t-statistics for testing the individual hypothesis $H_0 : \alpha_i = 0$ over 12 five-year time periods, where $\hat{\sigma}_i^2$ is the residual variance of the ith portfolio.

From Figure 5.4, most estimated α's are smaller than half of the idiosyncratic noises, making *statistical arbitrages* hard. The alternating signs make the task even harder. The individual tests for the null hypothesis $H_0 : \alpha_i = 0$ are largely insignificant over each 6 portfolios across 12 five-year time periods. However, the combined tests using T_0, T_1 and T_3 provide evidence of the deviation from the CAPM, though the deviations are not very large. For instance, if the significance level of 1% is used, many of the tests are indeed statistically insignificant.

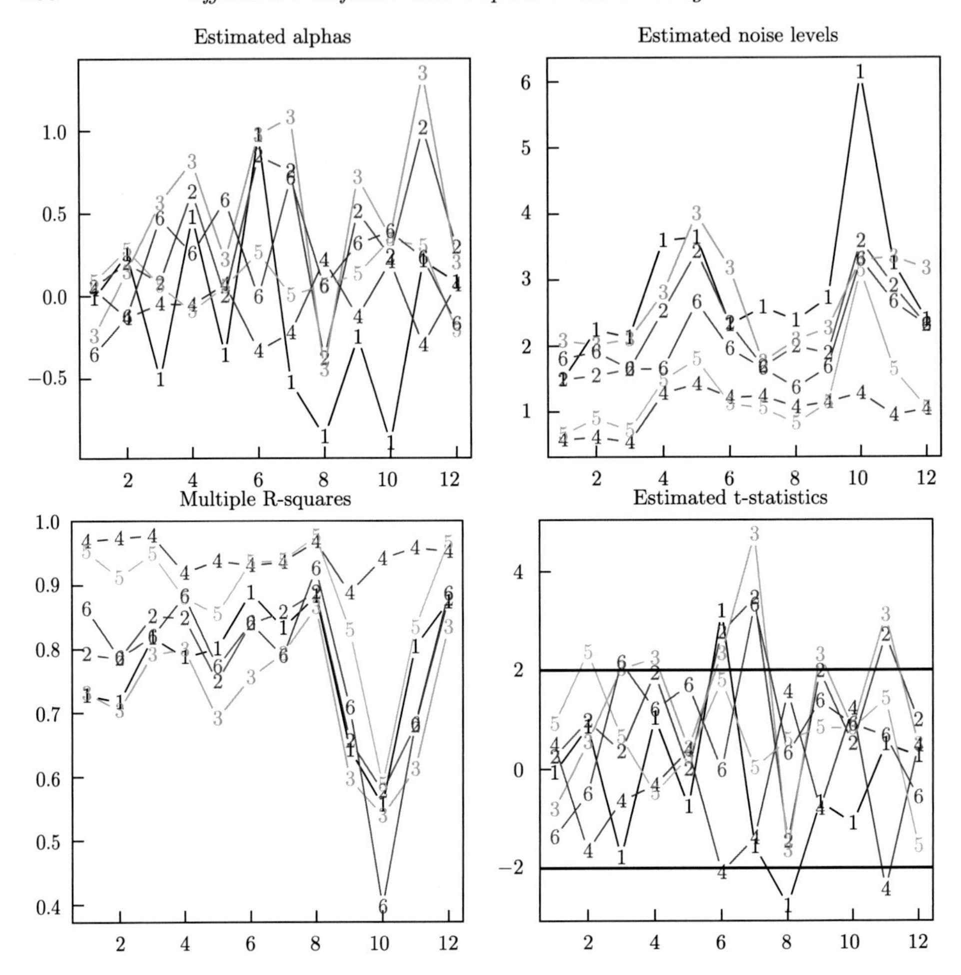

Figure 5.4 The estimated α's, standard deviations of idiosyncratic noises, multiple R-squares, and t-statistics for 6 Fama–French portfolios over 12 five-year periods from January 1951 to December 2010. They are obtained from the marginal regression (5.21) for each i over a five-year time window. The horizontal axis is time period and vertical axis is monthly returns or volatilities in percents for the top panel. Portfolios 1–6 are respectively small-value, small-neutral, small-growth, large-value, large-neutral, large-growth.

5.5.3 Further remarks

We would like to conclude this section with a few technical notes. First of all, the variances for individual stocks are usually much larger than those for portfolios. Individual stocks are not usually powerful enough to detect the small deviations from CAPM due to large variances. For this reason, portfolios are frequently used to validate CAPM. In addition, portfolio returns typically behave more like a normal distribution than individual stocks, particularly for monthly returns. For monthly returns of portfolios, the independence and normality assumptions are reasonable. This makes our test statistically valid. Secondly, the constant coefficients $\boldsymbol{\alpha}$ and $\boldsymbol{\beta}$ cannot be expected to hold over a long period of time (see Figure 5.3). Therefore, we can not use a lengthy time series since model (5.24) is an assumption that holds for multiple periods. Because of the length of the time series, usually $T = 60$ months, we use portfolios to reduce noises. Now, for the asymptotic distributions, such as (5.35), to hold with a reasonable accuracy, we need $T - N/2$ to be large [see also (5.30)]. This limits the choice of N, usually in the range from 5 to 20 [see also Table 5.2]. With this limitation, we do not want to use individual stocks, but use more broadly represented portfolios. Since the portfolios are used, even when the idiosyncratic noises are independent for individual stocks, $\boldsymbol{\Sigma}$ in (5.24) is usually not diagonal.

New econometric techniques are needed to accommodate the case $N > T$. When such techniques are introduced, individual stocks should be used to test CAPM. Pesaran and Yamagata (2012) and Fan, Liao and Yao (2015) introduced such new econometric techniques. The latter furthers the power of such large panel tests with *power enhancement* techniques.

Finally, we would like to remark that the P-values based on the last decade's data tend to be larger. This is again due to the large volatilities of the stocks in several sub-periods that reduce the power of the test.

5.6 Cross-sectional regression

Blume and Friend (1973) and Fama and MacBeth (1973) introduced the following cross-sectional regressions for testing the Sharpe–Lintner version of CAPM. An extension of the technique can be found in Gagliardini, Ossola, and Scaillet (2016). Note that under the CAPM, by (5.20), we have

$$\xi_j = \boldsymbol{E} Y_{j,t} = \lambda \beta_j, \quad \text{for } j = 1, \ldots, N, \tag{5.38}$$

where $\lambda = \boldsymbol{E} Y_t^m > 0$ is the *risk premium*. This simple relation leads to the following simpler method. Let $\hat{\xi}_j = T^{-1} \sum_{t=1}^{T} Y_{j,t}$ be the estimated excess return and

$$\hat{\beta}_j = \frac{\operatorname{cov}\{(Y_{jt}, Y_t^m), t = 1, \ldots, T\}}{\operatorname{var}(Y_t^m, t = 1, \ldots, T)}$$

be the estimate of market β for the j asset [see (5.19)]. Then, (5.38) suggests the following cross-sectional regression:

$$\hat{\xi}_j = c_0 + c_1 \hat{\beta}_j + \varepsilon_j, \quad j = 1, \ldots, N. \tag{5.39}$$

Let $\hat{c}_0$ and $\hat{c}_1$ be the fitted coefficients. If the CAPM holds, the following three properties should be true for cross-sectional fit:

1. the pairs $\{(\hat{\beta}_j, \hat{\xi}_j)\}$ should have a linear trend with a large multiple R^2.
2. $\hat{c}_0$ is statistically insignificant. The P-value for testing $H_0 : c_0 = 0$ should not be small;
3. the estimated risk premium $\hat{c}_1$ is statistically positive;

One of the drawbacks of the cross-sectional approach is that $\hat{\beta}_j$ is not true β_j that we need. The errors-in-variables in $\hat{\beta}_j$ create biases in estimating c_0 and c_1 (see Carroll *et al.*, 2006). In addition, the errors $\{\varepsilon_j\}_{j=1}^N$ might not be exogenous: the correlation between ε_j and $\hat{\beta}_j$ might not be zero. There are also issues such as the sector correlations that make P-value computation inaccurate.

5.7 Portfolio optimization without a risk-free asset

In Section 5.3, we derived the capital asset pricing model under the existence of a risk-free asset using a financial economics approach. In this section, we will derive a similar result without the assumption of the existence of risk-free interest using a mathematical approach. The result applies to any set of risky assets, not necessarily to the entire market. The results and approaches in Section 4.3 and here give a comprehensive view on the implications of CAPM.

We continue to use the notation in Section 4.1. The returns $\boldsymbol{R}$ of the p risky assets under consideration have the expected value $\boldsymbol{\mu}$ and covariance matrix $\boldsymbol{\Sigma}$. Let $\boldsymbol{\alpha}$ be the allocation vector, satisfying $\boldsymbol{\alpha}^{\mathrm{T}}\mathbf{1} = 1$, where $\mathbf{1} = (1, \ldots, 1)^{\mathrm{T}}$. The portfolio optimization can be formulated as follows. Given a target return μ, we wish to minimize the portfolio variance:

$$\min_{\boldsymbol{\alpha}} \boldsymbol{\alpha}^{\mathrm{T}} \boldsymbol{\Sigma} \boldsymbol{\alpha}, \qquad \text{s.t. } \boldsymbol{\alpha}^{\mathrm{T}} \boldsymbol{\mu} = \mu \text{ and } \boldsymbol{\alpha}^{\mathrm{T}} \mathbf{1} = 1. \tag{5.40}$$

The essential difference of the problem (5.40) from that in Section 4.1 is the possible absence of the risk-free asset. The existence of a risk-free interest asset allows us to remove one constraint [see (5.1)] so that the problem becomes simpler. The problem was further simplified when target (5.4) is used so that only unconstrained quadratic optimization (5.4) is considered.

When there exists a risk-free asset, say the first one, then by definition, the variance of its return $\mathrm{var}(R_1) = 0$. Hence, the first row and first column of $\boldsymbol{\Sigma}$ must be zero and the portfolio risk does not depend on α_1. The constraint on expected return can now be expressed as the constraint on the excess return

$$\sum_{i=2}^{p} \alpha_i(\mu_i - \mu_1) = \mu - \mu_1$$

so that the first constraint does not depend on α_1 either. Hence, the constraint $\boldsymbol{\alpha}^{\mathrm{T}}\mathbf{1} = 1$ can be removed and problem (5.40) reduces to the problem in Section 5.1 (see Remark 5.1). In short, problem (5.40) includes problem (5.4) as a specific case.

The constrained optimization is usually handled by the *Lagrange multiplier* method. Introducing multipliers λ_1 and λ_2, the optimization problem (5.40) becomes the unconstrained multiplier one: minimizing

$$\frac{1}{2}\boldsymbol{\alpha}^{\mathrm{T}}\boldsymbol{\Sigma}\boldsymbol{\alpha} + \lambda_1(\mu - \boldsymbol{\alpha}^{\mathrm{T}}\boldsymbol{\mu}) + \lambda_2(1 - \boldsymbol{\alpha}^{\mathrm{T}}\mathbf{1}).$$

Taking derivative with respect to $\boldsymbol{\alpha}$ and setting it to zero, we obtain

$$\boldsymbol{\Sigma}\boldsymbol{\alpha}^* - \lambda_1\boldsymbol{\mu} - \lambda_2\mathbf{1} = 0,$$

or

$$\boldsymbol{\alpha}^* = \lambda_1\boldsymbol{\Sigma}^{-1}\boldsymbol{\mu} + \lambda_2\boldsymbol{\Sigma}^{-1}\mathbf{1}. \tag{5.41}$$

The multipliers λ_1 and λ_2 are determined by

$$\boldsymbol{\alpha}^{*\mathrm{T}}\boldsymbol{\mu} = \mu \quad \text{and} \quad \boldsymbol{\alpha}^{*\mathrm{T}}\mathbf{1} = 1. \tag{5.42}$$

The solution to the above linear equations is easy to obtain.

It is more instructive to express solution (5.41) explicitly in terms of the target expected return μ. It can be computed (see the computation in Section 5.9) that

$$\boldsymbol{\alpha}^* = \mathbf{g} + \mu\boldsymbol{h}, \tag{5.43}$$

where

$$\mathbf{g} = D^{-1}[B\boldsymbol{\Sigma}^{-1}\mathbf{1} - A\boldsymbol{\Sigma}^{-1}\boldsymbol{\mu}]$$

and

$$\boldsymbol{h} = D^{-1}[C\boldsymbol{\Sigma}^{-1}\boldsymbol{\mu} - A\boldsymbol{\Sigma}^{-1}\mathbf{1}],$$

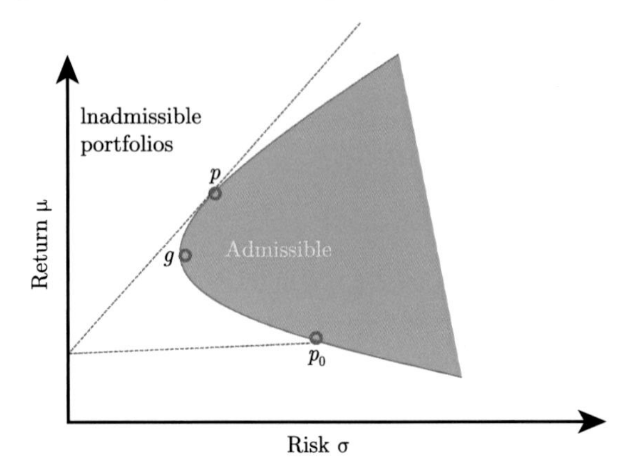

Figure 5.5 Minimum-variance portfolios without a risk-free asset. g is the minimum variance portfolio; p is a tangent portfolio and p_0 is its zero-beta portfolio.

with $A = \mathbf{1}^{\mathrm{T}}\boldsymbol{\Sigma}^{-1}\boldsymbol{\mu}, B = \boldsymbol{\mu}^{\mathrm{T}}\boldsymbol{\Sigma}^{-1}\boldsymbol{\mu}, C = \mathbf{1}^{\mathrm{T}}\boldsymbol{\Sigma}^{-1}\mathbf{1}$ and $D = BC - A^2$. Hence, the portfolio variance of the optimal portfolio is

$$\sigma^2 = (\mathbf{g} + \mu \boldsymbol{h})^{\mathrm{T}}\boldsymbol{\Sigma}(\mathbf{g} + \mu \boldsymbol{h}). \tag{5.44}$$

The right hand side is a simple quadratic function in μ. Simple algebra shows that (5.44) can be written as (this is a special case of (5.48); see Section 5.9 for computation)

$$C\sigma^2 - C^2/D(\mu - A/C)^2 = 1. \tag{5.45}$$

This parabola defines an *efficient frontier* in the risk-return space of (σ, μ). For any portfolio with expected return μ, its variance cannot be smaller than σ. See Figures 5.5.

Example 5.6 Let us consider the three risky assets in Example 5.1 (taking away the riskless asset). The covariance matrix is still the same as the one given in Example 5.1. With the expected returns and covariance matrix, it can easily be computed that

$$A = 6.4757, \quad B = 0.7275, \quad C = 92.5366, \quad D = 25.3932$$

and

$$\mathbf{g} = \begin{pmatrix} 1.5052 \\ -0.4244 \\ -0.0807 \end{pmatrix}, \qquad \boldsymbol{h} = \begin{pmatrix} -4.7307 \\ 3.7627 \\ 0.9680 \end{pmatrix}.$$

Therefore, for any given target expected return μ, the optimal portfolio allocation is

$$\boldsymbol{\alpha}^* = \begin{pmatrix} 1.5052 \\ -0.4244 \\ -0.0807 \end{pmatrix} + \mu \begin{pmatrix} -4.7307 \\ 3.7627 \\ 0.9680 \end{pmatrix}.$$

It is easy to verify that $g_1 + g_2 + g_3 = 1$ and $h_1 + h_2 + h_3 = 0$ so that the above is a valid portfolio (sum of components of $\boldsymbol{\alpha}$ is one) for any μ. By (5.44), its associated risk is given by

$$\begin{aligned}\sigma^2 &= \mathbf{g}^{\mathrm{T}}\boldsymbol{\Sigma}\mathbf{g} + 2\mu\boldsymbol{h}^{\mathrm{T}}\boldsymbol{\Sigma}\mathbf{g} + \mu^2\boldsymbol{h}^{\mathrm{T}}\boldsymbol{\Sigma}\mathbf{g} \\ &= 0.0287 - 0.5110\mu + 3.6442\mu^2.\end{aligned}$$

This parabola defines the efficient frontier given by (5.45). For example, if the target return is $\mu = 0$, then the minimum portfolio variance is 0.0287 and its associated standard deviation 16.94%.

Similarly, for $\mu = 18.48\%$, the same expected return as the optimal portfolio in Example 5.1, then the allocation vector is

$$\alpha^* = (0.6310, 0.2709, 0.0982)^{\mathrm{T}},$$

i.e. 63.10% on Asset 1, 27.09% on Asset 2, and 9.82% on Asset 3. Its associated variance is

$$0.0287 - 0.5110 * .1848 + 3.6442 * .1848^2 = 0.0589$$

and its standard deviation is 24.27%. Comparing with the portfolio standard deviation of 24.15% given in Example 5.1 with the same expected return of 18.48%, the optimal portfolio here has a larger volatility.

The last result is expected, since Example 5.1 is a specific case of our portfolio optimization problem with the fourth asset whose variance is zero. The optimization on 4-dimensional space always yields no worse results than the optimization on the 3-dimensional space. It also shows that adding a risk-free bond can improve the portfolio efficiency.

We now highlight some properties of the efficient frontier. First of all, by (5.45),

$$\sigma^2 = C^{-1} + C(\mu - A/C)^2/D \geqslant C^{-1}. \tag{5.46}$$

Thus, the *global minimum variance* is $\sigma_g^2 = C^{-1}$. It corresponds to the point (σ_g, μ_g) in the efficient frontier with $\mu_g = A/C$. This can easily be seen from (5.46) that when $\mu = A/C$, the last inequality therein becomes an equality. The allocation vector of the global minimum variance portfolio is $\alpha_g = \mathbf{g} + \mu_g \boldsymbol{h}$. See Figure 5.5. It is the solution to the risk minimization problem:

$$\min_{\boldsymbol{\alpha}} \boldsymbol{\alpha}^{\mathrm{T}} \boldsymbol{\Sigma} \boldsymbol{\alpha}, \qquad \text{s.t. } \boldsymbol{\alpha}^{\mathrm{T}} \mathbf{1} = 1. \tag{5.47}$$

The expected return of the global minimum portfolio is $\mu_g = A/C$.

Any efficient portfolio whose expected return is less than μ_g is practically inefficient. It is dominated by the global minimum variance portfolio, which has a smaller risk and higher expected return. From Figure 5.5, such a portfolio exists and solves mathematically problem (5.40). Therefore, when the parameter μ is smaller than μ_g, while the optimal portfolio still exists with the targeted return, such a portfolio is not practically admissible. Thus, in application, we need to determine the expected return μ_g of the global minimum portfolio, from which, admissible target returns $\mu \geqslant \mu_g$ can be set.

Example 5.6 (continued) When $\mu = 0.05$, the portfolio standard deviation is 11.10%. This is more efficient than the optimal portfolio with $\mu_g = 0$, since the latter has a larger risk. However, the latter portfolio is still optimal among the portfolios with expected return 0, though it is practically inadmissible. This raises the question whether the target return 5% or 18.48% is large enough to be practically admissible. They should be compared with the expected return of the global minimum portfolio, which is

$$A/C = 7.00\%.$$

Therefore, the optimal portfolio with expected return 5% is not admissible, but the optimal portfolio with expected return 18.48% is admissible. The standard deviation of the global minimum variance portfolio is $\sqrt{1/C} =$ 10.40%. This is the portfolio with the minimum possible risk constructed based on the three risky assets in Example 5.1. We do not know the Sharpe ratios of those portfolios, since there is no riskless portfolio in this example.

We summarize the above result and other properties in the following theorem. The theorem reveals that for any portfolio p on the efficient frontier, there must exist another portfolio on the efficient frontier p_0 such that these two portfolios are uncorrelated. The latter can be obtained (see Figure 5.5) by drawing a tangent line along the efficient frontier at the point p, intersecting the line with the vertical axis, and moving horizontally to the efficient frontier. The portfolio p is called a *tangent portfolio* and p_0 is called its associated *zero-beta portfolio*.

Theorem 5.4 *For any given p risky assets with $\boldsymbol{\Sigma} > 0$,*

(i) *there exists a minimum global portfolio* g *with the allocation vector* $\alpha_g = \mathbf{g} + \mu_g \boldsymbol{h}$ *vector and expected return* $\mu_g = C/A$ *that has the smallest possible variance* $\sigma_g^2 = 1/C$*;*

(ii) *for any portfolio* p *on the efficient frontier* (5.45)*, there exists a zero-beta portfolio* p_0 *that is uncorrelated with* p*. Such a zero-beta portfolio can be obtained by the geometry presented in Figure 5.5, whose expected return is given by* (5.49)*.*

Proof We need only to prove the second result. First of all, the covariance of the returns for any two portfolios is

$$\text{cov}(\boldsymbol{\alpha}_1^{\text{T}} \boldsymbol{R}, \boldsymbol{\alpha}_2^{\text{T}} \boldsymbol{R}) = \boldsymbol{\alpha}_1^{\text{T}} \boldsymbol{\Sigma} \boldsymbol{\alpha}_2.$$

In particular, the covariance of the returns R_p and R_q of two frontier portfolios with allocation vectors $\boldsymbol{\alpha}_1 = \mathbf{g} + \mu \boldsymbol{h}$ and $\boldsymbol{\alpha}_2 = \mathbf{g} + \mu_q \boldsymbol{h}$ is

$$\text{cov}(R_p, R_q) = (\mathbf{g} + \mu \boldsymbol{h})^{\text{T}} \boldsymbol{\Sigma} (\mathbf{g} + \mu_q \boldsymbol{h}).$$

Substituting the definition of $\mathbf{g}$ and $\boldsymbol{h}$ and unfoiling the above terms, we can easily see that each of these individual terms can be expressed in terms of A or B or C. After some tedious algebra (see Section 5.9 for details), we obtain

$$\text{cov}(R_p, R_q) = C/D \cdot (\mu - A/C)(\mu_q - A/C) + C^{-1}. \tag{5.48}$$

For each frontier portfolio p, there exists a p_0 with the expected return

$$\mu_{p_0} = \frac{A}{C} - \frac{D}{C^2(\mu_p - A/C)} \tag{5.49}$$

that is uncorrelated with portfolio p. This can easily be obtained by setting (5.48) to zero and solving for μ_q. The solution is given by (5.49).

It remains to establish the geometric interpretation of the zero-beta portfolio. Let us first determine the slope. From (5.45), we have

$$\sigma d\sigma - C/D \cdot (\mu - A/C) d\mu = 0.$$

Thus, the slope at point p, whose expected return and volatility are denoted as μ_p and σ_p, is given by

$$\frac{d\mu_p}{d\sigma_p} = \frac{\sigma_p D}{C\mu_p - A}.$$

It can easily be verified that by (5.46),

$$\begin{aligned}\mu_p - \frac{d\mu_p}{d\sigma_p}\sigma_p &= \mu_p - \frac{\sigma_p^2 D}{C\mu_p - A} \\ &= \mu_p - \frac{D\{C^{-1} - C/D \cdot (\mu_p - A/C)^2\}}{C\mu_p - A} \\ &= \mu_{p0}.\end{aligned}$$

This establishes the geometry of the zero-beta portfolio and completes the proof. □

Example 5.6 (continued) For the optimal portfolio with expected return $\mu_p = 18.48\%$, its zero-beta portfolio has expected return

$$\mu_{p0} = \frac{A}{C} - \frac{D}{C^2(\mu_p - A/C)} = 4.415\%.$$

Its associated portfolio allocation vector is $\boldsymbol{\alpha}_{p0}^* = (1.2963, -0.2583, -0.0380)^{\mathrm{T}}$ with portfolio risk 11.53%. It is easy to verify that the zero-beta portfolio is uncorrelated with the optimal portfolio with expected return $\mu_p = 18.48\%$:

$$(0.6310, 0.2709, 0.0982)\boldsymbol{\Sigma}(1.2963, -0.2583, -0.0380)^T = 0.$$

Similarly to (5.16), any random variable Y can be decomposed as

$$Y = \beta_0 + \beta_1 X_1 + \beta_2 X_2 + \varepsilon, \qquad E\varepsilon = 0,$$

in which ε is idiosyncratic noise, uncorrelated with X_1 and X_2:

$$\mathrm{cov}(X_1, \varepsilon) = 0, \qquad \mathrm{cov}(X_2, \varepsilon) = 0.$$

By taking $X_1 = R_{p0}$ and $X_2 = R_p$, applying the above decomposition to each portfolio with return R_a, constructed based on the p risky assets with return vector $\boldsymbol{R}$, we have the following result.

Theorem 5.5 *The return of any portfolio R_a, constructed based on the risky assets $\boldsymbol{R}$, can be decomposed as*

$$R_a = \beta_1 + \beta_2 R_{p0} + \beta_3 R_p + \varepsilon, \tag{5.50}$$

where ε is idiosyncratic noise, uncorrelated to the tangent portfolio R_p and its associated zero-beta portfolio R_{p0}, provided that $\mu \neq \mu_g$, i.e., R_p is not the global minimum portfolio. Furthermore,

$$\beta_1 = 0, \quad \beta_2 = 1 - \beta_3, \quad \beta_3 = \mathrm{cov}(R_a, R_p)/\sigma_p^2,$$

where $\sigma_p^2 = \mathrm{var}(R_p)$. *In other words,*

$$ER_a - ER_{p0} = \beta_{ap} E(R_p - R_{p0}), \tag{5.51}$$

where $\beta_{ap} = \mathrm{cov}(R_a, R_p)/\sigma_p^2$ *is the beta of portfolio* R_a *with respect to the tangent portfolio* R_p.

Proof Let $\mu_a = ER_a$ and $R_a^* = (\mathbf{g} + \mu_a \boldsymbol{h})^{\mathrm{T}} \boldsymbol{R}$ be the tangent portfolio with expected return μ_a. Set $\varepsilon = R_a - R_a^*$. Since both R_a and R_a^* are portfolios constructed based on $\boldsymbol{R}$, we can write $\varepsilon = \boldsymbol{b}^{\mathrm{T}} \boldsymbol{R}$ for some coefficient vector $\boldsymbol{b}$. The coefficient vector must satisfy

$$\boldsymbol{b}^{\mathrm{T}} \mathbf{1} = 0, \qquad \boldsymbol{b}^{\mathrm{T}} \boldsymbol{\mu} = 0,$$

as ε is a long-short portfolio with zero expected return $E\varepsilon = 0$. Therefore, using the definition of $\mathbf{g}$ and $\boldsymbol{h}$, we have

$$\mathrm{cov}(\varepsilon, \mathbf{g}^{\mathrm{T}} \boldsymbol{R}) = \boldsymbol{b}^{\mathrm{T}} \boldsymbol{\Sigma} \mathbf{g} = \boldsymbol{b}^{\mathrm{T}} (B\mathbf{1} - A\boldsymbol{\mu})/D = 0$$

and

$$\mathrm{cov}(\varepsilon, \boldsymbol{h}^{\mathrm{T}} \boldsymbol{R}) = \boldsymbol{b}^{\mathrm{T}} \boldsymbol{\Sigma} \boldsymbol{R} = \boldsymbol{b}^{\mathrm{T}} (C\boldsymbol{\mu} - A\mathbf{1})/D = 0.$$

Consequently, by (5.43), ε is uncorrelated with all tangent portfolios, including R_p and R_{p0}.

Next, it is easy to see that $R_a^* = w^* R_p + (1 - w^*) R_{p0}$, by taking w^* to satisfy

$$w^* \mu + (1 - w^*) \mu_{p0} = \mu_a, \quad \text{or} \quad w^* = (\mu_a - \mu_{p0})/(\mu - \mu_{p0}).$$

Combining the above results, we have

$$R_a = R_a^* + \varepsilon = w^* R_p + (1 - w^*) R_{p0} + \varepsilon,$$

where ε is uncorrelated with R_p and R_{p0}. Therefore,

$$\beta_1 = 0, \quad \beta_2 = w^*, \qquad \beta_3 = 1 - w^*.$$

This proves the first conclusion.

An alternative way of expressing β_3 is as follows. By the decomposition (5.50), recalling the zero correlation between R_p and R_{p0}, we have

$$\mathrm{cov}(R_a, R_p) = \mathrm{cov}(\beta_1 + \beta_2 R_{p0} + \beta_3 R_p + \varepsilon, R_p) = \beta_3 \sigma_p^2,$$

which yields $\beta_3 = \mathrm{cov}(R_a, R_p)/\sigma^2$. Similarly, we have

$$\beta_2 = \mathrm{cov}(R_a, R_{p0})/\sigma_{p0}^2.$$

The above result shows also that $\beta_2 = w^*$ and $\beta_3 = 1 - w^*$. This completes the proof. □

Let us now interpret the result of Theorem 5.5. First of all, Theorem 5.5 can be written as

$$R_a - R_{p_0} = \beta_{ap}(R_p - R_{p_0}) + \varepsilon. \tag{5.52}$$

The excess return of any portfolio over the zero-beta portfolio equals its beta multiplied by the excess return of the tangent portfolio over its zero-beta portfolio plus idiosyncratic noise. Decomposition (5.52) is an extension of CAPM to the case without a risk-free interest bond, when the p risky assets consist of all tradable assets at that time. The tangent portfolio plays a similar role to the market portfolio, and its zero-beta portfolio acts like the risk-free bond. For any portfolio a, define its

$$\text{Sharpe ratio} = E(R_a - R_{p_0})/\sigma_a, \tag{5.53}$$

where $\sigma_a = \mathrm{SD}(R_a)$. With this definition, the slope at the tangent portfolio p is the Sharpe ratio of the portfolio p. It maximizes the Sharpe ratio among all admissible portfolios (see Figure 5.5). This can easily be seen from (5.51):

$$\begin{aligned}\frac{E(R_a - R_{p_0})}{\sigma_a} &= \frac{\beta_{ap}(ER_p - ER_{p_0})}{\sigma_a} \\ &= \frac{\mathrm{cov}(R_a, R_p)(ER_p - ER_{p_0})}{\sigma_p^2 \sigma_a} \\ &= \mathrm{Corr}(R_a, R_p) \cdot \text{Sharpe ratio of } p.\end{aligned}$$

5.8 CAPM with unknowing risk-free rate

When (5.52) is applied to all tradable assets, let us call a tangent portfolio of this set of risky assets the market portfolio R^m. This results in the Black (1972) version of CAPM in absence of the risk-free asset [see (5.51)]:

$$\boldsymbol{ER} - \gamma\mathbf{1} = \boldsymbol{\beta}(ER^m - \gamma), \tag{5.54}$$

where $\boldsymbol{R}$ is the vector of returns of individual stocks or portfolios, R^m is the return of the market portfolio, and $\gamma = ER_{p_0}$ is the expected return of the zero-beta portfolio p_0, uncorrelated with the market portfolio. Regarding γ as the risk-free rate, this is the same as the Sharpe–Lintner version of CAPM. However, unlike the Sharpe–Linter version of CAPM, the expected return γ of the zero-beta rate is unknown.

The model (5.54) is an application of (5.51) to all assets on the market. When the number of assets on the market is large, it is expected that the zero-beta portfolio of the market portfolio is approximately risk-free, i.e.,

$R_{p_0} \approx ER_{p_0}$, which is denoted by γ. In this case, the decomposition (5.52) becomes

$$\boldsymbol{R} - \gamma \mathbf{1} = \boldsymbol{\beta}(R^m - \gamma) + \boldsymbol{\epsilon}, \tag{5.55}$$

where $\boldsymbol{\epsilon}$ is idiosyncratic noise, uncorrelated with the market portfolio. This is the Black version of CAPM.

5.8.1 Validating the Black version of CAPM

To validate the Black version of CAPM, we need to assume that the model (5.55) holds for multiple periods of time. Let $\{\boldsymbol{R}_t\}$ and $\{R_t^m\}$ be respectively observed return vectors of N given assets and of the market portfolio. We assume the statistical model:

$$\boldsymbol{R}_t = \boldsymbol{\alpha} + \boldsymbol{\beta} R_t^m + \boldsymbol{\epsilon}_t, \qquad \boldsymbol{E}\boldsymbol{\epsilon}_t = 0 \tag{5.56}$$

with $\operatorname{cov}(R_t^m, \boldsymbol{\epsilon}_t) = 0$ and $\operatorname{var}(\boldsymbol{\epsilon}_t) = \boldsymbol{\Sigma}$. This is an extension of model (5.55) from one-period to multi-period, assuming $\boldsymbol{\alpha}$, $\boldsymbol{\beta}$ and $\boldsymbol{\Sigma}$ remain constant over the time period.

When the Black version of CAPM holds, (5.54) shows that the expected return

$$E\boldsymbol{R} = \gamma \mathbf{1} + \boldsymbol{\beta}(ER^m - \gamma).$$

On the other hand, model (5.56) shows that $E\boldsymbol{R} = \boldsymbol{\alpha} + \boldsymbol{\beta} ER^m$. Equating these two quantities, the CAPM imposes the following restrictions

$$\boldsymbol{\alpha} = \gamma(\mathbf{1} - \boldsymbol{\beta}).$$

This is the object that we wish to validate. Hence, the hypothesis testing problem becomes

$$H_0 : \boldsymbol{\alpha} = \gamma(\mathbf{1} - \boldsymbol{\beta}) \quad \text{vs.} \quad H_1 : \boldsymbol{\alpha} \neq \gamma(\mathbf{1} - \boldsymbol{\beta}) \tag{5.57}$$

The form looks different from that of the Sharpe–Lintner version of CAPM (5.25). This is due to the different data that was used for the regression: raw returns versus the excess returns.

5.8.2 Testing statistics

If $\boldsymbol{\epsilon}_t \sim_{i.i.d} N(\mathbf{0}, \boldsymbol{\Sigma})$, then as in (5.26), the likelihood function is given by

$$\ell(\boldsymbol{\alpha}, \boldsymbol{\beta}, \boldsymbol{\Sigma}) = -\frac{NT}{2}\log(2\pi) - \frac{T}{2}\log|\boldsymbol{\Sigma}| \\ -\frac{1}{2} \times \sum_{t=1}^{T} (\boldsymbol{R}_t - \boldsymbol{\alpha} - \boldsymbol{\beta} R_t^m)^{\mathrm{T}} \boldsymbol{\Sigma}^{-1} (\boldsymbol{R}_t - \boldsymbol{\alpha} - \boldsymbol{\beta} R_t^m). \quad (5.58)$$

The maximum likelihood ratio test can be derived. In particular, the MLE under the full model (no restrictions) is the same as the Sharpe-Lintner version (5.27) except the data are now the raw returns $\{\boldsymbol{R}_t\}$ and $\{\boldsymbol{R}_t^m\}$. This results in an estimated covariance $\hat{\boldsymbol{\Sigma}}$.

The MLE under the null hythothesis $H_0 : \boldsymbol{\alpha} = \gamma(\mathbf{1} - \boldsymbol{\beta})$ can not be found explicitly. This is due to the estimation of parameter γ, which makes the constraint nonlinear. It solves iteratively the following equations. For a given $\hat{\boldsymbol{\Sigma}}_0$ and $\hat{\boldsymbol{\beta}}_0$, γ is estimated by

$$\hat{\gamma} = (\mathbf{1} - \hat{\boldsymbol{\beta}}_0)^T \hat{\boldsymbol{\Sigma}}_0^{-1} (\bar{\boldsymbol{R}} - \hat{\boldsymbol{\beta}}_0 \bar{R}_m) / (\mathbf{1} - \hat{\boldsymbol{\beta}}_0)^T \hat{\boldsymbol{\Sigma}}_0^{-1} (\mathbf{1} - \hat{\boldsymbol{\beta}}_0). \quad (5.59)$$

This can easily be shown by substituting $\boldsymbol{\alpha} = \gamma(\mathbf{1} - \boldsymbol{\beta})$ into (5.58) and then taking derivative with respect to γ and setting it to zero. We leave this to readers to derive (Exercise 5.14).

With an estimate of the risk-free rate $\hat{\gamma}$, the problem now becomes the Sharpe–Lintner version. Specifically, for given $\hat{\gamma}$, we compute [see (5.32)]

$$\hat{\boldsymbol{\beta}}_0 = \sum_{t=1}^{T} (R_t^m - \hat{\gamma})(\bar{\boldsymbol{R}} - \hat{\gamma}\mathbf{1}) \Big/ \sum_{t=1}^{T} (R_t^m - \hat{\gamma})^2. \quad (5.60)$$

For each given $\hat{\gamma}$ and $\hat{\boldsymbol{\beta}}_0$, the estimated covariance $\hat{\boldsymbol{\Sigma}}_0$ is given by

$$\hat{\boldsymbol{\Sigma}}_0 = T^{-1} \sum_{t=1}^{T} \hat{\boldsymbol{\epsilon}}_t^0 (\hat{\boldsymbol{\epsilon}}_t^0)^T, \quad (5.61)$$

where $\hat{\boldsymbol{\alpha}}_0 = \hat{\gamma}(\mathbf{1} - \hat{\boldsymbol{\beta}}_0)$ and

$$\hat{\boldsymbol{\epsilon}}_t^0 = \boldsymbol{R}_t - \hat{\boldsymbol{\alpha}}_0 - \hat{\boldsymbol{\beta}}_0 R_t^m = (\boldsymbol{R}_t - \hat{\gamma}\mathbf{1}) - \hat{\boldsymbol{\beta}}_0 (R_t^m - \hat{\gamma}).$$

Similar to (5.34), the maximum likelihood ratio statistic is

$$T_4 = T[\log|\hat{\boldsymbol{\Sigma}}_0| - \log|\hat{\boldsymbol{\Sigma}}|] \overset{a}{\sim}_{H_0} \chi^2_{N-1}, \quad (5.62)$$

where N is the number of assets used to run the regression. The asymptotic null distribution of T_4 is approximately χ^2 with degree of freedom $N-1$. This is due to the fact that the number of restrictions under the null hypothesis is $N - 1$ (recalling that γ is also a free parameter). Like the test statistic T_3, a correction can be made to improve the accuracy of approximation. Replacing the factor T in T_4 by $(T - N/2 - 2)$ yields

$$T_5 = (T - N/2 - 2)[\log|\hat{\boldsymbol{\Sigma}}_0| - \log|\hat{\boldsymbol{\Sigma}}|] \overset{a}{\sim}_{H_0} \chi^2_{N-1}.$$

The implementation of the above maximum likelihood ratio test is very simple. Starting from the MLE $\hat{\boldsymbol{\beta}}$ and $\hat{\boldsymbol{\Sigma}}$ of the full model, we can compute $\hat{\gamma}$ by taking $\hat{\boldsymbol{\beta}}_0 = \hat{\boldsymbol{\beta}}$ and $\hat{\boldsymbol{\Sigma}}_0 = \hat{\boldsymbol{\Sigma}}$ as the initial value. We can then iteratively compute $\hat{\boldsymbol{\beta}}_0$, $\hat{\boldsymbol{\Sigma}}_0$ and $\hat{\gamma}$. Note that the MLE $\hat{\boldsymbol{\Sigma}}$ is estimated based on the full model. It should be a consistent estimate of $\boldsymbol{\Sigma}$ whether the null hypothesis holds or not. Therefore, it should be close to $\hat{\boldsymbol{\Sigma}}_0$ when the null hypothesis is true. In other words, $\hat{\boldsymbol{\Sigma}}$ is a good initial value of $\hat{\boldsymbol{\Sigma}}_0$. A few iterations suffice.

We summarize the implementation of testing the Black version of CAPM by the following few steps.

1. Fit the linear model (5.56) to obtain $\hat{\boldsymbol{\alpha}}$, $\hat{\boldsymbol{\beta}}$ and $\hat{\boldsymbol{\Sigma}}$.
2. Using these estimates as the initial value, compute $\hat{\gamma}$ by (5.59), $\hat{\boldsymbol{\beta}}_0$ by (5.60), and $\hat{\boldsymbol{\Sigma}}_0$ by (5.61)
3. Iterate steps 2 a couple of times to obtain the estimate $\hat{\boldsymbol{\Sigma}}_0$ under the null hypothesis.
4. Compute the test statistics T_5 and its associated P-value by using χ^2_{N-1} distribution.

We now illustrate the above method by the following toy example.

Example 5.7 We now test the Black version of CAPM using the data in Example 5.5. The essential difference is that the risk-free rate is unknown and has to be estimated from the given data. Using the monthly returns of XOM, GE and JNJ, the covariance matrix of the residuals regressed on the returns of S&P 500 over the period February 2001 and January 2011 is given by

$$\hat{\boldsymbol{\Sigma}} = \begin{pmatrix} 21.097 & -2.699 & -0.844 \\ -2.699 & 34.570 & 2.656 \\ -0.844 & 2.656 & 14.301 \end{pmatrix}.$$

Using this and the estimate $\hat{\boldsymbol{\beta}} = (0.543, 1.387, 0.493)^T$ as the initial value in (5.59), we obtain $\hat{\gamma}_0 = 1.041$ as the market implied risk-free rate. With this risk-free rate, we can now run the Sharpe–Lintner version of CAPM, yielding

$$\hat{\boldsymbol{\Sigma}}_0 = \begin{pmatrix} 21.166 & -2.721 & -0.894 \\ -2.721 & 34.579 & 2.670 \\ -0.894 & 2.669 & 14.333 \end{pmatrix}, \qquad \hat{\boldsymbol{\beta}}_0 = \begin{pmatrix} 0.531 \\ 1.390 \\ 0.501 \end{pmatrix}.$$

With the estimated $\hat{\boldsymbol{\Sigma}}_0$ and $\hat{\boldsymbol{\beta}}_0$, we can now compute γ_0 again using (5.59). We iterate this two more times, and obtain

$$\hat{\boldsymbol{\Sigma}}_0 = \begin{pmatrix} 21.164 & -2.720 & -0.894 \\ -2.720 & 34.579 & 2.670 \\ -0.894 & 2.669 & 14.333 \end{pmatrix}, \qquad \hat{\boldsymbol{\beta}}_0 = \begin{pmatrix} 0.531 \\ 1.390 \\ 0.501 \end{pmatrix}.$$

and $\gamma_0 = 1.049\%$. The speed of convergence is very fast. The test statistic T_5 is then computed, yielding $T_5 = 0.665$ with degree of freedom 2. This gives a P-value of 0.7172 under χ^2_2. Hence, we have no strong evidence against the Black version of CAPM.

The above example yields that the market implied risk-free interest rate is 1.0487% per month. This is much higher than the average monthly yield rate 0.17% of the three 3-month Treasury Bills. There are several reasons for the discrepancy. First of all, we use only three stocks rather than the stocks from the entire market to compute $\hat{\gamma}_0$. It would be better to use several portfolios constructed based on the market to estimate $\hat{\gamma}_0$. Secondly, the varying market conditions, starting from the burst of the high technology bubble (2001), to the housing bubble (2006), subprime crisis (2007), and financial crisis (2008), make it hard to believe that the time homogeneous model (5.56) holds (see for example the time varying β in Figure 5.3). Thirdly, one factor is probably insufficient to capture all cross-sectional risk. See Example 5.5 in Chapter 5 for further analysis. We have no intention to give a better estimate of the market implied risk interest rate. This serves purely as a pedagogical example on how to use the Black version of CAPM.

5.9 Complements

We now provide detailed calculations (5.43) and (5.45) in the next two subsections.

5.9.1 Proof of (5.43)

Substituting the expression (5.41) of $\boldsymbol{\alpha}^*$ into (5.42), we obtain

$$\lambda_1 \boldsymbol{\mu}^{\mathrm{T}} \boldsymbol{\Sigma}^{-1} \boldsymbol{\mu} + \lambda_2 \boldsymbol{\mu}^{\mathrm{T}} \boldsymbol{\Sigma}^{-1} \mathbf{1} = \mu,$$
$$\lambda_1 \mathbf{1}^{\mathrm{T}} \boldsymbol{\Sigma}^{-1} \boldsymbol{\mu} + \lambda_2 \mathbf{1}^{\mathrm{T}} \boldsymbol{\Sigma}^{-1} \mathbf{1} = 1.$$

With the notation introduced in Section 5.7, the last two equations can be written as

$$\lambda_1 B + \lambda_2 A = \mu,$$
$$\lambda_1 A + \lambda_2 C = 1.$$

Solving for λ_1 and λ_2, we obtain

$$\lambda_1 = D^{-1}(C\mu - A), \quad \lambda_2 = D^{-1}(-A\mu + B).$$

Substituting λ_1 and λ_2 into (5.41), we conclude that

$$\begin{aligned}\boldsymbol{\alpha}^* &= D^{-1}(C\mu - A)\boldsymbol{\Sigma}^{-1}\boldsymbol{\mu} + D^{-1}(-A\mu + B)\boldsymbol{\Sigma}^{-1}\mathbf{1}\\ &= D^{-1}(-A\boldsymbol{\Sigma}^{-1}\boldsymbol{\mu} + B\boldsymbol{\Sigma}^{-1}\mathbf{1}) + \mu D^{-1}(C\boldsymbol{\Sigma}^{-1}\boldsymbol{\mu} - A\boldsymbol{\Sigma}^{-1}\mathbf{1})\\ &= \mathbf{g} + \mu\boldsymbol{h}.\end{aligned}$$

The proof is completed.

5.9.2 Proof of (5.48)

With the notation introduced in Section 5.7, it follows that

$$\begin{aligned}\mathbf{g}^{\mathrm{T}}\boldsymbol{\Sigma}^{-1}\mathbf{g} &= D^{-2}\big[B\boldsymbol{\Sigma}^{-1}\mathbf{1} - A\boldsymbol{\Sigma}^{-1}\boldsymbol{\mu}\big]\boldsymbol{\Sigma}\big[B\boldsymbol{\Sigma}^{-1}\mathbf{1} - A\boldsymbol{\Sigma}^{-1}\boldsymbol{\mu}\big]\\ &= D^{-2}\big[B^2\mathbf{1}^{\mathrm{T}}\boldsymbol{\Sigma}^{-1}\mathbf{1} - AB\boldsymbol{\mu}^{\mathrm{T}}\boldsymbol{\Sigma}^{-1}\mathbf{1} - AB\mathbf{1}^{\mathrm{T}}\boldsymbol{\Sigma}^{-1}\boldsymbol{\mu} + A^2\boldsymbol{\mu}\boldsymbol{\Sigma}^{-1}\boldsymbol{\mu}\big]\\ &= D^{-2}(B^2C - 2A^2B + A^2B)\\ &= D^{-2}(B^2C - A^2B) = B/D,\end{aligned}$$

and that

$$\begin{aligned}\mathbf{g}^{\mathrm{T}}\boldsymbol{\Sigma}^{-1}\boldsymbol{h} &= D^{-2}\big[B\boldsymbol{\Sigma}^{-1}\mathbf{1} - A\boldsymbol{\Sigma}^{-1}\boldsymbol{\mu}\big]\boldsymbol{\Sigma}\big[C\boldsymbol{\Sigma}^{-1}\boldsymbol{\mu} - A\boldsymbol{\Sigma}^{-1}\mathbf{1}\big]\\ &= D^{-2}(ABC - ACB - ABC + A^3)\\ &= -A/D.\end{aligned}$$

Furthermore,

$$\begin{aligned}\boldsymbol{h}^{\mathrm{T}}\boldsymbol{\Sigma}^{-1}\boldsymbol{h} &= D^{-2}\big[C\boldsymbol{\Sigma}^{-1}\boldsymbol{\mu} - A\boldsymbol{\Sigma}^{-1}\mathbf{1}\big]\boldsymbol{\Sigma}\big[C\boldsymbol{\Sigma}^{-1}\boldsymbol{\mu} - A\boldsymbol{\Sigma}^{-1}\mathbf{1}\big]\\ &= D^{-2}(C^2B - 2ACA + A^2C)\\ &= D^{-2}(C^2B - A^2C)\\ &= C/D.\end{aligned}$$

Therefore, we conclude that

$$\begin{aligned}\text{cov}(R_p, R_q) &= \mu_p\mu_q \boldsymbol{h}^{\mathrm{T}}\boldsymbol{\Sigma}^{-1}\boldsymbol{h} + (\mu_p + \mu_q)\mathbf{g}^{\mathrm{T}}\boldsymbol{\Sigma}^{-1}\boldsymbol{h} + \mathbf{g}^{\mathrm{T}}\boldsymbol{\Sigma}^{-1}\mathbf{g} \\ &= D^{-1}\big[C\mu_p\mu_q - A(\mu_p + \mu_q) + B\big] \\ &= D^{-1}\big[C(\mu_p - A/C)(\mu_q - A/C) - A^2/C + B\big] \\ &= C/D \cdot (\mu_p - A/C)(\mu_q - A/C) + C^{-1}.\end{aligned}$$

This completes the proof.

5.10 Exercises

5.1 Suppose that Jack holds a portfolio with expected return 10% and volatility 20% and that Jill has a portfolio with expected return 15% and volatility 30%. Furthermore the risk-free interest rate is 5%.

1. Whose investments are more efficient?
2. Suppose that Jack is willing to increase his leverage so that his holding is the same as Jill's, through the self-financing scheme. If his initial investment is $1000. What is the expected returns and volatility, in terms of dollars, after his increase of leverage?

5.2 Let s_A and s_B be the Sharpe ratio of portfolios A and B, respectively. Let r_A and r_B be the expected returns of these two portfolios, with standard deviation denoted by σ_A and σ_B, respectively. Assume that through self financing, portfolio A borrows $(\sigma_B/\sigma_A - 1)$ at risk-free rate r_f to leverage (or de-leverage) on portfolio A so that its risk is now the same as that of Portfolio B. Show that the excess return of leveraged investment in portfolio A is $s_A\sigma_B$ and that the expected return of the leveraged portfolio on A is larger than the expected return of portfolio B if $s_A > s_B$. This shows that the Sharpe ratio measures the efficiency of a portfolio.

5.3 Suppose that three mutual funds (conservative, growth and aggressive) have annual log-returns respectively 15% and 20% and 30% with volatility (standard deviation) respectively 20%, 30% and 50%. The correlation between any of the two funds is 0 and the risk-free interest rate is 5%.

1. What is the minimum variance portfolio with these three mutual funds?
2. Find the optimal portfolio allocation among the three mutual funds, if the expected return is set at 15%. Give the associated standard deviation of this portfolio.
3. Compute the Sharpe ratio for the portfolio in (a). How does it compare with that in portfolio (b)?

5.4 Assets 1 and 2 have the following covariance matrix:

$$\boldsymbol{\Sigma} = \begin{pmatrix} 0.01 & -0.02 \\ -0.02 & 0.05 \end{pmatrix}.$$

1. What is the covariance between portfolio A, which has 10% in asset 1 and 90% in asset 2, and portfolio B, which has 60% in asset 1 and 40% in asset 2?
2. Suppose that the excess returns of Assets 1 and 2 are 7% and 10%, respectively, what are the Sharpe ratios of portfolios A and B?

3. What is the variance portfolio with w on asset 1 and $1-w$ on asset 2?
4. Derive the minimum variance portfolio.

5.5 Suppose that two mutual funds (conservative and aggressive) have an annual expected return respectively 8% and 20% with volatility (standard deviation) 20% and 50%. In addition, the two mutual funds have correlation 0.6 and the risk free interest rate is 4%.

5.6 Which fund is more efficient in terms of the Sharpe ratio? Show your work.

5.7 Suppose that the mutual funds have market *beta*'s 0.5 and 1.5 respectively and the strict one-factor model holds. Deduce from the model on the variances of idiosyncratic noises and market portfolio.

5.8 Suppose that a portfolio consists of 10% riskless bond, 50% of conservative mutual fund and 40% of aggressive fund. What is the market beta of the portfolio?

5.9 Suppose that the excess returns of n risky assets follow CAPM with uncorrelated noise:

$$Y_i = \boldsymbol{\beta}_i Y^m + \epsilon_i$$

They together with a risk-free bond are used to track the market portfolio. Construct the optimal portfolio allocation problem in terms of minimizing the variance of the tracking error.

5.10 Let $\boldsymbol{Y}$ be the excess returns of risky assets. Let $X = \boldsymbol{a}^{\mathrm{T}}\boldsymbol{Y}$ be a portfolio with allocation vector $\boldsymbol{a}$. Denote by $\boldsymbol{\Sigma} = \mathrm{var}(\boldsymbol{Y})$ and $\boldsymbol{\mu} = E\boldsymbol{Y}$. Consider the following decomposition (regression)

$$\boldsymbol{Y} = \boldsymbol{\alpha} + \boldsymbol{\beta} X + \boldsymbol{\epsilon}, \quad E\boldsymbol{\epsilon} = 0, \quad \mathrm{cov}(\boldsymbol{\epsilon}, X) = 0.$$

1. Show that if $\boldsymbol{a} = c\boldsymbol{\Sigma}^{-1}\boldsymbol{\mu}$ (optimal portfolio in the mean-variance efficiency), then $\boldsymbol{\alpha} = 0$.
2. Conversely, if $\boldsymbol{\alpha} = 0$, there exists a constant c such that $\boldsymbol{a} = c\boldsymbol{\Sigma}^{-1}\boldsymbol{\mu}_0$.

5.11 Using the notation in Section 5.7, show

1. The vector $\boldsymbol{g}$ is an allocation vector, namely $\mathbf{1}^{\mathrm{T}}\boldsymbol{g} = 1$.
2. The portfolio formed by $\boldsymbol{h}$ is a long-short portfolio, namely $\mathbf{1}^{\mathrm{T}}\boldsymbol{h} = 0$.

5.12 Derive directly the minimum variance portfolio using the Lagrange multiplier method, namely

$$\boldsymbol{\alpha}^{\mathrm{T}}\boldsymbol{\Sigma}\boldsymbol{\alpha}, \quad \text{suchthat} \quad \boldsymbol{\alpha}^{\mathrm{T}}\mathbf{1} = 1.$$

Give the minimum variance and its associated allocation vector.

5.13 Consider the following portfolio optimization problem with a risk-free asset having return r_0:

$$\min \boldsymbol{\alpha}^{\mathrm{T}}\boldsymbol{\Sigma}\boldsymbol{\alpha}, \qquad \text{s.t. } \boldsymbol{\alpha}^{\mathrm{T}}\boldsymbol{\mu} + (1 - \boldsymbol{\alpha}^{\mathrm{T}}\mathbf{1})r_0 = \mu.$$

That is, we minimize the variance of the portfolio consisting of allocation vector $\boldsymbol{\alpha}$ on risky assets with return vector $\boldsymbol{\mu}$ and allocation $(1 - \boldsymbol{\alpha}^{\mathrm{T}}\mathbf{1})$ on the risk-free bond with return r_0, subject to the constraint that the portfolio's expected return is μ.

1. The optimal solution is

$$\boldsymbol{\alpha} = P^{-1}(\mu - r_0)\boldsymbol{\Sigma}^{-1}\boldsymbol{\mu}_0,$$

where $P = \boldsymbol{\mu}_0^{\mathrm{T}}\boldsymbol{\Sigma}^{-1}\boldsymbol{\mu}_0$ is the squared Sharpe ratio, and $\boldsymbol{\mu}_0 = \boldsymbol{\mu} - r_0\mathbf{1}$ is the vector of excess returns.

2. The variance of this portfolio is $\sigma^2 = (\mu - r_0)^2/P$.
3. When $r_0 < \mu$, show that $r_0 + P^{1/2}\sigma = \mu$, namely, the optimal allocation for the risky asset $\boldsymbol{\alpha}$ is the tangent portfolio.

5.14 Show that given $\boldsymbol{\Sigma}_0$ and $\boldsymbol{B}_0$, $\hat{\gamma}_0$ given by (5.59) is the maximum likelihood estimator.

1. Download the monthly data of 8 stocks: Dell, Ford, GE, IBM, Intel Johnson & Johnson, Merck, Microsoft from January 2001 to December 2014. Use the three-month treasury bill rates as a proxy for the risk-free rate and the S&P 500 index as a proxy of the market portfolio.
2. Construct the optimal allocations of the 8 stocks, if an investor is willing to invest 20% in riskless asset, using the the monthly data between 2001 and 2011.
3. If the allocation is fixed over the next two years (invested in December 2011), compare the performance of the portfolio over the next 6-month, one-year, two-year and three-year with the S&P 500 stock, in terms of return, volatility (standard deviation), and Sharpe ratio.
4. Create a value-weighted portfolio of the 8 stocks using the data in year 2011. As a proxy, the weight for Dell computer, for example, is proportional to the sum of the volume times closing price (un-adjusted) over the year. Report the percentage of allocation, if 20% of the asset is allocated to the 3-month treasury bills. Compare the performance of the portfolio over the next 6-month, one-year two-year, and three-year with the S&P 500 stock, in terms of gain, volatility (standard deviation) and Sharpe ratio.
5. Create a portfolio with 20% invested on risk-free bond and 10% over each of the 8 stocks. Compare the performance of the portfolio over the last 6-month, one-year, two-year and three-year with the S&P 500 stock, in terms of gain, volatility (standard deviation) and Sharpe ratio.

5.15 Verify the results in Table 5.4.

6

Factor Pricing Models

In chapter 5, we have provided some empirical evidence that the proxy of the market portfolio does not completely explain cross-sectional risks. Other factors such as firm characteristics can have additional explanatory power for cross-sectional risks. Therefore, more factors are needed to better account for the cross-sectional risks.

To this end, in this chapter, we introduce multiple factor models, which is an extension of the CAPM model, to better capture the cross-sectional risks. Econometric techniques are then introduced to validate these factor models. Statistical approaches to constructing factors will also be discussed.

6.1 Multifactor pricing models

6.1.1 Multifactor models

Let $f_1, \ldots, f_K$ be the K common factors that influence market returns and $R_1, \ldots, R_p$ be the returns of p risky assets. As in Chapter 5, for any given return of a risky asset R, consider the following least-squares problem: Minimize

$$E(R - \beta_0 - \beta_1 f_1 - \cdots - \beta_K f_K)^2$$

with respect to $\{\beta_j\}_{j=0}^K$. Taking derivative with respect β_j, we get the partial derivative

$$-E(R - \beta_0 - \beta_1 f_1 - \cdots - \beta_K f_K) f_j. \tag{6.1}$$

Let a and $\{b_j\}_{j=1}^p$ be the least-square solutions and

$$\varepsilon = R - a - b_1 f_1 - \cdots - b_K f_K.$$

Then, it follows by setting (6.1) to zero that

$$E\varepsilon = 0 \quad \text{and} \quad E f_j \varepsilon = 0.$$

Note that

$$R = a + b_1 f_1 + \cdots + b_K f_K + \varepsilon.$$

Applying the above decomposition to each of the risky assets, we obtain the following *multifactor model*:

$$R_i = a_i + b_{i1} f_1 + \cdots + b_{iK} f_K + \varepsilon_i, \tag{6.2}$$

with

$$E\varepsilon_i = 0, \quad E\varepsilon_i f_j = 0, \quad j = 1, \ldots, K. \tag{6.3}$$

Here, the *factor loadings* $\{b_{ij}\}$ are an extension of the market beta to the multi-factor model, i.e. b_{ij} indicates the sensitivity of the return of asset i on the jth factor. The multiple factor model (6.2) allows us to see how the risks are decomposed.

Let $\boldsymbol{a} = (a_1, \ldots, a_p)^{\mathrm{T}}$ and $\boldsymbol{B} = (b_{ij})$ be the $p \times K$ loading matrix. Then, (6.2) and (6.3) can be written in the matrix form:

$$\mathbf{R} = \boldsymbol{a} + \boldsymbol{B}\boldsymbol{f} + \boldsymbol{\epsilon}, \qquad \text{with } \boldsymbol{E}(\boldsymbol{\epsilon}) = 0 \text{ and } \operatorname{cov}(\boldsymbol{\epsilon}, \boldsymbol{f}) = 0, \tag{6.4}$$

where $\mathbf{R} = (R_1, \ldots, R_p)^{\mathrm{T}}$ is the vector of returns of p traded assets, $\boldsymbol{\epsilon} = (\varepsilon_1, \ldots, \varepsilon_p)^{\mathrm{T}}$ be the vector of idiosyncratic noise, and

$$\operatorname{cov}(\boldsymbol{\epsilon}, \boldsymbol{f}) = E(\boldsymbol{\epsilon} - E\boldsymbol{\epsilon})(\boldsymbol{f} - E\boldsymbol{f})^{\mathrm{T}} = E(\boldsymbol{\epsilon}\boldsymbol{f}^{\mathrm{T}}).$$

Commonly-used factors are the proxy of market portfolio, firm characteristics, tradable market variables, and macroeconomic variables. Examples of firm characteristics are "size", "value" (measured by book-to-market ratio), "Price-earning ratio" (PE), and "Price-sale ratio"(PS). For example, the size effect can be measured as the difference of returns of a portfolio with small capitalization (e.g. bottom decile) and high capitalization (top decile); the value effect and PE effect can be defined similarly with the market capitalization replaced respectively by the *book-to-market* ratio and *price-to-earning* ratio. Tradable market variables include "*maturity premium*", defined as the yield spread between long and short rates, "*default risk*", measured by the yield spread between high and low grade bonds, "liquidity spread" (differences between the 1-month repo rates and the 1-month treasury bill rates, measuring the short-term counterparty liquidity risk), exchange rates, and volatility index such as *VIX*, which is the *implied volatility* for the one-month at-money option of S&P 500 index calculated and disseminated by the Chicago Board Options Exchange Market Volatility Index. Figure 6.1 depicts those factors for the 10 year period from January 29, 2001 to February 28, 2011. Macroeconomic variables include inflation rate, unemployment

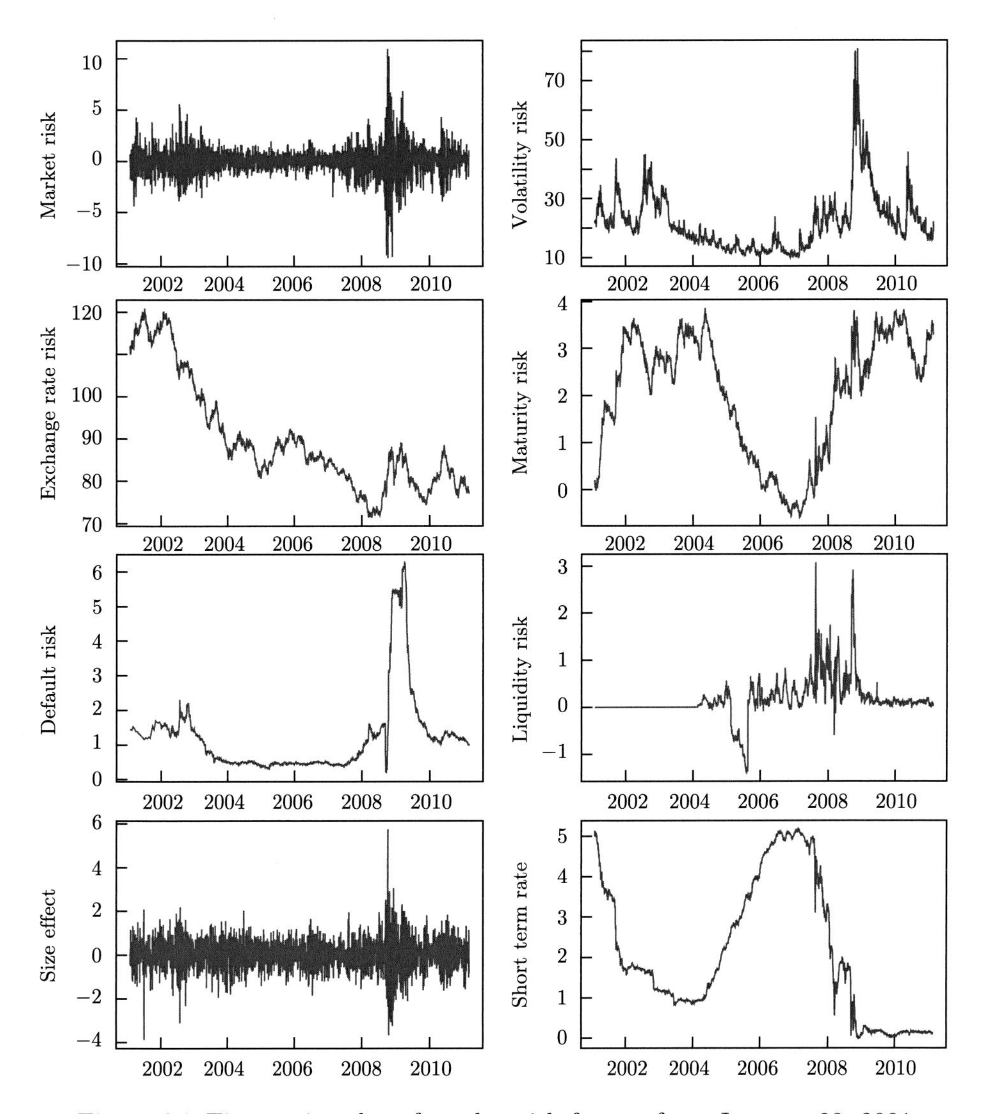

Figure 6.1 Time series plot of market risk factors from January 29, 2001 to February 28, 2011 at daily frequency. Market risk is the returns of S&P 500; volatility risk is the VIX index; exchange rate risk is the index of US dollar value; maturity risk is the yield spread between 10-year US treasury notes and 3-month treasury bills; default risk is the yield spread between AAA and BBB corporate bonds; liquidity risk is the differences between 1 months repo rates and 1 month treasury bill rates (from Feb. 4, 2004), and size effect is the difference of returns between S&P 500 and Russell 2000. The yields of 3-month treasury bills, marked as short term rate, are also presented.

rate, industrial production growth, housing price index, among others. These are typically nontradable. The correlation matrix of these factors over the period from February 23, 2004 to Feb. 28, 2011 is summarized as follows:

	SPX	Vol	ExR	Mat	Def	Liq	Size	Tbill
SPX	1.00	−0.13	0.00	0.00	0.01	−0.05	0.34	0.00
Vol	−0.13	1.00	−0.18	0.46	0.81	0.32	−0.06	−0.56
ExR	0.00	−0.18	1.00	−0.26	0.10	−0.34	−0.01	0.31
Mat	0.00	0.46	−0.26	1.00	0.41	−0.04	0.01	−0.96
Def	0.01	0.81	−0.10	0.41	1.00	0.06	0.01	−0.54
Liq	−0.05	0.32	−0.34	−0.04	0.06	1.00	−0.02	0.04
Size	0.34	−0.06	−0.01	0.01	0.01	−0.02	1.00	−0.02
Tbill	0.00	−0.56	0.31	−0.96	-0.54	0.04	−0.02	1.00

Example 6.1 Fama and French (1993) consider the following factors: excess returns of CRSP value-weighted stock index, difference of returns between large and small capitalization (size effect), and difference of returns between high and low book-to-market ratios (value effect). Specifically, they sort all stocks in the CRSP database by the market capitalization (large and small) and book equity to market equity (value, neutral and growth); see Section 5.5.2. They construct the following three factors:

1. SMB (Small Minus Big) is the average return on the three small portfolios minus the average return on the three big portfolios:

$$\begin{aligned} \text{SMB} = & \frac{1}{3}(\text{Small Value} + \text{Small Neutral} + \text{Small Growth}) \\ & -\frac{1}{3}(\text{Big Value} + \text{Big Neutral} + \text{Big Growth}). \end{aligned}$$

2. HML (High Minus Low) is the average return on the two value portfolios minus the average return on the two growth portfolios,

$$\begin{aligned} \text{HML} = & \frac{1}{2}(\text{Small Value} + \text{Big Value}) \\ & -\frac{1}{2}(\text{Small Growth} + \text{Big Growth}). \end{aligned}$$

3 The excess returns of the market portfolio, which are the value-weighted returns on all NYSE, AMEX, and NASDAQ stocks minus the one-month Treasury bills rates.

See Section 5.5.2 for these portfolios. The model is usually referred to as the *Fama–French three-factor model*. Note that SMB and HML are a long-short portfolio and hence are expected to have a small correlation with the market portfolio. The SMB and HML are so constructed that they have

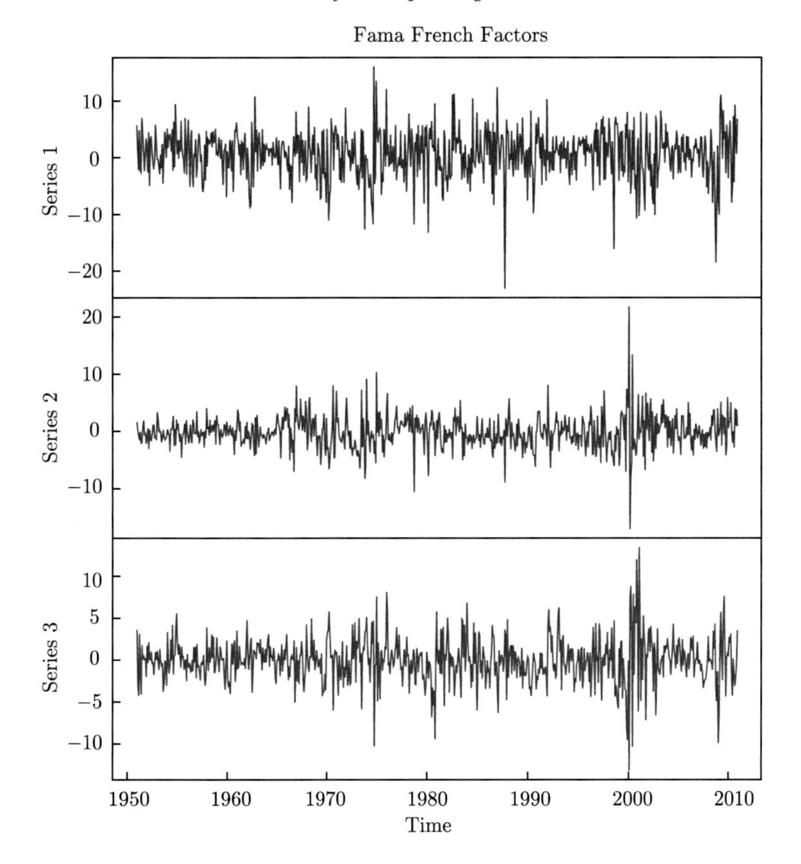

Figure 6.2 Time series plot of Fama–French three factors from January 1951 to December 2010. Series 1, 2 and 3 are respectively the excess returns of the market, and returns of SMB, and HML.

low correlation too. The correlation matrix of these three factors, computed based on monthly data from January 1951 to December 2010, is given by

$$\begin{pmatrix} 1.000 & 0.266 & -0.257 \\ 0.266 & 1.000 & -0.213 \\ -0.257 & -0.213 & 1.000 \end{pmatrix}.$$

Figure 6.2 shows the realization of these three factors over the same period.

From a statistical point of view, with more factors, the market portfolio can be better approximated and pricing errors can be reduced. In addition, ϵ becomes more exogenous and idiosyncratic. This is the advantage and flexibility of the multifactor model.

6.1.2 Factor pricing models

Like CAPM, financial economic theory postulates the returns of multifactor models. Let $\boldsymbol{\mu} = \boldsymbol{E}\boldsymbol{R}$ be the expected return and γ_0 be the risk-free interest rate. According to the Intertemporal CAPM (ICAPM) of Merton (1973), the expected returns of the risky assets should admit the form:

$$\boldsymbol{\mu} = \gamma_0 \mathbf{1} + \boldsymbol{B}\boldsymbol{\lambda}_K, \tag{6.5}$$

where $\boldsymbol{\lambda}_K$ is the vector of *risk premia*. Compare with (5.51) and (5.54). When the factors are tradable, it is the vector of excess returns of the risk factors over the risk-free interest rate, i.e. $\boldsymbol{\lambda}_K = \boldsymbol{E}(\boldsymbol{f} - \gamma_0 \mathbf{1})$. The *multifactor pricing model* (6.5) is clearly an extension of the CAPM. It postulates that the expected excess return of any asset is equal to its loadings on the risk factor times the risk premia.

The above multifactor pricing model was also derived by Ross (1976) using Arbitrage Pricing Theory (APT), which does not require the identification of the market portfolio and accommodates multiple risk factors. In the absence of arbitrage in large economies, Ross (1976) shows that (6.5) holds approximately. Connor (1984) introduces a competitive equilibrium version of APT that allows the exact pricing formula (6.5).

Several questions arise naturally: How to use the multifactor pricing model to forecast the expected return and volatility of a firm? How to empirically test the above pricing theory? How to select factors?

Before closing this section, we would like to differentiate between the multifactor model (6.4) and multifactor pricing model (6.5). The multifactor model (6.4) is a statistical decomposion and is always valid. It decomposes the return of an asset into its dependence on the returns of factors and the return of the idiosyncratic component. On the other hand, the factor pricing model (6.5) is a financial economics model. It is derived from financial economics theory. Its validity depends on the factors being used and needs validation.

6.2 Applications of multifactor models

Having provided theoretical insights on the factor pricing models, we proceed to two important applications of the factor model: to forecast the expected return of a firm, and to estimate the covariance matrix of assets.

For simplicity, assume that the factors are tradable and that there exists a risk-free asset with rate r_f. Then, according to (6.5), for any asset

$$ER = r_f + \beta_1(r_{F,1} - r_f) + \cdots + \beta_K(r_{F,K} - r_f),$$

since the risk premia is now $\boldsymbol{\lambda}_K = (E\boldsymbol{f} - r_f\mathbf{1})$. Thus, its expected return can be estimated as

$$\hat{\mu} = \bar{r}_f + \hat{\beta}_1(\bar{r}_{F,1} - \bar{r}_f) + \cdots + \hat{\beta}_K(\bar{r}_{F,K} - \bar{r}_f), \tag{6.6}$$

where $\bar{r}_f$ is the average risk-free rate, $\bar{r}_{F,i}$ is the average return of the ith factor, and $\hat{\beta}_i$ is the associated factor loading with the ith factor. This expected return can be used for capital budgeting. It helps determine whether a new project would enhance the value of a firm. See Copeland, Weston, and Shastri (2005). It can also be applied for performance evaluation of a fund or a strategy. A fund outperforming the market can be due to its correlation with the other risk factors that have positive returns. Thus, its return should be compared with the factor-model based return (6.6) and the volatility of the fund.

The challenges of estimating a large covariance matrix have already been discussed in Section 5.3. It involves too many parameters when no parametric model is imposed. From the factor model (6.4), it follows that

$$\text{var}(\boldsymbol{R}) = \boldsymbol{B}\text{var}(\boldsymbol{f})\boldsymbol{B}^{\mathrm{T}} + \text{var}(\boldsymbol{\epsilon}). \tag{6.7}$$

If the factors capture the cross-sectional risks, we may assume that $\text{var}(\boldsymbol{\epsilon})$ is either diagonal or sparse. This significantly reduces the number of parameters in $\boldsymbol{\Sigma} = \text{var}(\boldsymbol{R})$. It allows us to utilize recent developments on the estimation of a sparse high-dimensional covariance matrix; see Fan, Fan and Lv (2008) and Fan, Liao and Mincheva (2011). We defer the details to Section 7.2. Note that $\text{var}(\boldsymbol{f})$ is a low-dimensional matrix (e.g. 3×3 for a three-factor model). Formula (6.7) always holds since it utilizes the multifactor model (6.4).

Multiple factor models can also be used for portfolio hedging. Using the correlation between portfolios and factors, a hedging portfolio can be constructed based on tradable factors. In addition, one can also impose shocks on the factors and see how portfolios perform with the stress tests. An example of this is given in Section 6.5, in which we construct a principal factor that summarizes the 7 risk factors given in Figure 6.1 that are uncorrelated with the market factor. We can give a stress test based on the shocks to the market factor and the summarized factor to see how the portfolios handle those shocks.

As mentioned at the end of Section 6.2, the multifactor model (6.4) always holds, yet the multifactor pricing model (6.5) is a theory that needs to be tested. In the first application (6.6), we utilize the multifactor pricing model (6.5), but in the last three applications, we need only the multifactor model (6.4).

6.3 Model validation with tradable factors

The multifactor pricing model (6.5) postulates a beautiful theory on where the returns come from. It was derived based on several theoretical assumptions, and it needs to be checked whether it is consistent with market data. To validate such a model, we need to assume that model (6.2) holds over multiple periods of time. In other words, the return of the ith portfolio is governed by

$$R_{it} = a_i + b_{i1} f_{1t} + \cdots + b_{iK} f_{Kt} + \varepsilon_{it}, \quad t = 1, \ldots, T,$$

or putting in the matrix form,

$$\boldsymbol{R}_t = \boldsymbol{a} + \boldsymbol{B}\boldsymbol{f}_t + \boldsymbol{\epsilon}_t, \qquad E\boldsymbol{\epsilon}_t = 0, \quad \mathrm{cov}(\boldsymbol{R}_t, \boldsymbol{f}_t) = 0. \tag{6.8}$$

Direct application of (6.4) over each time period yields time-varying coefficients $\boldsymbol{a}_t$ and $\boldsymbol{B}_t$. Yet, model (6.8) assumes that they are time independent. This is an assumption to make econometric inference feasible.

Assuming $E\boldsymbol{R}_t$ and $E\boldsymbol{f}_t$ are constant in the testing period, model (6.8) implies that $\boldsymbol{\mu} = E\boldsymbol{R} = \boldsymbol{a} + \boldsymbol{B}E\boldsymbol{f}$. Equating this with (6.5), noticing that the risk premia for tradable factors are $\boldsymbol{\lambda}_K = E(\boldsymbol{f} - \gamma_0 \mathbf{1})$, the testable object now becomes

$$H_0 : \boldsymbol{a} = \gamma_0(\mathbf{1} - \boldsymbol{B}\mathbf{1}). \tag{6.9}$$

6.3.1 *Existence of a risk-free asset*

Let us now assume that there exists a risk-free asset which earns risk-free rate $r_{f,t}$ at time t. In this case, it is easier to express the problem in terms of the excess returns. Let

$$\boldsymbol{Y}_t = \boldsymbol{R}_t - r_{f,t}\mathbf{1} \quad \text{and} \quad \boldsymbol{X}_t = \boldsymbol{f}_t - r_{f,t}\mathbf{1},$$

in which $\boldsymbol{Y}_t$ are the excess returns of N testing portfolios (not necessarily all of the p assets). Then, the multiple factor model can be parameterized as

$$\boldsymbol{Y}_t = \boldsymbol{\alpha} + \boldsymbol{B}\boldsymbol{X}_t + \boldsymbol{\epsilon}_t. \tag{6.10}$$

The hypothesis (6.9) under this parametrization becomes $H_0 : \boldsymbol{\alpha} = 0$.

6.3.2 *Estimation of risk premia*

When the risk-free asset exists, the risk premia are particularly easy to estimate. They are estimated by the sample average of the observed risk premia over the same time period. More precisely,

$$\hat{\boldsymbol{\lambda}}_K = \bar{\boldsymbol{X}} = T^{-1} \sum_{t=1}^{T} \boldsymbol{X}_t. \tag{6.11}$$

The variance of the estimator is $\text{var}(\boldsymbol{\lambda}_K) = T^{-1}\text{var}(\boldsymbol{X})$, which can be estimated as

$$\hat{\text{var}}(\hat{\boldsymbol{\lambda}}_K) = T^{-1} \hat{\boldsymbol{\Sigma}}_X, \tag{6.12}$$

where $\hat{\boldsymbol{\Sigma}}_X$ is the sample covariance matrix of $\{\boldsymbol{X}_t\}_{t=1}^T$ given by

$$\hat{\boldsymbol{\Sigma}}_X = T^{-1} \sum_{t=1}^{\mathrm{T}} (\boldsymbol{X}_t - \bar{\boldsymbol{X}})(\boldsymbol{X}_t - \bar{\boldsymbol{X}})^{\mathrm{T}}.$$

Example 6.2 Using 60-year monthly data from January 1951 to December 2010, the estimated annualized risk premia for the Fama–French factors in Example 6.1 and their estimated standard errors are

$$\hat{\boldsymbol{\lambda}}_K = \begin{pmatrix} 6.736 \\ 2.309 \\ 4.368 \end{pmatrix}, \qquad \hat{\text{SE}} = \begin{pmatrix} 1.946 \\ 1.310 \\ 1.231 \end{pmatrix}.$$

The standard errors are obtained by taking the square root of the diagonal elements of the estimated covariance, computed from (6.12), which is given by

$$\text{var}(\hat{\boldsymbol{\lambda}}_K) = \begin{pmatrix} 3.788 & 0.678 & -0.617 \\ 0.678 & 1.715 & -0.343 \\ -0.617 & -0.343 & 1.514 \end{pmatrix}.$$

The result indicates that the excess return of the market portfolio over the last 60 years is about 6.736% give or take 1.946%, in spite of the recent financial crisis in 2008, the subprime crisis in 2007 and the technology bubble in 2000. This is the risk premium for taking risks in the equity market over the risk-free bonds. Similarly, the risk premia for SMB and HML are respectively estimated as 2.309% and 4.368%, with estimated standard errors 1.310% and 1.231%, respectively. This also shows how hard it is to estimate the expected returns. Indeed, even with 60-year data, the standard errors are still large. This is due to large stochastic errors in comparison with the expected returns. In the computing of the risk premia of SMB and HML, we compute directly the raw returns of these factors over the last 60 years, without subtracting the risk-free rates. This is due to the fact that they are basically long-short portfolios without capital cost (other than margin requirements). Indeed, the risk-free rates cancel themselves in the long-short portfolios.

6.3.3 Testing statistics

To derive the maximum likelihood ratio test, we assume further that $\boldsymbol{\epsilon}_t \sim_{i.i.d} N(0, \boldsymbol{\Sigma})$ in (6.10). Under this assumption, the likelihood function can easily be derived. It is analogous to (5.26) and MLE is similar to (5.27). For example, the conditional log-likelihood is the same as (5.26) except that $\boldsymbol{\beta} Y_t^m$ is now replaced by $\boldsymbol{B}\boldsymbol{X}_t$, namely

$$\ell(\boldsymbol{\alpha}, \boldsymbol{B}, \boldsymbol{\Sigma}) = -\frac{NT}{2}\log(2\pi) - \frac{T}{2}\log|\boldsymbol{\Sigma}| - \frac{1}{2} \times \sum_{t=1}^{\mathrm{T}} (\boldsymbol{Y}_t - \boldsymbol{\alpha} - \boldsymbol{B}\boldsymbol{X}_t)^{\mathrm{T}} \boldsymbol{\Sigma}^{-1} (\boldsymbol{Y}_t - \boldsymbol{\alpha} - \boldsymbol{B}\boldsymbol{X}_t). \quad (6.13)$$

Instead of following the same mathematical routine, we provide an alternative way from the least-squares point of view, or more precisely, the *method of moment* which replaces theoretical moments by their empirical moments. Note that it follows from (6.10) that

$$\mathrm{cov}(\boldsymbol{Y}_t, \boldsymbol{X}_t) = \boldsymbol{B}\mathrm{cov}(\boldsymbol{X}_t, \boldsymbol{X}_t) = \boldsymbol{B}\mathrm{var}(\boldsymbol{X}_t),$$

which implies that

$$\boldsymbol{B} = \mathrm{cov}(\boldsymbol{Y}_t, \boldsymbol{X}_t)\mathrm{var}(\boldsymbol{X}_t)^{-1}.$$

Substituting them by the empirical moments, noticing that

$$\boldsymbol{\alpha} = E\boldsymbol{Y}_t - \boldsymbol{B}E\boldsymbol{X}_t, \qquad \boldsymbol{\Sigma} = \mathrm{var}(\varepsilon_t),$$

we have the conditional maximum likelihood estimator:

$$\begin{aligned}
\hat{\boldsymbol{B}} &= \Big[\sum_{t=1}^{\mathrm{T}} (\boldsymbol{Y}_t - \bar{\boldsymbol{Y}})(\boldsymbol{X}_t - \bar{\boldsymbol{X}})^{\mathrm{T}}\Big]\Big[\sum_{t=1}^{\mathrm{T}} (\boldsymbol{X}_t - \bar{\boldsymbol{X}})(\boldsymbol{X}_t - \bar{\boldsymbol{X}})^{\mathrm{T}}\Big]^{-1}, \\
\hat{\boldsymbol{\alpha}} &= \bar{\boldsymbol{Y}} - \hat{\boldsymbol{B}}\bar{\boldsymbol{X}}, \qquad (6.14) \\
\hat{\boldsymbol{\Sigma}} &= T^{-1}\sum_{t=1}^{\mathrm{T}} \hat{\boldsymbol{\epsilon}}_t \hat{\boldsymbol{\epsilon}}_t^{\mathrm{T}},
\end{aligned}$$

where $\hat{\boldsymbol{\epsilon}}_t = \boldsymbol{Y}_t - \hat{\boldsymbol{\alpha}} - \hat{\boldsymbol{B}}\boldsymbol{X}_t$ is the residual vector at time t.

Remark 6.1 Let $\hat{\alpha}_i$ and $(\hat{b}_{i1}, \ldots, \hat{b}_{ik})$ be the marginal multiple regression fit of the ith asset:

$$Y_{it} = \alpha_i + b_{i1}X_{1t} + \cdots + b_{iK}X_{Kt} + \varepsilon_{it}, \qquad t = 1, \ldots, T, \tag{6.15}$$

with the tth residual denoted by $\hat{\varepsilon}_{it}$. Then, $\hat{\boldsymbol{\alpha}}$, $\hat{\boldsymbol{B}}$ and $\hat{\boldsymbol{\epsilon}}_t$ in (6.14) can be constructed from the marginal regression fits as follows:

$$\hat{\boldsymbol{\alpha}} = \begin{pmatrix} \hat{\alpha}_1 \\ \vdots \\ \hat{\alpha}_N \end{pmatrix}, \quad \hat{\boldsymbol{B}} = \begin{pmatrix} \hat{b}_{11} & \cdots & \hat{b}_{1K} \\ \vdots & & \vdots \\ \hat{b}_{N1} & \cdots & \hat{b}_{NK} \end{pmatrix}, \quad \hat{\boldsymbol{\epsilon}}_t = \begin{pmatrix} \hat{\varepsilon}_{1t} \\ \vdots \\ \hat{\varepsilon}_{Nt} \end{pmatrix}.$$

In other words, the MLE for model (6.10) can be obtained by the marginal least-squares fit to the regression model (6.15), with proper arrangements in the matrix form. When K is large in comparison with T, it is easy to overfit the model. Often, one takes approximately 3 to 5 factors with monthly returns of approximately 5 to 10 years approximately.

The maximum likelihood estimates under the null hypothesis $H_0 : \boldsymbol{\alpha} = 0$ can be derived analogously. They are given by

$$\hat{\boldsymbol{B}}_0 = \left[\sum_{t=1}^{\mathrm{T}} \boldsymbol{Y}_t \boldsymbol{X}_t^{\mathrm{T}}\right] \left[\sum_{t=1}^{\mathrm{T}} \boldsymbol{X}_t \boldsymbol{X}_t^{\mathrm{T}}\right]^{-1},$$

$$\hat{\boldsymbol{\Sigma}}_0 = T^{-1} \sum_{t=1}^{\mathrm{T}} \hat{\boldsymbol{\epsilon}}_t^o \hat{\boldsymbol{\epsilon}}_t^{oT},$$

where $\hat{\boldsymbol{\epsilon}}_t^o = \boldsymbol{Y}_t - \hat{\boldsymbol{B}}_0 \boldsymbol{X}_t$. As in Remark 6.1, the elements in $\hat{\boldsymbol{B}}_0$ and residuals in $\hat{\boldsymbol{\epsilon}}_t^o$ can be obtained via the marginal regression model (6.15) without using the intercept term $\boldsymbol{\alpha}$.

As in (5.34), it can be easily derived that the maximum likelihood ratio test is given by

$$T(\log|\hat{\boldsymbol{\Sigma}}_0| - \log|\hat{\boldsymbol{\Sigma}}|).$$

According to Wilks' theorem, the asymptotic null distribution is χ_N^2, since the number of restrictions under the null hypothesis is N. To improve the accuracy of the approximation, we recommend using the corrected version of the maximum likelihood ratio test

$$T_0 = (T - N/2 - K - 1)(\log|\hat{\boldsymbol{\Sigma}}_0| - \log|\hat{\boldsymbol{\Sigma}}|) \overset{a}{\sim}_{H_0} \chi_N^2, \tag{6.16}$$

in which "$\overset{a}{\sim}_{H_0}$" means "distributed approximately under the null hypothesis as". This is an extension of (5.35) to the K-factor model.

The Wald test based on the estimator $\hat{\boldsymbol{\alpha}}$ in (6.14) can easily be derived.

Table 6.1 *Summary of testing results using the Wald test*

K	2	3	5
P-value	0.010	0.039	0.025

Since the estimator $\hat{\boldsymbol{\alpha}}$ is linear in $\boldsymbol{Y}_t$, its covariance matrix can easily be derived and so can the Wald test statistic. We omit the detail here. The Wald test is given by

$$T_1 = \frac{T-N-K}{N}[1+\bar{\boldsymbol{X}}^{\mathrm{T}}\hat{\boldsymbol{\Sigma}}_X^{-1}\bar{\boldsymbol{X}}]^{-1}\hat{\boldsymbol{a}}^{\mathrm{T}}\hat{\boldsymbol{\Sigma}}^{-1}\hat{\boldsymbol{a}} \sim_{H_0} F_{N,T-N-K}, \tag{6.17}$$

where $\hat{\boldsymbol{\Sigma}}_X$ is the sample covariance matrix of $\{\boldsymbol{X}_t\}$ in (6.12). This null distribution is exact rather than the approximate one if $\boldsymbol{\epsilon} \sim N(0, \boldsymbol{\Sigma})$. It can be shown that this test is equivalent to the maximum likelihood ratio test T_0. This an extension of the test statistic (5.30) for the single-factor model, CAPM.

Example 6.3 Fama and French (1993) consider the following 5 possible factors. In addition to the three factors given in Example 6.1, they considered further two additional factors

- a term structure factor (yield spread between long and short bonds);
- a default risk (yield spread between high and low grade bonds)

They used 25 stock portfolios and 7 bond portfolios, namely $N = 32$. The stock portfolios are created using a two-way sort based on the market capitalization and book-to-equity ratio, each factor is grouped into 5 categories according according to their quintiles. This results in 5×5 portfolios (see Section 5.52). The bond portfolios include five US government ones and two corporate ones. The period of study is 1963/07–1991/12, i.e. $T = 354$. They considered the two-factor model (SMB and HML), three factor model (SMB, HML, and market portfolio) and the five factor model. The P-values are summarized in Table 6.1 based on the Wald test statistic T_1. Fama and French found that there are some improvements going from two factors to five factors, that three factors are necessary when testing portfolios consisting only of stocks, and that five factors are needed when bond portfolios are included.

6.3.4 An empirical study using Fama–French portfolios

In the following example, we apply the maximum likelihood ratio test to validate the Fama–French model over a more extensive period.

Example 6.4 We now consider validating the Fama–French three factor model over 12 five-year periods, from January 1951 to December 2010. The same 6 Fama–French portfolios as those in Section 5.5.2 are used as the testing portfolios. For each portfolio, we first run the marginal regression (6.15). The relevant summary is presented in Figure 6.3. They are similar quantities as defined in Section 5.5.2. The difference is that we now use three factors rather than just one factor.

Comparing with the CAPM model, the residual variances are now significantly smaller. Compare with Figure 5.4. This means that the multiple R-squares are now significantly larger, giving much better fits. Indeed, the average multiple R^2 is now 0.972, compared with 0.816 in the CAPM. The magnitudes of estimated α's are now significantly smaller in comparison with those from CAPM. The smaller estimated α's and estimated residual variances indicate that the three-factor model fits the cross-sectional risk much better. However, their ratios cancel each other out. As a result, the t-test statistics are only slightly smaller and the P-values for testing $\alpha_i = 0$ improve only somewhat. From Figure 6.3, it is clear that portfolios 2 and 4 (small-neutral, large-growth) have statistically significant α's during the periods 1986-1990 and 1991-1995, making a rejection of the Fama–French model in those periods. See Table 6.2 for test results based on the modified maximum likelihood ratio test (6.16), using the corrected maximum likelihood ratio test T_0.

Note that the Fama–French factors were constructed based on those 6 portfolios. One may argue that they are not good testing portfolios. Indeed, we include them here so that the results can be compared with those using the CAPM, presented in Section 5.5.2. To mitigate the problem, we now use the Fama–French 25 portfolios in Example 6.3. The results are somewhat better, as far as the testing results are concerned. Other characteristics are very similar. For example, the multiple R-squares are in the similar order of magnitude (see Figure 6.4). This provides, once more, evidence that the Fama–French three-factor model captures the cross-sectional risk better than the CAPM does.

We have also applied the Wald test to the above three data sets to validate the Fama–French three-factor model over the 12 five-year periods. The results are about the same and hence are omitted. Instead, we present the results based the Wald test for six 10-year time periods. The P-values for

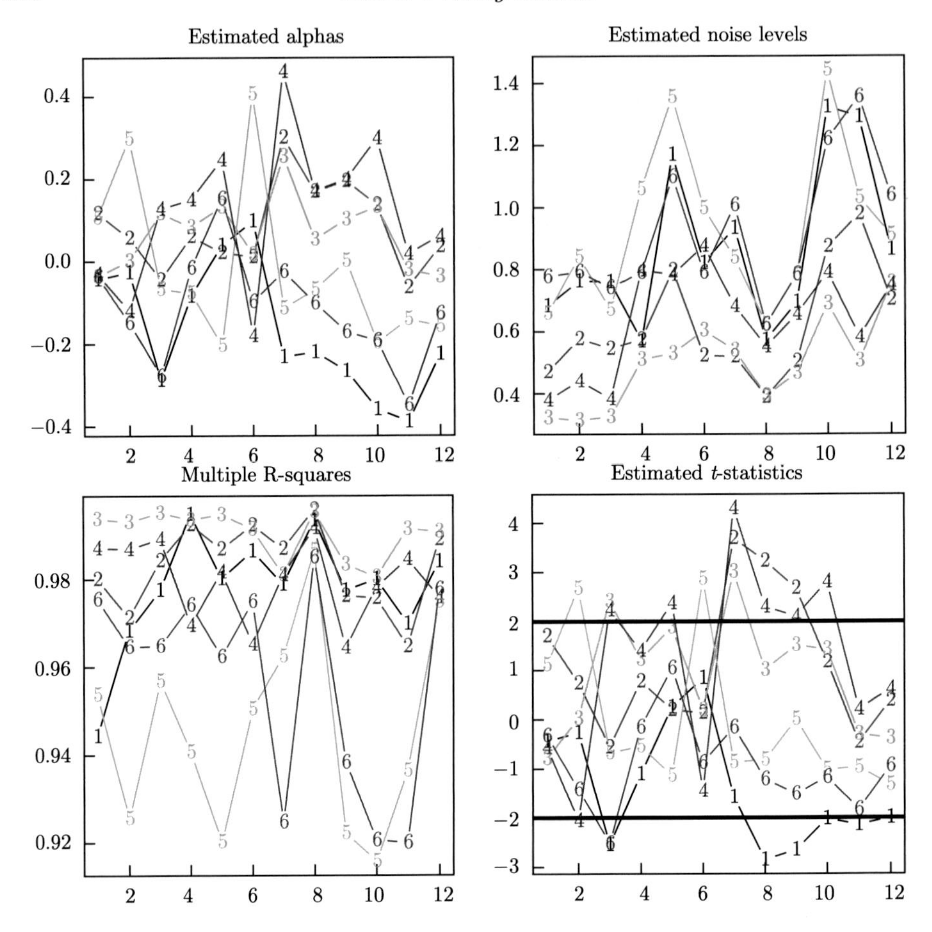

Figure 6.3 Estimated α's, SDs of noises, multiple R-squares, and t-statistics for 6 Fama–French portfolios over 12 five-year periods from January 1951 to December 2010, using the Fama–French 3 factor model. Portfolios 1–6 are respectively small-value, small-neutral, small-growth, large-value, large-neutral, large-growth.

the periods 1981–1990 and 1991–2000 show that there exist small but statistically significantly α's during the periods.

In summary, the Fama–French three-factor fits the market data better than the CAPM. They yield smaller estimated α's and larger multiple R-squares. The three-factor model captures cross-sectional risks very well. However, there seems to be small but statistically significant α's for some specific periods; they are too small to use for meaningful statistical arbitrage.

Table 6.2 *Empirical results for testing the Fama–French model*

Time	6 portfolios		25 portfolios	
	T_0	p-value	T_0	p-value
1/51-12/55	5.540	47.662	23.334	55.809
1/56-12/60	14.329	2.617	40.620	2.516
1/61-12/65	13.193	4.008	26.865	36.267
1/66-12/70	7.857	24.877	21.672	65.459
1/71-12/75	15.818	1.477	28.404	28.958
1/76-12/80	11.584	7.193	27.922	31.146
1/81-12/85	21.191	0.170	39.871	3.006
1/86-12/90	18.917	0.431	49.111	0.274
1/91-12/95	21.008	0.183	52.873	0.093
1/96-12/00	16.601	1.087	38.276	4.347
1/01-12/05	7.512	27.609	27.292	34.145
1/06-12/10	12.133	5.907	39.530	3.257

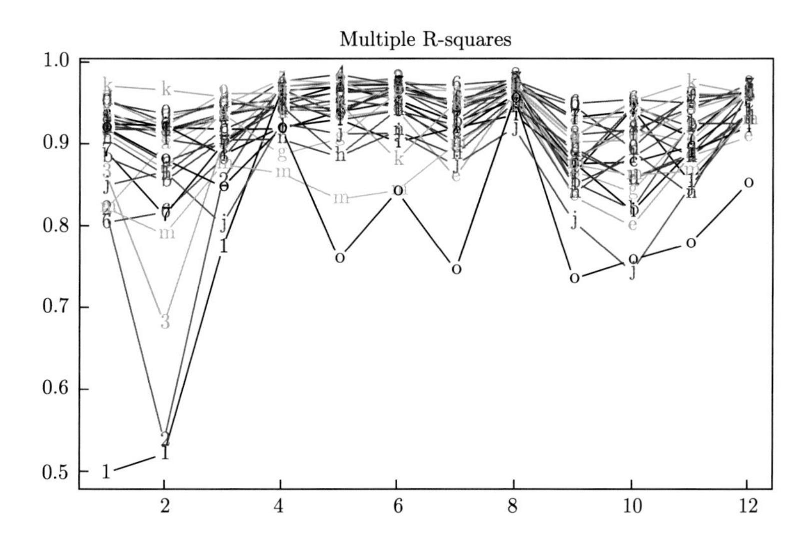

Figure 6.4 Multiple R^2 for the 25 Fama–French portfolios over 12 five-year periods from January 1951 to December 2010, using the Fama–French 3-factor model.

6.3.5 *Absence of a risk-free asset**

When the risk-free asset is unknown, we have to regress directly the raw return $\boldsymbol{R}_t$ on the tradable risk factors and infer directly the expected risk-free rate from the data. This is an extension of the Black version of CAPM, but now with more than one factor. The null hypothesis is now (6.9), since we are now working in terms of the raw returns instead of the excess returns.

The maximum likelihood estimator under model (6.8) is the same as that

Table 6.3 *Empirical results for testing the Fama–French model (Wald test)*

Time	6 portfolios		25 portfolios	
	T_0	p-value	T_0	p-value
1/51-12/60	2.243	4.416	1.745	2.972
1/61-12/70	3.150	0.685	0.803	72.901
1/71-12/80	1.070	38.468	0.571	94.413
1/81-12/90	6.796	0.000	3.158	0.003
1/91-12/00	5.135	0.011	2.690	0.032
1/01-12/10	2.442	2.960	1.579	6.101

under model (6.10) with appropriate changes of notation. This enables us to obtain $\hat{\boldsymbol{\Sigma}}$ under model (6.10). The estimate under the null hypothesis requires some additional effort. It cannot be found explicitly due to the nonlinear constraint (6.9).

As in Section 5.8.2, the maximum likelihood estimator under the null hypothesis can be solved iteratively. Given the estimated $\hat{\boldsymbol{B}}_0$ and $\hat{\boldsymbol{\Sigma}}_0$, the coefficient $\hat{\gamma}_0$ is determined by

$$\hat{\gamma}_0 = \left[(\mathbf{1}-\hat{\boldsymbol{B}}_0\mathbf{1})^{\mathrm{T}}\hat{\boldsymbol{\Sigma}}_0^{-1}(\mathbf{1}-\hat{\boldsymbol{B}}_0\mathbf{1})\right]^{-1}\left[(\mathbf{1}-\hat{\boldsymbol{B}}_0\mathbf{1})^{\mathrm{T}}\hat{\boldsymbol{\Sigma}}_0^{-1}(\bar{\boldsymbol{R}}-\hat{\boldsymbol{B}}_0\bar{\boldsymbol{R}}_K)\right], \quad (6.18)$$

where $\bar{\boldsymbol{R}}$ and $\bar{\boldsymbol{f}}$ are the averages of $\{\boldsymbol{R}_t\}_{t=1}^{\mathrm{T}}$ and $\{\boldsymbol{f}_t\}_{t=1}^{\mathrm{T}}$. It maximizes the log-likelihood function (6.13) with respect to γ_0 when $\boldsymbol{\alpha} = \gamma_0(\mathbf{1}-\boldsymbol{B}\mathbf{1})$ is plugged in. It is an extension of (5.59). We leave this derivation to the reader.

Now, given $\hat{\gamma}_0$, the situation is the same as that in Section 5.2.1 and $\hat{\boldsymbol{B}}_0$ and $\hat{\boldsymbol{\Sigma}}_0$ can easily be obtained. Specifically,

$$\hat{\boldsymbol{B}}_0 = \left[\sum_{t=1}^{\mathrm{T}}(\boldsymbol{R}_t-\hat{\gamma}_0\mathbf{1})(\boldsymbol{f}_t-\hat{\gamma}_0\mathbf{1})^{\mathrm{T}}\right]\left[\sum_{t=1}^{\mathrm{T}}(\boldsymbol{f}_t-\hat{\gamma}_0\mathbf{1})(\boldsymbol{f}_t-\hat{\gamma}_0\mathbf{1})^{\mathrm{T}}\right]^{-1},$$

and

$$\hat{\boldsymbol{\Sigma}}_0 = T^{-1}\sum_{i=1}^{\mathrm{T}}\hat{\boldsymbol{\epsilon}}_t^o\hat{\boldsymbol{\epsilon}}_t^{o\mathrm{T}},$$

where $\hat{\boldsymbol{\epsilon}}_t^o = (\boldsymbol{R}_t-\mathbf{1}\hat{\gamma}_0)-\hat{\boldsymbol{B}}_0(\boldsymbol{f}_t-\mathbf{1}\hat{\gamma}_0)$. We can now iterate these two steps until convergence.

The above process suggests a very simple algorithm for computing the maximum likelihood ratio test. Starting from the unconstrained model, we have the initial values of $\hat{\boldsymbol{B}}_0$ and $\hat{\boldsymbol{\Sigma}}_0$. Iterate the above steps several times. The convergence is very fast. The corrected maximum likelihood ratio test statistic is now given by

$$T_0 = (T - N/2 - K - 1)(\log|\hat{\boldsymbol{\Sigma}}_0| - \log|\hat{\boldsymbol{\Sigma}}|) \overset{a}{\sim}_{H_0} \chi^2_{N-1}.$$

which is an extension of (5.62). The degree of freedom is now $N - 1$, since this is the number of restrictions under (6.9).

The risk premia for the tradable risk factors are $\boldsymbol{\lambda}_K = E\boldsymbol{f} - \gamma_0$. This can be estimated as

$$\hat{\boldsymbol{\lambda}}_K = T^{-1}\sum_{t=1}^{\mathrm{T}} \boldsymbol{f}_t - \hat{\gamma}_0\mathbf{1}, \tag{6.19}$$

where the first part is just the average return of the returns of the risk factors and the second part is the estimated risk-free interest rate. Therefore, the error in the estimation comes from both parts. Indeed, it can be derived that

$$\widehat{\mathrm{var}}(\hat{\boldsymbol{\lambda}}_K) = T^{-1}\hat{\boldsymbol{\Sigma}}_X + \widehat{\mathrm{var}}(\hat{\gamma}_0)\mathbf{1}\mathbf{1}^{\mathrm{T}}, \tag{6.20}$$

where

$$\hat{\mathrm{var}}(\hat{\gamma}_0) = \left[(\mathbf{1} - \hat{\boldsymbol{B}}_0\mathbf{1})^{\mathrm{T}}\hat{\boldsymbol{\Sigma}}_0^{-1}(\mathbf{1} - \hat{\boldsymbol{B}}_0\mathbf{1})\right]^{-1}.$$

We omit the details. The standard errors for estimation of the risk premia are simply the square root of the diagonal elements in (6.20).

Example 6.5 We now test the Fama–French model in absence of the risk-free rate. The 25 Fama–French portfolios from January 2006 to December 2010 are used as our testing portfolios, i.e. $N = 25$. The MLE under the null hypothesis was computed with the algorithm above. The convergence is extremely fast. After two iterations, the algorithm basically converges. The maximum likelihood ratio statistic was computed to be $T_0 = 33.401$ with degree of freedom 24, giving a P-value of 9.59%. The evidence against the model is indeed weak.

The estimated risk-free rate $\hat{\gamma}_0 = 0.117\%$ with SE 1.623% per annum. This compares with the average of the 1-month treasury bill rates 2.204% per annum with a standard deviation of 2.131% during the same time period. The estimated risk premia is $\hat{\boldsymbol{\lambda}}_K = (3.170\%, 4.142\%, 0.044\%)^{\mathrm{T}}$ with estimated SE 8.579%, 4.146%, and 4.871% per annum, respectively. The standard errors are computed based on the square root of the estimated covariance matrix (6.20), which is given by

$$\hat{\mathrm{var}}(\hat{\boldsymbol{\lambda}}) = \begin{pmatrix} 73.601 & 15.209 & 19.540 \\ 15.209 & 17.191 & 7.600 \\ 19.540 & 7.600 & 23.723 \end{pmatrix}.$$

Note that the risk premium of the market portfolio is low during this period, because of the subprime crisis in 2007 and the financial crisis in 2008.

6.4 Macroeconomic variables as factors*

We will now briefly discuss the econometric techniques for validating the multiple factor pricing model when factors are non-tradable macroeconomic variables. Examples of such macroeconomic variables are growth in gross domestic product, changes in bond yields, and unanticipated inflation. These factors are observable, albeit at a low frequency, but they are not traded. So the risk premia is unknown and cannot be written as $\boldsymbol{\lambda}_K = E\boldsymbol{f} - \gamma_0 \mathbf{1}$.

The basic assumption is still the factor model (6.8) and the MLE under the factor model can be computed similarly to that in Section 6.3.5. The main difference here is that the risk premia $\boldsymbol{\lambda}_K$ are not the same as $E\boldsymbol{f}_K - \gamma_0\mathbf{1}$, and they are treated as unknown parameters. From the multifactor pricing model (6.5) and regression model (6.8), we have

$$\boldsymbol{\mu} = \gamma_0 \mathbf{1} + \boldsymbol{B}\boldsymbol{\lambda} = \boldsymbol{a} + \boldsymbol{B}E\boldsymbol{f}_K.$$

Thus, the multifactor pricing model puts constraint on the regression coefficient

$$\boldsymbol{a} = \gamma_0 \mathbf{1} + \boldsymbol{B}\boldsymbol{\gamma}_1, \tag{6.21}$$

where $\boldsymbol{\gamma}_1 = \boldsymbol{\lambda}_K - E\boldsymbol{f}$. The testable object now becomes (6.21).

The MLE under the null hypothesis (6.21) has now K extra parameters $\boldsymbol{\gamma}_1$. Nevertheless, the idea in Section 6.3.5 can be used to obtain the MLE under the null hypothesis. We omit the details.

With the estimated $\hat{\boldsymbol{\Sigma}}$ and $\hat{\boldsymbol{\Sigma}}_0$, we can now form the modified maximum likelihood ratio statistic T_0. The asymptotic null distribution is still χ^2-distribution. The degree of freedom is either $N-K$ or $N-K-1$, depending on whether the risk-free assets are known or not. For example, when γ_0 is unknown, the number of constraints in (6.21) is $N-K-1$, which is the degree of freedom under the null hypothesis.

6.5 Selection of factors

There are two main approaches to choose factors. One is based on the financial economic theory. The other is a statistical approach powered by arbitrage pricing theory, which builds factors from a comprehensive set of asset returns via *factor analysis* and *principal component analysis*. For example, variables such as the firm characteristics and macroeconomic and financial market variables do have an economic basis. These are known factors and can be selected either manually or via statistical variable selection techniques.

Regardless of how factors are selected, economic/market variables or statistically built ones, linear combinations of the selected factors are equivalent. For example, in the Fama–French model, one could have defined the two factors such as

$$f_1 = \text{SMB} + \text{HML}, \qquad f_2 = \text{SMB} - 2 * \text{HML}.$$

The linear space spanned by the factors f_1 and f_2 is the same as that spanned by the factors SMB and HML. In general, the linear space spanned by the K factors $\boldsymbol{f}$ is equivalent to the linear space spanned by $\boldsymbol{A}\boldsymbol{f}$ for any nonsingular matrix $\boldsymbol{A}$.

Principal components analysis (PCA) and factor analysis are classical statistical tools, prominently featured in statistical multivariate analysis. They can be found in many multivariate books. Many statistical and econometric software packages contain both PCA and factor analysis. Therefore, we just briefly touch the subject. These two topics are very different when the number of variables (assets) is small, but are approximately the same when the number of variables are large. We will elaborate this further in Section 7.2.7. For a thorough account of principal component analysis and factor analysis, we refer readers to Anderson (2003).

6.5.1 Principal component analysis

Given p time series in $\{\boldsymbol{X}_t\}_{t=1}^T$, they are usually correlated and can be summarized by their main co-movement directions, called principal compoments. This can be the risk factors summarized in Section 5.1, in which we attempt to summarize multiple factors by a few principal components. These can also be a large set of returns of assets, where we try to identify key factors that explain the cross-sectional risks. Regardless of the purpose, our starting point is the sample covariance matrix $\hat{\boldsymbol{\Sigma}}$, computed based on the time series $\{\boldsymbol{X}_t\}$.

Note that the sample covariance of $\{\boldsymbol{b}^{\mathrm{T}}\boldsymbol{X}_t\}_{t=1}^T$ is $\boldsymbol{b}^{\mathrm{T}}\hat{\boldsymbol{\Sigma}}\boldsymbol{b}$. The principal direction is to find $\boldsymbol{b}$ that maximizes the sample variance, subject to the normalization $\|\boldsymbol{b}\|^2 = \boldsymbol{b}^{\mathrm{T}}\boldsymbol{b} = 1$. In other words, the first principal component $\boldsymbol{\xi}_1$ is defined as

$$\boldsymbol{\xi}_1 = \operatorname{argmax}_{\|\boldsymbol{b}\|=1} \boldsymbol{b}^{\mathrm{T}}\hat{\boldsymbol{\Sigma}}\boldsymbol{b}. \tag{6.22}$$

Note that the sign of $\boldsymbol{\xi}_1$ can not be determined uniquely. If $\boldsymbol{\xi}_1$ is the principal component, so is $-\boldsymbol{\xi}_1$.

To appreciate why the maximum instead of minimum variability direction is taken, let us image that we would like to summarize the homework assignment scores, the results of midterm and final exams into one overall performance index. Such a summary is a linear combination of homework, midterm and final results. Such a linear combination should be chosen to maximize the variability of overall performance, subject to normalization constraint, so that we can differentiate one student from another. Therefore, we use the principal component as a measure. In other words, the principal component summarizes the performance factor of the class. In a similar application, the principal component of the returns of the components of S&P500 index is highly correlated to the return of S&P500 index. It summarizes the performance of the market factor. See, for example, Fan, Liao and Mincheva (2013) for a theoretical justification of this.

For bivariate normal data whose scatter plot is of oval shape, the principal component direction is the major axis of the oval. The minor axis direction is the second principal component, which we now define.

The second principal component is defined to maximize the variance in that direction, subject to the constraint that it should be orthogonal to the first principal component, i.e.

$$\boldsymbol{\xi}_2 = \mathrm{argmax}_{\boldsymbol{b}^{\mathrm{T}}\boldsymbol{x}i_1=0, \|\boldsymbol{b}\|=1} \boldsymbol{b}^{\mathrm{T}} \hat{\boldsymbol{\Sigma}} \boldsymbol{b}. \tag{6.23}$$

Note that the constraint that $\boldsymbol{b}^{\mathrm{T}}\boldsymbol{\xi}_1 = 0$ is equivalent to $\boldsymbol{b}^{\mathrm{T}}\hat{\boldsymbol{\Sigma}}\boldsymbol{\xi}_1 = 0$ so that $\{\boldsymbol{\xi}_1^{\mathrm{T}}\boldsymbol{X}_t\}$ and $\{\boldsymbol{\xi}_2^{\mathrm{T}}\boldsymbol{X}_t\}$ are uncorrelated. The mathematics of the first part of the statement follows from (6.24) that $\boldsymbol{b}^{\mathrm{T}}\hat{\boldsymbol{\Sigma}}\boldsymbol{\xi}_1 = \lambda_1 \boldsymbol{b}^{\mathrm{T}}\boldsymbol{\xi}_1$ and of the second part can be found in Exercise 6.2.

In general, the kth principal direction is defined as the maximizer of the variance subject to its orthogonality to the first $k-1$ principal components.

Recall that any symmetric matrix admits the following *spectral decomposition*:

$$\hat{\boldsymbol{\Sigma}} = \lambda_1 \boldsymbol{\xi}_1 \boldsymbol{\xi}_1^{\mathrm{T}} + \cdots + \lambda_p \boldsymbol{\xi}_p \boldsymbol{\xi}_p^{\mathrm{T}}, \tag{6.24}$$

where $\{\boldsymbol{\xi}_j\}_{j=1}^p$ are othonormal, i.e.

$$\boldsymbol{\xi}_j^{\mathrm{T}} \boldsymbol{\xi}_k = 0, \text{ for } i \neq j, \quad \|\boldsymbol{\xi}_j\| = 1.$$

Assume that the eigenvalues $\{\lambda_j\}_{j=1}^p$ have been ordered in a decreasing order. Then, it is easy to show that the corresponding eigenvector $\boldsymbol{\xi}_k$ is the kth principal component:

$$\hat{\boldsymbol{\Sigma}} \xi_k = \lambda_k \boldsymbol{\xi}_k.$$

For any two vectors, it can be shown (see Exercise 6.2) that the sample covariance

$$\operatorname{cov}(\{\boldsymbol{\xi}_j^T \boldsymbol{X}_t\}, \{\boldsymbol{\xi}_k^T \boldsymbol{X}_t\}) = \boldsymbol{\xi}_j^T \hat{\boldsymbol{\Sigma}} \boldsymbol{\xi}_k, \quad j \neq k.$$

Using this and (6.24), the principal components are uncorrelated:

$$\operatorname{cov}(\{\boldsymbol{\xi}_j^T \boldsymbol{X}_t\}, \{\boldsymbol{\xi}_k^T \boldsymbol{X}_t\}) = \boldsymbol{\xi}_j^T \hat{\boldsymbol{\Sigma}} \boldsymbol{\xi}_k = \lambda_k \boldsymbol{\xi}_j^T \boldsymbol{\xi}_k = 0, \quad \text{for } j \neq k,$$

Furthermore, the variance of the jth principal component is

$$\operatorname{var}(\{\boldsymbol{\xi}_j^T \boldsymbol{X}_t\}) = \lambda_j.$$

Note that from (6.24),

$$\operatorname{tr}(\hat{\boldsymbol{\Sigma}}) = \lambda_1 + \cdots + \lambda_p,$$

which can be thought of as the total variance. The largest k principal direction $\{\boldsymbol{\xi}_k^T \boldsymbol{X}_t\}_{t=1}^T$ contributes to the total variance by the amount of $\lambda_1 + \cdots + \lambda_k$. Therefore, the proportion of total variability explained by the first k principal components is

$$p_k = \frac{\lambda_1 + \cdots + \lambda_k}{\lambda_1 + \cdots + \lambda_p}. \tag{6.25}$$

In particular, p_1 is the proportion of the total variance explained by the first principal component alone, and $p_2 - p_1$ is the proportion of variance explained by the second principal component, and so on. For many applications, a few principal components can explain over 80% of the total variability, achieving significant dimensionality reduction and interpretability.

PCA above is introduced based on the sample covariance matrix. It can also be based on the sample correlation matrix. The results are in general different, but the interpretation is similar. The correlation matrix based principal components are based on marginally standardized variables. They are independent of the units being used. Note that for the correlation matrix, since $\operatorname{tr}(\hat{\boldsymbol{\Sigma}}) = p$, (6.25) reduces to

$$p_k = p^{-1}(\lambda_1 + \cdots + \lambda_k).$$

Therefore, the jth principal component explains λ_j / p variability, and the average contribution by each principal component is

$$p^{-1} \sum_{j=1}^{p} \lambda_j / p = 1/p.$$

The spectral decomposition (6.24) and principal component analysis can easily carried out by the R-function:

Table 6.4 *Principal components of idiosyncratic factors to the market portfolio*

		Vol	ExR	Mat	Def	Liq	Size	Tbill
first PC	$\boldsymbol{\xi}_1$	0.479	−0.230	0.481	0.450	0.111	0.008	−0.521
second PC	$\boldsymbol{\xi}_2$	0.154	−0.492	−0.263	−0.049	0.779	−0.015	0.238

```
> res = eigen(S);  xi = res$vectors; lam = res$values
```

where S is a given sample covariance or correlation matrix.

We now use the principal component analysis to blend eight risk factors introduced in Section 5.1 into two risk factors. One is the market factor and the other is the principal component of the other 7 risk factors introduced in Figure 6.1. Of course, these two factors are highly correlated. But, we wish to construct an uncorrelated factor that summarizes the other 7 risk factors. We analyze the data in the following example.

Example 6.6 We now analyze the risk factors given in Figure 6.1. The data from February $23, 2004$ to February $28, 2011$ are used. To obtain an uncorrelated risk factor to this market portfolio, we first regress the other 7 risk factors on the returns of S&P 500 and obtain the residual vectors. The residual vectors are uncorrelated with the market portfolio. We then compute the correlation matrix of the residual vectors ($p = 7$) and do the spectral decomposition (6.24). The first two principal components are the linear combinations of the standardized factors (each of them has zero mean and unit variance) with weights given by Table 6.4. These weights are $\boldsymbol{\xi}_1$ and $\boldsymbol{\xi}_2$ in the decomposition (6.24), whose corresponding eigenvalues are $\lambda_1 = 3.002$ and $\lambda_2 = 1.010$. The first principal component explains about $\lambda_1/p = 42.86\%$ of variability, and the second component explains $\lambda_2/p = 14.44\%$, which is about the average $1/7$.

The above computation is based on the entire data set. To see how the proportion of variance explained by the first two principal components vary over time, for each given t, we restrict ourselves to the data only in the time window $[t - w, t]$ and repeat the above analysis. The results are depicted in Figure 6.6.

To see how much prediction power these two risk factors have, we use the Fama–French 100 portfolios as testing cases. We regress these 100 portfolios respectively on the returns of the SP 500 and the PCA during in the period January $23, 2004$ to December $31, 2010$. The multiple regression R^2 for those 100 portfolios are presented in Figure 6.7. We also presented the multiple R^2 using all of the 8 factors introduced in Section 5.1.

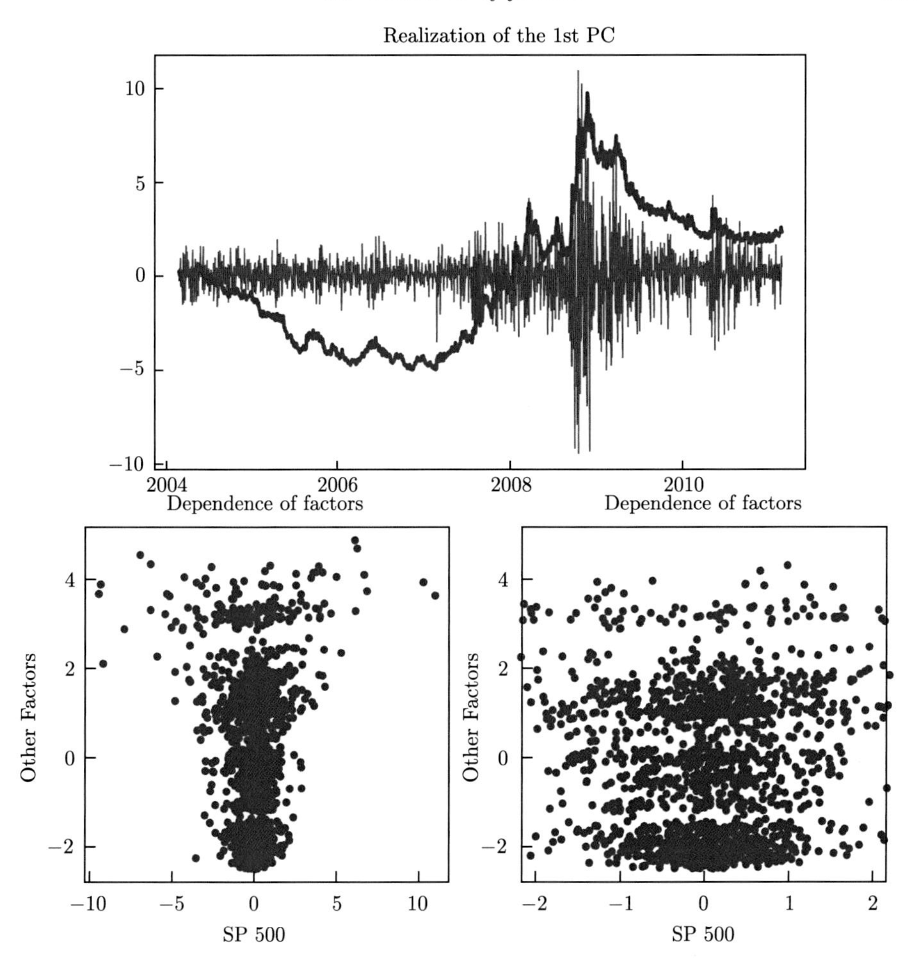

Figure 6.5 Top panel is the realization of the returns of S&P 500 and the first PCA from February 23, 2004 to February 28, 2011. The bottom panel shows that these two factors are nearly independent. The bottom right is a zoom version of the figure at bottom left.

The market portfolio explains a large portion of the variability of the returns. Unfortunately, the PCA does not add very much to the explanatory power, yet the other 7 factors do have additional prediction power.

There are many possible improvements on the analysis. The PCA looks relatively smooth and cannot be expected to have much prediction power on the returns of the market at daily frequency. One possibility is to consider monthly returns, and the other is to use the returns of the risk factors and lag variables. We do not pursue further the details of the analysis.

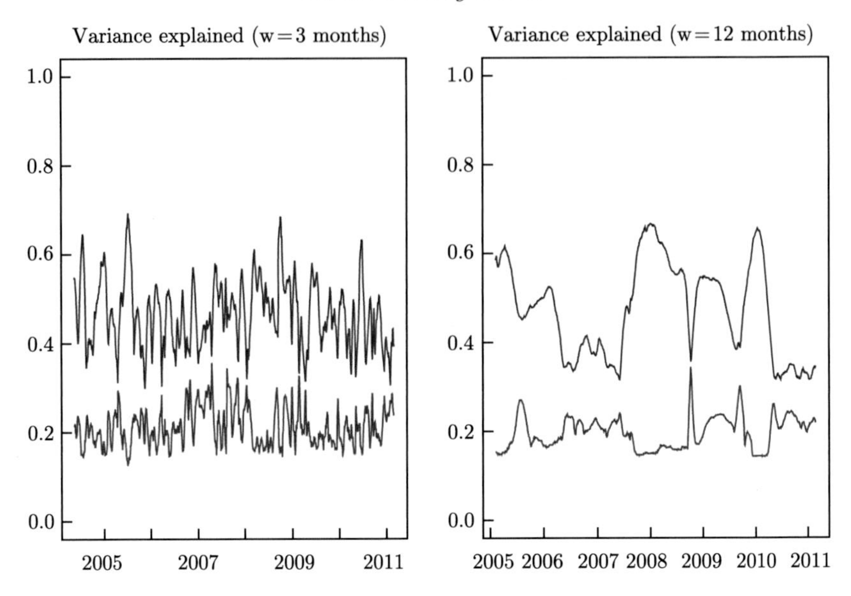

Figure 6.6 Proportion of variances explained by the first two principal components with moving window of 3 months (left) and 1 year (right) used to estimate the correlation matrix.

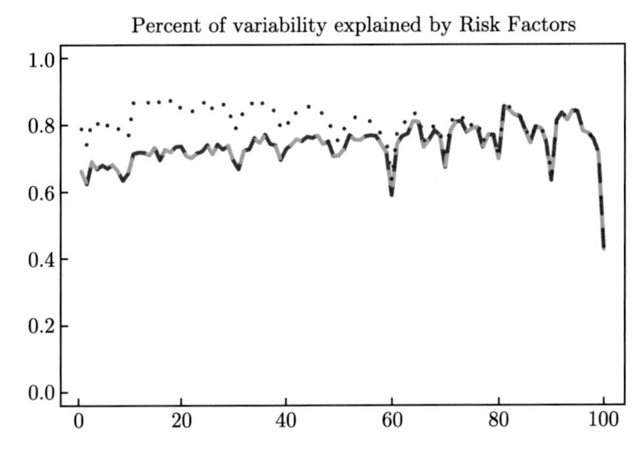

Figure 6.7 Multiple R^2 of the 100 Fama–French portfolios when regressed on the returns of S&P 500 (red), S&P 500 and PCA (green), and all 8 risk factors (blue).

6.5.2 Factor analysis*

The goal of factor analysis is to extract unknown factors as well as their factor loadings from the data. Assume that K factors account for all covariance of asset returns so that the covariance of the idiosyncratic noise $\boldsymbol{\Sigma}$

is diagonal. This is a strict factor structure, imposed by Ross (1973) in his development of the APT.

With the factor structure (6.4), the covariance matrix is given by (6.7). Let $\boldsymbol{\Sigma}_f = \text{var}(\boldsymbol{f})$ and $\boldsymbol{B}^* = \boldsymbol{\Omega}_K^{\frac{1}{2}}\boldsymbol{B}$. Then,

$$\text{var}(\boldsymbol{R}) = \boldsymbol{B}^{\mathrm{T}}\boldsymbol{\Sigma}_f\boldsymbol{B} + \boldsymbol{\Sigma}_\epsilon = \boldsymbol{B}^{*T}\boldsymbol{B}^* + \boldsymbol{\Sigma}_\epsilon.$$

In other words, $\boldsymbol{B}$, a $p \times K$ loading matrix, can be identified only up to a $K \times K$ nonsingular matrix, and $\boldsymbol{B}^*$ is unique up to an orthogonal transform, corresponding to the loading matrix of the orthonormal factors $\boldsymbol{f}^* = \boldsymbol{\Sigma}_f^{-1}\boldsymbol{f}$ with $\text{var}(\boldsymbol{f}^*) = I_K$. Without loss of generality, assume $E\boldsymbol{f} = 0$. Thus, $\boldsymbol{a} = \boldsymbol{E}\boldsymbol{R}_t$ in (6.4), which can be estimated as $\hat{\boldsymbol{a}} = \bar{\boldsymbol{R}}$, the sample mean vector.

Adapting the likelihood function (6.13) into the current notation, under the normality assumption, the maximum likelihood estimator is to find $\boldsymbol{B}$ and $\boldsymbol{f}$ that minimize

$$\sum_{t=1}^{T}(\boldsymbol{R}_t - \bar{\boldsymbol{R}} - \boldsymbol{B}\boldsymbol{f}_t)^{\mathrm{T}}\boldsymbol{\Sigma}^{-1}(\boldsymbol{R}_t - \bar{\boldsymbol{R}} - \boldsymbol{B}\boldsymbol{f}_t). \tag{6.26}$$

With estimated $\{\boldsymbol{B}\boldsymbol{f}_t\}$, the estimated $\boldsymbol{\Sigma}$ is the diagonal matrix, consisting of residual variances of individual assets, recalling the strict factor model assumption.

To solve (6.26), let us consider the ordinary least-squares instead of generalized least-squares (6.26):

$$\sum_{t=1}^{T}(\boldsymbol{R}_t - \bar{\boldsymbol{R}} - \boldsymbol{B}\boldsymbol{f}_t)^{\mathrm{T}}(\boldsymbol{R}_t - \bar{\boldsymbol{R}} - \boldsymbol{B}\boldsymbol{f}_t). \tag{6.27}$$

We will show at the end of the section that space spanned by the columns of $\boldsymbol{B}$ is the same as the space spanned by the first K principal components $\{\boldsymbol{\xi}_j\}_{j=1}^K$ of the data $\{\boldsymbol{R}_t\}$ using its sample covariance matrix. In other words, the projections of $\{\boldsymbol{R}_t - \bar{\boldsymbol{R}}\}_{t=1}^{\mathrm{T}}$ on principal directions

$$\boldsymbol{f}_t = (\boldsymbol{\xi}_1, \ldots, \boldsymbol{\xi}_K)^{\mathrm{T}}(\boldsymbol{R}_t - \bar{\boldsymbol{R}}), \quad t = 1, \ldots, T$$

are estimated realized factors from the data. This is an un-normalized version. With these given factors, from (6.14), we have the loading matrix

$$\begin{aligned}\boldsymbol{B} &= \sum_{t=1}^{T}(\boldsymbol{R}_t - \bar{\boldsymbol{R}})(\boldsymbol{R}_t - \bar{\boldsymbol{R}})^{\mathrm{T}}\boldsymbol{\Theta}_K\Big[\sum_{t=1}^{T}\boldsymbol{\Theta}_K^{\mathrm{T}}(\boldsymbol{R}_t - \bar{\boldsymbol{R}})(\boldsymbol{R}_t - \bar{\boldsymbol{R}})^{\mathrm{T}}\boldsymbol{\Theta}_K\Big] \\ &= \hat{\boldsymbol{\Sigma}}\boldsymbol{\Theta}_K[\boldsymbol{\Theta}_K^{\mathrm{T}}\hat{\boldsymbol{\Sigma}}\boldsymbol{\Theta}_K]^{-1} \\ &= (\lambda_1\boldsymbol{\xi}_1, \ldots, \lambda_K\boldsymbol{\xi}_K)[\mathrm{diag}(\lambda_1, \ldots, \lambda_K)]^{-1} \\ &= \boldsymbol{\Theta}_K,\end{aligned}$$

where $\boldsymbol{\Theta}_K = (\boldsymbol{\xi}_1, \ldots, \boldsymbol{\xi}_K)$. Since the least-squares estimator also provides a consistent estimator when the identifiability condition is imposed, we in fact obtain a good initial value for the problem (6.26):

$$\boldsymbol{f}_t = (\boldsymbol{\xi}_1, \ldots, \boldsymbol{\xi}_K)^{\mathrm{T}}(\boldsymbol{R}_t - \bar{\boldsymbol{R}}), \qquad \boldsymbol{B} = (\boldsymbol{\xi}_1, \ldots, \boldsymbol{\xi}_K).$$

One can indeed use these principal components as factors, if one does not intend to use MLE. Recently, Fan, Liao and Mincheva (2013) showed that when the portfolio size is large, the principal component analysis and the factor analysis are approximately the same.

The MLE (6.26) can be solved by iteratively finding the factors $\boldsymbol{f}$, factor loadings $\boldsymbol{B}$, and the residual variance matrix $\boldsymbol{\Sigma}$. For example, given $\boldsymbol{f}$, the loading matrix can be found via (6.14) with appropriate adaptation of notation. On the other hand, given the loading matrix $\hat{\boldsymbol{B}}$ and residual variance $\hat{\boldsymbol{\Sigma}}$, minimizing (6.26) is equivalent to minimizing separately for each t in the generalized least-squares:

$$(\boldsymbol{R}_t - \bar{\boldsymbol{R}} - \boldsymbol{B}\boldsymbol{f}_t)^{\mathrm{T}}\hat{\boldsymbol{\Sigma}}^{-1}(\boldsymbol{R}_t - \bar{\boldsymbol{R}} - \boldsymbol{B}\boldsymbol{f}_t),$$

which gives the solution

$$\hat{\boldsymbol{f}}_t = (\hat{\boldsymbol{B}}^{\mathrm{T}}\hat{\boldsymbol{\Sigma}}^{-1}\hat{\boldsymbol{B}})^{-1}\hat{\boldsymbol{B}}^{\mathrm{T}}\hat{\boldsymbol{\Sigma}}^{-1}(\boldsymbol{R}_t - \bar{\boldsymbol{R}}),$$

where $\hat{\boldsymbol{\Sigma}}$ is an estimate of $\boldsymbol{\Sigma}$. Given $\boldsymbol{B}$ and $\boldsymbol{f}$, recalling the strict factor model assumption, the MLE of $\boldsymbol{\Sigma}$ is a diagonal matrix, consisting of variance of residuals of each stock. The iteration stops when the target (6.26) fails to decrease.

In the above iterative process, we do not enforce the identifiability condition. This does not matter, since we aim at minimizing (6.26). Even though the factors may not converge without imposing identifiability conditions, the linear space spanned by the factors should converge.

When the normality assumption is removed, (6.26) becomes the least-squares estimator. It still provides a consistent estimate of the linear space spanned by the factors.

We now show that the space spanned by the first K principal vectors is the solution to the least squares problem (6.27). To this end, let $\boldsymbol{A}$ be a $p \times (p-K)$ matrix such that the space spanned by the columns of $\boldsymbol{A}$ is the orthogonal completement to the space spanned by the columns of $\boldsymbol{B}$. Then, $\boldsymbol{R}_t - \bar{\boldsymbol{R}}$ can be decomposed as

$$\boldsymbol{R}_t - \bar{\boldsymbol{R}} = \boldsymbol{A}\boldsymbol{\eta}_{1t} + \boldsymbol{B}\boldsymbol{\eta}_{2t}, \tag{6.28}$$

for some vectors $\boldsymbol{\eta}_{1t}$ and $\boldsymbol{\eta}_{2t}$. Substituting this into (6.27), and using the orthogonality $\boldsymbol{A}^{\mathrm{T}}\boldsymbol{B} = 0$, we have

$$\sum_{t=1}^{T} \boldsymbol{\eta}_{1t}^{\mathrm{T}}(\boldsymbol{A}^{\mathrm{T}}\boldsymbol{A})\boldsymbol{\eta}_{1t} + \sum_{t=1}^{T}(\boldsymbol{\eta}_{2t} - \boldsymbol{f}_t)^{\mathrm{T}}\boldsymbol{B}^{\mathrm{T}}\boldsymbol{B}(\boldsymbol{\eta}_{2t} - \boldsymbol{f}_t).$$

Minimizing it with respect to $\boldsymbol{f}_t$, we obtain

$$\sum_{t=1}^{T} \boldsymbol{\eta}_{1t}^{\mathrm{T}}(\boldsymbol{A}^{\mathrm{T}}\boldsymbol{A})\boldsymbol{\eta}_{1t}.$$

Noticing that $\boldsymbol{\eta}_{1t} = (\boldsymbol{A}^{\mathrm{T}}\boldsymbol{A})^{-1}\boldsymbol{A}^{\mathrm{T}}(\boldsymbol{R}_t - \bar{\boldsymbol{R}})$, the above quantity can now be written as

$$\sum_{t=1}^{T}(\boldsymbol{R}_t - \bar{\boldsymbol{R}})^{\mathrm{T}}\boldsymbol{A}(\boldsymbol{A}^{\mathrm{T}}\boldsymbol{A})^{-1}\boldsymbol{A}^{\mathrm{T}}(\boldsymbol{R}_t - \bar{\boldsymbol{R}}) = (T-1)\mathrm{tr}(\hat{\boldsymbol{\Sigma}}\boldsymbol{C}),$$

where $\boldsymbol{C} = \boldsymbol{A}(\boldsymbol{A}^{\mathrm{T}}\boldsymbol{A})^{-1}\boldsymbol{A}^{\mathrm{T}}$, the projection matrix onto the space spanned by the columns of $\boldsymbol{A}$. Minimizing the above quantity with respect to $\boldsymbol{B}$ is equivalent to minimizing the above quantity with respect to projection matrix $\boldsymbol{C}$ on the $(n-K)$ dimensional space. From the spectral decomposition (6.24), it is easy to see that $\boldsymbol{C}$ corresponds to the $(p-K)$-dimensional space spanned by the eigenvectors of the smallest $(p-K)$ eigenvalues. Therefore, $\boldsymbol{B}$ corresponds to the eigenvectors of the first K largest eigenvalues.

6.6 Exercises

6.1 Consider the multi-factor model:

$$\boldsymbol{Y} = \boldsymbol{\alpha} + \boldsymbol{B}\boldsymbol{X} + \boldsymbol{\epsilon}, \qquad E\boldsymbol{\epsilon} = 0, \quad \mathrm{cov}(\boldsymbol{X}, \boldsymbol{\epsilon}) = E\boldsymbol{X}\boldsymbol{\epsilon}^{\mathrm{T}} = 0.$$

1. What is $\mathrm{var}(\boldsymbol{Y})$? Show the work.
2. What is $\mathrm{var}(\boldsymbol{A}\boldsymbol{Y})$?

6.2 For any two portfolios $\{\boldsymbol{a}^{\mathrm{T}}\boldsymbol{R}_t\}_{t=1}^{\mathrm{T}}$ and $\{\boldsymbol{a}^{\mathrm{T}}\boldsymbol{R}_t\}_{t=1}^{\mathrm{T}}$, show that its sample covariance is given by

$$\mathrm{cov}(\{\boldsymbol{a}^{\mathrm{T}}\boldsymbol{R}_t\}, \{\boldsymbol{b}^{\mathrm{T}}\boldsymbol{R}_t\}) = \boldsymbol{a}^{\mathrm{T}}\hat{\boldsymbol{\Sigma}}\boldsymbol{b},$$

where $\hat{\boldsymbol{\Sigma}}$ is the sample covariance matrix of $\{\boldsymbol{R}_t\}_{t=1}^{\mathrm{T}}$.

6.3 Let $\boldsymbol{Y}_t$ be a vector of excess returns of N assets. Consider the multivariate linear regression model

$$\boldsymbol{Y}_t = \boldsymbol{\alpha} + \boldsymbol{\beta} Y_t^m + \boldsymbol{\epsilon}_t,$$

where $\boldsymbol{\epsilon}_t \sim N(0, \boldsymbol{\Sigma})$ and $\text{cov}(Y_t^m, \boldsymbol{\epsilon}_t) = 0$.

1. Derive the maximum likelihood estimators for $\boldsymbol{\alpha}$ and $\boldsymbol{\beta}$. (You do not need to derive the MLE for $\boldsymbol{\Sigma}$, since this part is hard; you just take for granted that $\hat{\boldsymbol{\Sigma}}$ is the MLE).
2. Show that the maximum likelihood ratio test for the null hypothesis: H_0 : $\boldsymbol{\alpha} = 0$ is

$$T_2 = T[\log(|\hat{\boldsymbol{\Sigma}}_0|) - \log(|\hat{\boldsymbol{\Sigma}}|)]$$

where $\hat{\boldsymbol{\Sigma}}_0$ is the MLE under H_0. Give explicitly the expression for $\hat{\boldsymbol{\Sigma}}_0$.

6.4 Consider the multi-factor model

$$\boldsymbol{Y}_t = \boldsymbol{a} + \boldsymbol{B}\boldsymbol{X}_t + \boldsymbol{\epsilon}_t$$

with observable factor $\boldsymbol{X}_t$, where $E\boldsymbol{\epsilon}_t = 0$ and $\text{cov}(\boldsymbol{X}_t, \boldsymbol{\epsilon}_t) = 0$.

1. Based on 20 stock portfolios over a period of 60 months on the three factors, it was computed that $|\hat{\boldsymbol{\Sigma}}_0| = 2.375$ and $|\hat{\boldsymbol{\Sigma}}| = 1.624$. Test if the multifactor model is consistent with the empirical data, i.e. $H_0 : \boldsymbol{a} = 0$.
2. Suppose that the beta's of the GE stock over the S&P 500 index (X_1), the size effect X_2 and book-to-market effect X_3 are respectively 1.3, 0.3 and -0.4. Assume further that over the last 10 years the average risk-free interest is 4%, the average return of the S&P 500 is 11%, the average difference of returns between the small large capitalization is 3%, and the average difference of returns between the high and low book-to-market is 2%, what is the expected return of the GE stock using the Fama–French model?

6.5 Consider the multifactor model

$$\boldsymbol{Y}_t = \boldsymbol{a} + \boldsymbol{B}\boldsymbol{X}_t + \boldsymbol{\epsilon}_t$$

with observable factor $\boldsymbol{X}_t$.

1. Suppose that the CAPM holds and over the last five years, the average of the risk-free interest rate is 3.5% and the average return of the CRSP value-weighted index is 12.5%. If the market β of a stock (with respect to the index) is 1.3, what is the expected return of the stock?
2. Based on 15 stock portfolios over a period of 60 months regressed on five factors without knowing the risk-free interest rate, it is computed that $|\hat{\boldsymbol{\Sigma}}_0| = 2.425$ and $|\hat{\boldsymbol{\Sigma}}| = 1.742$. Test if the multifactor model is consistent with the empirical data, i.e. $H_0 : \boldsymbol{a} = 0$.
3. Suppose that the strict multi-factor model is correct so that $\text{var}(\boldsymbol{\epsilon}_t) = \boldsymbol{\Sigma}_0$ is a diagonal matrix and $\boldsymbol{X}_t$ and $\boldsymbol{\epsilon}_t$ are uncorrelated. Show how to estimate the covariance matrix of $\boldsymbol{Y}$ based on the past T days' data:

$$\{(\boldsymbol{X}_t, \boldsymbol{Y}_t) : t = 1, \ldots, T\}.$$

6.6 Consider a basket of bonds. Let $X_i = 1$ be the event that the ith bond will be default in two years. The probability of default $p_i = P(X_i = 1)$ is usually known (from bond rating). The challenge is to model the correlation of default. Suppose these bonds share one common unobserved risk factor $Z \sim N(0,1)$ and condition on Z, the events of default are independent with probability

$$P(X_i = 1|Z) = \exp(a_i + b_i Z),$$

where b_i is the factor-loading and a_i is the intercept. Show that

$$EX_iX_j = p_i p_j \exp(b_i b_j/2).$$

Hint: $EX_iX_j = E[P(X_i = 1, X_j = 1|Z)]$ and $E\exp(bZ) = \exp(b^2/2)$.

6.7 Verify the results in Table 6.2.

6.8 Use the Fama–French 100 portfolios in the last five years to construct three common factors via the principal component analysis based on the correlation matrix. Report the variance explained by each principal components. Now, regress each of the Fama–French 100 portfolios on these three principal components and report the distribution (histogram) of the residual variances. Report also the distribution of the variances of these 100 portfolios over the same time period.

7

Portfolio Allocation and Risk Assessment

The groundbreaking work on portfolio choice by Markowitz (1952, 1959) is the cornerstone of modern finance on which beautiful pricing theories are built upon. Yet, its applications face a number of challenges such as the accuracy of estimated expected returns and the stability of estimated volatility matrices. The problem becomes more challenging when the pool of candidate assets is large. For example, among the universe of 1000 stocks in the emerging markets, an estimate of the covariance matrix of order 1000 is needed in order to be able to efficiently select tens of stocks for investment and monitoring. Yet the sample covariance matrix based on the past one years' daily (say) data must be degenerate, making it impossible to directly use the formula for optimal portfolio allocations.

This chapter covers some empirical aspects of portfolio allocation and risk assessment. Emphasis will be given to the case where the pool of financial assets is large, since the issues of error accumulation is more severe, though the approaches are also applicable to the situation where the candidate pool is moderate or small.

How do we assess the risk of a large portfolio? Do the errors in estimating large covariance matrix accumulate? How do they impact portfolio optimization? How do we select a subset of portfolio to invest? How do we track a portfolio? Is a given portfolio efficient? These are among the topics to be addressed in this chapter.

7.1 Risk assessment of large portfolios

There are many measures of risks of a given portfolio with allocation $\boldsymbol{w}$ on the p risky assets with returns $\boldsymbol{R}$. See Artzner, Delbaen, Eber and Heath for a definition of coherent *risk measures*. Here, we specifically focus on the portfolio standard deviation $\sqrt{\mathrm{var}_t(\boldsymbol{w}^{\mathrm{T}}\boldsymbol{R}_{t+1})}$ at time t. Since the time t does not play a specific role, the dependence of current time t will be dropped. Thus, the risk of the portfolio is $\sqrt{R(\boldsymbol{w})}$, where

$$R(\boldsymbol{w}) = \boldsymbol{w}^{\mathrm{T}} \boldsymbol{\Sigma} \boldsymbol{w}, \quad \text{with} \quad \boldsymbol{\Sigma} = \mathrm{var}(\boldsymbol{R}). \tag{7.1}$$

The portfolio risk depends on unknown volatility matrix. From historical data, an estimate $\hat{\boldsymbol{\Sigma}}$ is obtained and the estimated risk $\sqrt{\hat{R}(w)}$ is computed, where

$$\hat{\boldsymbol{R}}(\boldsymbol{w}) = \boldsymbol{w}^{\mathrm{T}} \hat{\boldsymbol{\Sigma}} \boldsymbol{w}. \tag{7.2}$$

We will refer to this risk as the *perceived risk* or the *empirical risk*, since it is our estimated risk, not the actual risk. How different can they be? Let us first look at the number of elements being estimated.

Estimating high-dimensional covariance matrices is intrinsically challenging. Suppose we have a pool of 2000 stocks to be managed or selected. There are over 2 million free parameters in the covariance matrix! Yet, one year daily returns yield only about sample size $T = 252$. In this case, the sample covariance matrix is degenerate. Further extension of the sampling period will introduce more biases in the estimation of the covariance matrices, since the stochastic dynamics of asset returns evolve over time. For this reason, a shorter sampling period such as three months or half a year is used frequently to estimate the volatility matrix. Accurately estimating it poses significant challenges due to both contradicting demands of statistical biases and stochastic errors. Since the portfolio risk assessment involves millions of estimated covariances, accumulation of millions of estimation errors can have a devastating effect.

7.1.1 Stability of a portfolio

The question then arises what kind of portfolios can avoid the large noise accumulation. That leads us to define the stability of a portfolio. We call the L_1-norm

$$\|\boldsymbol{w}\|_1 = |w_1| + \cdots + |w_p|$$

the *gross exposure* of the portfolio with the allocation vector $\boldsymbol{w}$. It is innerly related with the total *short position*. This concept was independently introduced by Brodie et al.(2009), DeMiguel et al.(2009), Fan, Zhang and Yu (2012). Let

$$\boldsymbol{w}^{+} = \sum_{w_i \geqslant 0} w_i, \qquad \boldsymbol{w}^{-} = \sum_{w_i < 0} |w_i|$$

An earlier version of this paper appeared online in 2008.

be total long and short positions and $c = \|\boldsymbol{w}\|_1$. Then, $c = w^+ + w^-$. Since the total portfolio is 100%, we have $w^+ - w^- = 1$. Solving the last two equations yields

$$w^- = (c-1)/2, \quad w^+ = (c+1)/2.$$

Hence, the minimum gross exposure is $c \geqslant 1$. All portfolios with *no short* positions have the minimum gross exposure $c = 1$.

Example 7.1 Consider the following five portfolios of size 4:

$$\boldsymbol{w}_1 = \begin{pmatrix} 0.25 \\ 0.25 \\ 0.25 \\ 0.25 \end{pmatrix}, \boldsymbol{w}_2 = \begin{pmatrix} 0 \\ 0.50 \\ 0.25 \\ 0.25 \end{pmatrix}, \boldsymbol{w}_3 = \begin{pmatrix} -0.5 \\ 0.5 \\ 0.5 \\ 0.5 \end{pmatrix},$$

$$\boldsymbol{w}_4 = \begin{pmatrix} -0.5 \\ 0.5 \\ 1.5 \\ -0.5 \end{pmatrix}, \boldsymbol{w}_5 = \begin{pmatrix} -8 \\ 25 \\ -7 \\ -9 \end{pmatrix}.$$

Both $\boldsymbol{w}_1$ and $\boldsymbol{w}_2$ are no short-sale portfolios with $c = 1$. Portfolio $\boldsymbol{w}_1$ is more diversified and robust to withstand individual influence and modeling errors. The gross exposure for the third portfolio is $c = \|\boldsymbol{w}_3\|_1 = 2$, and hence the total short position is $w^- = (c-1)/2 = 0.5$ and long position is $w^+ = (c+1)/2 = 1.5$. It is less stable, since the gross exposure is large (The L_1-norm is similar to the standard deviation, but measures deviation from 0 rather than from the average). This can also be seen from the portfolio weights. Since $c = \|\boldsymbol{w}_4\|_1 = 3$, the total short position is $(c-1)/2 = 1$. It is even more aggressive and unstable. Clearly, the least stable and speculative portfolio is $c = \|\boldsymbol{w}_5\|_1 = 49$. The total short position is $w^- = (c-1)/2 = 24$ and long position is $w^+ = (c+1)/2 = 25$. While the portfolio is allowable theoretically, it is not allowable in practice due to its speculative nature. For this reason, it is not allowed by regulatory requirement and marginal requirement by financial brokage firms.

The above example reveals that the gross exposure is related to the stability and speculativeness of a portfolio. It is reasonable to expect that the estimation errors do not accumulate very rapidly unless the gross exposures are very large. This will be shown in Theorem 7.1. First of all, note that if $\hat{\boldsymbol{\Sigma}}$ is an unbiased estimate of $\boldsymbol{\Sigma}$, so is $\hat{R}(\boldsymbol{w})$. This follows simply from

$$E\hat{R}(\boldsymbol{w}) = \boldsymbol{w}^{\mathrm{T}}(E\hat{\boldsymbol{\Sigma}})\boldsymbol{w} = R(\boldsymbol{w}).$$

The *unbiasedness* implies that the procedure does not have any systematic biases: repeated use of the procedure many times on average gives the right estimate. However, the estimation error can be large.

7.1.2 Stability and risk approximations

The following theorem gives a bound on the *risk approximation*, which shows that the risk approximation error depends on the gross-exposure parameter c.

Theorem 7.1 (Risk Approximation) *For any portfolio of size p with allocation vector $\boldsymbol{w}$,*

$$|R(\boldsymbol{w}) - \hat{R}(\boldsymbol{w})| \leqslant e_{\max} c^2, \tag{7.3}$$

where $e_{\max} = \max_{i,j} |\hat{\sigma}_{ij} - \sigma_{ij}|$, denoted also by $|\hat{\boldsymbol{\Sigma}} - \boldsymbol{\Sigma}|_\infty$.

The proof of the theorem is surprisingly simple. The risk difference is $\left|\sum_{i,j}(\hat{\sigma}_{i,j} - \sigma_{i,j}) w_i w_j\right|$. It is bounded by

$$\sum_{i,j}\left|\hat{\sigma}_{i,j} - \sigma_{i,j}\right| |w_i||w_j| \leqslant e_{\max} \sum_{i,j} |w_i||w_j| = e_{\max}\Big(\sum_{i=1}^{p} |w_i|\Big)^2.$$

This completes the proof.

Let us discuss the implications of the theorem. First of all, the result holds for any portfolio size even when p is much larger than the sample size T, and for any matrices $\hat{\boldsymbol{\Sigma}}$ (even non semi-definite matrix). There is little noise accumulation effect when c is modest ($\leqslant 2$ or 3), since $e_{\max}$ increases with portfolio size p only in logarithmic order (see Theorem 7.4 in Section 7.4). In particular, among the universe of no-short sale portfolios, the upper bound in Theorem 7.1 is the tightest. The risk difference is bounded by the maximum componentwise estimation error.

To gain additional insights, we consider a simulated example in which we know what the true risks are.

Example 7.2 Consider 4 assets all with the expected returns 10% and volatility 30% per annum. They are equally correlated, with $\rho = 0.5$. For each simulation, data are sampled from the multivariate normal distribution at daily frequency for 3 months ($T = 63$ days). We repeat the simulation 1000 times. Figure 7.1 depicts one of the simulated returns of the four assets over a period of three months. The daily expected return is $0.1/252$ or 4 basis points, which is not visible in the presence of daily volatility of $0.3/\sqrt{252}$ or 189 basis point. The co-movements of the four asset returns can be seen due to the positive correlation of $\rho = 0.5$.

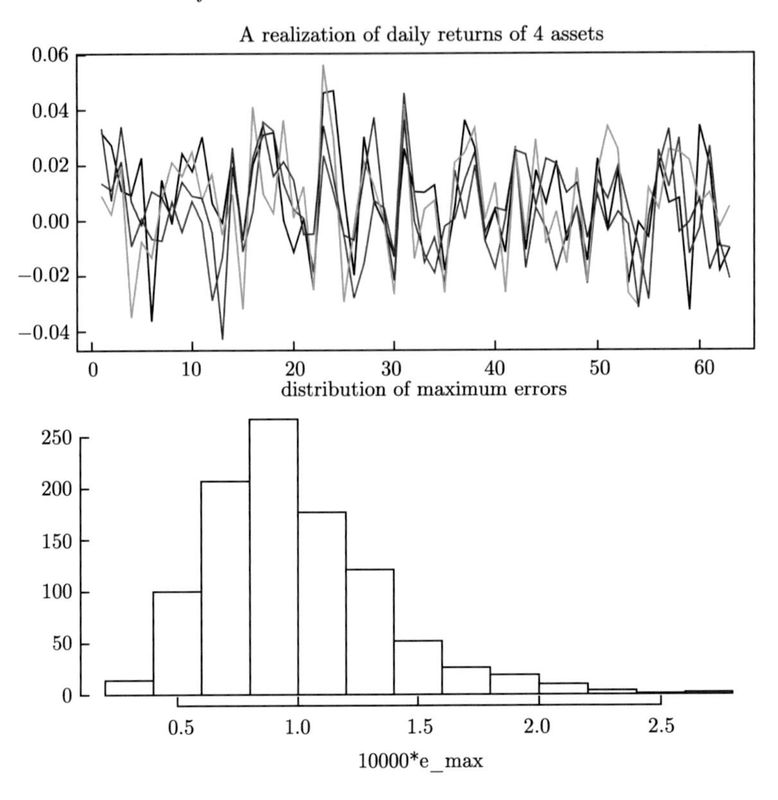

Figure 7.1 A realization of simulated daily returns of the four assets over a three-month period and the distribution of the maximum componentwise estimation errors $e_{\max}$ of the sample covariance over 1000 simulations.

Consider specifically the risk assessment of the five portfolios given in Example 7.1. The risks are assessed by using the sample variance based on the past 3-month daily returns. This is repeated 1000 times. Then, as noted before, portfolio risks are estimated unbiasedly. That is, among 1000 estimates $\hat{R}(\boldsymbol{w})$, their average should be approximately $R(\boldsymbol{w})$ for any $\boldsymbol{w}$. This is indeed the case for the five specific portfolios constructed in Example 7.1. Compare the first two rows of Table 7.1.

Unbiasedness refers to the overall average of the estimated portfolio variance. The average of estimated portfolio risks is indeed close to the true risk. It does not say the typical size of the estimation errors. The average of the mean absolute errors $|R(\boldsymbol{w})^{1/2} - \hat{R}(\boldsymbol{w})^{1/2}|$ is summarized in Table 7.1. This should also be compared to the portfolio true risks $R(\boldsymbol{w})^{1/2}$. As shown in Theorem 7.1, the upper bound of estimation error depends on the maximum componentwise estimation error $e_{\max}$. Figure 7.1 summarizes the distribu-

Table 7.1 *True risks and average estimation errors for five portfolios (in percent, annualized)*

Portfolios	$\boldsymbol{w}_1$	$\boldsymbol{w}_2$	$\boldsymbol{w}_3$	$\boldsymbol{w}_4$	$\boldsymbol{w}_5$
True Risks	23.72	24.88	30.00	42.43	607.45
Average Estimated Risks	23.73	24.86	29.95	42.33	606.04
Average Estimation Error	1.73	1.75	2.14	2.91	43.00

tion of $e_{\max}$. True risks increase with the gross exposures and so do the patterns of estimation errors.

In practice, the choice of portfolio is often driven by recent market conditions. When $\boldsymbol{w}$ is chosen by data (e.g. via optimization), the resulting allocation is denoted by $\hat{\boldsymbol{w}}$. It is helpful to distinguish a few concepts. The true risk of the portfolio $\hat{\boldsymbol{w}}$ is $R(\hat{\boldsymbol{w}})^{1/2} = \sqrt{\hat{\boldsymbol{w}}^T \boldsymbol{\Sigma} \hat{\boldsymbol{w}}}$. This depends on unknown volatility matrix $\boldsymbol{\Sigma}$. The perceived risk is $\hat{R}(\hat{\boldsymbol{w}})$, which is known. It is not an unbiased estimate of $R(\hat{\boldsymbol{w}})$ any more, since $\hat{\boldsymbol{w}}$ depends also on the data. They can be very different: the former can be as small as zero, while the latter is usually far away from zero. This is illustrated in the following studies by Fan, Zhang and Yu (2012).

Example 7.3 We focus on 1000 stocks with least missing data from the components of Russell 3000 index in the five-year period 2003–2007. *Russell 3000 index* consists of 3000 publicly held US companies based on total market capitalization, which represents approximately 98% of the US equity market. It divides further into the *Russell* 1000 *index* and *Russell* 2000 *index*. The former consists of the largest 1000 companies, representing 92% of the market capitalization of the Russell 3000 index and the latter comprises of the smaller 2000 companies, accounting for approximately 8% of the total market capitalization of the Russell 3000 Index.

To mitigate selection bias, 600 stocks are randomly selected from those 1000 stocks. These 600 stocks are the pool of stocks that are used to form various optimal portfolios below ($p = 600$). At any given time, the sample covariance matrix from past two years ($T = 504$) is used to estimate the volatility matrix.

Perceived minimum variance portfolio is defined by

$$\hat{\boldsymbol{w}}_{\text{opt}} = \text{argmin}_{\boldsymbol{w}^T \mathbf{1}=1} \hat{R}(\boldsymbol{w}), \tag{7.4}$$

minimizing the perceived portfolio variance. Let us focus on the case with sample covariance matrix $\hat{\boldsymbol{\Sigma}}$. Since the rank of the volatility matrix $\hat{\boldsymbol{\Sigma}}$ is no greater than $T - 1$, we have the minimum perceived risk $\hat{R}(\hat{\boldsymbol{w}}_{\text{opt}}) = 0$, since

$p > T$. This can also be understood as follows. Given the past T days of the data, let us choose the weight $\hat{\boldsymbol{w}}$ such that the returns of the portfolio on each of the past T days are zero. Such a solution exists since we have $p - 1 = 599$ undetermined weights $\boldsymbol{w}$ to choose and $T = 504$ equations (returns are zero) to satisfy. Hence, the perceived risk of such a portfolio is zero. Of course, the actual risk $R(\hat{\boldsymbol{w}}_{\text{opt}})$ is far from zero. It behaves like a random portfolio selected from the pool, having risks around 20%-30%, the typical readings of VIX during the period. Hence, the perceived risk $\hat{R}(\hat{\boldsymbol{w}}_{\text{opt}})$ is not an unbiased estimate of the actual risk.

The above large difference is due to a lack of control of gross exposure in the optimization problem (7.4). Thus, Fan, Zhang and Yu (2012) consider the following portfolio optimization under gross exposure constraint:

$$\hat{\boldsymbol{w}}_{\text{opt},c} = \text{argmin}_{\boldsymbol{w}^\top \mathbf{1}=1, \|\boldsymbol{w}\|_1 \leqslant c} \hat{R}(\boldsymbol{w}), \tag{7.5}$$

a specific example of the portfolio choice. For real data, the actual risks are unknown. We use the following measurement as a proxy. At the begining of each month, we optimize the portfolio according to (7.5) for each given c and we then hold the portfolio for a month. The daily actual returns of such a portfolio are recorded. This is done for the three-year testing period from 2005 to 2007 or the actual returns of approximately 756 trading days. The standard deviations of these returns are regarded as the actual risk with gross-exposure constraint c. Figure 7.2 depicts the actual risks for various choices of c. The volatility matrix can also be estimated by RiskMetrics or the Fama–French three-factor model. We will furnish details of these methods in Section 7.2.

First of all, the optimal no-short sale portfolio has a risk of approximately 9.3%, whereas the perceived optimal portfolio (7.4) has an average risk 23% (corresponding to the picture with $c = \infty$). This demonstrates the phenomenon observed by Jagannathan and Ma (2003) that the optimal no-short sale portfolio outperforms the perceived optimal portfolio. Clearly, the risk decreases first and then increases steadily with gross exposure c after $c \geqslant 2$. This is consistent with Theorem 7.1. When c is small, the risk approximation error is small. On the other hand, when c increases, the space of allowable portfolios increases and hence the risks of the theoretical optimal portfolios decrease. For a small c, the benefit of increasing c outpaces the noise accumulation. When c is sufficiently large, the benefit obtained from increasing the space of portfolios gradually vanishes (recalling the law of diminishing returns), yet the noise accumulation grows steadily. That explains the shape of the risks of the perceived optimal portfolios. It also answers the

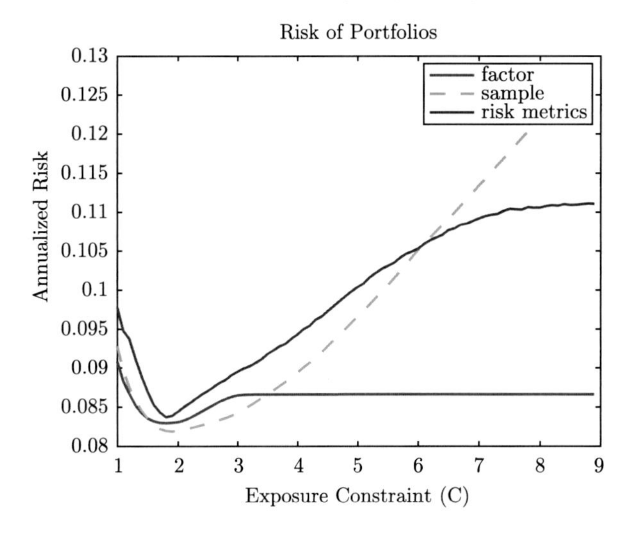

Figure 7.2 The actual risks of the minimum variance portfolios with gross-exposure constraint c, using the 600 randomly selected assets from the 1000 stocks with least missing data in Russel 3000 in the period 2003–2007. The covariance matrices are estimated by the sample covariance, RiskMetrics, and the Fama–French three-factor model. Adapted from Fan, Zhang and Yu (2012).

question posed by Jagannathan and Ma why imposing a wrong constraint helps.

Figure 7.2 also reveals that the estimators of large covariance matrices play a role in the risk profile. In particular, the covariance matrix estimated by using the factor model appears very stable and the risk profile does not increase with c at $c \geqslant 3$

7.1.3 Errors in risk assessments

For a given portfolio $\boldsymbol{w}$, its risk is estimated by $\hat{R}(\boldsymbol{w})^{1/2}$. Is estimation error negligible? How to quantify the estimation error in the risk assessment? Of course, (7.3) provides an upper bound. However, this upper bound is too crude to be useful in many applications. For example, for portfolio $\boldsymbol{w}_3$, the upper bound for errors in estimating daily portfolio variance is $e_{\max}c^2 = 4e_{max}$, which is around $4 * 10^{-4}$ (see Figure 7.1). The typical estimate of portfolio variance before annualization (see Table 7.1) is $(0.2995/\sqrt{252})^2 = 0.00036$. Using the upper bound, the true daily portfolio variance is in the interval $[0.0000, 0.00076]$ or the annualized portfolio standard deviation in the interval $[0, 0.4376]$. Such a wide interval, while covering the true risk of 0.3000, is useless. Using the same calculation, for $\boldsymbol{w}_1$, the

Table 7.2 *Averages and standard deviations of RE over 10,000 portfolios*

$c = 1$	$c = 1.2$	$c = 1.4$	$c = 1.6$	$c = 1.8$	$c = 2$
		$T = 200$			
4.84%	4.81%	4.66%	4.75%	4.73%	4.77%
(1.00%)	(1.01%)	(0.94%)	(1.00%)	(0.99%)	(0.96%)
		$T = 400$			
3.48%	3.48%	3.48%	3.49%	3.52%	3.48%
(0.51%)	(0.49%)	(0.48%)	(0.53%)	(0.51%)	(0.56%)

typical estimated risk is 23.73% per annum. However, the error bound in (7.3) says that the true annualized portfolio risk for the portfolio $\boldsymbol{w}_1$ is in the interval $[0.176, 0.286]$, still too wide for many applications.

The above observations lead to Fan, Liao and Shi (2015) to introduce the concept of a *high-confidence level upper bound* (H-CLUB). For each proposed portfolio variance estimator $\hat{R}(\boldsymbol{w}) = \boldsymbol{w}^T \hat{\boldsymbol{\Sigma}} \boldsymbol{w}$ of the true variance $R(\boldsymbol{w}) = \boldsymbol{w}^T \boldsymbol{\Sigma} \boldsymbol{w}$ and a given $\tau \in (0, 1)$, an *H-CLUB* $\hat{U}(\tau)$ is the one such that

$$P(|\hat{R}(\boldsymbol{w}) - R(\boldsymbol{w})| \leqslant \hat{U}(\tau)) \to 1 - \tau, \tag{7.6}$$

as the sample size $T \to \infty$. Typically, we take $\tau = 0.05$ so that $\hat{U}(\tau)$ is similar to twice the standard error of the estimator. The upper bound $e_{\max}\|\boldsymbol{w}\|_1^2$ in (7.3) corresponds to H-CLUB with $\tau = 0$. This is typically 5 to 20 times as large as $\hat{U}(0.05)$ for gross exposure constraint $\|\boldsymbol{w}\|_1$ ranges from 1 to 2, according to Fan, Liao and Shi (2015) (see Table 3 there).

Fan, Liao and Shi (2015) construct H-CLUB for several commonly used risk estimators in the next section. For simplicity, we introduce here only the method based on the sample covariance matrix. For a given portfolio $\boldsymbol{w}$, define the autoregressive function of its squared returns

$$\gamma_T(h) = \operatorname{cov}((\boldsymbol{w}'\boldsymbol{R}_t)^2, (\boldsymbol{w}'\boldsymbol{R}_{t+h})^2).$$

This can be estimated by its sample version:

$$\hat{\gamma}(h) = T^{-1} \sum_{t=1}^{T-h} ((\boldsymbol{w}'\boldsymbol{R}_t)^2 - \boldsymbol{w}'\boldsymbol{S}\boldsymbol{w})((\boldsymbol{w}'\boldsymbol{R}_{t+h})^2 - \boldsymbol{w}'\boldsymbol{S}\boldsymbol{w}),$$

where $\boldsymbol{S}$ is the sample covariance matrix and $\boldsymbol{w}'\boldsymbol{S}\boldsymbol{w}$ is the mean of the squared returns $\{(\boldsymbol{w}'\boldsymbol{R}_t)^2\}_{t=1}^T$. Let $z_{\tau/2}$ denote the upper $\tau/2$ quantile of the standard normal distribution. For some large L, which diverges as T

Table 7.3 *Summary of parameters used for the simulation of the Fama–French three-factor model*

$\boldsymbol{\mu}_B$	$\boldsymbol{\Sigma}_B$			$\boldsymbol{\mu}_f$	$\boldsymbol{\Sigma}_f$		
0.783	0.0291	0.0239	0.0102	0.0236	1.251	−0.035	−0.204
0.518	0.0239	0.0540	−0.0070	0.0140	−0.035	0.316	−0.002
0.410	0.0102	−0.0070	0.0869	0.0207	−0.204	−0.002	0.193

does but at slower rate than T, let

$$\hat{\sigma}^2 = \hat{\gamma}(0) + 2\sum_{h=1}^{L} \hat{\gamma}(h), \quad \hat{U}(\tau) = \frac{z_{\tau/2}\hat{\sigma}}{\sqrt{T}}. \tag{7.7}$$

See Exercise 7.2 for an explanation. Fan, Liao and Shi (2015) show that $\hat{U}(\tau)$ is the H-CLUB for the sample covariance matrix, namely, it satisfies (7.6). Translating this into portfolio risk (instead of portfolio variance), by using the Delta method (2.51), we have

$$P\left(|\hat{R}(\boldsymbol{w})^{1/2} - R(\boldsymbol{w})^{1/2}| \leqslant \hat{U}(\tau)/\sqrt{4\boldsymbol{w}^{\mathrm{T}}\boldsymbol{S}\boldsymbol{w}}\right) \to 1 - \tau. \tag{7.8}$$

In other words, $\hat{U}(\tau)/\sqrt{4\boldsymbol{w}^{\mathrm{T}}\boldsymbol{S}\boldsymbol{w}}$ is the H-CLUB for the portfolio risk.

Are estimation errors in risk assessments negligible? Let us compare the relative error

$$\mathrm{RE} = 0.5(\hat{U}(0.05)/\sqrt{4\boldsymbol{w}^{\mathrm{T}}\boldsymbol{S}\boldsymbol{w}})/R(\boldsymbol{w})^{1/2} \approx \frac{\hat{\sigma}}{2\sqrt{T}R(\boldsymbol{w})},$$

which is half H-CLUB divided by the portfolio risk, where $\hat{\sigma}$ is given by (7.7). This RE represents the coefficient of variation for estimating portfolio risk. Through extensive simulated data based on the Fama–French three-factor model with parameters calibrated to the daily returns of S&P 500's top 100 constituents, Fan, Liao and Shi (2015) obtained the results, depicted in Table 7.3, based on 10,000 testing portfolios with different portfolio sizes. The good piece of news is that the relative error is around 3—5%, which is negligible for many applications.

7.1.4 Representative portfolios with a given exposure

In order to produce Table 7.2, Fan, Liao and Shi (2015) have to select representative portfolios with a given exposure c. In other words, they have to sample uniformly the portfolio weight $\boldsymbol{w}$ from the set

$$\Big\{\boldsymbol{w} \in R^p : \sum_{i=1}^{p} w_i = 1 \text{ and } \sum_{i=1}^{p} |w_i| = c\Big\}. \tag{7.9}$$

This is not an easy task and they provide some simple heuristics. Such a simulation can have applications to portfolio optimization and other risk assessment problems.

First of all, given the gross exposure c, the total long and short positions are respectively $w^+ = (c+1)/2$ and $w^- = (c-1)/2$. For $c = 1$, there are no-short positions, whereas for $c > 1$, there are both long and short positions. The identities (or indices) of long and short positions are hard to identify, but the following sampling scheme is a reasonable approximation: The positive positions are determined by a Bernoulli trial (p times) with probability of success $w^+/(w^+ + w^-) = (c+1)/(2c)$. Once the identities are determined, we can normalize them and the problem reduces to the case with $c = 1$. For the case with $c = 1$, the uniform distribution on the set $\{\boldsymbol{w} \in R^p : \sum_{i=1}^p w_i = 1, w_i \geqslant 0\}$ can be generated from a normalized exponential distribution:

$$w_i = \zeta_i \Big/ \sum_{i=1}^{p} \zeta_i, \qquad \zeta_i \sim_{i.i.d.} \text{standard exponential.} \tag{7.10}$$

It can be shown that (7.10) produces a uniform distribution on the $\{w_i : \sum_{i=1}^p w_i = 1, w_i \geqslant 0\}$.

The above discussion leads to the following simple method.

1. Generate a positive integer k, the number of stocks with positive weights in $\boldsymbol{w}$, from a binomial distribution $\text{Bin}\left(p, \dfrac{c+1}{2c}\right)$.
2. Generate independently $\{\zeta_i\}_{i=1}^k$ from the standard exponential distribution and set $w_i^+ = (c+1)\zeta_i/(2\sum_{j=1}^k \zeta_j)$. This yields $\boldsymbol{w}_+ = (w_1^+, \ldots, w_k^+)$, a temporary vector of the positive weights in $\boldsymbol{w}$.
3. The temporary negative weights in $\boldsymbol{w}_- = (w_1^-, \ldots, w_{N-k}^-)$ are generated analogously with $w_i^- = (1-c)\zeta_i^*/2\sum_{j=1}^{N-k} \zeta_j^*$, where $\{\zeta_j^*\}_{j=1}^{N-k}$ are obtained independently from the standard exponential distribution.
4. Take the portfolio weights $\boldsymbol{w}$ as a random permutation of the vector $(\boldsymbol{w}_+, \boldsymbol{w}_-)$.

In the algorithm above, only step 1 is heuristic; the rest are based on sound mathematical principles.

7.2 Estimation of a large volatility matrix

How do we estimate a time-varying volatility matrix from financial returns? How can we obtain a nondegenerate covariance matrix even when $p \gg T$? How well does it perform? In this section, we will introduce progressively more sophisticated methods to answer the above questions.

We always assume that we have return vectors $\{\boldsymbol{R}_t\}_{t=1}^{T}$ for p assets over T different time periods. Our approaches focus on low-frequency such as daily financial returns. For high-frequency financial data analysis for large portfolios, see, for example, Barndorff-Nielsen, Hansen, Lunde and Shephard (2011), and Fan, Li and Yu (2012). For a full and comprehensive account of high-frequency financial econometrics, see Ait-Sahalia and Jacod (2014).

7.2.1 Exponential smoothing

The *exponential smoothing* method is a time varying estimate of conditional variance matrix $\boldsymbol{\Sigma}_t = \mathrm{var}_t(\boldsymbol{R}_{t+1})$. The essential idea to obtain an estimate $\boldsymbol{\Sigma}_t$ is to localize the data in time by considering a period of data right before the time t, namely using only the data $\{X_{t-j}\}_{j=0}^{h}$ for a given window size h. An example of this is the local sample covariance matrix using only the data $\{X_{t-j}\}_{j=0}^{h}$. This is a moving window with width h, as t varies. As t varies, we obtain a time-varying estimate of the covariance matrix. This method was used in the computation of $\hat{\boldsymbol{\Sigma}}_t$ in Figure 7.2 with $h = 501$. The number of assets of p can even be larger than the window size, as in Example 7.3. Therefore, this kind of localization method does not have any intention to solve the aforementioned ill-conditioned problems: The resulting covariance matrix can still be rank-deficient.

The local sample covariance method assigns a radical weighting scheme: whether a data point is used or not depends on whether it is within h periods from time t. An improvement of the idea is to use a smooth weighting scheme such as the *exponential smoothing*: For a given smoothing parameter $\lambda < 1$, it estimates $\hat{\boldsymbol{\Sigma}}_t$ by

$$\hat{\boldsymbol{\Sigma}}_t = (1-\lambda)(\boldsymbol{R}_t\boldsymbol{R}_t^{\mathrm{T}} + \lambda \boldsymbol{R}_{t-1}\boldsymbol{R}_{t-1}^{\mathrm{T}} + \lambda^2 \boldsymbol{R}_{t-2}\boldsymbol{R}_{t-2}^{\mathrm{T}} + \cdots). \tag{7.11}$$

Note that the total weight

$$(1-\lambda)\sum_{j=0}^{\infty} \lambda^j = 1.$$

The volatility of each component is the exponential smoothing of its squared returns, as in Section 3.4:

$$\hat{\sigma}_{i,t}^2 = (1-\lambda)(r_{i,t}^2 + \lambda r_{i,t-1}^2 + \lambda^2 r_{i,t-2}^2 + \cdots).$$

The remote observations were exponentially down weighted. The effective data points are quite recent. For example, if $\lambda = 0.94$, $\lambda^{40} = 0.084$, which is much smaller than one. It is a time-domain smoothing: the larger λ, the wider the effective data windows and the smoother the estimated volatility matrix (See Figure 3.16). The estimator can be computed recursively by

$$\hat{\boldsymbol{\Sigma}}_t = \lambda \hat{\boldsymbol{\Sigma}}_{t-1} + (1-\lambda)\boldsymbol{R}_t \boldsymbol{R}_t^{\mathrm{T}}. \tag{7.12}$$

In practical implementation, one can take the sample covariance matrix of the data as the initial value $\hat{\boldsymbol{\Sigma}}_0$ and discard some initial estimates such as $\hat{\boldsymbol{\Sigma}}_1, \ldots, \hat{\boldsymbol{\Sigma}}_{10}$.

The smoothness of the resulting estimate $\hat{\boldsymbol{\Sigma}}_t$ as a function of t is governed by the *smoothing parameter* λ. The larger the value λ, the smoother the resulting estimate is. A larger value of λ increases the number of effective local data points and hence reduces the variance in the estimate, but it also increases the approximation error or biases in the estimate. Therefore, an appropriate choice of λ is important for estimating a volatility matrix. RiskMetrics recommends $\lambda = 0.94$ for computing a daily volatility matrix and $\lambda = 0.97$ for computing a monthly volatility. Note that the resulting monthly volatility matrix is $5\hat{\boldsymbol{\Sigma}}_{t,0.97}$ when a month is defined as 25 trading days, where $\hat{\boldsymbol{\Sigma}}_{t,0.97}$ is computed based on (7.12) with $\lambda = 0.97$ and daily returns. The RiskMetrics estimator in Figure 7.1 was constructed by using such a method.

Example 7.4 We now consider the volatility matrix of the returns of the S&P 500 index and the changes of the VIX on the 10-year window from January 29, 2001 to February 28, 2011 presented in Figure 6.1. We use the exponential smoothing (7.12) with $\lambda = 0.94$ and initial value $\hat{\boldsymbol{\Sigma}}_0 = 0$. This gives an estimate of the daily volatility matrix. The result is presented in Figure 7.3. The estimates at the initial 3 months (63 days) are influenced by the initial value since $\lambda^{62} = 0.022$. The estimates afterwards are not influenced much by the initial choices. If the sample covariance is used as the initial value, the initial effect is smaller. See Exercise 7.4. The correlation here is the measure of the leverage effect, which hovers around -0.8. The volatility of the S&P 500 index increases dramatically during the financial crisis, so does the change of the volatility in the same period.

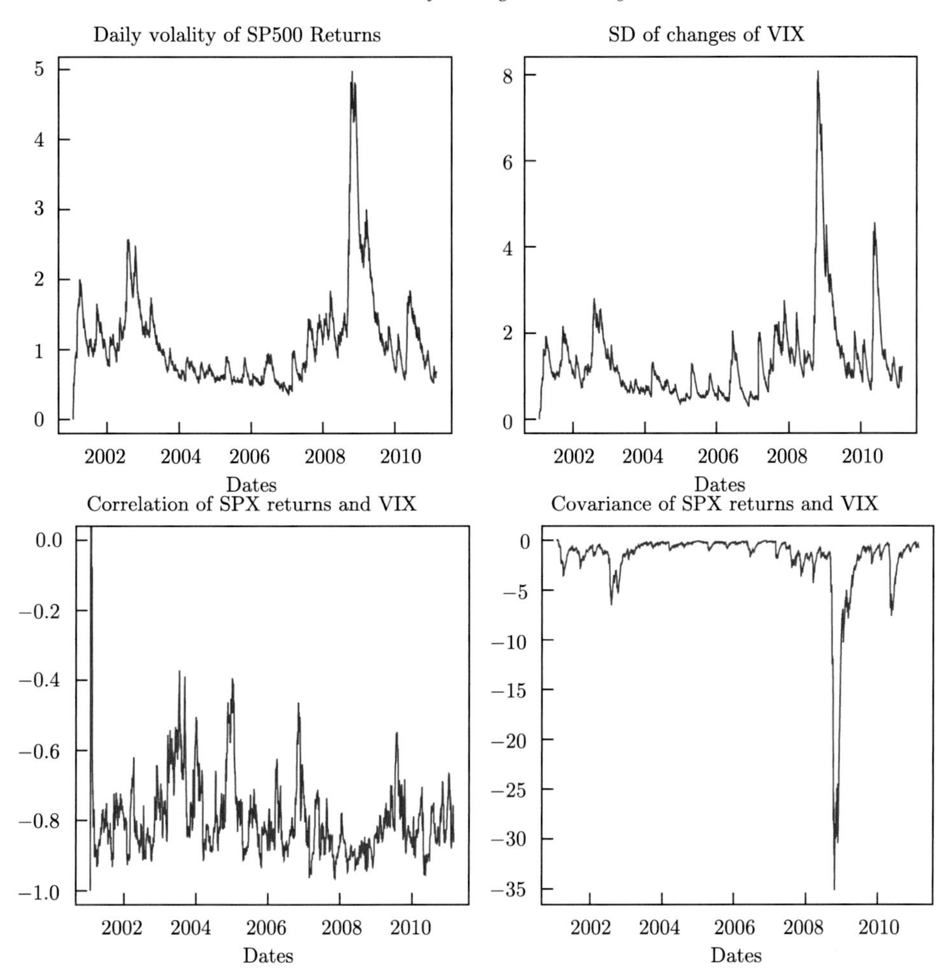

Figure 7.3 The volatility matrix between the returns of S&P 500 and the changes of VIX are estimated by the exponential smoothing with $\lambda = 0.94$.

7.2.2 Regularization by thresholding

Exponential smoothing does not solve the ill-conditioned problem. It merely aggregates local volatility into a time-varying estimate. The simplest method to solve the ill-conditioned problem is *thresholding*. The essential assumption is that $\boldsymbol{\Sigma}$ is sparse; a majority of its off-diagonal elements are nearly zero. While this assumption might not be valid for many financial applications, it can be combined with the multifactor model to yield a powerful method.

The sparsity assumption is most conveniently explored by thresholding, which sets small estimated elements to zero (Bickel and Levina, 2008). For a given thresholding parameter λ, it is defined as

$$\hat{\boldsymbol{\Sigma}}_\lambda = \Big(\hat{\sigma}_{i,j} I(|\hat{\sigma}_{i,j}| \geqslant \lambda)\Big). \tag{7.13}$$

This method is simple but powerful. By setting small estimated values to zero, they are not estimated and do not contribute to the variance of the estimator. The sparsity ensures that the biases due to thresholding are small too.

There are a number of useful variations of the simple thresholding estimator (7.13). *Generalized thresholding* rules of Antoniadis and Fan (2001) can also be employed (Rothman, Levina and Zhu, 2009). The simple thresholding (7.13) or its generalization does not even take the varying scales of the covariance into account. One way to account this is to threshold on t-type statistics. For example, using the simple thresholding, we can define the *adaptive thresholding* estimator (Cai and Liu, 2011) as

$$\hat{\boldsymbol{\Sigma}}_\lambda = \Big(\hat{\sigma}_{i,j} I(|\hat{\sigma}_{i,j}/\mathrm{SE}(\hat{\sigma}_{i,j})| \geqslant \lambda)\Big), \tag{7.14}$$

where $\mathrm{SE}(\hat{\sigma}_{i,j})$ is the estimated standard error of $\hat{\sigma}_{i,j}$. For example, if $\hat{\boldsymbol{\Sigma}}$ is the sample covariance matrix, then $\hat{\sigma}_{ij} = T^{-1}\sum_{t=1}^T r_{it}r_{jt}$ and

$$\mathrm{SE}(\hat{\sigma}_{i,j}) = T^{-1/2}\mathrm{SD}(\{r_{it}r_{jt}\}_{t=1}^T),$$

when returns are martingale differences.

A simpler method to take the scale into account is to apply thresholding on the correlation matrix. Let $\hat{\boldsymbol{\Psi}}_\lambda$ be the thresholded correlation matrix with thresholding parameter λ. Then,

$$\hat{\boldsymbol{\Sigma}}_\lambda^* = \mathrm{diag}(\hat{\boldsymbol{\Sigma}})^{1/2}\hat{\boldsymbol{\Psi}}_\lambda \mathrm{diag}(\hat{\boldsymbol{\Sigma}})^{1/2} \tag{7.15}$$

is an estimate of the volatility matrix. In particular, when $\lambda = 0$, it is the sample covariance matrix, whereas when $\lambda = 1$, it is the diagonal matrix, consisting of the sample variances. This form is more appropriate since it is thresholded on the standardized scale. The estimator (7.15) is equivalent to applying the *entry dependent thresholding*

$$\lambda_{ij} = \sqrt{\hat{\sigma}_{i,i}\hat{\sigma}_{j,j}}\;\lambda$$

to the original covariance matrix $\hat{\boldsymbol{\Sigma}}$.

We now illustrate the thresholding methods by using the following example.

Example 7.5 Consider the eight risk factors presented in Example 6.1.

The correlation matrix of these factors over the period from Feb. 23, 2004 to Feb. 28, 2011 are summarized as follows:

	SPX	Vol	ExR	Mat	Def	Liq	Size	Tbill
SPX	1.00	−0.127	0.001	0.004	0.007	−0.053	0.342	0.003
Vol	−0.127	1.00	−0.184	0.456	0.805	0.324	−0.057	−0.555
ExR	0.001	−0.184	1.00	−0.258	−0.104	−0.342	−0.011	0.312
Mat	0.004	0.456	−0.258	1.00	0.409	−0.036	0.013	−0.960
Def	0.007	0.805	−0.104	0.409	1.00	0.059	0.010	−0.545
Liq	−0.053	0.324	−0.342	−0.036	0.059	1.00	−0.018	0.039
Size	0.342	−0.057	−0.011	0.013	0.010	−0.018	1.00	−0.017
Tbill	0.003	−0.555	0.312	−0.960	−0.545	0.039	−0.017	1.00

The minimum and maximum eigenvalues are

$$\lambda_{\min}(\hat{\boldsymbol{\Sigma}}) = 0.024 \quad \text{and} \quad \lambda_{\max}(\hat{\boldsymbol{\Sigma}}) = 2.994$$

so that the *conditioning number* is $2.994/0.024 = 124.75$. The problem is well posed since $T \gg p$. Applying the thresholding with $\lambda = 0.25$, we obtain

	SPX	Vol	ExR	Mat	Def	Liq	Size	Tbill
SPX	1.00	0.00	0.00	0.00	0.00	0.00	0.34	0.00
Vol	0.00	1.00	−0.18	0.46	0.81	0.32	0.00	−0.56
ExR	0.00	−0.18	1.00	−0.26	0.00	−0.34	0.00	0.31
Mat	0.00	0.46	−0.26	1.00	0.41	0.00	0.00	−0.96
Def	0.00	0.81	0.00	0.41	1.00	0.00	0.00	−0.54
Liq	0.00	0.32	−0.34	0.00	0.00	1.00	0.00	0.00
Size	0.34	0.00	0.00	0.00	0.00	0.00	1.00	0.00
Tbill	0.00	−0.56	0.31	−0.96	−0.54	0.00	0.00	1.00

Many small elements of the sample correlation matrix are now replaced by zero. Remark that $\lambda_{\min}(\hat{\boldsymbol{\Sigma}}^*_\lambda) = 0.023$ and $\lambda_{\max}(\hat{\boldsymbol{\Sigma}}^*_\lambda) = 2.974$. The thresholding operator does not alter, very much, the spectral property in this case. This is mainly due to the fact that the problem is well posed: $T \gg p$.

The advantage of thresholding is to avoid estimating small elements so that noise does not accumulate. The decision of whether an element should be estimated is much easier than the attempt to estimate it accurately. Indeed, under some regularity conditions, Bickel and Levina (2008) show that

$$\|\hat{\boldsymbol{\Sigma}}_\lambda - \boldsymbol{\Sigma}\| = O_p\left(\sqrt{\frac{\log(p)s_1}{T}}\right), \qquad \|\hat{\boldsymbol{\Sigma}}_\lambda^{-1} - \boldsymbol{\Sigma}^{-1}\| = O_p\left(\sqrt{\frac{\log(p)s_1}{T}}\right)$$

and

$$\|\hat{\boldsymbol{\Sigma}}_\lambda - \boldsymbol{\Sigma}\|_F = O_p\left(\sqrt{\log(p)s_2/T}\right),$$

where s_1 the maximum number of nonzero elements in each row, and s_2 is the total number of non-zero elements. Here, the norms refer to the *operator norm* defined by

$$\|\boldsymbol{A}\| = \lambda_{\max}(\boldsymbol{A}^{\mathrm{T}}\boldsymbol{A})^{1/2},$$

the square root of the largest eigenvalue of the symmetric matrices $\boldsymbol{A}^{\mathrm{T}}\boldsymbol{A}$, and the *Frobenius norm* defined by

$$\|\boldsymbol{A}\|_F = \mathrm{tr}(\boldsymbol{A}^{\mathrm{T}}\boldsymbol{A})^{1/2},$$

which is the square-root of the sum of all squared entries of the matrix $\boldsymbol{A}$. It is well known that

$$p^{-1/2}\|\boldsymbol{A}\|_F \leqslant \|\boldsymbol{A}\| \leqslant \max_i \sum_j |a_{ij}| \leqslant \|A\|_F.$$

When the matrix $\boldsymbol{\Sigma}$ is sparse, s_1 and s_2 are small. The above theoretical result reveals that the impact of dimensionality p is limited, entering into the equation only at the logarithmic order, and that the volatility matrix can be estimated accurately when s_1 and s_2 are small. Since each element in the volatility matrix can be estimated with an error of order $O_p(T^{-1/2})$, the total sum of squared errors for non-zero elements is of $O_p(s_2/T)$, if their positions are known. Thus, it only costs us a $\log(p)$ order to learn the unknown locations of the non-zero elements.

7.2.3 Projections onto semi-positive and positive definite matrix spaces

Despite nice theoretical properties, the thresholding estimator $\hat{\boldsymbol{\Sigma}}_\lambda$ is not necessarily positive definite. Here, we first offer two simple methods for the projection of a symmetric matrix into the space of semi-definite matrices. The first method is to perform the eigenvalue decomposition

$$\hat{\boldsymbol{\Sigma}}_\lambda = \boldsymbol{\Gamma}^{\mathrm{T}}\mathrm{diag}(\lambda_1, \ldots, \lambda_p)\boldsymbol{\Gamma} \tag{7.16}$$

and compute

$$\hat{\boldsymbol{\Sigma}}_\lambda^+ = \boldsymbol{\Gamma}^{\mathrm{T}}\mathrm{diag}(\lambda_1^+, \ldots, \lambda_p^+)\boldsymbol{\Gamma}, \tag{7.17}$$

where $\lambda_j^+ = \max(\lambda_j, 0)$ is the positive part of λ_j. This L_2-projection is the solution to the least-squares problems: Minimize, with respect $\boldsymbol{S} \geqslant 0$, to the least-squares problem:

$$\|\hat{\boldsymbol{\Sigma}}_\lambda - \boldsymbol{S}\|_F^2, \quad \text{such that} \quad \boldsymbol{S} \geqslant 0. \tag{7.18}$$

The second method is to compute

$$\hat{\boldsymbol{\Sigma}}_\lambda^+ = (\hat{\boldsymbol{\Sigma}}_\lambda + \lambda_{\min}^- I_p)/(1 + \lambda_{\min}^-).$$

where $\lambda_{\min}^-$ is the negative part of the minimum eigenvalue of $\hat{\boldsymbol{\Sigma}}_\lambda$, which equals $-\lambda_{\min}$ when $\lambda_{\min} < 0$ and 0 otherwise. Both projections do not alter eigenvectors, using the same matrix $\boldsymbol{\Gamma}$ as the original decomposition (7.16). When applied to the correlation matrix, the second method still gives a correlation matrix.

The above projection gives a simple and analytic solution. However, it is not necessarily optimal. For example, Qi and Sun (2006) introduced an algorithm for computing the *nearest correlation matrix*: For a given symmetric matrix $\boldsymbol{A}$, find a correlation matrix $\boldsymbol{R}$ to minimize

$$\|\boldsymbol{A} - \boldsymbol{R}\|_F^2, \quad \text{s.t.} \quad \boldsymbol{R} \geqslant 0, \text{diag}(\boldsymbol{R}) = \boldsymbol{I}_p. \tag{7.19}$$

To apply the algorithm (7.19) to our threhsolding estimator $\hat{\boldsymbol{\Sigma}}_\lambda$, we compute the standardized input

$$\boldsymbol{A} = \text{diag}(\hat{\boldsymbol{\Sigma}}_\lambda)^{-1/2} \hat{\boldsymbol{\Sigma}}_\lambda \text{diag}(\hat{\boldsymbol{\Sigma}}_\lambda)^{-1/2}$$

first and then apply (7.19) to obtain $\hat{\boldsymbol{R}}_\lambda$ and transform back to the covariance matrix as

$$\boldsymbol{\Sigma}_\lambda^* = \text{diag}(\hat{\boldsymbol{\Sigma}}_\lambda)^{1/2} \hat{\boldsymbol{R}}_\lambda \text{diag}(\hat{\boldsymbol{\Sigma}}_\lambda)^{1/2}. \tag{7.20}$$

Of course, if the thresholding is applied to the correlation matrix, as in Example 7.5, we can then take the input $\boldsymbol{A} = \hat{\boldsymbol{\Psi}}_\lambda$. After the nearest correlation matrix projection, we can apply (7.15) to transform the problem back.

As a generalization of (7.19), one can also solve the following optimization problem

$$\|\boldsymbol{A} - \boldsymbol{R}\|_F^2, \quad \text{such that} \quad \lambda_{\min}(\boldsymbol{R}) \geqslant \delta, \quad \text{diag}(\boldsymbol{R}) = \boldsymbol{I}_p \tag{7.21}$$

for a given $\delta \geqslant 0$. Namely, we wish to find covariance matrix $\boldsymbol{S}$ with minimum eigenvalue δ such that it is closest to $\boldsymbol{A}$. When $\delta = 0$, it reduces to problem (7.19).

There is a R-package called *nearPD* that computes (7.19). The function is also called `nearPD`.

*7.2.4 Regularization by penalized likelihood**

Penalized likelihood, introduced by Fan and Li (2001), is a powerful method for exploring sparsity. Let $\ell_T(\boldsymbol{\theta})$ be the log-likelihood function based on a sample size T, with p parameters. If $\boldsymbol{\theta}$ is sparse, it can be explored by the penalized likelihood

$$\operatorname{argmin}_{\boldsymbol{\theta}} \left\{ -\frac{2}{T}\ell_T(\boldsymbol{\theta}) + \sum_{j=1}^{p} p_\lambda(|\theta_j|) \right\}, \tag{7.22}$$

where p_λ is a penalty function, depending on the regularization parameter λ. An ideal choice is L_0*-penalty*: $p_\lambda(|\theta|) = \lambda I(\theta \neq 0)$. This results in the *best subset* selection: among the subsets of variables of size m whose penalty part in (7.22) being λm, the minimization of (7.22) is obtained by the subset of m variables with the maximum likelihood value among all $\binom{p}{m}$ subsets. This needs to be computed for all integer values m. The computation of this is impossible even for moderate dimensionality p. A convex relaxation of the problem is to use the L_1-penalty $p_\lambda(\theta) = \lambda|\theta|$, called *LASSO* by Tibshirani (1996). However, it introduces biases due to the shrinkage of LASSO. For this reason, Antoniadis and Fan (2001) and Fan and Li (2001) introduced a family of *folded-concave penalty functions* such as the smoothly clipped absolute deviation (*SCAD*). The SCAD is a quadratic spline function (see Figure 7.4), whose derivative function is given by

$$p'_\lambda(t) = \lambda \left\{ I\left(t \leqslant \lambda\right) + \frac{(a\lambda - t)_+}{(a-1)\,\lambda} I\left(t > \lambda\right) \right\} \tag{7.23}$$

where the constant $a = 3.7$ was suggested by Fan and Li (2001) from a Bayesian argument. The essential idea is that when an estimate exceeds a value a, it should not penalized any further.

The penalized likelihood (7.22) can be computed by an iterated reweighted LASSO by a local linear approximation. Given the estimate $\hat{\boldsymbol{\theta}}^{(k)}$ at the kth iteration, approximate

$$p_\lambda(|\theta_j|) \approx p_\lambda(|\hat{\theta}_j^{(k)}|) + p'_\lambda(|\hat{\theta}_j^{(k)}|)(|\theta_j| - |\hat{\theta}_j^{(k)}|).$$

Figure 7.4 shows such an approximation to the SCAD function at a given point. The symmetric local linear approximation (dashed line) is the convex majorant of the folded convex function. Other approximations are possible such as the local quadratic approximation in the picture, but it is not as accurate. The local quadratic approximation was used in Fan and Li (2001) and the local linear approximation was introduced by Zou and Li (2008). With the local linear approximation, the target becomes

$$Q_{k+1}(\theta) = -\frac{2}{T}\ell_T(\boldsymbol{\theta}) + \sum_{j=1}^{p} w_{k,j}|\theta_j| + c, \tag{7.24}$$

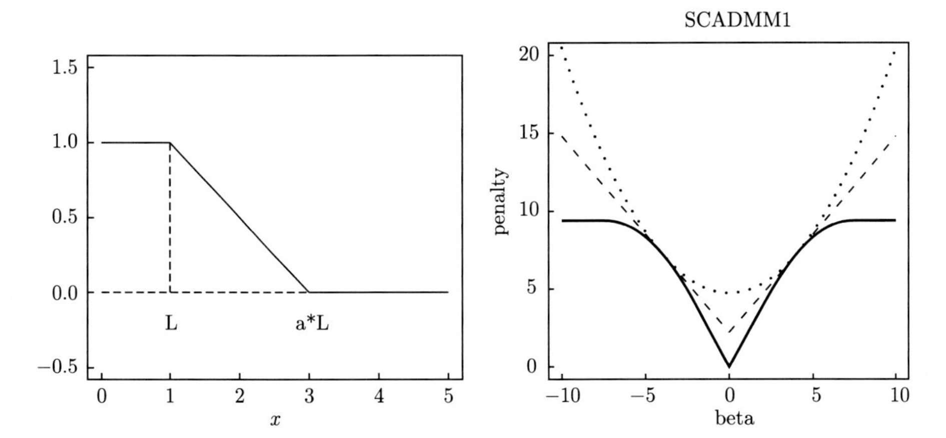

Figure 7.4 The derivative function of SCAD (left panel) and SCAD function (right panel) and its symmetric local linear and local quadratic approximation at a given point.

where $w_{k,j} = p'_\lambda(|\hat{\theta}_j^{(k)}|)$ and

$$c = \sum_{j=1}^{p} \left[p_\lambda(|\hat{\theta}_j^{(k)}|) - p'_\lambda(|\hat{\theta}_j^{(k)}|)|\hat{\theta}_j^{(k)}| \right]$$

is a constant, independent of $\boldsymbol{\theta}$. (7.24) is the weighted penalized L_1-likelihood.

Starting with $\hat{\boldsymbol{\theta}}^{(0)} = 0$, $Q_1(\boldsymbol{\theta})$ is the same target as LASSO by noticing $p_\lambda(0) = \lambda$ and $\hat{\boldsymbol{\theta}}^{(1)}$ is the LASSO estimator. The next iteration reduces biases with a larger component receiving smaller weights or even zero weights. See the left panel of Figure 7.4. The algorithm is a majorization-minimization algorithm and each iteration reduces the target value. To see this, let

$$Q(\boldsymbol{\theta}) = -\frac{2}{T}\ell_T(\boldsymbol{\theta}) + \sum_{j=1}^{p} p_\lambda(|\theta_j|)$$

be the target function. Then, since the penalty function is approximated from above, we have

$$Q(\hat{\boldsymbol{\theta}}^{(k+1)}) \leqslant Q_{k+1}(\hat{\boldsymbol{\theta}}^{(k+1)}) \leqslant Q_{k+1}(\hat{\boldsymbol{\theta}}^{(k)}) = Q(\hat{\boldsymbol{\theta}}^{(k)}).$$

The second inequality follows since $\hat{\boldsymbol{\theta}}^{(k+1)}$ minimizes the function $Q_{k+1}(\cdot)$, and the last equality holds by the definition of Q_{k+1}. In other words, the target value $\{Q(\hat{\boldsymbol{\theta}}^{(k)})\}$ is a decreasing sequence and converges.

As a specific example, we now apply the penalized likelihood method to the covariance regularization in the following example.

Example 7.6 Let $\boldsymbol{R}_t \sim N(\boldsymbol{\mu}, \boldsymbol{\Sigma})$. Then, its density is

$$(2\pi)^{-p/2}|\boldsymbol{\Sigma}|^{-1/2}\exp(-(\boldsymbol{R}_t - \boldsymbol{\mu})^{\mathrm{T}}\boldsymbol{\Sigma}^{-1}(\boldsymbol{R}_t - \boldsymbol{\mu})/2).$$

Regarding $\boldsymbol{\mu}$ and $\boldsymbol{\Sigma}$ as $\boldsymbol{\theta}$, the log-likelihood function

$$\ell_T(\boldsymbol{\mu}, \boldsymbol{\Sigma}) = -\frac{1}{2}\Big\{Tp\log(2\pi) + T\log|\boldsymbol{\Sigma}| + \sum_{t=1}^{\mathrm{T}}(\boldsymbol{R}_t - \boldsymbol{\mu})^{\mathrm{T}}\boldsymbol{\Sigma}^{-1}(\boldsymbol{R}_t - \boldsymbol{\mu})\Big\}.$$

See (5.26) for a similar expression (our case corresponds to the special case there with $\boldsymbol{\beta} = 0$ and $N = p$, and $\boldsymbol{\alpha} = \boldsymbol{\mu}$). Note that

$$\sum_{t=1}^{\mathrm{T}}(\boldsymbol{R}_t - \boldsymbol{\mu})^{\mathrm{T}}\boldsymbol{\Sigma}^{-1}(\boldsymbol{R}_t - \boldsymbol{\mu}) = \mathrm{tr}\Big(\boldsymbol{\Sigma}^{-1}\sum_{t=1}^{\mathrm{T}}(\boldsymbol{R}_t - \boldsymbol{\mu})(\boldsymbol{R}_t - \boldsymbol{\mu})^{\mathrm{T}}\Big)$$

where $tr(\boldsymbol{A})$ is the summation of the diagonal elements of a square matrix $\boldsymbol{A}$, called the *trace* of the matrix $\boldsymbol{A}$ and that

$$\sum_{t=1}^{\mathrm{T}}(\boldsymbol{R}_t - \boldsymbol{\mu})(\boldsymbol{R}_t - \boldsymbol{\mu})^{\mathrm{T}} = T\hat{\boldsymbol{\Sigma}} + (\bar{\boldsymbol{R}} - \boldsymbol{\mu})(\bar{\boldsymbol{R}} - \boldsymbol{\mu})^{\mathrm{T}},$$

where $\hat{\boldsymbol{\Sigma}}$ be the sample covariance matrix (divided by T, rather than $(T-1)$) and $\bar{\boldsymbol{R}}$ is the sample average. It is then clear that the maximum likelihood is obtained at $\hat{\boldsymbol{\mu}} = \bar{\boldsymbol{R}}$. Using these, after dropping the constant term, the log-likelihood becomes

$$-\frac{2}{T}\ell_T(\hat{\boldsymbol{\mu}}, \boldsymbol{\Sigma}) = \log|\boldsymbol{\Sigma}| + \mathrm{tr}(\boldsymbol{\Sigma}^{-1}\hat{\boldsymbol{\Sigma}}), \tag{7.25}$$

which can be regarded as a *divergence* between $\hat{\boldsymbol{\Sigma}}$ and $\boldsymbol{\Sigma}$.

We are now ready to apply the penalized likelihood (7.22) to estimate the sparse covariance matrix. If $\boldsymbol{\Sigma} = (\sigma_{ij})$ is sparse, by (7.25), the penalized likelihood (7.22) becomes

$$\log|\boldsymbol{\Sigma}| + \mathrm{tr}(\boldsymbol{\Sigma}^{-1}\hat{\boldsymbol{\Sigma}}) + \sum_{i<j}p_\lambda(|\sigma_{ij}|). \tag{7.26}$$

Here, we do not penalize on the diagonal elements since they are not sparse. Minimization of (7.26) among the class of positive definite matrix yields a sparse estimate of the covariance matrix.

An advantage of the penalized likelihood approach is that it allows us to explore other aspects of sparsity. For example, if $\boldsymbol{\Sigma}^{-1}$ is sparse, then by letting $\boldsymbol{\Omega} = \boldsymbol{\Sigma}^{-1}$, the likelihood function (7.25) can be written as

$$-\log|\boldsymbol{\Omega}| + \mathrm{tr}(\boldsymbol{\Omega}\hat{\boldsymbol{\Sigma}}).$$

As the sparsity is now on the elements of $\Omega = (\omega_{ij})$, the penalized likelihood function (7.22) reduces to

$$-\log|\boldsymbol{\Omega}| + \operatorname{tr}(\boldsymbol{\Omega}\hat{\boldsymbol{\Sigma}}) + \sum_{i<j} p_\lambda(|\omega_{ij}|). \tag{7.27}$$

Theoretical properties have been thoroughly studied by Rothman, Bickel, Levina and Zhu (2008) and Lam and Fan (2009).

7.2.5 Factor model with observable factors

Sparsity is not a reasonable assumption in many financial applications. It can be combined with a factor model where the covariance matrix of idiosyncratic noises is sparse. One of the earliest references on this topic is Fan, Fan and Lv (2008) who use the strict factor model to deal with large covariance matrix estimation. This imposes a structure on the covariance matrix: it can be decomposed as a sum of a low-rank matrix plus a diagonal matrix.

We have shown in Section 6.1 [see (6.2)] that the returns of asset i at time period t can be decomposed into

$$R_{it} = a_i + b_{i1} f_{1t} + \cdots + b_{iK} f_{Kt} + \varepsilon_{it}, \quad t = 1, \ldots, T, \tag{7.28}$$

or putting it into the matrix form [see (6.4)]

$$\boldsymbol{R}_t = \boldsymbol{a} + \boldsymbol{B}\boldsymbol{f}_t + \boldsymbol{\epsilon}_t, \tag{7.29}$$

where $\boldsymbol{\epsilon}_t$ is the vector of idiosyncratic noise with

$$E\boldsymbol{\epsilon}_t = 0, \qquad \operatorname{cov}(\boldsymbol{f}_t, \boldsymbol{\epsilon}_t) = 0. \tag{7.30}$$

The volatility matrix implied by (7.29) and (7.30) is

$$\boldsymbol{\Sigma} \equiv \operatorname{var}(\boldsymbol{R}_t) = \boldsymbol{B}\operatorname{var}(\boldsymbol{f}_t)\boldsymbol{B}^{\mathrm{T}} + \operatorname{var}(\boldsymbol{\epsilon}_t). \tag{7.31}$$

We have not yet imposed any assumption so far. The model (7.29) relies merely on the decomposition of random variables. The strict factor model assumes that $\operatorname{var}(\boldsymbol{\epsilon}_t)$ is diagonal. In this case, the model-implied volatility matrix admits a low-rank (the rank of $\boldsymbol{B}\operatorname{var}(\boldsymbol{f}_t)\boldsymbol{B}^{\mathrm{T}}$ is K) plus a diagonal decomposition. It has $K \cdot p$ parameters in the matrix $\boldsymbol{B}$, p parameters in matrix $\operatorname{var}(\boldsymbol{\epsilon})$, and $K(K+1)/2$ in covariance $\operatorname{var}(\boldsymbol{f}_t)$. This gives a total of $(p+1)K + K(K+1)/2$ parameters. To put this in perspective, we assume that the three factor model is applied to $p = 2000$ stocks. In this case, the factor model implied covariance matrix (7.31) has $8,006$ free parameters, whereas the original covariance matrix has $2,001,000$ parameters!

Expression (7.31) suggests a simple substitution method. Assume further that the factors are observable. Then, for each asset i, run multiple regression (7.28) to get factor loadings $\{\hat{b}_{ij}\}_{j=1}^K$ and residual variance $\hat{\sigma}_i^2$. Obtain the sample covariance $\hat{\boldsymbol{\Sigma}}_f$ of factors $\{\boldsymbol{f}_t\}_{t=1}^{\mathrm{T}}$ and compute the strict factor model based estimator:

$$\hat{\boldsymbol{\Sigma}}^S = \hat{\boldsymbol{B}}\hat{\boldsymbol{\Sigma}}_f\hat{\boldsymbol{B}}^{\mathrm{T}} + \mathrm{diag}(\hat{\sigma}_1^2, \ldots, \hat{\sigma}_p^2). \tag{7.32}$$

The resulting estimate is always positive definite even when $p \gg T$, because the last part is positive definite. The factor-model based covariance matrix in Figure 7.2 refers to this approach.

What are the benefits of using (7.32)? Compared with the sample covariance matrix, Fan, Fan and Lv (2008) showed that the factor-model based estimator has a better rate for estimating $\boldsymbol{\Sigma}^{-1}$ and the same rate for estimating $\boldsymbol{\Sigma}$. The implication of this is that the factor-model based covariance matrix is better in portfolio allocation (see also Figure 7.2), but not much better for risk assessment.

To gain insight into this kind of result, let us look at the following toy example.

Example 7.7 We consider a one-factor model in which the factor loadings are known with $\mathbf{B} = \mathbf{1}$. We also assume ideally that the levels of idiosyncratic noises are also known with $\mathrm{var}(\boldsymbol{\epsilon}) = I_{p_n}$. In this case, by (7.31),

$$\boldsymbol{\Sigma} = \sigma_f^2 \mathbf{1}\mathbf{1}^{\mathrm{T}} + I_{p_n},$$

with only one unknown element $\sigma_f^2 = \mathrm{var}(f)$. After substituting in the sample version, we obtain the substitution estimator

$$\hat{\boldsymbol{\Sigma}} = \hat{\sigma}_f^2 \mathbf{1}\mathbf{1}^{\mathrm{T}} + I_{p_n}.$$

Now it can easily be seen that the estimator error of $\hat{\sigma}_f^2$ is propagated to all entries of the covariance matrix. For example,

$$\|\hat{\boldsymbol{\Sigma}} - \boldsymbol{\Sigma}\|_2^2 = p^2 |\hat{\sigma}_f^2 - \sigma_f^2|^2.$$

This will be of the same order as the sample covariance matrix, which estimates p^2 entries of elements. Note that the eigenvalues of $\boldsymbol{\Sigma}$ are $(1 + \sigma_f^2 p), 1, \ldots, 1$. Therefore, the largest eigenvalue is hard to estimate accurately in absolute terms, when p is large. On the other hand, the eigenvalues of $\boldsymbol{\Sigma}^{-1}$ are $(1 + \sigma_f^2 p)^{-1}, 1, \ldots, 1$, which can be estimated precisely. That explains why the inversion of the matrix can be better estimated with the factor structure.

To demonstrate the above theory numerically, Fan, Fan and Lv (2008) performed an extensive simulation study. To report the result, we need some measures of distance or divergence between two matrices. There are three commonly used distances or *divergence of matrices*: For a given estimator $\hat{\boldsymbol{\Sigma}}$ of $\boldsymbol{\Sigma}$,

- *Quadratic loss*: $\|\hat{\boldsymbol{\Sigma}} - \boldsymbol{\Sigma}\|_Q^2 = \mathrm{tr}[\hat{\boldsymbol{\Sigma}}\boldsymbol{\Sigma}^{-1} - \boldsymbol{I}_p]^2/p$;
- *Entropy loss*: $\mathrm{tr}(\hat{\boldsymbol{\Sigma}}\boldsymbol{\Sigma}^{-1}) - \log|\hat{\boldsymbol{\Sigma}}\boldsymbol{\Sigma}^{-1}| - p$
- *Norm-based losses*: $\|\hat{\boldsymbol{\Sigma}} - \boldsymbol{\Sigma}\|$, where $\|\cdot\|$ can be the operator norm or the Frobenius norm.

Note that the quadratic loss is a relative loss in matrix estimation, as it can be written as

$$\|\boldsymbol{\Sigma}^{-1/2}\hat{\boldsymbol{\Sigma}}\boldsymbol{\Sigma}^{-1/2} - \boldsymbol{I}_p\|_F^2/p.$$

See Exercise 7.8.

Example 7.8 Fan, Fan and Lv (2008) simulated returns of p assets from a 3-factor model. To determine factor loadings, they calibrated the model parameters to market data. Specifically, they used the 3-year daily data of 30 industrial portfolios during 5/1/02–8/29/05 to fit the Fama–French three-factor model. The estimated 30 factor loading vectors are summarized in Table 7.3 (the left panel), which gives the sample mean vector and covariance matrices of these 30 three-dimensional factor loadings. The residual variance of the 30 idiosyncratic errors have average 0.6608 and SD 0.3275, with a minimum of 0.195. The sample mean and sample variance of the Fama–French factors are also summarized in Table 7.3 (right panel).

With the parameters above, the daily returns of p assets from the three-factor model can be simulated as follows. For a given p and T,

- generate the factors $\boldsymbol{f}_t$ of size T independently from $\mathcal{N}\left(\boldsymbol{\mu}_{\boldsymbol{f}}, \boldsymbol{\Sigma}_{\boldsymbol{f}}\right)$;
- generate factor loadings $\boldsymbol{b}_1, \ldots, \boldsymbol{b}_p$ from $\mathcal{N}(\boldsymbol{\mu}_{\boldsymbol{B}}, \boldsymbol{\Sigma}_{\boldsymbol{B}})$;
- generate the levels of idiosyncratic noises $\sigma_1, \ldots, \sigma_p$ from a gamma distribution Gamma(α, β) conditioned on $[0.195, \infty)$. The parameters $\alpha = 3.3586$ and $\beta = .1876$ are chosen to match the first two moments of the empirical data. In short, we generate $\sigma_1, \ldots, \sigma_p$ from Gamma$(3.3586, 0.1876)$ and reject the outcome whenever the results is less than 0.195.
- generate T idiosyncratic noises $\{\varepsilon_{it}\}_{t=1}^{\mathrm{T}}$ from $N(0, \sigma_i^2)$ for each asset i;
- compute the daily returns $\{R_{it}\}_{t=1}^{\mathrm{T}}$ by using (7.28).

Based on 500 simulations of three years' data ($T = 756$), Fan, Fan and Lv

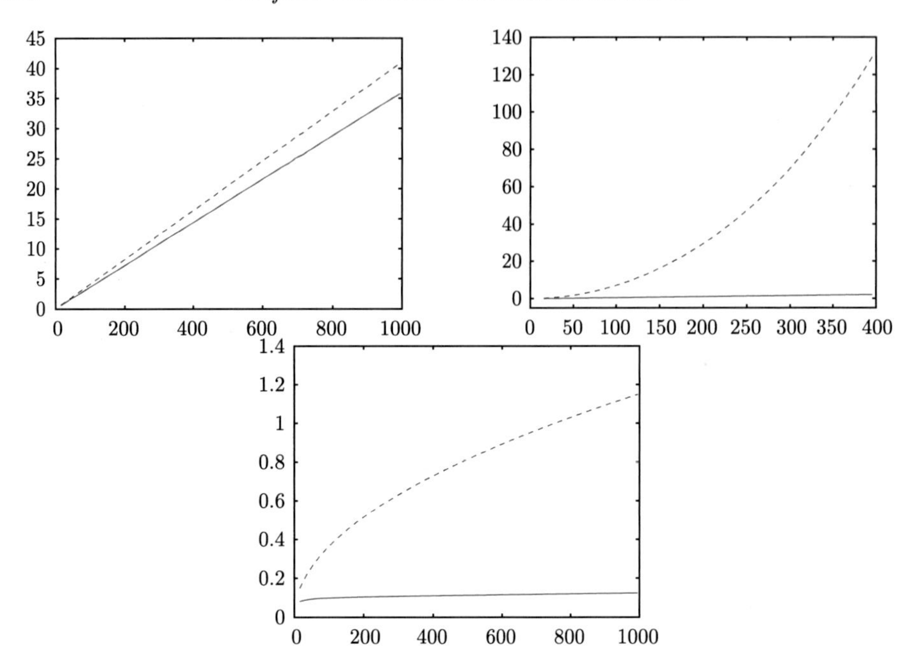

Figure 7.5 Average losses of estimating covariance matrices Average losses of estimating $\hat{\boldsymbol{\Sigma}}$ based on 500 simulations. Dashed lines represent the sample covariance matrix and solid lines are for the factor model-based estimator. From the left to right panel: Average loses under the Frobenius norm, the entropy loss.

(2008) examine the impact of dimensionality for both factor-based covariance matrix compared with the sample one. Figure 7.5 show the performance for estimating $\boldsymbol{\Sigma}$. For the Frobenius norm, the sample covariance and the factor-model based estimation have comparable performance. The performance deteriorates as p increases. In terms of the quadratic loss or entropy loss, the factor-based estimation outperforms the sample covariance matrix. This is due to the involvement of the inverse of the covariance matrix.

Figure 7.6 summarizes the performance on the estimation of $\boldsymbol{\Sigma}^{-1}$ based on both estimation methods. Clearly, the results show the better performance of the factor-model based method and are consistent with the theory reported above.

7.2.6 Approximate factor models with observable factors

Sparsity and exact factor models rarely occur in financial practice. But they can be combined to yield powerful methods. Instead of assuming uncor-

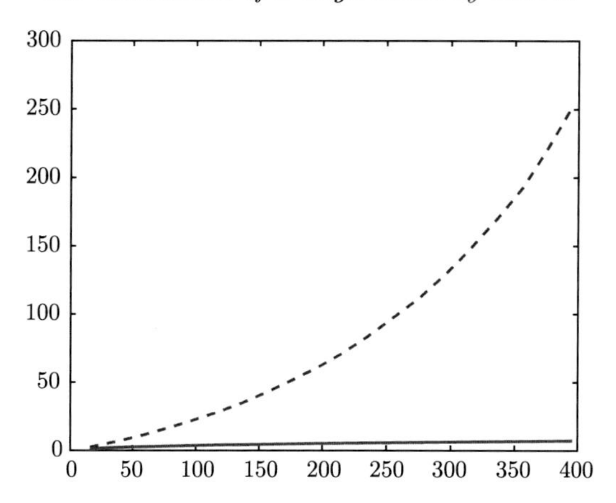

Figure 7.6 Average loss $\|\hat{\boldsymbol{\Sigma}}^{-1} - \boldsymbol{\Sigma}^{-1}\|_F$ for estimating $\hat{\boldsymbol{\Sigma}}^{-1}$ based on 500 simulations. Dashed lines are for the sample covariance matrix and solid line for the factor model-based estimator.

relatedness of idiosyncratic noises, an *approximate factor model* imposes only sparsity on $\boldsymbol{\Sigma}_\epsilon = \mathrm{var}(\boldsymbol{\epsilon})$. The sparsity can be explored by thresholding as shown in Section 7.2.2 or penalized likelihood as in Section 7.2.4. For simplicity, we only employ the simple thresholding rule on the correlation matrix: Compute $\hat{\boldsymbol{\Sigma}}_\epsilon$, the sample covariance matrix of residuals $\{\hat{\boldsymbol{\epsilon}}_t\}_{t=1}^T$ and construct the thresholded estimator $\hat{\boldsymbol{\Sigma}}_{\epsilon,\lambda}^*$ as in (7.15), which applies the thresholding rule to the correlation matrix.

Following (7.31), the estimation of the volatility matrix based on an approximate factor model is now given by

$$\hat{\boldsymbol{\Sigma}}_\lambda^A = \hat{\boldsymbol{B}}\hat{\boldsymbol{\Sigma}}_f\hat{\boldsymbol{B}}^T + \hat{\boldsymbol{\Sigma}}_{\epsilon,\lambda}^*, \tag{7.33}$$

where $\hat{\boldsymbol{B}}$ and $\hat{\boldsymbol{\Sigma}}_f$ are defined as (7.32). This simple estimator combines the techniques of the factor model with the thresholding estimator, which was introduced by Fan, Liao and Mincheva (2011). They show that the nice theoretical properties continue to hold. In the regularization step, many alternative approaches can be also used for estimating $\mathrm{var}(\boldsymbol{\epsilon})$ (e.g., penalized likelihood, adaptive thresholding). We would like to note that (7.33) is not necessarily a semi-positive definite matrix. The projection techniques in Section 7.2.3 can be employed.

With the thresholding parameter $\lambda = 1$ in computing (7.15), the resulting $\hat{\boldsymbol{\Psi}}_\lambda$ there is the identity matrix and $\hat{\boldsymbol{\Sigma}}_{\epsilon,1}^*$ is just the diagonal matrix with residual variances as its diagonal elements. Therefore, the method reduces to the exact factor model in the last subsection, i.e. the same estimator as

(7.32):

$$\hat{\boldsymbol{\Sigma}}_1^A = \hat{\boldsymbol{\Sigma}}^S.$$

On the other hand, at the other extreme that $\lambda = 0$, $\hat{\boldsymbol{\Sigma}}_{\epsilon,0}^*$ is just the sample covariance $\hat{\boldsymbol{\Sigma}}_\epsilon$ of residual vectors. Then, it follows from (7.33) that

$$\hat{\boldsymbol{\Sigma}}_0^A = \hat{\boldsymbol{B}}\hat{\boldsymbol{\Sigma}}_f\hat{\boldsymbol{B}}^{\mathrm{T}} + \hat{\boldsymbol{\Sigma}}_\epsilon$$

is the sample covariance matrix of the returns $\{\boldsymbol{R}_t\}_{t=1}^{\mathrm{T}}$, by recalling the decomposition (7.31). In summary, the family of the covariance estimator (7.33) includes the sample covariance matrix and the estimator based the strict factor model as specific examples, corresponding to, respectively, the choice $\lambda = 0$ and $\lambda = 1$.

We now illustrate the method by the following example.

Example 7.9 Consider the returns of the Fama–French 100 portfolios in the last quarter of 2010. We wish to estimate the covariance matrix based on 63 daily returns. The largest eigenvalue of the sample covariance is 81.68, explaining about 79.59% of the variability (See Section 6.5.1 for these concepts). This principal direction is basically the market portfolio. Since the sample size is 63, the rank of the sample covariance matrix is at most 62 (usually exactly 62). That means that there are 38 zero eigenvalues. The distribution of the eigenvalues, called the *spectral distribution*, is presented in Figure 7.7. The strict factor model estimation of the covariance matrix is always positive definite, whose spectral distribution is also presented in Figure 7.7.

The approximate factor model estimate of covariance matrix is more flexible. When $\lambda = 1$, it is the same estimate as that by the strict factor model. We take $\lambda = 0.1$ that sets any sample correlations of idiosyncratic errors to zero if their magnitudes are less than 0.1. This results in 2249 zeros. However, the resulting estimate is not semi-positive definite. The minimum eigenvalue is -0.447. We project the resulting estimate further into the space of semipositive definite matrices, using method 2 in Section 7.2.2.

Figure 7.7 summarizes the spectral distributions of the estimated covariance matrices. The largest eigenvalues of all estimated covariance matrices are around 80 and are not presented in the figure. Each of the spectral distributions has a few distinct second, third, and fourth eigenvalues. They provide evidence of the existence of the second, third and possibly fourth factors that explains the market common risks. Interestingly, despite various thresholdings ($\lambda = 0, 0.1$ and 1), these salient features remain in all spectral estimates. The regularization changes only small eigenvalues and their cor-

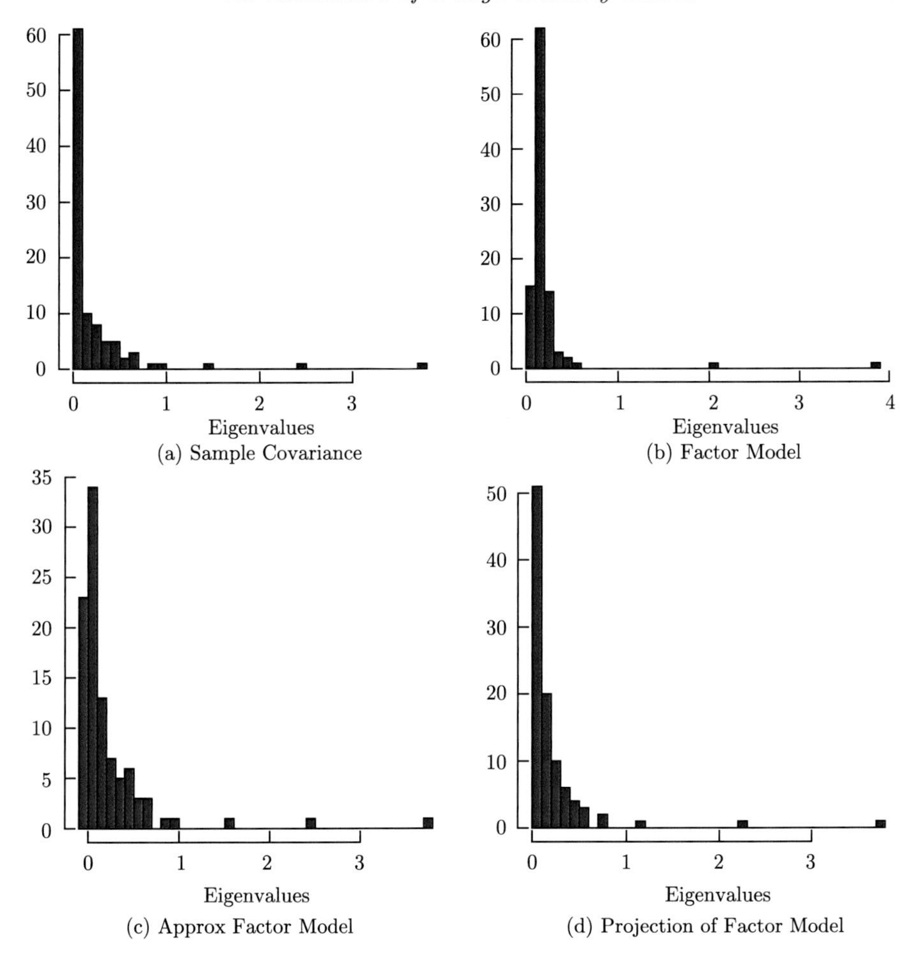

Figure 7.7 Spectral distributions of the covariance matrices estimated (the largest eigenvalue around 80 is excluded in the plot) by (a) sample covariance matrix; (b) strict factor model; (c) approximate factor model; (d) projection of estimate in (c) onto the space of semi-positive definite matrices.

responding eigenvectors, which cannot be reliably estimated. For example, the sample covariance corresponds to the regularized estimator with $\lambda = 0$ and the covariance matrix estimated by the strict factor corresponds to the regularized estimator with $\lambda = 1$. These two estimators result from two very different estimators for the correlation matrix $\mathbf{\Psi}$ of the idiosyncratic noises. In addition, the projection method shrinks all eigenvalues by a factor of $(1 - \lambda_{\min}^{-}) = 1.447$. These still do not affect the main features presented in the figure.

7.2.7 Approximate factor models with unobservable factors

In many applications such as foreign investments or exotic asset classes, the common factors that drive the correlation of returns are unknown. These *latent factors* have to be estimated from the given asset returns. As discussed in Section 6.5.1, the principal components are the natural candidate for estimating these eigenvectors.

For estimating large volatility $\boldsymbol{\Sigma}$, Fan, Liao and Mincheva (2013) introduced the following simple method, called *Principal Orthogonal complEment Thresholding*, or *POET* for short. For a given number of factors K, the method runs the following steps.

1. Obtain the sample covariance matrix $\hat{\boldsymbol{\Sigma}}$ based on T period returns.
2. Run a singular value decomposition for the sample covariance to obtain $\hat{\boldsymbol{\Sigma}} = \sum_{j=1}^{p} \hat{\lambda}_j \hat{\boldsymbol{\xi}}_j \hat{\boldsymbol{\xi}}_j^T$.
3. Compute the residual covariance matrix $\hat{\boldsymbol{Q}} = \sum_{j=K+1}^{p} \hat{\lambda}_j \hat{\boldsymbol{\xi}}_j \hat{\boldsymbol{\xi}}_j^T$.
4. Apply a regularization method to $\boldsymbol{Q}$ with a regularization parameter λ to obtain $\hat{\boldsymbol{Q}}_\lambda$. For example, we can apply the thresholding on the correlation matrix as in (7.15) or adaptive thresholding as in (7.14).
5. Compute the POET estimator as

$$\hat{\boldsymbol{\Sigma}}_\lambda^P = \sum_{j=1}^{K} \hat{\lambda}_j \hat{\boldsymbol{\xi}}_j \hat{\boldsymbol{\xi}}_j^T + \hat{\boldsymbol{Q}}_\lambda. \tag{7.34}$$

First of all, the POET estimator is a nonparametric approach. It works the best when the underlying covariance matrix admits the approximate factor structure, namely, $\text{var}(\boldsymbol{\epsilon}_t)$ is sparse in (7.31). Secondly, the approach is very versatile. When $\lambda = 0$, $\hat{\boldsymbol{\Sigma}}_0^P$ is the sample covariance matrix, whereas when $\lambda = 1$, it reduces to the covariance matrix based on the strict factor models: $\hat{\boldsymbol{\Sigma}}_1^P = \hat{\boldsymbol{\Sigma}}^S$. When $K = 0$, the approach reduces to the usual thresholding estimator such as Bickel and Levina (2008) or Cai and Liu (2011).

Second, overestimating the number of factor K does very little harm. Fan, Liao and Mincheva (2013) show that if true $K = 3$, even when you use $K = 10$, the resulting estimator performs robustly with the choice of K. On the other hand, under estimating K creates a lot of biases in the POET estimator. For example, with $K = 1$ or $K = 2$, the estimator is not even consistent. Data-driven choice of K is possible. See, for example, Bai and Ng (2002, 2008) and Onatski (2010), Ahn, Lam and Yao (2012), and Horenstein (2013).

Third, the POET estimator is relatively easy to compute. As long as we have software to compute the singular-value decomposition, the rest can be computed analytically. There is a R-package called `POET` that computes the POET estimator.

Finally, Fan, Liao and Mincheva (2013) showed when the number of assets p is sufficiently large, the latent can be sufficiently accurately estimated and the POET estimator behaves as if the latent factors are observable. More precisely, when $p \log p \geqslant T$, $\hat{\boldsymbol{\Sigma}}_\lambda^P$ has the same performance as $\hat{\boldsymbol{\Sigma}}_\lambda^A$.

The remaining materials of this section can be skipped for first-time readers.

High-dimensional PCA and factor analysis* To provide additional insights as to why principal component analysis and factor analysis are approximately the same when the number of assets or variables is large, let us begin with the factor model (7.29). We assume for simplicity that $\boldsymbol{a} = 0$; otherwise, we can subtract the averages from the returns and the factors. Then, model (7.29) can be written as

$$\boldsymbol{R}_t = \boldsymbol{B}\boldsymbol{f}_t + \boldsymbol{\epsilon}_t = (\boldsymbol{B}\boldsymbol{H})(\boldsymbol{H}^{-1}\boldsymbol{f}_t) + \boldsymbol{\epsilon}_t, \tag{7.35}$$

for any $K \times K$ nonsingular matrix $\boldsymbol{H}$. Since both $\mathbf{B}$ and $\mathbf{f}_t$ are unknown, this creates the issue of *identifiability*. To resolve the ambiguity between $\mathbf{B}$ and $\mathbf{f}_t$, we impose the identifiability constraint that $\text{var}(\mathbf{f}_t) = \mathbf{I}_K$ and the columns of $\mathbf{B}$ are orthogonal. Under this canonical form, it follow from (7.31), we have

$$\boldsymbol{\Sigma} = \boldsymbol{B}\boldsymbol{B}^{\mathrm{T}} + \boldsymbol{\Sigma}_\varepsilon, \qquad \boldsymbol{\Sigma}_\varepsilon = \text{var}(\boldsymbol{\epsilon}). \tag{7.36}$$

Let $\boldsymbol{b}_1, \ldots, \boldsymbol{b}_K$ be the K orthogonal columns of $\boldsymbol{B}$, sorting according to the norms $\{\|\boldsymbol{b}_j\|\}_{j=1}^K$, from the largest to the smallest. For simplicity, we assume that there is no tie in $\{\|\boldsymbol{b}_j\|\}_{j=1}^K$. Then, it is clear that the normalized vector $\boldsymbol{b}_j/\|\boldsymbol{b}_j\|$ is the j-th eigenvector of $\boldsymbol{B}\boldsymbol{B}^{\mathrm{T}}$. This can easily be seen from $\boldsymbol{B} = (\boldsymbol{b}_1, \ldots, \boldsymbol{b}_K)$ and the orthogonality assumption:

$$(\boldsymbol{B}\boldsymbol{B}^{\mathrm{T}})\boldsymbol{b}_j/\|\boldsymbol{b}\| = (\sum_{i=1}^K \boldsymbol{b}_i\boldsymbol{b}_i^{\mathrm{T}})\boldsymbol{b}_j/\|\boldsymbol{b}_j\| = \|\boldsymbol{b}_j\|^2\boldsymbol{b}_j/\|\boldsymbol{b}_j\|.$$

The above identity also shows that $\|\boldsymbol{b}_j\|^2$ is the jth largest eigenvalue of $\boldsymbol{B}\boldsymbol{B}^{\mathrm{T}}$. How large is it? From the definition, $\|\boldsymbol{b}_j\|^2 = \sum_{i=1}^p b_{ij}^2$, which is usually the order of p, if the factor j is *pervasive*. Namely, the jth factor influences a non-negligible fraction of the asset returns among p assets, meaning, a non-negligible fraction of $|b_{ij}|$ is bounded away from zero.

From linear algebra, it is well known that the first K eigenvalues of matrix $\boldsymbol{B}\boldsymbol{B}^{\mathrm{T}}$ are the same as the first K eigenvalues of matrix $\boldsymbol{B}^{\mathrm{T}}\boldsymbol{B}$. Let $\widetilde{\boldsymbol{b}}_i$ be the transpose of the ith row of $\boldsymbol{B}$. Then, $\boldsymbol{B}^{\mathrm{T}} = (\widetilde{\boldsymbol{b}}_1, \ldots, \widetilde{\boldsymbol{b}}_p)$ and

$$\boldsymbol{B}^{\mathrm{T}}\boldsymbol{B} = p \cdot \frac{1}{p}\sum_{i=1}^{p} \widetilde{\boldsymbol{b}}_i \widetilde{\boldsymbol{b}}_i^{\mathrm{T}}.$$

If we assume that the factors are *pervasive* in the sense that

$$\frac{1}{p}\sum_{i=1}^{p} \widetilde{\boldsymbol{b}}_i \widetilde{\boldsymbol{b}}_i^{\mathrm{T}} \longrightarrow \boldsymbol{\Sigma}_B,$$

as $p \to \infty$, where $\boldsymbol{\Sigma}_B$ is $K \times K$ positive definite matrix with distinct and positive eigenvalues $\lambda_1(\boldsymbol{\Sigma}_B), \ldots, \lambda_k(\boldsymbol{\Sigma}_B)$. An example of this that $\widetilde{\boldsymbol{b}}_i$ is a random sample from a population with mean zero and covariance $\boldsymbol{\Sigma}_B$, as in Example 7.8. Therefore, the jth eigenvalue of $\boldsymbol{B}\boldsymbol{B}^{\mathrm{T}}$ in (7.36) is approximately $p \cdot \lambda_j(\boldsymbol{\Sigma}_B)$ for $j \leqslant K$. This can make $\boldsymbol{\Sigma}_\varepsilon$ there negligible under some mild assumption $\|\boldsymbol{\Sigma}_\varepsilon\| = o(p)$.

Let λ_j be the jth largest eigenvalue of $\boldsymbol{\Sigma}$ with the corresponding eigenvector $\boldsymbol{\xi}_j$, for $j = 1, \ldots, p$. To gain further intuition, let us assume for a moment that $\boldsymbol{\Sigma}_\varepsilon = 0$ so that $\boldsymbol{\Sigma} = \boldsymbol{B}\boldsymbol{B}^{\mathrm{T}}$ and

$$\lambda_j = \|\boldsymbol{b}_j\|^2, \quad \text{for } j \leqslant K, \qquad \lambda_j = 0, \quad j > K. \tag{7.37}$$

and

$$\boldsymbol{\xi}_j = \boldsymbol{b}_j / \|\boldsymbol{b}_j\|. \tag{7.38}$$

In this case, the POET estimator is just the sample version of the spectral decomposition

$$\boldsymbol{\Sigma} = \sum_{j=1}^{K} \lambda_j \boldsymbol{\xi}_j \boldsymbol{\xi}_j^{\mathrm{T}}.$$

Furthermore, from (7.35) and the orthogonality of $\{\boldsymbol{b}_j\}$, we have

$$(\boldsymbol{b}_j/\|\boldsymbol{b}_j\|)^{\mathrm{T}}\boldsymbol{R}_t = \|\boldsymbol{b}_j\| f_{jt} + (\boldsymbol{b}_j/\|\boldsymbol{b}_j\|)^{\mathrm{T}}\boldsymbol{\epsilon}_t,$$

where f_{jt} is the realized jth factor at time t. The last term is the weighted average of noise $\boldsymbol{\epsilon}_t$ over all p assets and hence typically negligible when p is large. Therefore,

$$f_{jt} \approx \boldsymbol{\xi}_j^{\mathrm{T}}\boldsymbol{R}_t / \lambda_j^{1/2}. \tag{7.39}$$

Namely, the jth realized factor can be recovered by projecting the returns $\boldsymbol{R}_t$ onto the jth principal component, divided by the square root of its corresponding eigenvalue. This shows that the principal component analysis and factor analysis for large portfolios are approximately the same.

Let us now consider more general case of $\boldsymbol{\Sigma}_\varepsilon$. In this case, (7.37) and (7.38) holds only approximately. Indeed, by Wely's Theorem, the differences between eigenvalues of $\boldsymbol{\Sigma}$ and $\boldsymbol{B}\boldsymbol{B}^{\mathrm{T}}$ are bounded by $\|\boldsymbol{\Sigma}_\varepsilon\|$. That is

$$\left|\lambda_j - \|\boldsymbol{b}_j\|^2\right| \leqslant \|\boldsymbol{\Sigma}_\varepsilon\|, \text{ for } j \leqslant K, \qquad |\lambda_j| \leqslant \|\boldsymbol{\Sigma}_\varepsilon\|, \text{ for } j > K. \tag{7.40}$$

Since $\|\boldsymbol{b}_j\|^2 \approx p\lambda_j(\boldsymbol{\Sigma}_B)$ for $j \leqslant K$, the approximation error is negligible when $\|\boldsymbol{\Sigma}_\varepsilon\| = o(p)$. By applying the $\sin(\theta)$-theorem of Davis and Kahan (1970), it can be shown that

$$\left\|\boldsymbol{\xi}_j - \boldsymbol{b}_j/\|\boldsymbol{b}_j\|\right\| = O(p^{-1}\|\boldsymbol{\Sigma}_\varepsilon\|), \quad \text{for } j \leqslant K. \tag{7.41}$$

Expressions (7.40) and (7.41) are the generalizations of (7.37) and (7.38). They entail that

$$\sum_{j=1}^{K} \lambda_j \boldsymbol{\xi}_j \boldsymbol{\xi}_j \approx \sum_{j=1}^{K} \boldsymbol{b}_j \boldsymbol{b}_j^{\mathrm{T}} = \boldsymbol{B}\boldsymbol{B}^{\mathrm{T}}.$$

This is estimated by the first part of the POET estimator (7.34). The remaining part of spectral estimator $\hat{\boldsymbol{R}}$ estimates $\boldsymbol{\Sigma}_\varepsilon$. Since it is assumed $\boldsymbol{\Sigma}_\varepsilon$ is sparse, regularization method applies to $\boldsymbol{\Sigma}_\varepsilon$. The operator norm of $\|\boldsymbol{\Sigma}_\varepsilon\|$ can easily be bounded by its sparsity assumption. See 7.9.

The above method is designed for pervasive factors or strong factors. The asymptotic theory of the POET and related method has been established by Fan, Liao and Mincheva (2013). For theory and methods for weak factors, see Onatski (2012). For other important treatments of factor models, including using principal components as estimated factors and *dynamic factor models*, see Stock and Watson (2002, 2005), Bai (2003, 2009), and Forni, and Hallin, Lippi, and Reichlin (2000, 2005).

7.3 Portfolio allocation with gross-exposure constraints

Markowitz's mean-variance optimal portfolio solves the following optimization problem

$$\min_{\boldsymbol{w}} \boldsymbol{w}^{\mathrm{T}}\boldsymbol{\Sigma}\boldsymbol{w}, \quad \text{s.t. } \boldsymbol{w}^{\mathrm{T}}\mathbf{1} = 1 \text{ and } \boldsymbol{w}^{\mathrm{T}}\boldsymbol{\mu} = r_0.$$

The optimal solution is given by (5.43), which depends on a few quantities such as

$$\boldsymbol{\Sigma}^{-1}\boldsymbol{\mu}, \quad \boldsymbol{\Sigma}^{-1}\mathbf{1}, \qquad \boldsymbol{\mu}^{\mathrm{T}}\boldsymbol{\Sigma}^{-1}\boldsymbol{\mu}$$

that involve unknown $\boldsymbol{\mu}$ and $\boldsymbol{\Sigma}$. While the work is the cornerstone of modern finance, its practical applications face a number of challenges. It is not clear whether the aforementioned unknown quantities can be estimated accurately enough in application. The optimal allocation depends very sensitively on the estimated mean and volatility matrix as well as their estimation errors. It can result in extremely large short positions.These problems become more severe when the pool of candidate assets is large. Therefore, it is reasonable to seek more stable and robust optimal portfolios.

7.3.1 Portfolio selection with gross-exposure constraint

Let c be the maximum gross exposure allowed. Let $U(\cdot)$ be the utility function. An extension of Markowitz' work is to add constraints on the gross exposure, searching optimal portfolios among the class of stable ones. This leads to the following utility optimization with the gross-exposure constraint:

$$\begin{aligned} &\max_{\boldsymbol{w}} E[U(\boldsymbol{w}^T \boldsymbol{R})] \\ &\text{s.t.} \quad \boldsymbol{w}^T \mathbf{1} = 1, \ \ \|\boldsymbol{w}\|_1 \leqslant c, \ \ \boldsymbol{A}^T \boldsymbol{w} = \boldsymbol{a}. \end{aligned} \tag{7.42}$$

The gross exposure constraint is equivalent to total short position $\boldsymbol{w}^- \leqslant (c-1)/2$. The equality constraints can be on the expected return such as $\boldsymbol{A} = \boldsymbol{\mu}$. However, the expected returns $\boldsymbol{\mu}$ are usually very hard to estimate accurately. Instead, one might put constraints on the allocations in different sectors. One can also optimize the portfolio subject to the constraint that the resulting portfolio is neutral to certain known factors such as the market portfolio. These can be accommodated in the framework (7.42) by an appropriate choice of the matrix $\boldsymbol{A}$.

When c is small or moderate, the constraint $\|\boldsymbol{w}\|_1 \leqslant c$ is usually binding and the solution is typically sparse. This achieves the purpose of asset selection. For example, when $c = 1$, and the utility is the portfolio risk as in Figure 7.2, the average number of stocks in the optimal no-short portfolios there are about 53, among 600 stocks. One can replace the utility function in (7.42) by any risk measure (Artzner, Delbaen, Eber and Heath, 1999)

When $c = \infty$, the gross-exposure constraint is no longer binding. It becomes Markowitz's utility optimization problem. When $c = 1$, no short-sales are allowed in the portfolios. These are the most conservative but most stable portfolios. As c ranges from 1 to ∞, it creates a continuous solution path from most conservative to most aggressive portfolios, with different utility profiles. See Figure 7.2.

As in Section 5.2, when the return distribution is normal $\boldsymbol{R} \sim N(\boldsymbol{\mu}, \boldsymbol{\Sigma})$ and the utility function is exponential $U(w) = 1 - \exp(-Aw)$, then the expected utility is given by

$$E[U(\boldsymbol{w}^{\mathrm{T}}\boldsymbol{R})] = 1 - \exp(-AM(\boldsymbol{\mu}, \boldsymbol{\Sigma})) \tag{7.43}$$

where

$$M(\boldsymbol{\mu}, \boldsymbol{\Sigma}) = \boldsymbol{w}^{\mathrm{T}}\boldsymbol{\mu} - A\boldsymbol{w}^{\mathrm{T}}\boldsymbol{\Sigma}\boldsymbol{w}/2.$$

Therefore, utility maximization is equivalent to mean variance analysis. Thus, we focus only on the function M. Following the same argument as in Theorem 7.1, we can easily obtain the *utility approximation*:

$$|M(\hat{\boldsymbol{\mu}}, \hat{\boldsymbol{\Sigma}}) - M(\boldsymbol{\mu}, \boldsymbol{\Sigma})| \leqslant \|\hat{\boldsymbol{\mu}} - \boldsymbol{\mu}\|_\infty \|\boldsymbol{w}\|_1 + Ae_{\max}\|\boldsymbol{w}\|_1^2/2, \tag{7.44}$$

for any estimated expected return $\hat{\boldsymbol{\mu}}$ and estimated volatility matrix $\hat{\boldsymbol{\Sigma}}$. No short-sale portfolios give the tightest upper bound with $c = 1$. However, this usually yields an optimal portfolio that is not diversified enough. For the empirical study summarized in Figure 7.2, the average number of stocks in the optimal no-short portfolios is only 31 when RiskMetrics is used.

From now on, like many papers in the literature, we focus on the risk profile. Theory and methods are readily extendable to utility optimization. However, it is very hard to estimate expected returns that make empirical studies inconclusive.

Define the theoretical optimal and perceived optimal allocation vectors respectively by

$$\boldsymbol{w}_{\mathrm{opt}} = \mathrm{argmin}_{\|\boldsymbol{w}\|_1 \leqslant c} R(\boldsymbol{w}), \qquad \hat{\boldsymbol{w}}_{\mathrm{opt}} = \mathrm{argmin}_{\|\boldsymbol{w}\|_1 \leqslant c} \hat{R}(\boldsymbol{w}). \tag{7.45}$$

This is a quadratic program and can easily be solved by a quadratic solver. The allocation vector $\boldsymbol{w}_{\mathrm{opt}}$ is also referred to as the *oracle* allocation as it can only be done by an oracle who knows $\boldsymbol{\Sigma}$. The best possible actual risk is $\sqrt{R(\boldsymbol{w}_{\mathrm{opt}})}$. In practice, the allocation vector $\hat{w}_{\mathrm{opt}}$ is used, which will be referred to as the *perceived optimal* or *empirical optimal* allocation vector. The perceived optimal risk is $\sqrt{\hat{R}(\hat{\boldsymbol{w}}_{\mathrm{opt}})}$, which is computable and is often smaller than the oracle risk. The actual risk of the perceived optimal portfolio $\sqrt{R(\hat{\boldsymbol{w}}_{\mathrm{opt}})}$ is unknown and is of our primary interest. The following theorem, obtained by Fan, Zhang and Yu (2012), reveals that all those three quantities are very close when c and $e_{\max}$ are not large.

Theorem 7.2 ((Optimal Risk Approximations)) *Let* $e_{\max} = |\hat{\boldsymbol{\Sigma}} - \boldsymbol{\Sigma}|_\infty$. *Then, we have*

$$\begin{aligned}
|R(\hat{\boldsymbol{w}}_{opt}) - R(\boldsymbol{w}_{opt})| &\leqslant 2e_{\max}\, c^2 \\
|R(\hat{\boldsymbol{w}}_{opt}) - \hat{R}(\hat{\boldsymbol{w}}_{opt})| &\leqslant e_{\max}\, c^2 \\
|R(\boldsymbol{w}_{opt}) - \hat{R}(\hat{\boldsymbol{w}}_{opt})| &\leqslant e_{\max}\, c^2.
\end{aligned}$$

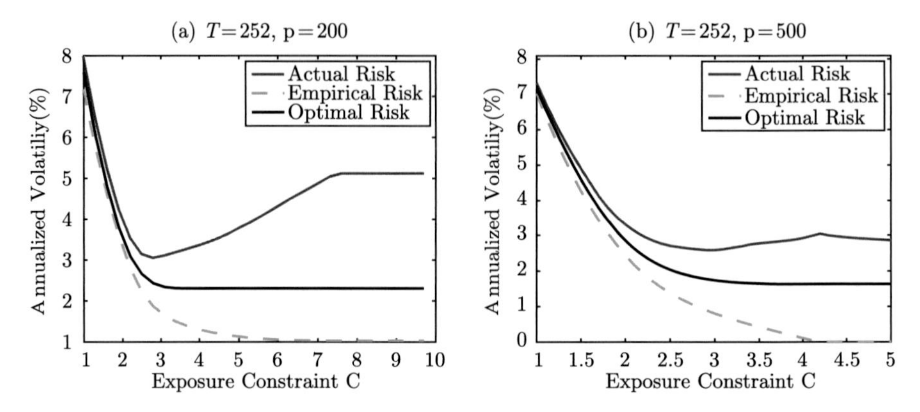

Figure 7.8 The risks of theoretically optimal portfolios, and the actual risks of the empirically optimal portfolios, and the perceived risks of the empirically optimal portfolios under gross-exposure constraints are plotted against the gross-exposure parameter c. Taken from Fan, Zhang and Yu (2012).

The proof is straightforward, but we defer it to Section 7.6. For bounding in terms of relative errors, see Exercise 7.10. Theorem 7.2 reveals that when c and e_{max} are small, the three aforementioned optimal risks are in fact very close. Since $\hat{R}(\hat{\boldsymbol{w}}_{\text{opt}})$ is known, the unknown oracle risk or the actual risk of the perceived optimal portfolio can be estimated for a range of c. To illustrate the point, we consider the following simulation studies, in which we know the oracle risk.

Example 7.10 We simulate daily returns of $p = 200$ and $p = 500$ assets over 252 days from a three-factor model with parameters given by Table 7.3. For simulation studies, since we know the true covariance matrix, the three different risks can be computed. The estimated covariance matrix is based on the sample covariance matrix. When $p = 500$, it is degenerate. Nevertheless, the perceived optimal risk is well behaved when c is small (e.g. $c \leqslant 1.5$). See Figure 7.8. This shows that our constrained optimization problem can handle the case with $p \gg T$.

In consistency with Theorem 7.2, the differences among those three risks are small when c is modest and the discrepancies generally increase with c. The figure also explains the question raised by Jagannathan and Ma (2003) why using the wrong constraint helps. As c increases, the oracle risk decreases. However, when c is sufficiently large, the constraint is no longer binding and the oracle risk is the same as that of the global minimum variance portfolio, regardless of c. See the flat part of the actual risk profile in Figure 7.8. On the other hand, as c increases the portfolio becomes less stable and its risk increases. This is why a large c does not necessarily produce the smallest actual risk. Indeed, they are far from optimal in Figure 7.8.

We would like to note that the perceived optimal risks are usually below the actual risks. For $p = 500$, due to the singularity of the sample covariance matrix, the perceived optimal risk can even reach zero when c is large enough. The gross-exposure constraints prevents this from happening. We would like also to note that the actual risk profile does not increase as rapidly as those presented in Figure 7.2 when c is large. This is mainly due to the fact that the parameters used are calibrated from 30 industrial portfolios. Hence, our simulated data behave more closely to portfolios than to the individual stocks used in Figure 7.2.

Theorem 7.2 shows that the approximation error depends on $e_{\max}$, the maximum componentwise estimation error of the covariance matrix. How fast does this quantity grow with the size p of the pool of candidate assets? The following theorem gives an answer to this question.

Theorem 7.3 *If for a sufficiently large x,*

$$\max_{i,j} P\{b_T|\sigma_{ij} - \hat{\sigma}_{ij}| > x\} < \exp(-Cx^{1/a}),$$

for some positive constants a and C and rate b_T, then

$$|\boldsymbol{\Sigma} - \hat{\boldsymbol{\Sigma}}|_\infty = O_P\left(\frac{(\log p)^a}{b_T}\right).$$

The proof of theorem is simple, but we defer it to Section 7.6. The theorem states that if each element can be estimated with the rate of convergence b_T and it has an exponential tail, the impact of the size of the portfolio p is limited. It enters into the risk approximation only in the logarithmic order.

Fan, Zhang and Yu (2012) give several situations under which the conditions of Theorem 7.3 hold. We state one of their results as an example.

Theorem 7.4 *Suppose that $\|\boldsymbol{R}_t\|_\infty$ is bounded and that the returns are weakly dependent in that its α mixing coefficient $\alpha(q)$ decays exponentially, namely $\alpha(q) = O(\exp(-Cq^{1/b}))$, for some $b < 2a-1$. If $\log p = o(n^{1/(2b+1)})$, then conclusion in Theorem 7.3 holds.*

*7.3.2 Relation with covariance regularization**

To better understand why the gross-exposure constraint helps on risk approximation, we make connections with covariance regularization.

By the Lagrange multiplier method, problem (7.45) is to minimize

$$\boldsymbol{w}^{\mathrm{T}}\hat{\boldsymbol{\Sigma}}\boldsymbol{w}/2 + \lambda_1(\|\boldsymbol{w}\|_1 - c) + \lambda_2(1 - \boldsymbol{w}^{\mathrm{T}}\mathbf{1}).$$

Let $\mathbf{g}$ be the *subgradient* vector of the function $\|\boldsymbol{w}\|_1$, whose ith element is -1, 1 or any values in $[-1, 1]$ depending on whether w_i is positive, negative or zero, respectively. Then, the first order condition is

$$\hat{\boldsymbol{\Sigma}}\boldsymbol{w} + \lambda_1\mathbf{g} - \lambda_2\mathbf{1} = 0, \tag{7.46}$$

$$\lambda_1(c - \|\boldsymbol{w}\|_1) = 0, \qquad \lambda_1 \geqslant 0, \tag{7.47}$$

in addition to the constraints $\boldsymbol{w}^{\mathrm{T}}\mathbf{1} = 1$ and $\|\boldsymbol{w}\|_1 \leqslant c$. Let $\tilde{\boldsymbol{w}}$ be the solution to (7.46) and (7.47). We will show in Section 7.6 that it is also the solution to the unconstrained portfolio optimization problem

$$\min_{\boldsymbol{w}^{\mathrm{T}}\mathbf{1}=1} \boldsymbol{w}^{\mathrm{T}}\tilde{\boldsymbol{\Sigma}}_c\boldsymbol{w}, \tag{7.48}$$

in which, with $\tilde{\mathbf{g}}$ being the gradient evaluated at $\tilde{\boldsymbol{w}}$,

$$\tilde{\boldsymbol{\Sigma}}_c = \hat{\boldsymbol{\Sigma}} + \lambda_1(\tilde{\mathbf{g}}\mathbf{1}^{\mathrm{T}} + \mathbf{1}\tilde{\mathbf{g}}^{\mathrm{T}}). \tag{7.49}$$

This is in a similar spirit to Jagannathan and Ma (2003) and DeMiguel et al. (2008). The covariance matrix (7.49) is a regularized covariance estimator. In other words, the gross-exposure constrained portfolio optimization (7.45) is equivalent to the global minimum variance portfolio (7.48) using the covariance matrix $\tilde{\boldsymbol{\Sigma}}_c$.

7.4 Portfolio selection and tracking

The gross-exposure constraint can also be applied to the problem of portfolio selection and portfolio tracking. It limits the gross exposure of the solutions.

7.4.1 Relation with regression

Risk minimization is innately related to multiple regression. Letting $Y = R_p$ and $X_j = R_p - R_j$, we have

$$\begin{aligned}\mathrm{var}(\boldsymbol{w}^{\mathrm{T}}\boldsymbol{R}) &= \min_b E(\boldsymbol{w}^{\mathrm{T}}\boldsymbol{R} - b)^2 \\ &= \min_b E(Y - w_1X_1 - \cdots - w_{p-1}X_{p-1} - b)^2\end{aligned} \tag{7.50}$$

Therefore, the portfolio risk optimization is equivalent to the regression problem, minimizing $E(Y - w_1X_1 - \cdots - w_{p-1}X_{p-1} - b)^2$ with respect to $b, w_1, \ldots, w_{p-1}$. In this case, the portfolio weights become the regression coefficients $\boldsymbol{w}^* = (w_1, \ldots, w_{p-1})^{\mathrm{T}}$. Once it is found, $w_p = 1 - \mathbf{1}^{\mathrm{T}}\boldsymbol{w}^*$.

The gross exposure of the portfolio is

$$\|\boldsymbol{w}\|_1 = \|\boldsymbol{w}^*\|_1 + |1 - \mathbf{1}^{\mathrm{T}}\boldsymbol{w}^*|.$$

The constraint $\|\boldsymbol{w}\|_1 \leqslant c$ is not the same as the existence of a constant d such that $\|\boldsymbol{w}^*\|_1 \leqslant d$. For example, when $d = 0$, only one stock R_p is allowed, yet $c = 1$ picks multiple stocks.

The above usage creates asymmetry among stocks. A more typical use is to let $Y = \boldsymbol{w}_0^{\mathrm{T}}\boldsymbol{R}$, for a given portfolio with $\boldsymbol{w}_0^{\mathrm{T}}\mathbf{1} = 1$. This can be the equally weighted portfolio or an index based on those stocks or the optimal no-short portfolio. Then, we can also write

$$\operatorname{var}(\boldsymbol{w}^{\mathrm{T}}\boldsymbol{R}) = \min_b E(Y - \boldsymbol{w}^{\mathrm{T}}\boldsymbol{X} - b)^2 \tag{7.51}$$

where $\boldsymbol{X} = \boldsymbol{w}_0^{\mathrm{T}}\boldsymbol{R}\mathbf{1} - \boldsymbol{R}$. The problem becomes the constrained least-squares problem: minimizing (7.51) with respect to $\boldsymbol{w}$ subject to $\boldsymbol{w}^{\mathrm{T}}\mathbf{1} = 1$. The constraint $\boldsymbol{w}^{\mathrm{T}}\mathbf{1} = 1$ suggests that one needs to drop one of the X-variables. The one that is least correlated with Y is a natural candidate to drop. Let us say that it is the pth variable.

Let $\boldsymbol{X}^\star = (X_1, \ldots, X_{p-1})$. The constrained least-squares problem

$$\min_{b,\boldsymbol{w}^\star} E(Y - \boldsymbol{w}^{\star\mathrm{T}}\boldsymbol{X}^\star - b)^2, \qquad \text{s.t. } \|\boldsymbol{w}^*\|_1 \leqslant d, \tag{7.52}$$

namely, LASSO, can be solved rapidly by using the *Least-Angle Regression* (LARS, Efron, et al., 2003). The algorithm allows one to quickly solve the optimization problem for all $d \geqslant 0$. The results, as a function of d, are called a *solution path*: $\{\boldsymbol{w}^\star(d) : d \geqslant 0\}$. For example, $d = 0$ selects no X variable and a very small d should select the X variable that is most correlated with Y. In general, for given d, the algorithm selects basically the optimal subset to predict Y within the constraint.

LARS also provides an approximate solution to the optimization problem (7.45) with gross exposure constraint. Take Y to be the optimal no-short sale portfolio; use LARS to obtain $\{\boldsymbol{w}^\star(d) : d \geqslant 0\}$; compute the gross exposure of the portfolio $\{Y, -\boldsymbol{w}^\star(d)^{\mathrm{T}}\boldsymbol{X}^\star\}$ in terms of original assets $\{R_j\}_{j=1}^p$. Let the resulting gross exposure be $c(d)$. Regard the portfolio $\{Y, -\boldsymbol{w}^\star(d)^{\mathrm{T}}\boldsymbol{X}^\star\}$ as the solution to (7.45) with the gross exposure constraint $c(d)$. Fan, Zhang and Yu (2012) show that such a method turns out to work very well and produces a very close solution to the optimization problem (7.45).

7.4.2 Portfolio selection and tracking

If Y is the target portfolio to be tracked or replicated, and $\{X_j\}_{j=1}^p$ are candidate assets, the penalized least-squares (7.52) can be regarded as finding a portfolio to minimize the expected tracking error. Let the solution be $\boldsymbol{w}(d)$. The selected portfolio allocates $\boldsymbol{w}(d)$ on $\{X_j\}_{j=1}^p$ and the rest on a riskless bond. As d increases, the number of selected assets increases and the tracking error decreases. The gross exposure d is always under control.

The penalized least-squares (7.52) can also be regarded as an effort to improve the efficiency of the existing portfolio Y. Given a portfolio Y constructed from the financial assets $\{X_j\}_{j=1}^p$, is it efficient? Can it be improved? To answer the question, run LASSO with variables $\{X_j\}_{j=1}^p$ and Y. The penalized least-squares can be interpreted as modifying weights to improve the performance of Y in terms of portfolio risks. The resulting portfolio is now $\{Y, -\boldsymbol{w}^*(d)^T\boldsymbol{X}^*\}$. If the portfolio Y is efficient, the regression coefficients $\boldsymbol{w}^*(d)$ should be small for a range of value d. The perceived risk path $\hat{R}(d)$ helps decision making.

We illustrate the above idea by the following example.

Example 7.11 Let Y be the returns of the CRSP index and $\boldsymbol{X}$ be ten industrial portfolios from the Data Library of Kenneth French. They are "Consumer Non-durables", "Consumer durables", "Manufacturing", "Energy", "Business equipment", "Telecommunications", "Shops", "Health", "Utilities", and "Others", labeled 1 to 10 in the left panel of Figure 7.9. Suppose that our goal is to improve the risk of CRSP by using those 10 industrial portfolios on January 8, 2005. This is of course a difficult task. Thus, we use one-year daily returns of those 11 portfolios before the date and run LASSO (7.52) for various d. The result is summarized on in Figure 7.9.

When $d = 0$, only the CRSP index Y is allowed. As d increases, it first recruits portfolio 1 ("Consumer Non-durables") to reduce the perceived risk (blue curve on the right panel of Figure 7.9). Further increase of d recruits portfolio 9 to the pool. At any given value of d, the non-vanishing weights of those 11 portfolios are given and their sum must be one (see the vertical line in the left panel). Ex-ante risks (perceived risks) of these portfolios are presented in the right panel of Figure 7.9.

To evaluate the performance of these portfolios, they are held for one year. Their actual risks or ex-post risks are computed based on their daily returns from Jan. $9, 2005$ to Jan. $8, 2006$. The results are presented also in the right panel of Figure 7.9 (the red curve). It is interesting to note that ex-ante and ex-post have the same decreasing pattern until 6 stocks are added. Using

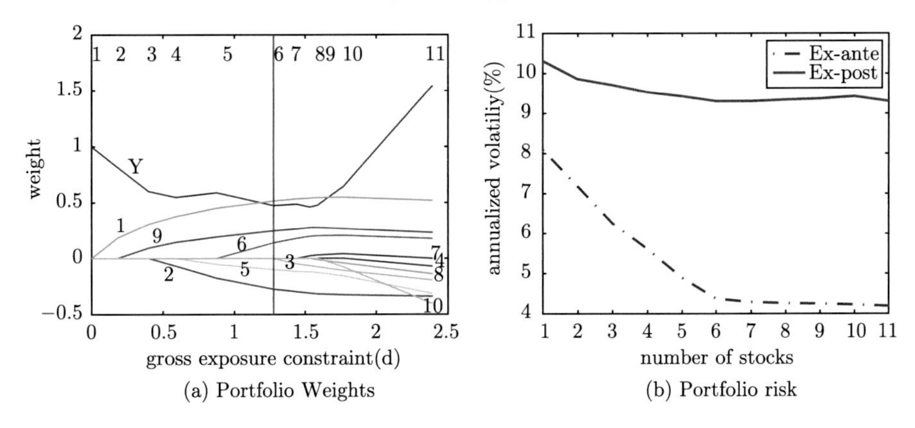

Figure 7.9 Illustration of risk improvement by using the penalized least-squares. (a) The solution paths for (7.52) as a function of d. The number of assets recruited are shown at the top of graph for each given d. (b) The ex-ante and ex-post risks (annualized volatility) of the selected portfolios. Adapted from Fan, Zhang and Yu (2012).

ex-ante as a guide, one would have selected 4–6 portfolios and the risk is improved (smaller than the CRSP index).

7.5 Empirical applications

We now illustrate the effectiveness of the methods by using two data sets:

- 600 stocks randomly selected from the 1000 stocks with least missing data in Russell 3000 during 2003–2007;
- 100 Fama Portfolios from January 1998 to December 2007.

To evaluate the performance, the portfolio is always optimized monthly and returns of the selected portfolios are recorded daily. The results are taken from Fan, Zhang and Yu (2012) and we do not further acknowledge.

7.5.1 Fama–French 100 portfolios

The data of the Fama–French 100 portfolios formed by size and book-to-market ratios (10×10) are downloaded from the Data Library from Kenneth French. Their daily returns over the 10-year period from 1998–2007 are used.

Three methods are used to estimate the covariance matrix that is needed for risk optimization: The sample covariance matrix based on past 252 days, the strict factor model using last one-year daily data, and RiskMetrics ($\lambda = 0.97$). The testing period is the 9 years since January 1999 and evaluation is conducted on a rolling basis. The results are summarized in Figure 7.10 and Table 7.4.

The optimal exposure c is around 2, independent of the method used to estimate the covariance matrix. Interestingly, the expected returns and Sharpe ratios peak around there too. This must be a coincidence, since our method merely optimizes the portfolio risk. The average number of assets in the optimal non-short sale portfolio is around 6 (see Table 7.4). It is not diversified enough. As a result, their risks can be improved as the gross exposure increases. Interestingly, the optimal no-short sale portfolio has a lower risk than the equally weighted portfolio, even though the former is not diversified enough. This is consistent with the theory. After all, the equally weighted portfolio is one of the no-short portfolios and should be outperformed by the optimal no-short portfolio.

RiskMetrics performs the best due to shorter time-window used to estimate volatility, which induces less biases. Since the individual assets are portfolios themselves, they are less volatile and a smaller amount of time-domain smoothing yields a better estimated volatility matrix. On the other hand, the factor model performs the worst due to modeling biases: we cannot expect the strict factor model to hold approximately due to the constructions of the portfolios. But it results in the most stable estimation of covariance matrices. Indeed, the factor model provides the most diversified portfolios, since the maximum allocations on individual portfolios have the smallest weights (see lower left panel of Figure 7.10).

The global minimum portfolios, namely, the optimal portfolios without gross exposure constraint, exist when the covariance matrix is estimated by the sample covariance matrix or by the strict factor model. Hence, the risks of these portfolios can also be computed. This corresponds to the rows with $c = \infty$ in Table 7.4. The result shows that even though the global minimum portfolios exist in our study, they are outperformed by the portfolio with gross exposure constraints due to the error accumulation in estimating large covariance matrices. Imposing gross exposure constraints regularizes the estimated covariance matrix.

Finally, we would like to note that when the RiskMetrics is used, since the effective number of data points is only about 3 months, the global minimum variance portfolio cannot be stably computed.

7.5.2 Russell 3000 stocks

We now apply the portfolio optimization technique under the gross-exposure constraint to the universe of 600 individual stocks from Russell 3000, de-

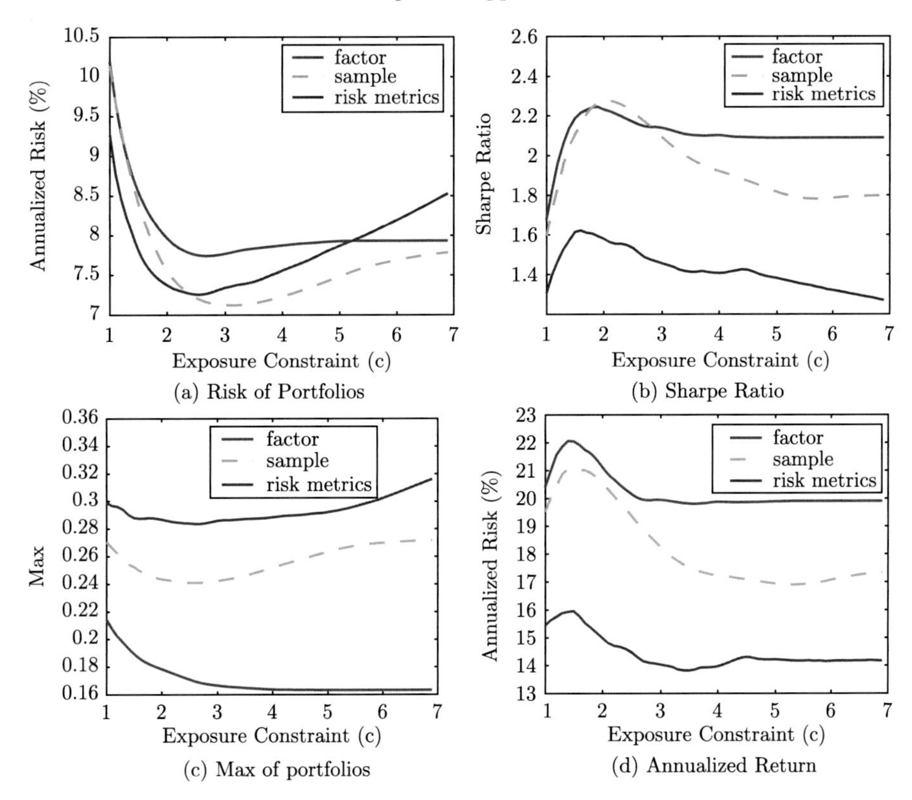

(a) Risk of Portfolios

(b) Sharpe Ratio

(c) Max of portfolios

(d) Annualized Return

Figure 7.10 Characteristics of invested portfolios as a function of gross exposure Characteristics of invested portfolios as a function of exposure constraint c from the Fama–French 100 industrial portfolios. (a) Annualized risk of portfolios. (b) Sharpe ratio of portfolios. (c) Max weight of allocations. (d) Annualized return of portfolios.

scribed in Example 7.3. Since the individual stocks are more volatile than the portfolios studied in the last section, we increase the window size from one year to two years for estimating the covariance matrix. This affects the sample covariance matrix and the factor-model based estimator, but does not affect the RiskMetrics ($\lambda = 0.97$).

The results were shown in Figure 7.2. Table 7.5 furnishes additional details. Again, the optimal no-short portfolio is not diversified enough, picking only 53 stocks on average. Its risk can be improved by increasing the value of the gross exposure, and the optimal one is obtained also around $c = 2$. Again, the factor-model provides the most stable estimates. The RiskMetrics, on the other hand, uses a too small time window to estimate the volatility, resulting in large variance in the estimation. Hence, it performs the worst among the three methods.

Table 7.4 *Ex-post risks and other characteristics of constrained optimal portfolios for 100 Fama–French portfolios*

Methods	Mean	Std	Sharpe-R	Max-W	Min-W	Long	Short
Sample Covariance Matrix Estimator							
c = 1	19.51	10.14	1.60	0.27	−0.00	6	0
c = 1.5	21.04	8.41	2.11	0.25	−0.07	9	6
c = 2	20.55	7.56	2.28	0.24	−0.09	15	12
c = 3	18.26	7.13	2.09	0.24	−0.11	27	25
$c = \infty$	17.55	7.82	1.82	0.66	−0.32	52	48
Factor-Based Covariance Matrix Estimator							
c = 1	20.40	10.19	1.67	0.21	−0.00	7	0
c = 1.5	22.05	8.56	2.19	0.19	−0.05	11	8
c = 2	21.11	7.96	2.23	0.18	−0.05	17	18
c = 3	19.95	7.77	2.14	0.17	−0.05	35	41
$c = \infty$	19.90	7.93	2.09	0.43	−0.14	45	55
Covariance Estimation from Risk Metrics							
c = 1	15.45	9.27	1.31	0.30	−0.00	6	0
c = 1.5	15.96	7.81	1.61	0.29	−0.07	9	5
c = 2	14.99	7.38	1.58	0.29	−0.10	13	9
c = 3	14.03	7.34	1.46	0.29	−0.13	21	18
Unmanaged Index							
Equal-W	10.86	16.33	0.46	0.01	0.01	100	0
CRSP	8.2	17.9	0.26				

Since $T = 504$ and $p = 600$, the global minimum variance portfolio does not exist when the covariance matrix is estimated by the sample covariance matrix or RiskMetrics. They behave like random portfolios. We can use $c = 8$ as a proxy of the global optimal portfolio. Due to noise accumulation, it performs the worst. On the other hand, when the factor model is used, the estimated covariance matrix is well regularized. Hence, the global minimum variance portfolio (corresponding to $c = \infty$) exists. It is not optimal in the performance possibly due to the bias induced by the factor model.

7.6 Complements

We now furnish the proofs of Theorems 7.2, 7.3, and (7.48). The derivations of these results are simple, but provides mathematical insights to those conclusions.

Table 7.5 *Ex-post risks and other characteristics of constrained optimal portfolios*

Methods	Std	Max-W	Min-W	Long	Short
Sample Covariance Matrix Estimator					
No short	9.28	0.14	0.00	53	0
c = 2	8.20	0.11	−0.06	123	67
c = 3	8.43	0.09	−0.07	169	117
c = 4	8.94	0.10	−0.08	201	154
c = 5	9.66	0.12	−0.10	225	181
c = 6	10.51	0.13	−0.10	242	201
c = 7	11.34	0.14	−0.11	255	219
c = 8	12.20	0.17	−0.12	267	235
Factor-Based Covariance Matrix Estimator					
No short	9.08	0.12	0.00	54	0
c = 2	8.31	0.06	−0.03	188	120
c = 3	8.65	0.05	−0.03	314	272
c = 4	8.66	0.05	−0.03	315	273
c = 5	8.66	0.05	−0.03	315	273
c = 6	8.66	0.05	−0.03	315	273
c = 7	8.66	0.05	−0.03	315	273
c = 8	8.66	0.05	−0.03	315	273
Covariance Estimation from Risk Metrics					
No short	9.78	0.40	0.00	31	0
c = 2	8.44	0.12	−0.06	119	63
c = 3	8.95	0.11	−0.07	191	133
c = 4	9.43	0.12	−0.09	246	192
c = 5	10.04	0.12	−0.10	279	233
c = 6	10.53	0.12	−0.11	300	258
c = 7	10.92	0.13	−0.11	311	272
c = 8	11.06	0.13	−0.10	315	277

7.6.1 Proof of Theorem 7.2

First of all, $R(\hat{\boldsymbol{w}}_{\text{opt}}) - R(\boldsymbol{w}_{\text{opt}}) \geqslant 0$, since $\boldsymbol{w}_{\text{opt}}$ minimizes the function R. Similarly, we have $\hat{R}(\hat{\boldsymbol{w}}_{\text{opt}}) - \hat{R}(\boldsymbol{w}_{\text{opt}}) \leqslant 0$. Using this, we have

$$\begin{aligned} & R(\hat{\boldsymbol{w}}_{\text{opt}}) - R(\boldsymbol{w}_{\text{opt}}) \\ &= R(\hat{\boldsymbol{w}}_{\text{opt}}) - \hat{R}(\hat{\boldsymbol{w}}_{\text{opt}}) + \hat{R}(\hat{\boldsymbol{w}}_{\text{opt}}) - \hat{R}(\boldsymbol{w}_{\text{opt}}) + \hat{R}(\boldsymbol{w}_{\text{opt}}) - R(\boldsymbol{w}_{\text{opt}}) \\ &\leqslant R(\hat{\boldsymbol{w}}_{\text{opt}}) - \hat{R}(\hat{\boldsymbol{w}}_{\text{opt}}) + \hat{R}(\boldsymbol{w}_{\text{opt}}) - R(\boldsymbol{w}_{\text{opt}}) \end{aligned}$$

Since both $\|\boldsymbol{w}_{\text{opt}}\|_1 \leqslant c$ and $\|\hat{\boldsymbol{w}}_{\text{opt}}\|_1 \leqslant c$, the above two terms are no larger than the maximum deviation. In other words,

$$R(\hat{\boldsymbol{w}}_{\text{opt}}) - R(\boldsymbol{w}_{\text{opt}}) \leqslant 2 \sup_{\|\boldsymbol{w}\|_1 \leqslant c} |\hat{R}(\boldsymbol{w}) - R(\boldsymbol{w})|.$$

By Theorem 7.1, we conclude that

$$R(\hat{\boldsymbol{w}}_{\text{opt}}) - R(\boldsymbol{w}_{\text{opt}}) \leqslant 2e_{\max}c^2.$$

This proves the first conclusion of Theorem 7.2. The second inequality therein follows directly from Theorem 7.1.

To prove the third inequality, by using $R(\boldsymbol{w}_{\text{opt}}) - R(\hat{\boldsymbol{w}}_{\text{opt}}) \leqslant 0$, we have that

$$\begin{aligned} R(\boldsymbol{w}_{\text{opt}}) - \hat{R}(\hat{\boldsymbol{w}}_{\text{opt}}) &= R(\boldsymbol{w}_{\text{opt}}) - R(\hat{\boldsymbol{w}}_{\text{opt}}) + R(\hat{\boldsymbol{w}}_{\text{opt}}) - \hat{R}(\hat{\boldsymbol{w}}_{\text{opt}}) \\ &\leqslant R(\hat{\boldsymbol{w}}_{\text{opt}}) - \hat{R}(\hat{\boldsymbol{w}}_{\text{opt}}) \\ &\leqslant e_{\max}c^2, \end{aligned}$$

where the last inequality follows from Theorem 7.1. Similarly, it follows from $\hat{R}(\boldsymbol{w}_{\text{opt}}) - \hat{R}(\hat{\boldsymbol{w}}_{\text{opt}}) \geqslant 0$ that

$$\begin{aligned} R(\boldsymbol{w}_{\text{opt}}) - \hat{R}(\hat{\boldsymbol{w}}_{\text{opt}}) &= R(\boldsymbol{w}_{\text{opt}}) - \hat{R}(\boldsymbol{w}_{\text{opt}}) + \hat{R}(\boldsymbol{w}_{\text{opt}}) - \hat{R}(\hat{\boldsymbol{w}}_{\text{opt}}) \\ &\geqslant R(\boldsymbol{w}_{\text{opt}}) - \hat{R}(\boldsymbol{w}_{\text{opt}}) \\ &\geqslant -e_{\max}c^2 \end{aligned}$$

Combining the last two results, the third inequality follows.

7.6.2 Proof of Theorem 7.3

Note that by the union bound of probability, we have for any $D > 0$,

$$P\{\sqrt{n}\|\boldsymbol{\Sigma} - \hat{\boldsymbol{\Sigma}}\|_\infty > D(\log p)^a\} \leqslant p^2 \max_{i,j} P\{\sqrt{n}|\sigma_{ij} - \hat{\sigma}_{ij}| > D(\log p)^a\}.$$

By the assumption, the above probability is bounded by

$$p^2 \exp\left(-C[D(\log p)^a]^{1/a}\right) = p^2 p^{-CD^{1/a}},$$

which tends to zero when D is large enough. This completes the proof of the theorem.

7.6.3 Proof of (7.48)

First of all, note that the solution to problem (7.48) is given by

$$\boldsymbol{w}_{\text{opt}} = \tilde{\boldsymbol{\Sigma}}_c^{-1}\mathbf{1}/\mathbf{1}^{\mathrm{T}}\tilde{\boldsymbol{\Sigma}}_c^{-1}\mathbf{1}.$$

By the definition of $\tilde{\boldsymbol{\Sigma}}_c$ and $\tilde{\boldsymbol{w}}^{\mathrm{T}}\mathbf{1} = 1$, we have

$$\begin{aligned} \tilde{\boldsymbol{\Sigma}}_c\tilde{\boldsymbol{w}} &= \hat{\boldsymbol{\Sigma}}\tilde{\boldsymbol{w}} + \lambda_1\tilde{\mathbf{g}} + \lambda_1\tilde{\mathbf{g}}^{\mathrm{T}}\tilde{\boldsymbol{w}}\mathbf{1}. \\ &= \lambda_2\mathbf{1} + \lambda_1\tilde{\mathbf{g}}^{\mathrm{T}}\tilde{\boldsymbol{w}}\mathbf{1}, \end{aligned}$$

in which the last equality utilizes (7.46). Noting that $\tilde{\mathbf{g}}^T\tilde{\boldsymbol{w}} = \|\tilde{\boldsymbol{w}}\|_1$ and using (7.47), we conclude that

$$\tilde{\boldsymbol{\Sigma}}_c\tilde{\boldsymbol{w}} = (\lambda_2 + \lambda_1 c)\mathbf{1}.$$

Thus, $\tilde{\boldsymbol{w}} = (\lambda_2+\lambda_1 c)\tilde{\boldsymbol{\Sigma}}_c^{-1}\mathbf{1}$, which has the same direction $\boldsymbol{w}_{\text{opt}}$. Since $\mathbf{1}^T\tilde{\boldsymbol{w}} = 1$, they must be equal. This completes the proof.

7.7 Exercises

7.1 What is the gross exposure of the portfolio with weights

$$\boldsymbol{w} = (-0.2, 0.3, 0.4, -0.2, 0.1, 0.2, 0, 0.4)?$$

What is the risk of this portfolio invested on "Dell", "Ford", "GE", "IBM", "Johnson & Johnson", "Merck", "3-month Treasury Bill", "S&P 500 index" in the past ten years (January 1, 2005 to January 1, 2015, using daily data). Compare it with the portfolio with equal weight.

7.2 Let $X_1, \ldots, X_T$ be a sequence of stationary time series with the autocovariance function $\gamma(h) = \text{cov}(X_t, X_{t+h})$ and $\bar{X} = T^{-1}\sum_{t=1}^T X_t$. Show that

$$\text{var}(\bar{X}) = T^{-2}[T\gamma(0) + 2(T-1)\gamma(1) + \cdots + 2\gamma(T-1)],$$

and

$$\lim_{T\to\infty}[T\text{var}(\bar{X})] = \gamma(0) + 2\sum_{h=1}^{\infty}\gamma(h).$$

In other words,

$$\text{var}(\bar{X}) \approx T^{-1}[\gamma(0) + 2\sum_{h=1}^{L}\gamma(h)]$$

for a sufficient large integer L.

7.3 Draw 10,000 random portfolios of size $p = 100$ with gross exposure $c = 1.6$ from (7.9). Plot the weights (w_1, w_2) and (w_{99}, w_{100}).

7.4 Estimate the time-varying volatility matrix between the returns of S&P 500 and the percentage changes of the VIX over the last 1 year using the exponential smoothing (7.12) with $\lambda = 0.94$ and initial value $\hat{\boldsymbol{\Sigma}}_0 = 0$ and respectively $\hat{\boldsymbol{\Sigma}}_0 =$ the sample covariance matrix. Present the results as in Figure 7.3 and compare them.

7.5 Estimate the time-varying volatility matrix of the eight risk factors presented in Example 7.5 using the last 3-month data and the estimator (7.15) with $\lambda = 0.1$ (No estimate is needed for the initial three months). Present the results for the exchange rates and the S&P 500 index, similar to Figure 7.3. Repeat the exercise by using the second projection method in Section 7.2.2 and compare the results.

7.6 Prove (7.12).

7.7 Show that the solution to (7.18) is given by (7.17). As a further generalization, what is the solution to the following problem: Minimizing with respect the symmetric matrix $\boldsymbol{S}$

$$\|\hat{\boldsymbol{\Sigma}}_\lambda - \boldsymbol{S}\|_F^2, \quad \text{s.t.} \quad \lambda_{\min}(\boldsymbol{S}) \geqslant \delta$$

for a given $\delta \geqslant 0$. Namely, we wish to find covariance matrix $\boldsymbol{S}$ with minimum eigenvalue no smaller than δ such that it is closest to $\hat{\boldsymbol{\Sigma}}_\lambda$. When $\delta = 0$, it reduces to problem (7.18).

7.8 Show that $\text{tr}[\hat{\boldsymbol{\Sigma}}\boldsymbol{\Sigma}^{-1} - \boldsymbol{I}_p]^2 = \|\boldsymbol{\Sigma}^{-1/2}\hat{\boldsymbol{\Sigma}}\boldsymbol{\Sigma}^{-1/2} - \boldsymbol{I}_p\|_F^2$.

7.9 For any symmetric matrix $\boldsymbol{\Sigma}$, its operator norm is bounded by its L_1-norm: $\|\boldsymbol{\Sigma}\| \leqslant \max_i \sum_{j=1}^p |\sigma_{ij}|$. Use this to show that if $\sigma_{ii} \leqslant C$ for all i, then

$$\|\boldsymbol{\Sigma}\| \leqslant C^{1-q} \max_i \sum_{j=0}^p |\sigma_{ij}|^q,$$

for all $q \in [0, 1]$. The last factor is a *generalized sparsity measure* with $q = 0$ being the maximum number of non-vanishing components per column.

7.10 Let $\hat{\boldsymbol{\Sigma}}$ be an estimated volatility matrix of true volatility $\boldsymbol{\Sigma}$. Show that for any portfolio allocation $\boldsymbol{w}$, the relative estimation error is bounded by

$$\left| \frac{\boldsymbol{w}^T \hat{\boldsymbol{\Sigma}} \boldsymbol{w}}{\boldsymbol{w}^T \boldsymbol{\Sigma} \boldsymbol{w}} - 1 \right| \leqslant \|\boldsymbol{\Sigma}^{-1/2}\hat{\boldsymbol{\Sigma}}\boldsymbol{\Sigma}^{-1/2} - \boldsymbol{I}_p\|.$$

7.11 Suppose that we have 100 investable stocks, labeled as 1 through 100 and classified as “Consumer Non-durables”, “Consumer durables”, “Manufacturing”, “Energy”, “Business equipment”, “Telecommunications”, “Shops”, ”Health”, “Utilities”, and “Others”. Let $w_1, \ldots, w_{100}$ be the portfolio weights. If the first 10 stocks are labeled as “Consumer Non-durables”, the second 10 stocks are in “Consumer durables” and so on, write down the constraints of the portfolios:

1. the “health stocks” are no more than 15% and “energy stocks” are no more than 30%;
2. no exposure to “Telecommunications”;
3. exposure to “Consumer durables”, but gross exposure to “Consumer durables” is zero.

7.12 Let the study period be January 2001 to January 2015. Apply the sample covariance matrix, the Fama–French 3 factor model, and the RiskMetrics with $\lambda = 0.94$ to obtain the time-varying covariance matrix for ‘Dell”, “Ford”, “GE”, “IBM”, “Intel”, “Johnson & Johnson”, “Merck”, “3-month Treasury Bill” and“S&P 500 index” at the beginning of each month (defined as every 21 days after the initial 252 days). Optimize the portfolio and holds for the next 21 days. Compute the risk of such a portfolio and compare it with the equally weighted portfolio.

8

Consumption based CAPM

The capital asset pricing modeland its extension, the multifactor pricing model, assume a myopic behavior of investors, who optimize the portfolio value at the next period only. In spite of their popularity in practice, the models determine asset prices by the portfolio choices at one period, making investors aim at a moving target. This is a serious drawback.

This chapter relates asset prices to the consumption, investment and saving decisions of investors. In contrast with pricing financial derivatives (Hull, 2014; Karatzas and Shreve, 1998; Mikosch, 1998); Yan, 2014), which derives prices from those of the underlying assets without answering the question of where the underlying asset prices come from, this chapter briefly touches on absolute pricing, answering the question of what key factors determine the prices of underlying assets, from a financial economics viewpoint. More thorough discussions on this kind of absolute asset pricing can be found in Cochrane (2005). This together with relative pricing on financial derivatives provide a comprehensive understanding on the returns of financial assets.

8.1 Utility optimization

Suppose that an investor consumes a representative good with quantity C_t at time t with price p_t. A representative good refers to real *consumption* of good and services, including food, clothing, gasoline, automobiles, electronics, telecommunications, child care, laundry and other services. The price p_t refers to the weighted average of the aggregated prices. Suppose further that the investor has an external income of I_t, which can be a salary among others, and $\boldsymbol{\alpha}_t$ shares on stocks with prices $\boldsymbol{S}_t$. Here, $\boldsymbol{S}_t$ is a vector of prices of all tradable assets at time t and $\boldsymbol{\alpha}_t$ is their associated allocation vector. Then his total budget at time t is the sum of the sale of financial stocks and external incomes. This imposes the budget constraint on consumption:

$$p_t C_t = I_t + (\boldsymbol{\alpha}_{t-1} - \boldsymbol{\alpha}_t)^{\mathrm{T}} \boldsymbol{S}_t. \tag{8.1}$$

Note that this formulation allows savings. When the second term of (8.1) is negative, it actually saves some external income and invests it in the financial market.

Let δ be the subjective *discount factor* of future consumption, and $U(\cdot)$ be the utility function of consumption. The *intertemporal choice problem* assumes that individuals' consumption, saving and investment objective is to maximize the discounted expected utility

$$E_t \left(\sum_{j=0}^{\infty} \delta^j U(C_{t+j}) \right), \tag{8.2}$$

with respect to the allocation vectors $\{\boldsymbol{\alpha}_t\}$, subject to the budget constraint (8.1). In (8.2), E_t is the expectation given the information up to time t, which varies from person to person. The uncertainty of the future includes the prices of goods and services, external incomes, and the prices of stocks. Substituting (8.1) into (8.2), we wish to maximize

$$\max E_t \Big[\sum_{j=0}^{\infty} \delta^j U \Big(\frac{I_{t+j} + (\boldsymbol{\alpha}_{t+j-1} - \boldsymbol{\alpha}_{t+j})^{\mathrm{T}} \boldsymbol{S}_{t+j}}{p_{t+j}} \Big) \Big]$$

with respect to portfolio allocation vectors $\{\boldsymbol{\alpha}_{t+j}\}_{j=0}^{\infty}$. The first-order condition can be obtained by taking derivatives with respect to $\{\boldsymbol{\alpha}_{t+j}\}$ and setting them to zero. Note that $\boldsymbol{\alpha}_{t+j}$ appears only in both $(j+1)$th and $(j+2)$th terms. By (8.2), the sum of these two terms is $\delta^j U(C_{t+j}) + \delta^{j+1} U(C_{t+j+1})$, whose partial derivative with respect to $\boldsymbol{\alpha}_{t+j}$ is

$$\delta^j E_t \Big[U'(C_{t+j}) \frac{\partial C_{t+j}}{\partial \boldsymbol{\alpha}_{t+j}} + \delta U'(C_{t+j+1}) \frac{\partial C_{t+j+1}}{\partial \boldsymbol{\alpha}_{t+j}} \Big]. \tag{8.3}$$

By (8.1), it is easy to see that

$$\frac{\partial C_{t+j}}{\partial \boldsymbol{\alpha}_{t+j}} = -\frac{\boldsymbol{S}_{t+j}}{p_{t+j}}, \quad \frac{\partial C_{t+j+1}}{\partial \boldsymbol{\alpha}_{t+j}} = \frac{\boldsymbol{S}_{t+j+1}}{p_{t+j+1}}.$$

Using this and setting (8.3) to zero, we obtain the *optimal consumption* and *investment decision* should satisfy the following system of equations:

$$-E_t \left[U'(C_{t+j}) \cdot \frac{\boldsymbol{S}_{t+j}}{p_{t+j}} \right] + \delta E_t \left[U'(C_{t+1+j}) \frac{\boldsymbol{S}_{t+j+1}}{p_{t+j+1}} \right] = 0,$$

In particular, considering the case $j = 0$ leads to the following *Euler condition*:

$$\boldsymbol{S}_t = E_t[M_{t+1}\boldsymbol{S}_{t+1}], \tag{8.4}$$

where M_{t+1} is a *stochastic discount factor* given by

$$M_{t+1} = \frac{\delta U'(C_{t+1})p_t}{U'(C_t)p_{t+1}}. \tag{8.5}$$

Model (8.4) is the *stochastic discount pricing* model. It has many profound implications in asset pricing, as to be seen. For more comprehensive treatment, see Cochrane (2005).

Applying (8.5) to each tradable asset, the stochastic discount pricing model implies that the present price of any individual asset i is its next period price discounted by the same stochastic discount factor M_{t+1}, i.e.

$$S_{i,t} = E_t[M_{t+1}S_{i,t+1}]. \tag{8.6}$$

Dividing both sides by $S_{i,t}$ and using the fact that $S_{i,t}$ is known at time t, we obtain that the return of the ith asset satisfies

$$E_t[M_{t+1}(1 + R_{i,t+1})] = 1, \tag{8.7}$$

where $R_{i,t+1}$ is the simple return at period $t+1$. The price or return depends on the *inflation rate* p_{t+1}/p_t, which can sometimes be ignored when the time unit is small and the *intertemporal rate of substitution* $\delta U'(C_{t+1})/U'(C_t)$, which is the ratio of the discounted marginal utility of the consumption in the next period to that of the marginal utility of the consumption at present period.

If there exists a risk-free asset with return $r_{f,t+1}$ in the next period, then by (8.7), we have

$$E_t M_{t+1} = (1 + r_{f,t+1})^{-1}. \tag{8.8}$$

An extension of this is for any *"zero-beta"* asset $S_{0,t}$, which is uncorrelated with the stochastic discount factor $\text{cov}_t(S_{0,t+1}, M_{t+1}) = 0$, we have

$$S_{0,t} = E_t M_{t+1} S_{0,t+1} = (E_t M_{t+1})(E_t S_{0,t+1}),$$

where we use the fact that for any uncorrelated random variables $E(XY) = (EX)$
(EY). Dividing both sides of $S_{0,t}$, we obtain

$$E_t M_{t+1} = (1 + E_t R_{0,t+1})^{-1}, \tag{8.9}$$

a generalization of (8.8), where $R_{0,t+1}$ is the return of a zero-beta asset.

Remark* 8.1 The stochastic discount factor model has an interpretation in terms of the "*risk-neutral*" pricing. By (8.6) and (8.8), the price of the ith stock is

$$S_{i,t} = E_t \left[\frac{S_{i,t+1}}{1 + r_{f,t+1}} M_{t+1} \right] / E_t M_{t+1} = E_t^* \frac{S_{i,t+1}}{1 + r_{f,t+1}}. \tag{8.10}$$

where E_t^* is the expectation with the conditional probability density defined by

$$\frac{dP_t^*}{dP_t} = \frac{M_{t+1}}{E_t M_{t+1}}.$$

This follows from the change of variables in probability theory. Regarding probability measure P^* as the risk-neutral probability, (8.10) can be interpreted as that the current asset price is the expected discounted future price with respect to the risk-neutral probability. It shares a very similar interpretation to *derivative pricing* (Hull. 2014), which is the expected discounted payoff under the risk-neutral world.

8.2 Consumption-based CAPM

If assets and goods prices are identical for all individuals, an application of (8.4) to each individual yields

$$\boldsymbol{S}_t = E_{j,t}[M_{j,t+1} \boldsymbol{S}_{t+1}],$$

in which $E_{j,t}$ and $M_{j,t+1}$ are the expectation and stochastic discount factor of the jth individual. Note that individuals have different expectations and access to different information. They also have different time preference, utility and different income patterns. Therefore, the stochastic discount factors vary across individuals. It is hard to aggregate individual demands for financial assets and consumption of goods to derive the condition of the equilibrium of the aggregated demand and supply, as we did in Section 5.3.1.

8.2.1 CCAPM

One way to avoid the above mathematical difficulty is to assume that there is a *representative investor* who has rational expectation so that the Euler condition

$$\boldsymbol{S}_t = E_t[M_{t+1} \boldsymbol{S}_{t+1}]$$

can be applied at the aggregate level. This is the *consumption-based CAPM* (CCA-

PM). In other words,

$$\boldsymbol{S}_t = E_t\Big[\frac{p_t}{p_{t+1}}\delta\frac{U'(C_{t+1})}{U'(C_t)}\boldsymbol{S}_{t+1}\Big] \tag{8.11}$$

in which E_t, δ and U are respectively the expectation, discount factor, and utility of the representative investor, p_t is a retail price index of goods and services, and C_t is the aggregate consumption of physical goods and services. The model admits the same form as (8.4) but is interpreted at the aggregated level. For example, E_t is conditional expectation given the information collected by the representative agent before time t and the consumption is now referred to as the average consumption of all individuals.

Figure 8.1 presents the *consumer price index* and inflation rates from January 1881 to February 2011. The former can be regarded as p_t, while the latter inflation rate can be regarded as $\log(p_t/p_{t-1})$ for the monthly time series. The inflation rates after 1980 are less volatile, thanks to the changes in the monetary policy since October 6, 1979, when its newly appointed chairman, Paul Volcker, initiated money supply targeting and abandoned interest rate targeting. The *gross domestic product* (GDP) can be regarded as a proxy of the aggregated consumption C_t, which is also depicted in Figure 8.2.

Note that from the definition of the covariance, for any two random variables

$$EXY = (EX)(EY) + \mathrm{cov}(X, Y).$$

Using definition (8.5), by (8.11), we have

$$\boldsymbol{S}_t = (E_t M_{t+1})(E_t \boldsymbol{S}_{t+1}) + \mathrm{cov}_t(M_{t+1}, \boldsymbol{S}_{t+1}).$$

If there exists a risk-free asset with return $r_{f,t+1}$, then it follows from (8.8) that

$$\boldsymbol{S}_t = (1 + r_{f,t+1})^{-1} E_t \boldsymbol{S}_{t+1} + \mathrm{cov}_t(M_{t+1}, \boldsymbol{S}_{t+1}). \tag{8.12}$$

That is, the current price depends on the discounted *present value* and a *risk premium*. If an asset price evolves positively with consumption growth or more precisely the discounted factor, the asset price tends to be high. The more assets are demanded for intertemporal transfers, the higher their prices (recalling the inflation rate is negligible over a short term). Model (8.12) gives us insights into where the prices come from.

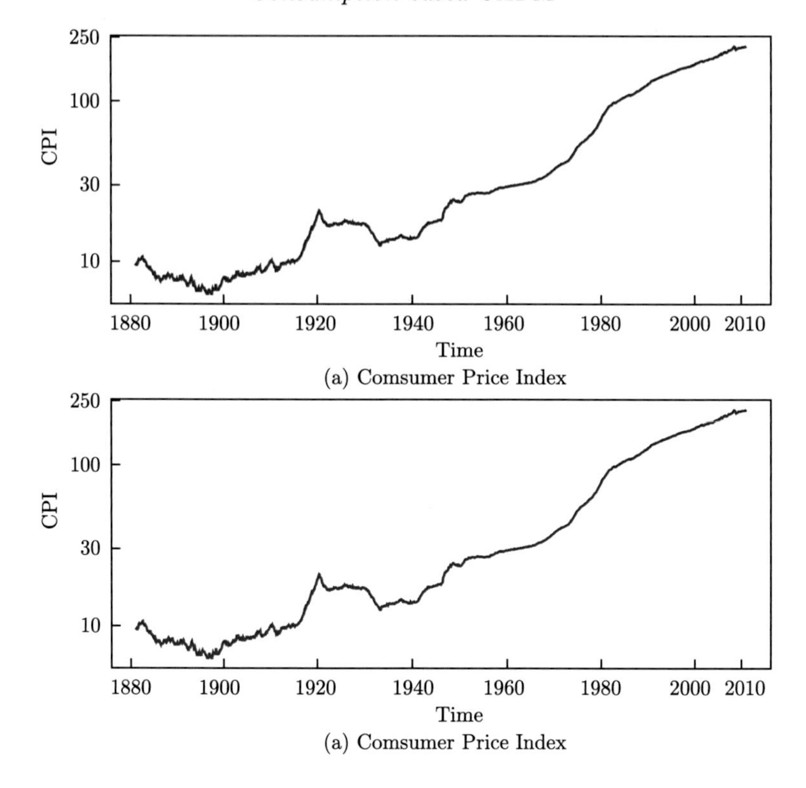

Figure 8.1 Consumer price index and inflation rate from 1881 to 2011 (monthly data).

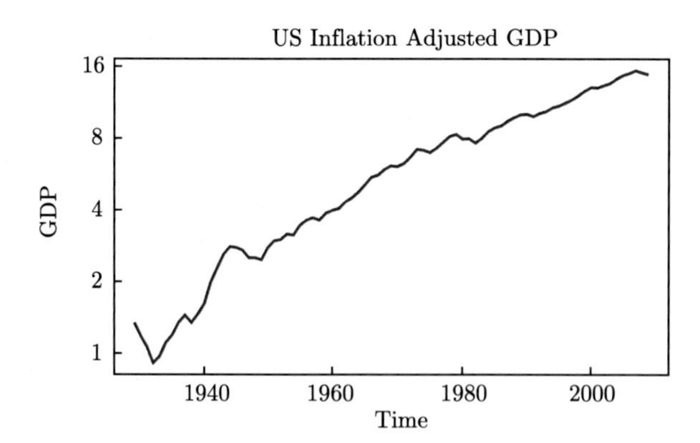

Figure 8.2 US inflation adjusted GDP from 1929 to 2009 in trillions (annual data in 2010 dollars).

8.2.2 Power utility

We will now further elaborate on how prices are determined by considering the *power utility* function. To facilitate the notation, we parameterize the power utility function as follows:

$$U(C) = \frac{C^{1-\gamma} - 1}{1 - \gamma}$$

where γ is the coefficient of *relative risk aversion*. This parametrization has the advantage that $U(C) = \log C$, as $\gamma \to 1$. For this utility function, $U'(C) = C^{-\gamma}$ and (8.11) reduces to

$$\boldsymbol{S}_t = E_t\left[\delta \frac{p_t}{p_{t+1}} \left(\frac{C_{t+1}}{C_t}\right)^{-\gamma} \boldsymbol{S}_{t+1}\right].$$

Let $\boldsymbol{R}_{t+1}$ be the vector of the inflation adjusted log-returns

$$\boldsymbol{R}_{t+1} = \log \boldsymbol{S}_{t+1}/\boldsymbol{S}_t - \log(p_{t+1}/p_t),$$

in which the vector division is interpreted as componentwise division. By dividing both sides by $\boldsymbol{S}_t$, we can write the pricing formula as

$$\delta E_t \exp(\boldsymbol{Y}_{t+1}) = 1, \tag{8.13}$$

where

$$\boldsymbol{Y}_{t+1} = \boldsymbol{R}_{t+1} - \gamma \Delta C_{t+1}, \tag{8.14}$$

and $\Delta C_{t+1} = \log C_{t+1}/C_t$ is the rate of the consumption growth.

To further simplify the above formula, we assume that $\boldsymbol{Y}_{t+1}$ has a normal distribution. Recall that

$$Ee^Y = \exp(\mu + \sigma^2/2) = \exp(EY + \mathrm{var}(Y)/2) \tag{8.15}$$

for $Y \sim \mathcal{N}(\mu, \sigma^2)$. Applying this formula to each component in (8.13), we obtain that

$$\delta \exp\{E_t Y_{i,t+1} + \mathrm{var}_t(Y_{i,t+1})/2\} = 1.$$

or equivalently

$$E_t Y_{i,t+1} = -\log \delta - \mathrm{var}_t(Y_{i,t+1})/2.$$

This together with the definition of $Y_{i,t+1}$ yields the *Hansen–Singleton formula* (Hansen and Singleton, 1983):

$$E_t R_{i,t+1} = -\log \delta + \gamma E_t \Delta C_{t+1} - \mathrm{var}_t(Y_{i,t+1})/2. \tag{8.16}$$

The above formula applies to all assets, including the risk-free asset. For the risk-free asset 0, by (8.14) with inflation rate ignored, we have

$$\mathrm{var}_t(Y_{0,t+1}) = \gamma^2 \mathrm{var}(\Delta C_{t+1}).$$

Hence, by (8.16), we have

$$r_{f,t+1} = -\log\delta + \gamma E_t \Delta C_{t+1} - \frac{\gamma^2}{2}\text{var}_t(\Delta C_{t+1}), \tag{8.17}$$

where $r_{f,t+1}$ is the inflation adjusted risk-free rate. It explains the factors that determine the risk-free rate: the time preference rate (log-discount factor) "$-\log\delta$", expected consumption growth (incentive to borrow for future consumption) and volatility of the growth (precautionary motive for saving).

Substracting (8.17) from (8.16), we obtain the expected excess return

$$\begin{aligned} E_t(R_{i,t+1} - r_{f,t+1}) &= -\frac{1}{2}\text{var}_t(Y_{i,t+1}) + \frac{1}{2}\text{var}_t(Y_{0,t+1}) \\ &= -\frac{1}{2}\text{var}_t(R_{i,t+1} - \gamma\Delta C_{t+1}) + \frac{1}{2}\text{var}_t(-\gamma\Delta C_{t+1}), \end{aligned}$$

after ignoring again the inflation. Expanding the first variance into variance and covariance, we obtain

$$E_t[R_{i,t+1} - r_{f,t+1}] = -\sigma_i^2/2 + \gamma\text{cov}_{ic}, \tag{8.18}$$

where $\sigma_i^2 = \text{var}_t(R_{i,t+1})$ and $\text{cov}_{ic} = \text{cov}(R_{i,t+1}, \Delta C_{t+1})$. It describes clearly where the returns come from. If an asset is positively correlated with the consumption growth, its expected return tends to be higher.

We conclude this section by the following illustrative example. It gives us an idea on the measurement of each quantity in the pricing formulas (8.17) and (8.18).

Example 8.1 We now illustrate the above pricing formula by using the following empirical data, from January, 1959 to December, 2010. The returns of the CRSP index are used as the returns of stocks. These, along with the risk-free interest rates, were obtained from the Data Library of Kenneth French's website. Consumption refers to that of nondurables and services. Thus, the consumption growth is the change in log real consumption of nondurables and services. The personal consumption expenditures were obtained from the Federal Research Bank of Saint Louis. The time series are summarized in Figure 8.3.

Based on the past 52 years' non-overlapping monthly data from December 1959 to December 2010, the summary of the four time series presented in Figure 8.3 is depicted in Table 8.1. The spike of the inflation adjusted risk-free rates corresponds to the change of the Federal Reserve's monetary policy on October 6, 1979.

First from Table 8.1, the expected excess returns $\log r_{E,t} = R_t - r_{f,t}$ per annum have mean 5.105% with a standard deviation 17.09%. If we assume

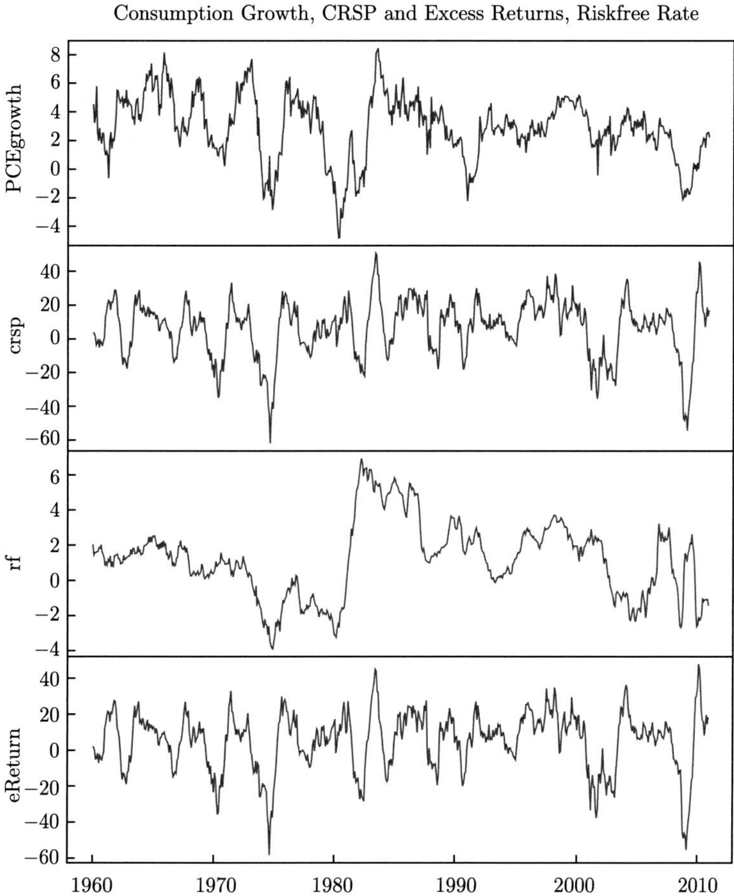

Figure 8.3 Consumption growth (top panel), inflation adjusted returns of the CRSP index (second panel), inflation adjusted risk-free interest rates (third panel), and the excess returns of the CRSP index (bottom) panel from January 1960 to December, 2010. The data are annualized and sampled at monthly frequency.

that they follow the normal distribution, i.e., $\log r_{E,t} \sim N(0.05105, 0.1709^2)$, then it follows from (8.15) that

$$Er_E = \exp(0.0511 + 0.1709^2/2) = 1.0598.$$

In other words, the real return has expected value about 5.98% with an SD 15.85% over this 52-year period.

Substituting empirical data into (8.18), we obtain

$$5.105 = -16.242^2/2 + 17.090\gamma,$$

yielding $\gamma = 8.541$.

Table 8.1 *Moments of consumption growth and asset returns*

Variable	Mean	Standard deviation	Corr with consumption growth	Covar with consumption growth
Consumption growth	2.925	2.343	1.000	5.489
Stock return	6.321	16.242	0.449	17.090
RF return	1.216	2.166	0.308	1.564
Stock − RF return	5.105	15.853	0.418	15.526

Similarly, substituting the above quantity into (8.17), we obtain

$$1.216 = -\log \delta + \gamma \cdot 2.925 - \frac{\gamma^2}{2} \cdot 2.343^2,$$

or $\log(\delta) = -154.17$.

Note that direct application of this does not yield stable estimates of the parameters γ and δ. The results are also very sensitive to the period of the study specified.

The above example is merely an illustration of the various quantities in the asset pricing model. Determining these parameters by the previous method is too crude to be useful. While the above theoretical models are insightful, they are too simple to be well fitted by the market data and consumption data. The misfit of the model gives rise to the *equity premium puzzle* of Mehra and Prescott (1985): the average excess return on the US stock market is too high to be easily explained by standard asset pricing model. This often refers to the fact that the calibrated risk aversion parameter γ is too large. In other studies, γ can be as large as 30. Also, in other fits, one can obtain a positive $\log(\delta)$, which gives the rise to the *risk-free rate puzzle* by Weil (1989): Given a positive average consumption growth, a low riskless interest rate and a positive rate of time preference ($\log \delta > 0$), such investors would have a strong desire to borrow from the future.

8.3 Mean-variance frontier*

We now show that the stochastic discount pricing model also implies CAPM. For a given portfolio with allocation vector $\boldsymbol{\alpha}_t$, its portfolio value is given by

$$W_t = \boldsymbol{\alpha}_t^{\mathrm{T}} \boldsymbol{S}_t.$$

This portfolio has value $W_{t+1} = \boldsymbol{\alpha}_t^{\mathrm{T}} \boldsymbol{S}_{t+1}$, with slight abuse of notation of W_{t+1}. From the stochastic discount model (8.4), we have

$$\begin{aligned} W_t &= \boldsymbol{\alpha}_t^{\mathrm{T}} E_t M_{t+1} \boldsymbol{S}_{t+1} \\ &= E_t M_{t+1} W_{t+1} \\ &= \mathrm{cov}_t(M_{t+1}, W_{t+1}) + (E_t M_{t+1})(E_t W_{t+1}). \end{aligned}$$

Recalling that $E_t M_{t+1} = (1 + r_{f,t+1})^{-1}$ from (8.8), we obtain

$$E_t W_{t+1} - (1 + r_{f,t+1}) W_t = \mathrm{cov}_t(M_{t+1}, W_{t+1}) / E_t M_{t+1}. \tag{8.19}$$

By the Cauchy-Schwartz inequality, we have

$$[E_t W_{t+1} - (1 + r_{f,t+1}) W_t]^2 \leqslant \mathrm{var}_t(M_{t+1}) \mathrm{var}(W_{t+1}) / (E_t M_{t+1})^2.$$

Hence, by dividing both sides by $\mathrm{var}(W_{t+1})$ and using (8.8), we have

$$\frac{(E_t W_{t+1} - (1 + r_{f,t+1}) W_t)^2}{\mathrm{var}_t(W_{t+1})} \leqslant \frac{\mathrm{var}_t(M_{t+1})}{(E_t M_{t+1})^2}. \tag{8.20}$$

It states that the excess gain (return) per unit risk, i.e. the Sharpe ratio, is bounded by $\mathrm{var}_t(M_{t+1})/(E_t M_{t+1})^2$. This upper bound is not always achievable. If there exists a portfolio allocation $\boldsymbol{\alpha}_t^*$ such that $\boldsymbol{\alpha}_t^{*\mathrm{T}} \boldsymbol{S}_{t+1} = M_{t+1}$, the upper bound is achievable by this portfolio.

Remark 8.1 For the power utility function, we have $M_{t+1} = p_t/p_{t+1} \cdot (C_{t+1}/C_t)^{-\gamma}$. The right hand side of (8.20) is

$$\frac{E_t (p_t/p_{t+1})^2 (C_{t+1}/C_t)^{-2\gamma}}{[E_t (p_t/p_{t+1})(C_{t+1}/C_t)^{-\gamma}]^2} - 1$$

This bound is not necessarily achievable by any tradable portfolio, since M_{t+1} is not a return of a portfolio.

Is there another stochastic discount factor M_{t+1}^*, constructed based on a portfolio, such that

$$\boldsymbol{S}_t = E_t[M_{t+1}^* \boldsymbol{S}_{t+1}]?$$

If so, the Sharpe ratio is attainable by such a portfolio. The answer is affirmative, as constructed by Hansen and Jagannathan (1991) as follows. Let $M_{t+1}^* = \boldsymbol{\alpha}_t^{*\mathrm{T}} \boldsymbol{S}_{t+1}$ where $\boldsymbol{\alpha}_t^*$ minimizes

$$E_t(M_{t+1} - \boldsymbol{\alpha}_t^{\mathrm{T}} \boldsymbol{S}_{t+1})^2.$$

This is the best approximation to the stochastic discount factor using portfolios $\boldsymbol{\alpha}_t^{\mathrm{T}} \boldsymbol{S}_{t+1}$. By taking the derivative with respect to $\boldsymbol{\alpha}_t$ and setting it to zero, we obtain the first order condition:

$$E_t(M_{t+1} - \boldsymbol{\alpha}_t^{*\mathrm{T}} \boldsymbol{S}_{t+1})\boldsymbol{S}_{t+1} = 0.$$

Thus,

$$E_t M_{t+1}^* \boldsymbol{S}_{t+1} = E_t M_{t+1} \boldsymbol{S}_{t+1} = \boldsymbol{S}_t.$$

Hence, by (8.20), for any portfolio, its Sharpe ratio is bounded by

$$\frac{(E_t W_{t+1} - (1 + r_{f,t+1})W_t)^2}{\mathrm{var}_t(W_{t+1})} \leqslant \frac{\mathrm{var}_t(M_{t+1}^*)}{(E_t M_{t+1}^*)^2}.$$

The portfolio M_{t+1}^* attains the maximum Sharpe ratio. This follows easily from (8.19) with $W_t = M_t^*$, which shows that

$$E_t M_{t+1}^* - (1 + r_{f,t+1})M_t^* = \mathrm{var}_t(M_{t+1}^*)/(EM_{t+1}^*).$$

For this reason, it is called the *benchmark portfolio*.

In terms of the stochastic discount factor M_{t+1}^*, it follows from (8.19) that

$$E_t W_{t+1} - (1 + r_{f,t+1})W_t = \mathrm{cov}_t(M_{t+1}^*, W_{t+1})(1 + r_{f,t+1}).$$

In particular, taking $W_t = M_t^*$, the excess gain of the benchmark portfolio is given by

$$E_t M_{t+1}^* - (1 + r_{f,t+1})M_t^* = \mathrm{var}_t(M_{t+1}^*)(1 + r_{f,t+1}).$$

The ratio of the above two equalities is then

$$E_t W_{t+1} - (1 + r_{f,t+1})W_t = \beta_t[E_t M_{t+1}^* - (1 + r_{f,t+1})M_t^*], \tag{8.21}$$

where $\beta_t = \mathrm{cov}_t(M_{t+1}^*, W_{t+1})/\mathrm{var}_t(M_{t+1}^*)$ is the regression coefficient of W_{t+1} on M_{t+1}^*. In other words, the excess gain of any portfolio is its β times the excess gain of the benchmark portfolio. In terms of the return, dividing both sides of (8.21) by W_t, it can be expressed as

$$E_t R_{t+1} - r_{f,t+1} = \beta_t^*[E_t R_{M,t+1}^* - r_{f,t+1}], \tag{8.22}$$

where $R_{M,t+1}^*$ is the return of the benchmark portfolio and

$$\beta_t^* = \beta_t M_t^* = \frac{\mathrm{cov}_t(R_{M,t+1}^*, R_{t+1})}{\mathrm{var}_t(R_{M,t+1}^*)}$$

is now the market beta in terms of the return. This is indeed the same as the Sharpe–Lintner version of CAPM, M_{t+1}^* regarded as the market portfolio.

8.4 Exercises

8.1 If an asset is positively correlated with the consumption growth, the stock price tends to be higher. Explain briefly using a pricing formula.

8.2 What is the consumption based CAPM? What is the expected value of the stochastic discount factor?

8.3 What is the price of an asset? What is its main difference from the derivative pricing?

8.4 What are the two key assumptions in deriving the Hansen–Singleton formula from the consumption based CAPM? According to the formula, what are the risk-free interest and the expected excess return?

8.5 What are the main factors that determine the risk-free interest rate?

8.6 What are the equity premium puzzle and risk-free rate puzzle?

8.7 If the log-return $r_t \sim N(8, 20^2)$ per annum, what is the expected real return $E\exp(r_t) - 1$?

9
Present-value Models

This chapter gives the stochastic discount pricing model, deriving asset prices from consumption, saving and investment. This chapter focuses on asset prices, calculating the prices implied by the models of returns. It gives us an idea of what is the fundamental price of a stock and how the prices are related to the dividend payments and short-term interest rates.

9.1 Fundamental price

We first infer the price of an asset from its expected return and dividend payment. Recall that the simple return of a stock is given by

$$R_{t+1} = \frac{S_{t+1} + D_{t+1}}{S_t} - 1.$$

Hence, the present value of the asset price satisfies

$$(1 + R_{t+1})S_t = S_{t+1} + D_{t+1}.$$

Taking conditional expectation, we have

$$(1 + E_t R_{t+1})S_t = E_t S_{t+1} + E_t D_{t+1},$$

or equivalently

$$S_t = (1 + E_t R_{t+1})^{-1}(E_t S_{t+1} + E_t D_{t+1}). \tag{9.1}$$

Assume that the expected return is a non-time-varying constant:

$$E_t[R_{t+1}] = R. \tag{9.2}$$

This is an unrealistic assumption, but nevertheless provides an actuarial valuation of the asset. Under model (9.2), we have

$$S_t = (1 + R)^{-1}(E_t S_{t+1} + E_t D_{t+1}). \tag{9.3}$$

Iterative application of the above formula yields

$$\begin{aligned} S_t &= (1+R)^{-1}E_tD_{t+1} + (1+R)^{-2}E_tD_{t+2} + (1+R)^{-2}E_tS_{t+2} \\ &= \sum_{i=1}^{K-1}(1+R)^{-i}E_tD_{t+i} + (1+R)^{-K}E_tS_{t+K}. \end{aligned}$$

If we further assume that

$$\lim_{K\to\infty}(1+R)^{-K}E_tS_{t+K} = 0, \tag{9.4}$$

then, by taking the limit, we have

$$S_t = \sum_{i=1}^{\infty}(1+R)^{-i}E_tD_{t+i}. \tag{9.5}$$

It says that the *fundamental price* of a stock is the discounted dividend paid in perpetuity by the stock. The discount rate is the expected return of the stock itself, which is assumed to be a constant over time. Since the dividend payment is the cash flow of the stock, (9.5) is called the *discounted cash flow* model, discounted by the long-run expected return of the stock. It is the present value of the dividend paid in the life time of the stock and is also referred to as the *present-value model*.

Suppose further that the dividends are expected to grow at a constant rate G:

$$E_tD_{t+i} = (1+G)E_tD_{t+i-1} = (1+G)^iD_t. \tag{9.6}$$

This assumption is unrealistic, but the formula provides useful intuition on the present value of a stock. It is referred to as the Gordon (1962) growth model. It is reasonable to assume that the stock return is expected to be higher than the dividend growth rate, i.e. $R > G$. Under assumption (9.6), we have

$$S_t = E_t\Big[\sum_{i=1}^{\infty}(1+R)^{-i}(1+G)^iD_t\Big] = \frac{(1+G)D_t}{R-G}. \tag{9.7}$$

It shows that the *dividend-to-price ratio* is given by

$$D_t/S_t = (R-G)/(1+G).$$

The dividend-to-price ratio is also referred to as the *dividend yield*.

Remark 9.1 It can be shown that the equity repurchases affect the expected future dividend, but they do not affect the validity of the price formula.

Let us now consider the co-movement between the price process $\{S_t\}$ and dividend process $\{D_t\}$. Let $\Delta D_{t+1+i} = D_{t+1+i} - D_{t+i}$. Then

$$\begin{aligned}
&\sum_{i=0}^{\infty}(1+R)^{-i}\Delta D_{t+1+i} \\
&= (1+R)\sum_{i=1}^{\infty}(1+R)^{-i}D_{t+i} - \sum_{i=0}^{\infty}(1+R)^{-i}D_{t+i} \\
&= R\sum_{i=1}^{\infty}(1+R)^{-i}D_{t+i} - D_t.
\end{aligned}$$

By using (9.5), the last equality is the same as $RS_t - D_t$. Hence,

$$S_t - D_t/R = R^{-1}E_t\sum_{i=0}^{\infty}(1+R)^{-i}\Delta D_{t+1+i}. \tag{9.8}$$

The price series $\{S_t\}$ and dividend series $\{D_t\}$ are co-integrated: If $\{\Delta D_t\}$ is stationary, so is $\{S_t - D_t/R\}$. The general definition of the *cointegration* of a multivariate time series $\boldsymbol{Y}_t$ is that each component is an integrated time series, but there exists a linear combination $\boldsymbol{a}^{\mathrm{T}}\boldsymbol{Y}_t$ ($\boldsymbol{a} \neq 0$ is called cointegrating vector) such that it is stationary (or more generally a lower order integration). Another example is the index of S&P 500 and its future. Both are a random walk, but their difference is stationary. The concept was introduced in Engle and Granger (1987). See Section 4.3 for further details.

9.2 Rational bubbles

We have shown that under condition (9.4),

$$S_{D,t} = \sum_{i=1}^{\infty}(1+R)^{-i}E_tD_{t+i}$$

is the unique solution to the pricing equation (9.3). What is the general solution to the pricing equation (9.3) without the growth assumption (9.4)?

Let us write $S_t = S_{D,t} + B_t$. Then, from (9.3), we have

$$S_{D,t} + B_t = (1+R)^{-1}E_t(S_{D,t+1} + B_{t+1} + D_{t+1}),$$

Since $S_{D,t}$ is a solution to the pricing equation (9.3), i.e.

$$S_{D,t} = (1+R)^{-1}E_t(S_{D,t+1} + D_{t+1}),$$

we have

$$B_t = (1+R)^{-1}E_tB_{t+1}. \tag{9.9}$$

In other words, the general solution to the pricing equation (9.3) consists of the sum of two parts. The first part $S_{D,t}$, which is called *fundamental value*. The second part B_t is called a *rational bubble*. The growth condition (9.4) rules out this possibility. That is, under the condition (9.4), the only solution to (9.9) is $B_t = 0$ or the solution to the pricing equation (9.3) is unique and is the fundamental value $S_{D,t}$.

Without the growth solution, the solution B_t in general can have an explosive pattern. In fact, it follows from (9.9) that

$$E_t B_{t+1} = (1+R)B_t.$$

Using the double expectation formula,

$$E_t B_{t+n} = E_t E_{t+n-1} B_{t+n} = (1+R) E_t B_{t+n-1}.$$

Iteratively applying this, we have

$$E_t B_{t+n} = (1+R)^n B_t, \qquad \text{for any } n > 0.$$

which is explosive if $B_t \neq 0$. The word "bubble" reminds us of some famous episodes in history in which asset prices rose higher than a level that can be easily explained by fundamentals: investors betting other investors would drive prices even higher in the future. For this reason, B_t is called the rational bubble. It is rational since the solution $S_t = S_{D,t} + B_t$ still satisfies the pricing equation (9.3).

An example of a rational bubble is given by Blanchard and Watson (1982), in which the bubble can grow at rate $\dfrac{1+R}{\theta} - 1$ with probability θ and burst with probability $1-\theta$. In other words,

$$B_{t+1} = \xi_{t+1} + \begin{cases} \dfrac{1+R}{\theta} B_t & \text{with probability } \theta \\ 0 & \text{with probability } 1-\theta, \end{cases} \tag{9.10}$$

where $\{\xi_t\}$ is a martingale difference, i.e. $E_t \xi_{t+1} = 0$. For such a bubble,

$$E_t B_{t+1} = E_t \xi_{t+1} + (1+R)B_t = (1+R)B_t.$$

Thus, it satisfies equation (9.9). Yet, it has an explosive behavior. For example, if $\theta = 0.9$ and $R = 0.1$, then with probability 90%, the bubble grows on average by $(1+R)/\theta = 22\%$. The probability of burst is 10% and the expected waiting time is therefore 10. Thus, the average growth length is 9 times, with a bubble of $1.22^9 \approx 6$ times of B_t. Furthermore, it is not a small probability to get even a consecutive growth of twenty times (such a probability is $\theta^{20} = 12.2\%$) and this can make the bubble very large (53.4

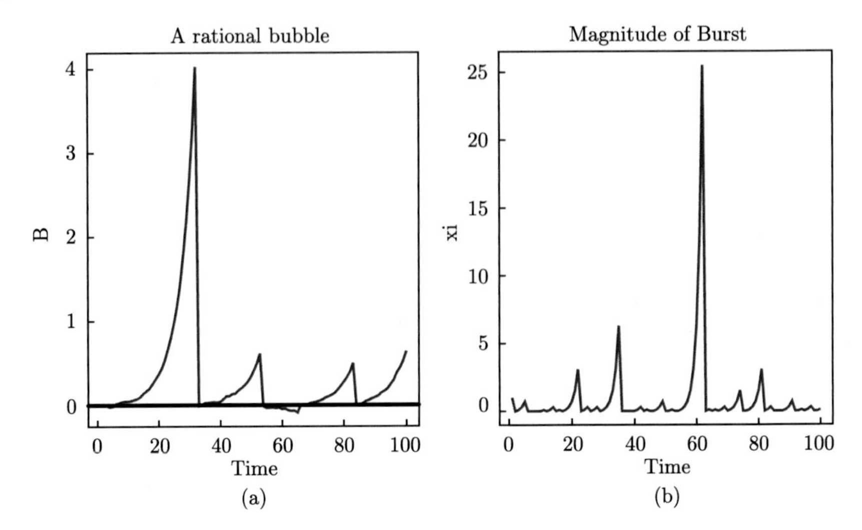

Figure 9.1 (a) Rational bubble B_t is plotted against t with $\theta = 0.9$ and $R = 0.1$, $\xi_t \sim N(0, 0.1^2)$; (b) $|\xi_t|$ for a martingale difference is plotted against t.

times of growth). The right panel of Figure 9.1 shows a realization of such a rational bubble.

Any martingale in (9.10) would make the rational bubble satisfy equation (9.9). Yet, the martingale itself can have a bubble kind of effect. These two kind of bubble behaviors would make B_t look even more like a bubble. An example of such a martingale is given by

$$\xi_{t+1} = \begin{cases} (2|\xi_t| + 1)\mathrm{sgn}(\varepsilon_{t+1}), & \text{with probability } 1/2 \\ 0 & \text{with probability } 1/2, \end{cases}$$

where ε_{t+1} is independent of ξ_t. Then, for symmetric random variable ε_t, ξ_t is a martingale difference:

$$E_t\xi_{t+1} = \frac{1}{2}(2|\xi_t| + 1)E_t\mathrm{sgn}(\varepsilon_{t+1}) = 0.$$

Let $\tau =$ first time that $\xi_{t+1} = 0$ (bubble burst). Then,

$$|\xi_{t+1}| = 2|\xi_t| + 1, \qquad \text{for } t < \tau.$$

and iteratively using this formula

$$|\xi_{\tau-1}| = 2^{\tau-1}|\xi_0| + 2^{\tau-1} - 1.$$

The magnitude grows very fast before the burst. Figure 9.1(b) shows the bubble effect from a simulated time series.

9.3 Time-varying expected returns

The pricing equation (9.3) is based on the unrealistic assumption of the constant rate of return (9.2). How do we price an asset if the expected returns are time varying? To generalize the pricing formula, let us introduce some notation.

Let $s_t = \log S_t$ and $d_t = \log D_t$ be the logarithm of the price of an asset and the logarithm of its dividend payment, respectively. Then, the log-return of the asset can be written as

$$r_{t+1} = \log(S_{t+1} + D_{t+1})/S_t = s_{t+1} - s_t + \log(1 + D_{t+1}/S_{t+1}).$$

Using $D_{t+1}/S_{t+1} = \exp(d_{t+1} - s_{t+1})$, we have

$$r_{t+1} = s_{t+1} - s_t + \log\{1 + \exp(d_{t+1} - s_{t+1})\}. \tag{9.11}$$

For a given time series $\{x_t\}$, the nonlinear time series $\{f(x_t)\}$ can naturally be linearized via a Taylor expansion at its center $\bar{x}$:

$$f(x_t) \approx f(\bar{x}) + f'(\bar{x})(x_t - \bar{x}). \tag{9.12}$$

To use the Taylor expansion (9.12), let $x_t = d_t - s_t$ and θ be the average of the time series, which is the average of the logarithms of the dividend-to-price ratios. Then, an application to the Taylor expansion (9.12) to $f(x_t) = \log(1 + \exp(x_t))$, we have

$$\begin{aligned} &\log\{1 + \exp(d_{t+1} - s_{t+1})\} \\ &\approx \log(1 + \exp(\theta)) + \frac{\exp(\theta)}{1 + \exp(\theta)}(d_{t+1} - s_{t+1}. - \theta) \end{aligned}$$

Substituting this into (9.11), after some simple algebra, we have

$$r_{t+1} = \kappa + \rho s_{t+1} + (1 - \rho)d_{t+1} - s_t,$$

where $\rho = (1+\exp(\theta))^{-1}$ and $\kappa = -(1-\rho)\log(1-\rho)$. Rearranging the above terms, we obtain

$$s_t \approx \kappa + \rho s_{t+1} + (1 - \rho)d_{t+1} - r_{t+1}. \tag{9.13}$$

This is the *approximate present-value model* of Campbell and Shiller (1988a, b). When the dividend-to-price ratio is constant, then the approximation becomes exact.

Figure 9.2 depicts the inflation adjusted index of S&P 500 as well as the dividend paid (multiplied by a factor of 20 to make the picture more visible) by the index. It also presents the dividend yield per annum. The dividend is computed based on the sum of those paid in the previous 12 months. This

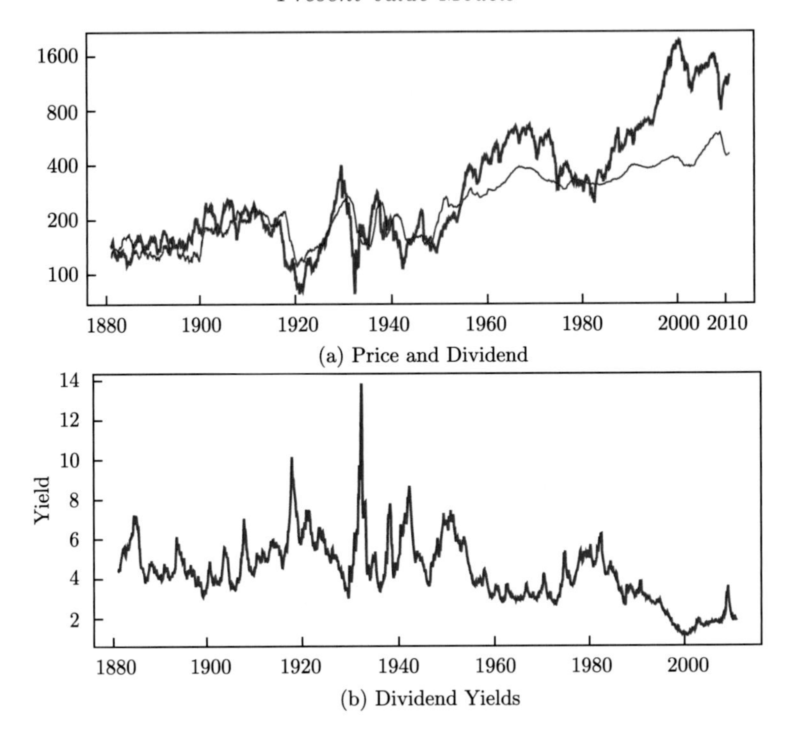

Figure 9.2 S&P 500 data. (a) The inflation adjusted S&P 500 index (in March 2011 dollars) and dividend paid (multiplied by a factor 20 to make it more visible). (b) The dividend yield of the S&P 500 index per annum (in percent).

reduces seasonality of the dividend payments. The average dividend-to-price ratio between 1881 and 2010 for the S&P 500 index has mean 4.35% per annum with standard deviation 1.61% , i.e., $\theta \approx \log(0.0435)$. This implies that $\rho = (1+0.0435)^{-1} \approx 0.958$ for annual data and $\rho = (1+0.0435/12)^{-1} \approx 0.996$ for monthly data.

How good is the approximation (9.13)? The following example gives an empirical study.

Example 9.1 Let us call s_t and $\kappa + \rho s_{t+1} + (1-\rho)d_{t+1} - r_{t+1}$ in (9.13) as the exactly and approximate log-prices. Using monthly nominal dividends and prices on the S&P 500 stock index over the period 1881-2010, the exact and approximate log-prices have mean of 0.136% and 0.136% and SDs and 4.21% and 4.20% per month and a correlation of 0.99999. Therefore, the approximation is very accurate in this case. Figure 9.3 depicts the exact and approximate log-prices over the period. The difference is indistinguishable.

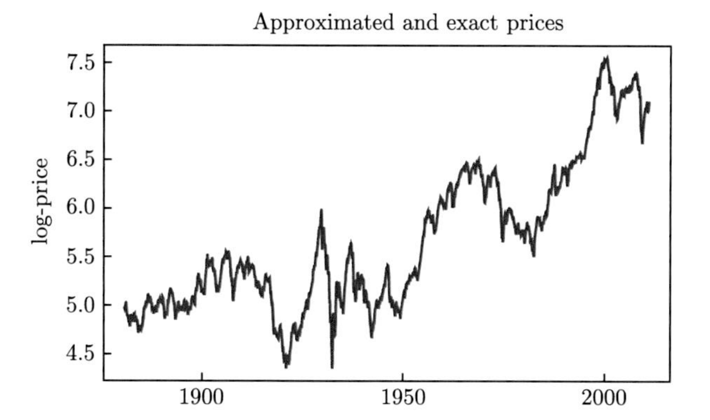

Figure 9.3 The log-prices (red) and approximated log-prices (9.13) based on the S&P 500 index from 1881 to 2010.

Regarding (9.13) as the exact formula, under the growth condition

$$\lim_{j\to\infty} \rho^j s_{t+j} = 0,$$

iterative application of (9.13) results in

$$s_t = \frac{\kappa}{1-\rho} + (1-\rho)P_{dt} - P_{rt}, \tag{9.14}$$

where $P_{dt} = \sum_{j=0}^{\infty} \rho^j d_{t+1+j}$ is the discounted value of the future log-dividends and $P_{rt} = \sum_{j=0}^{\infty} \rho^j r_{t+1+j}$ is the discounted value of future log-stock returns. The pricing formula (9.14) implies that the high stock price today must be some combination of high dividends and low stock returns in the future. Namely, large value s_t must imply either P_{dt} is large or P_{rt} is small or some combination of those.

In terms of the dividend-to-price ratio, by taking E_t on both sides, (9.14) can also be written as

$$d_t - s_t = -\frac{\kappa}{1-\rho} + E_t P_{rt} - \sum_{j=0}^{\infty} \rho^j E_t \Delta d_{t+1+j}, \tag{9.15}$$

where $\Delta d_{t+1+j} = d_{t+1+j} - d_t$. This can easilly be seen by using (9.14) and the following algebra:

$$\begin{aligned}
\sum_{j=0}^{\infty} \rho^j \Delta d_{t+1+j} &= \sum_{j=0}^{\infty} \rho^j d_{t+1+j} - d_t - \sum_{j=1}^{\infty} \rho^j d_{t+j} \\
&= \sum_{j=0}^{\infty} \rho^j d_{t+1+j} - d_t - \rho \sum_{j=0}^{\infty} \rho^j d_{t+j+1} \\
&= (1-\rho)P_{dt} - d_t.
\end{aligned}$$

The model (9.15) is called the *dynamic Gordon model* or the *dividend ratio model*. Recalling that $\rho \approx 1$, P_{rt} is approximately the same as the long-run return, and the last term is the discounted expected dividend growth rate. Model (9.15) indicates that the dividend price ratio is a good proxy of the expected long-run returns.

Example 9.2 Suppose that the return $\{r_t\}$ follows the AR(1) process with the expected return r:

$$r_{t+1} = (1-\gamma)r + \gamma r_t + \eta_{t+1}, \qquad \eta_{t+1} \sim \text{IID}(0, \sigma^2).$$

Write $r_t^* = r_t - r$. Then, $\{r_t^*\}$ follows the AR(1) model: $r_{t+1}^* = \gamma r_t^* + \xi_{t+1}$. By using $E_t r_{t+j}^* = \gamma^j r_t^*$ (see Example 2.10), we have

$$E_t P_{rt} = E_t \Big[\sum_{j=0}^{\infty} \rho^j (r + r_{t+j+1}^*) \Big] = \frac{r}{1-\rho} + \frac{\gamma r_t^*}{1-\rho\gamma}.$$

Since $\rho \approx 1$ as in Remark 9.1, a 1% increase in the expected return today reduces stock price by about 1% if $\gamma = 0.5$. To see this, if $r_t^* = 0.01$ (r_t is 1% above the expectation r) and $\gamma = 0.5$, then P_{rt} increases

$$0.01\gamma/(1-\gamma) \approx 0.01$$

and hence s_t decreases by 1% from (9.14). Similarly, the stock price decreases by about 3% if $\gamma = 0.75$ and by about 9% if $\gamma = 0.9$, when the expected return today increases by 1%. On the other hand, it can be shown (see Exercise 9.3)

$$\text{var}_t(P_{rt}) = \frac{\sigma^2}{(1-\gamma\rho)^2(1-\rho^2)},$$

which is usually large, making returns on assets nearly unforecastable. This is consistent with the empirical evidence in Chapter 2.

A similar expression can be obtained if we assume that the changes of dividend payments follows an AR(1) model. We leave this to an exercise (see Exercise 9.4).

9.4 Empirical evidence

According to the dividend-to-price ratio model (9.15), the dividend ratio is approximately the weighted average of future returns, plus the the discounted expected dividend growth rate. Typically, the expected change of the dividend is very small and as such variation of the term $E_t \Delta d_{t+1+j}$ is negligible. The value ρ close to 1 implies that the dividend ratio should have a higher predictive power for long-term return than short-term return.

Table 9.1 *Regression of log-returns of S&P 500 index on the logarithm of its dividend-to-price ratio over different time horizons*

	Forecast Horrizon(K)					
	1	3	12	24	36	48
1981/01–2010/12						
$\hat{\beta}(K)$	0.007	0.025	0.127	0.267	0.380	0.488
$R^2(K)$	0.006	0.020	0.102	0.232	0.317	0.420
$t(\hat{\beta}(K))$	1.441	1.904	2.316	3.049	4.044	5.882
1951/01–1980/12						
$\hat{\beta}(K)$	0.015	0.053	0.243	0.449	0.586	0.684
$R^2(K)$	0.012	0.039	0.170	0.315	0.462	0.552
$t(\hat{\beta}(K))$	2.408	3.139	3.114	2.783	4.206	6.793
1927/01–1950/12						
$\hat{\beta}(K)$	0.000	0.029	0.206	0.593	0.937	1.241
$R^2(K)$	0.000	0.003	0.035	0.140	0.270	0.413
$t(\hat{\beta}(K))$	−0.006	0.431	1.278	3.567	2.987	4.337
1927/01–2010/10						
$\hat{\beta}(K)$	0.001	0.009	0.057	0.126	0.171	0.211
$R^2(K)$	0.000	0.002	0.015	0.037	0.049	0.058
$t(\hat{\beta}(K))$	0.262	0.736	1.243	1.777	2.038	2.277

The t-statistics are computed based on the Newey and West (1987) method for computing standard errors, given in (9.25).

To provide some empirical evidence, we use the monthly data from S&P 500 index from 1926 to 2010. For returns of more than one month, overlapping monthly returns are used: $Y_t^K = r_{t+1} + \cdots + r_{t+K}$. This increases the sample size, but also introduces correlation. The dividend-to-price ratio is the sum of dividends paid in the previous year divided by current index level. This reduces seasonal patterns of dividends. We run the following simple linear regression:

$$Y_t^K = \alpha_K + \beta_K(d_t - s_t) + \eta_{t,K}, \tag{9.16}$$

where $\eta_{t,K}$ is the noise part that cannot be explained by the dividend-to-price ratio. Because of the dependence due to the overlapping, the standard formula for computing t-statistics no longer applies. Some adjustments are needed for computing the standard errors. Section 9.5 derives the standard error formula for the least-squares estimator $\hat{\alpha}_K$ and $\hat{\beta}_K$. In the application of the Newey and West (1987) formula (9.25), we take $L = K$.

We run regressions over the period from 1926 to 2010, also separately for the three subperiods: 1927–1950 (war period) and 1951-1980 (post war period) and 1981–2010 for different time horizons K. The subperiods have approximately the same length. The division of 1981 corresponds roughly to the change of the Federal Reserve monetary policy on October 6, 1979. The interest rate in the two years following October, 1979 was five times greater than that in the prior two years.

The results of regression analysis are summarized in Table 9.1. The multiple R^2 is very small over a short-term horizon and increases as the forecast horizon K does. This is consistent with the dividend ratio model. The scatter plots of the logarithm of dividend-to-price ratio (or dividend yield) are plotted against multiple-period returns to give us an idea on the strength of prediction power of dividend yields. See Figure 9.4–9.7. The multiple R^2 is relatively small even for long time horizon prediction when the entire period (1927-2010) is considered. This is not surprising. The static model (9.16) is not expected to hold for such a long time period. Indeed, Figure 9.2 suggests time-varying dividend yield over this period. In the period 1981–2010, the dividend yields decline; in the period 1950-1980, the dividend yield is relatively stable, and in the period 1927–1950, the dividend yields are volatile. It is hard to put these three subperiods into the same static model. That explains statistically why the multiple R^2 is so small when the model is fitted to the entire period.

There are many other variables that have some ability to predict the returns of a stock. These include the measures that are fundamental to a firm. In addition to the dividend yield, price-to-earning ratio and price-to-sale ratio are also important.
Other predictors include market variables such as various interest rate measures like yield spreads between long- and short-term, low- and high-grade corporate bonds or commercial paper.

As an example, let us take the changes on the yield of 3-month treasury bill rates as the regressor. Denote by z_t the three-month treasury bill rates. Let X_t be the changes of the yield over the previous year's average:

$$X_t = z_t - \sum_{i=0}^{11} z_{t-i}/12.$$

Figure 9.8 summarizes the changes from January 1935 to February 2011. There are barely any changes of interest rates in the 1930s and 1940s, as they were pegged by the Federal Reserve during most of the period. For this reason, we only consider the two subperiods: 1951–1980 and 1981–2010, both covering 30 years. We also considered the whole 60-year period, from 1951 to 2010.

To examine the impact of the short-term rate change on the future returns,

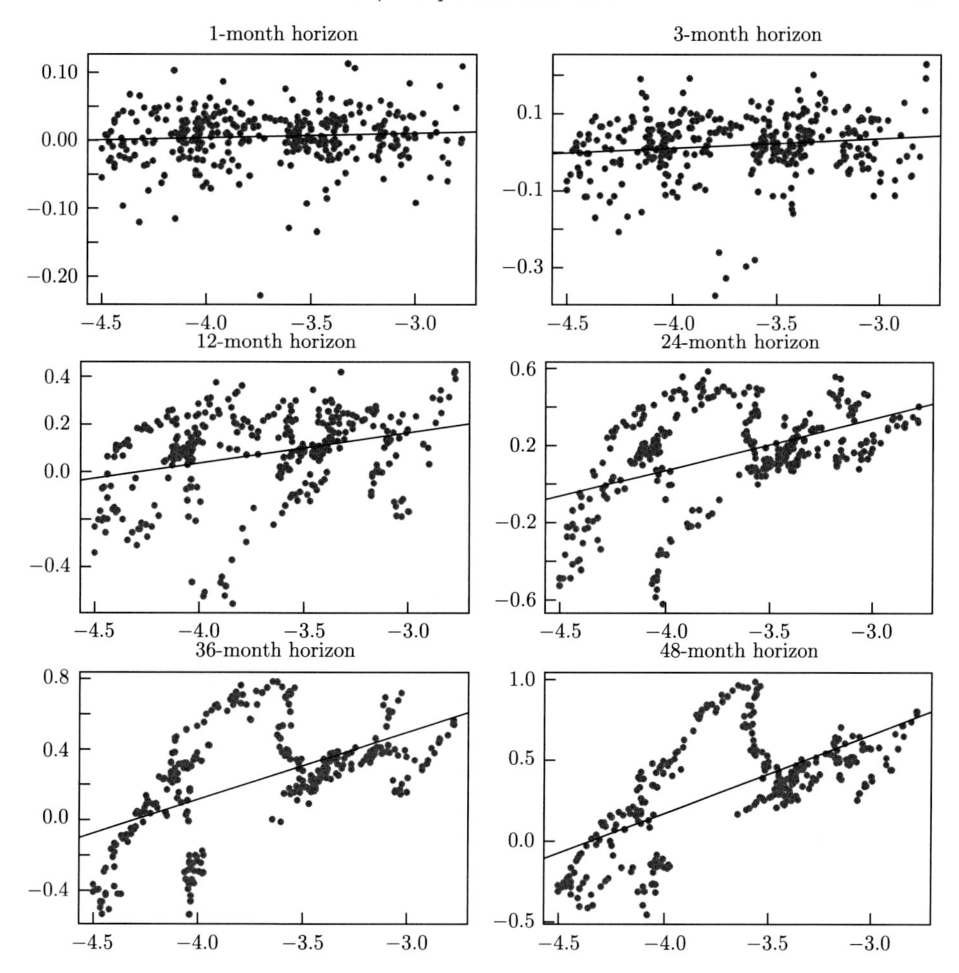

Figure 9.4 The scatter plot of the logarithm of the dividend-to-price ratio (the dividend yield) against the future 1-month, 3-month, 12-month, 24-month, 36-month, and 48-month returns. The correlation coefficient gets stronger when the time horizon gets larger. The time period is January 1981 to December 2010.

we consider the regression problem:

$$Y_t^K = a_K + b_K X_t + \eta_{t,K}, \tag{9.17}$$

in which the detrended series was used as a regressor, and $\eta_{t,K}$ is the regression error. Table 9.2 summarizes the results.

The detrended short rate X_t has some ability to forecast stock returns, in particular for short term returns. The forecasting power is concentrated in the first subsample, before the change of Federal Reserve's monetary policy in 1979. In that period, the multiple R^2 admits a hump structure. While the

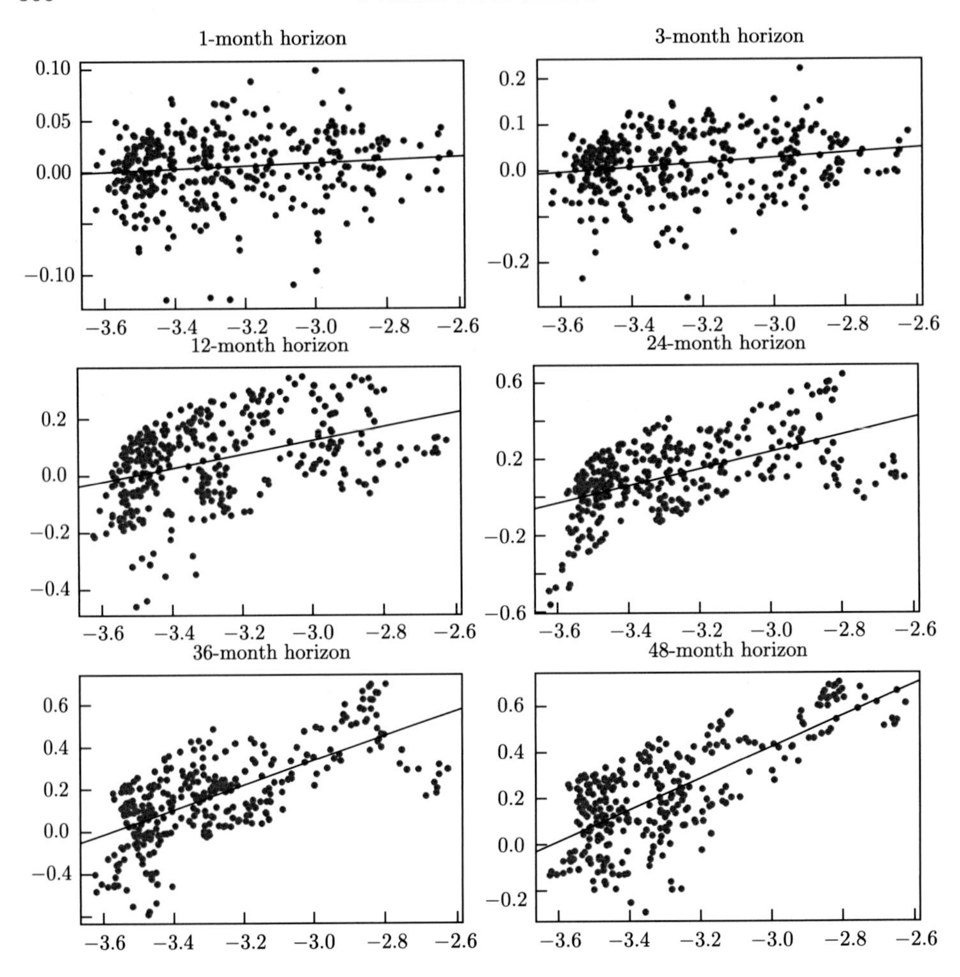

Figure 9.5 The scatter plot of the logarithm of the dividend-to-price ratio (the dividend yield) against the future 1-month, 3-month, 12-month, 24-month, 36-month, and 48-month returns. The correlation coefficient gets stronger when the time horizon gets larger. The time period is January 1951 to December 1980.

changes of short-term rates do have predictive power on the whole sample, the contribution is mainly due to that of the first subperiod. The predictive powers of the changes of interest rates are much smaller than those using the dividend yield. The signs of regression are mostly negative. This is expected. An increase of short-term rates has an adverse effect on the returns of stock. The changes of the short term rates tend to have a bigger impact on the short-term returns (holding less than a year) than the long-term returns.

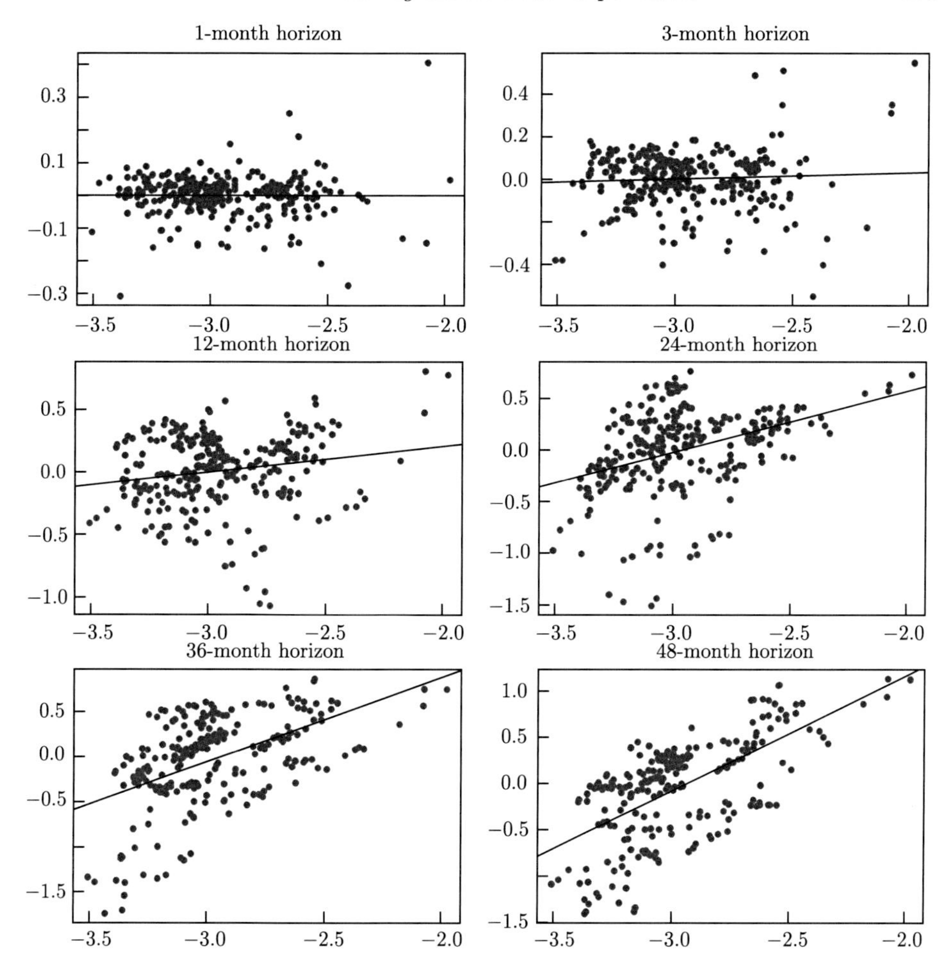

Figure 9.6 The scatter plot of the logarithm of the dividend-to-price ratio (the dividend yield) against the future 1-month, 3-month, 12-month, 24-month, 36-month, and 48-month returns. The correlation coefficient gets stronger when the time horizon gets larger. The time period is January 1981 to December 2010.

9.5 Linear regression under dependence

Consider the linear regression for time series

$$Y_t = \boldsymbol{X}_t^{\mathrm{T}} \boldsymbol{\beta} + \varepsilon_t.$$

Writing it in the matrix form, we have

$$\boldsymbol{Y} = \boldsymbol{X}\boldsymbol{\beta} + \boldsymbol{\epsilon}. \tag{9.18}$$

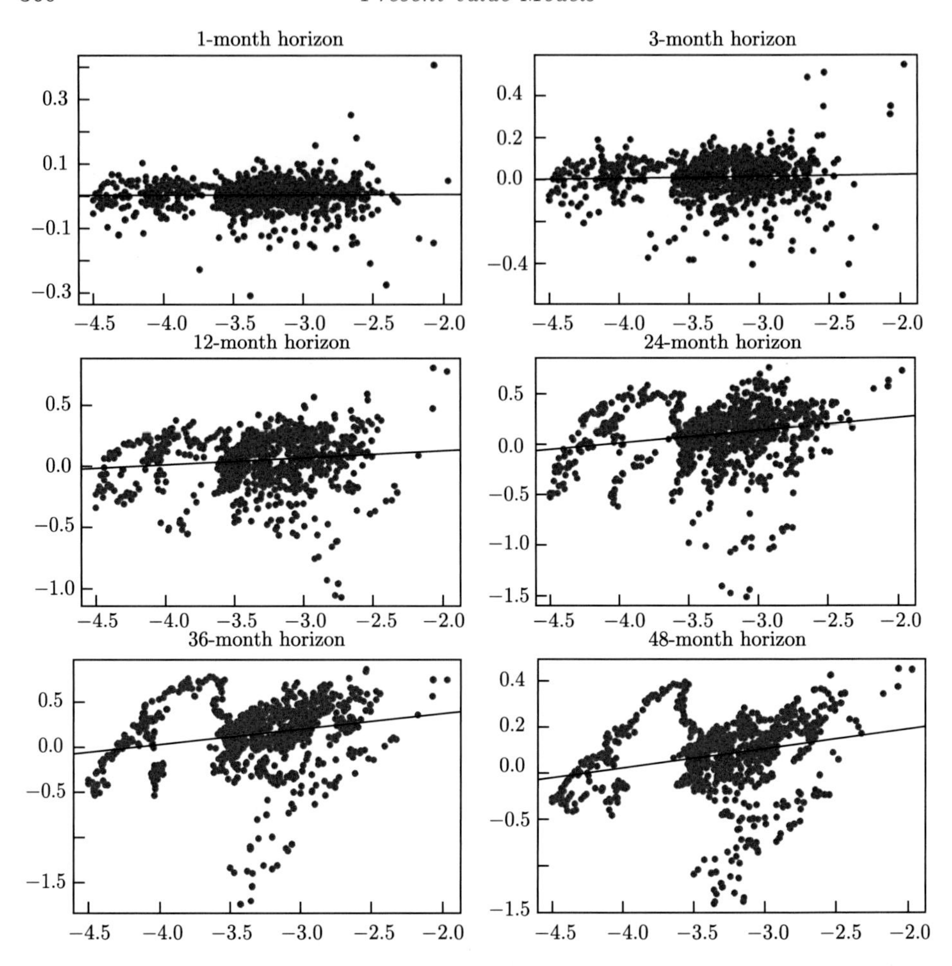

Figure 9.7 The scatter plot of the logarithm of the dividend-to-price ratio (the dividend yield) against the future 1-month, 3-month, 12-month, 24-month, 36-month, and 48-month returns. The correlation coefficient gets stronger when time horizon gets larger. The time period is January 1927 to December 1950.

Then, the ordinary least-squares estimator is to minimize

$$\sum_{t=1}^{T}(Y_t - \boldsymbol{X}_t^{\mathrm{T}}\boldsymbol{\beta})^2 = \|\boldsymbol{Y} - \boldsymbol{X}\boldsymbol{\beta}\|^2,$$

which gives the estimator

$$\hat{\boldsymbol{\beta}} = \boldsymbol{S}_T^{-1}\boldsymbol{X}^{\mathrm{T}}\boldsymbol{Y}/T,$$

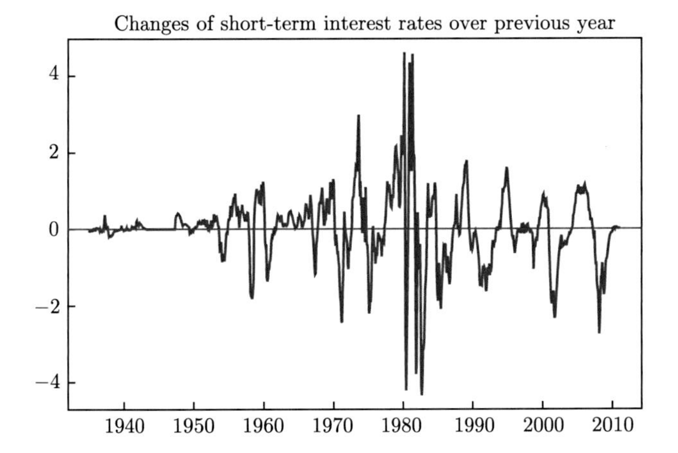

Figure 9.8 The changes of the yields of 3-month treasury bills over the previous year, from January 1935 to February 2011.

Table 9.2 *Regression of log stock returns on the short-term interest rate changes for different time horizons*

	Forecast Horrizon(K)					
	1	3	12	24	36	48
1981/01 – 2010/12						
$\hat{\beta}(K)$	−0.0014	−0.0001	0.0057	0.0119	−0.0113	−0.0403
$R^2(K)$	0.0017	0.0000	0.0014	0.0031	0.0018	0.0183
$t(\hat{\beta}(K))$	−0.7133	−0.0193	0.2087	0.5717	−0.4337	−1.3833
1951/01 – 1980/12						
$\hat{\beta}(K)$	−0.0084	−0.0190	−0.0711	−0.0651	−0.0251	−0.0267
$R^2(K)$	0.0559	0.0732	0.1652	0.0701	0.0089	0.0088
$t(\hat{\beta}(K))$	−5.0269	−3.8564	−3.4109	−2.1208	−0.9356	−0.8766
1951/01 – 2010/12						
$\hat{\beta}(K)$	−0.0042	−0.0080	−0.0187	−0.0098	−0.0157	−0.0288
$R^2(K)$	0.0154	0.0139	0.0154	0.0023	0.0042	0.0115
$t(\hat{\beta}(K))$	−3.1463	−2.0433	−1.0081	−0.6015	−1.0487	−1.4217

where

$$\boldsymbol{S}_T = T^{-1}\boldsymbol{X}^{\mathrm{T}}\boldsymbol{X} = T^{-1}\sum_{t=1}^{T}\boldsymbol{X}_t\boldsymbol{X}_t^{\mathrm{T}}.$$

Substituting (9.18) into the above expression, we obtain

$$\hat{\boldsymbol{\beta}} = \boldsymbol{\beta} + \boldsymbol{S}_T^{-1}\boldsymbol{X}^{\mathrm{T}}\boldsymbol{\epsilon}/T = \boldsymbol{\beta} + T^{-1}\boldsymbol{S}_T^{-1}\sum_{t=1}^{T}\boldsymbol{X}_t\varepsilon_t. \tag{9.19}$$

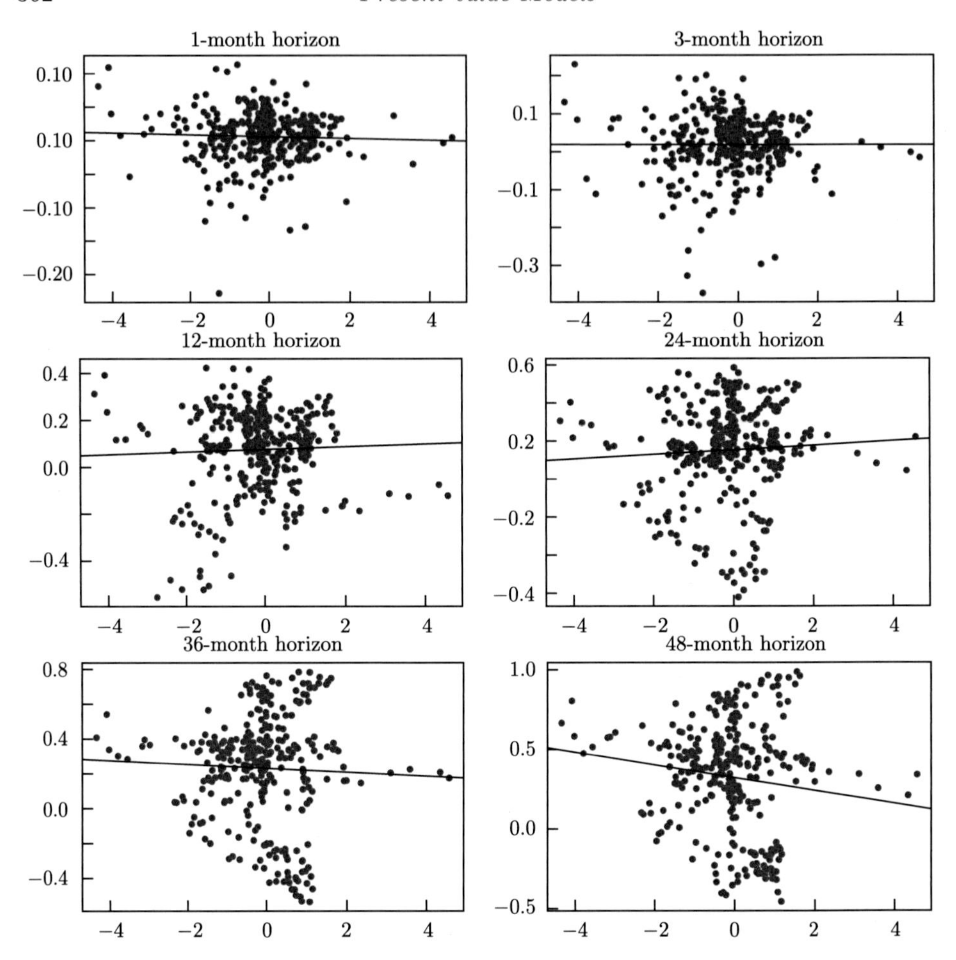

Figure 9.9 The changes of the yields of 3-month treasury bills against returns of S&P 500 index over the next 1 month, 3 months, 12 months, 24 months, 36 months, and 48 months. Time period: January 1981 to December 2010.

The ordinary least-squares estimator for the time series is the same as that for cross-sectional regression. The main difference lies in calculation of the variance–covariance matrix for $\hat{\boldsymbol{\beta}}$. This is similar to the calculation in Exercise 7.2. Specifically, the following term is calculated differently:

$$\begin{aligned} \operatorname{var}\Big(T^{-1/2}\sum_{t=1}^{T}\boldsymbol{X}_t\varepsilon_t\Big) &= T^{-1}\sum_{t,t'=1}^{T}\operatorname{cov}(\boldsymbol{X}_t\varepsilon_t, \boldsymbol{X}_{t'}\varepsilon_{t'}) \\ &= \sum_{j=0}^{T-1}\boldsymbol{\Sigma}_{T,j}, \end{aligned} \tag{9.20}$$

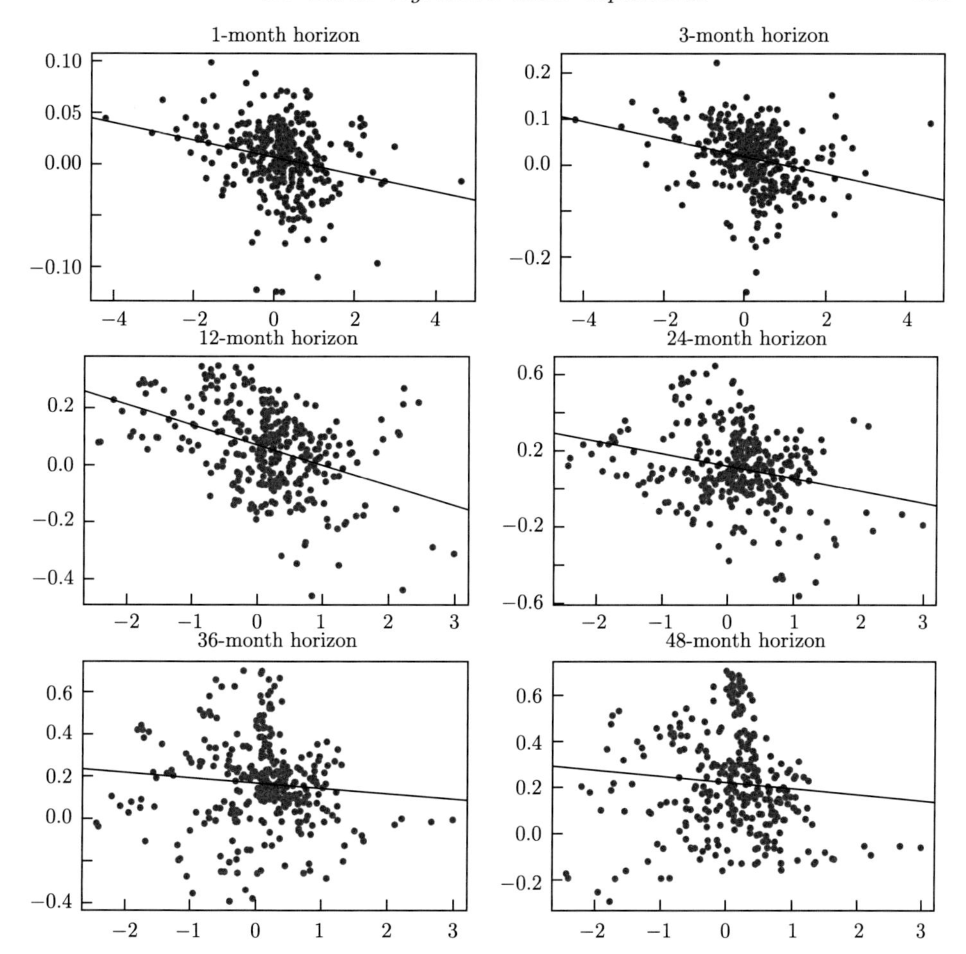

Figure 9.10 The changes of the yields of 3-month treasury bills against returns of S&P 500 index over the next 1 month, 3 months, 12 months, 24 months, 36 months, and 48 months. Time period: January 1951 to December 1980.

where

$$\boldsymbol{\Sigma}_{T,j} = T^{-1} \sum_{t=1}^{T-j} [\operatorname{cov}(\boldsymbol{X}_t \varepsilon_t, \boldsymbol{X}_{t+j} \varepsilon_{t+j}) + \operatorname{cov}(\boldsymbol{X}_t \varepsilon_t, \boldsymbol{X}_{t+j} \varepsilon_{t+j})']. \tag{9.21}$$

Under some stationarity and mixing conditions, it can be shown that

$$\lim_{T\to\infty} \boldsymbol{\Sigma}_{T,j} = \boldsymbol{\Sigma}_j, \quad \text{and} \quad \lim_{T\to\infty} \boldsymbol{S}_T = \boldsymbol{\Sigma}, \tag{9.22}$$

for some given limit matrices and hence the limit of (9.20) is $\sum_{j=0}^{\infty} \boldsymbol{\Sigma}_j$. Furthermore, it can be shown under additional conditions that the following asymptotic normality also holds:

$$T^{-1/2}\sum_{t=1}^{T}\boldsymbol{X}_t\varepsilon_t \xrightarrow{D} N\Big(0, \sum_{j=0}^{\infty}\boldsymbol{\Sigma}_j\Big).$$

Substituting the last two results into (9.19), it follows that

$$T^{1/2}(\hat{\boldsymbol{\beta}}-\boldsymbol{\beta}) \xrightarrow{D} N\Big(0, \boldsymbol{\Sigma}^{-1}\sum_{j=0}^{\infty}\boldsymbol{\Sigma}_j\boldsymbol{\Sigma}^{-1}\Big).$$

In other words, the asymptotic variance-covariance matrix of $\hat{\boldsymbol{\beta}}$ is given by

$$T^{-1}\boldsymbol{\Sigma}^{-1}\Big(\sum_{j=0}^{\infty}\boldsymbol{\Sigma}_j\Big)\boldsymbol{\Sigma}^{-1}. \tag{9.23}$$

Specifically, when the data are independent, $\boldsymbol{\Sigma}_j = 0$ for $j > 0$ and the asymptotic variance reduces to $\boldsymbol{\Sigma}^{-1}\boldsymbol{\Sigma}_0\boldsymbol{\Sigma}^{-1}$.

An obvious estimator of $\boldsymbol{\Sigma}$ is its sample average $\boldsymbol{S}_T$. By (9.21), a natural estimator for $\boldsymbol{\Sigma}_j$ is

$$S_{T,j} = T^{-1}\sum_{t=1}^{T-j}\hat{\varepsilon}_t\hat{\varepsilon}_{t+j}(\boldsymbol{X}_t\boldsymbol{X}'_{t+j} + \boldsymbol{X}_{t+j}\boldsymbol{X}'_t).$$

This cannot be applied to the case when j is very close to T, since there are not many data points in computing the average. Therefore, we have to truncate the infinity sum in (9.23) at a certain lag L. This leads to the estimator of the variance and covariance matrix for $\hat{\boldsymbol{\beta}}_j$ as follows:

$$\widehat{\text{var}}(\hat{\boldsymbol{\beta}}) = T^{-1}\boldsymbol{S}_T^{-1}\Big(\sum_{j=1}^{L}\boldsymbol{S}_{T,j}\Big)\boldsymbol{S}_T^{-1} \tag{9.24}$$

for a user choice parameter L. Note that while the matrix (9.23) is always semi-positive definite, the estimator (9.24) might not be so. The summation in (9.24) is radical, giving the weight zero or one, depending on whether $j \leqslant L$ or not. An alternative method is to taper the weight as follows

$$\widehat{\text{var}}(\hat{\boldsymbol{\beta}}) = T^{-1}\boldsymbol{S}_T^{-1}\Big(\sum_{j=0}^{L}\frac{p-j}{p}\boldsymbol{S}_{T,j}\Big)\boldsymbol{S}_T^{-1}. \tag{9.25}$$

The latter is the estimator due to Newey and West (1987). It is always semi-positive definite.

9.6 Exercises

9.1 What are the rational bubbles? Give mathematical definition and explain briefly.

9.2 What is the meaning of cointegration? Give two examples of cointegrated time series.

9.3 Assume that the mean-adjusted return follows the AR(1) model:

$$r_{t+1}^* = \gamma r_t^* + \eta_{t+1},$$

where η_{t+1} is a white noise series with mean zero and variance σ^2. Let $P_{Dt}^* = \sum_{j=0}^{\infty} \rho^j r_{t+1+j}^*$.

1. Show that $P_{Dt}^* = (1-\gamma\rho)^{-1}(\gamma r_t^* + \sum_{j=0}^{\infty} \rho^j \eta_{j+1+j})$.
2. Deduce from (a) that $E_t P_{Dt}^* = \gamma r_t^*/(1-\gamma\rho)$.
3. Deduce from (a) that $\text{var}_t(P_{Dt}^*) = \dfrac{\sigma^2}{(1-\gamma\rho)^2(1-\rho^2)}$.

9.4 Suppose that the log-dividend growth $\Delta d_{t+1} = d_{t+1} - d_t$ follows the $AR(1)$ model:

$$\Delta d_{t+1} = (1-\theta)d + \theta\Delta d_t + \varepsilon_{t+1}$$

where ε_t is a white noise series with mean zero. Assume further that the return follows the AR(1) model:

$$r_{t+1} = (1-\gamma)r + \gamma r_t + \eta_{t+1},$$

where η_{t+1} is a white noise series with mean zero.

1. What is $E_t d_{t+j}$?
2. What is the expected discounted log-dividend $P_{Dt} = \sum_{j=0}^{\infty} \rho^j E_t d_{t+1+j}$? Hint: An easier way is to show that

$$P_{Dt} = \sum_{j=0}^{\infty} \rho^j E_t \Delta d_{t+j+1} + d_t + \rho P_{Dt}$$

   with and then compute $E_t \Delta d_{t+j+1}$ or use Exercise 9.3.
3. What is the present value of the stock?

9.5 Determine the prediction power of the price to earning ratio on the returns of the S&P 500 index for the six time horizons: 1 month, 3 months, 12 months, 24 months, 36 months and 48 months for three subperiods: January 1927 to December 1950, January 1951 to December 1980, January 1981 to September 2010. Report the following in particular

1. A time series plot of PE ratios of the entire period.
2. Regression coefficients and multiple R^2.
3. Use both earning-to-price ratios and log-dividend-to-price ratios as the two regressors and report the results in (b). You are welcome to compute the t-statistics using the Newey–West estimator of the standard error, but this is optional.

References

Ahn, S.C. and Horenstein, A.R. (2013). Eigenvalue ratio test for the number of factors. *Econometrica*, **81**, 1203–1227.

Ait-Sahalia, Y., Fan, J. and Li, Y. (2013). The leverage effect puzzle: disentangling sources of bias in high. *Journal of Financial Economics*, **109**, 224–249.

Ait-Sahalia, Y. and Jacod, J. (2014). *High-Frequency Financial Econometrics.* Princeton University Press, Princeton, NJ.

Akaike, H. (1970). Statistical predictor identification. *Annals of the Institute of Statistical Mathematics*, **22**, 203-217.

Akaike, H. (1973). Information theory and an extension of the maximum likelihood principle. In *Second International Symposium in Information Theory*, B.N. Petroc and F. Caski (eds). Akademiai Kiado, Budapest, pp. 276–281.

Anderson, T.W. (2003). *An Introduction to Multivariate Statistical Analysis* (3rd edition). John Wiley & Sons, New York.

Antoniadis, A. and Fan, J. (2001). Regularized wavelet approximations (with discussion). *Journal of American Statistical Association*, **96**, 939-967.

Appel, G. (2009). *Technical Analysis: Power Tools for Active Investors.* FT Press, Upper Saddle River.

Artzner, P., Delbaen, F., Eber, J. and Heath, D. (1999). Coherent measures of risk. *Mathematical Finance*, **9**, 203–228.

Bai, J. (2003). Inferential theory for factor models of large dimensions. *Econometrica*, **71**, 135–171.

Bai, J. (2009). Panel data models with interactive fixed effects. *Econometrica*, **77**, 1229-1279.

Bai, J. and Ng, S. (2002). Determining the number of factors in approximate factor models. *Econometrica*, **70**, 191–221.

Bai, J. and Ng, S. (2008). Large dimensional factor analysis. *Foundations and Trends in Econometrics*, **3**, 89-163.

Banz, R.W. (1981). The relationship between return and market value of common stocks. *Journal of Financial Economics*, **9**, 3–18.

Barndorff-Nielsen, O., Hansen, P., Lunde, A. and Shephard, N. (2011). Multivariate realised kernels: consistent positive semi-definite estimators of the covariation of equity prices with noise and non-synchronous trading. *Journal of Econometrics*, **162**, 149–169.

Basu, S. (1977). Investment performance of common stocks in relation to their price-earnings ratios: a test of the efficient market hypothesis. *Journal of Finance*, **32**, 663–682.

Berkes, I., Horváth, L. and Kokoszka, P. (2003). GARCH processes: structure and estimation. *Bernoulli*, **9**, 183–371.

Berzuini, C., Bes, N.B., Gilks, W.R. and Larizza, C. (1997). Dynamic conditional independence models and Markov Chain Monte Carlo methods. *Journal of American Statistical Association*, **92**, 1403–1412.

Bickel, P.J. and Levina, E. (2008). Covariance regularization by thresholding. *Annals of Statistics*, **36**, 2577–2604.

Black, F. (1972). Capital market equilibrium with restricted borrowing. *Journal of Business*, **45**, 444–454.

Black, F. (1976). Studies of stock price volatility changes. *Proceedings of the 1976 meetings of the business and economics statistics section, American Statistical Association*, 177181.

Blanchard, O.J. and Watson, M.W. (1982). Bubbles, rational expectations and financial markets. In *Crises in the Economic and Financial Structure: Bubbles, Bursts, and Shocks*, P. Wachtel (ed). Lexington Press, Lexington, MA.

Blume, M.E. and Friend, I. (1973). A new look at the capital asset pricing model. *Journal of Finance*, **28**, 19–34.

Bollerslev, T. (1986). Generalized autoregressive conditional heteroscedasticity. *Journal of Econometrics*, **31**, 307–327.

Bougerol, P. and Picard, N. (1992). Strict stationarity of generalized autoregressive processes. *Annals of Probability*, **4**, 1714–1730.

Box, G.E.P. and Pierce, D.A. (1970). Distribution of residual autocorrelations in auto-regressive-integrated moving Average time series models, *Journal of American Statistical Association*, **65**, 1509–1526.

Brodie, J., Daubechies, I., De Mol, C., Giannoned, D. and Loris, I. (2009). Sparse and stable Markowitz portfolios. *Proceedings of the National Academy of Sciences USA*, **106**, 12267–12272.

Brown, C.M. (2012). *Technical Analysis for the Trading Professional* (2nd Edition). McGraw-Hill, New York.

Cai, T. and Liu, W. (2011). Adaptive thresholding for sparse covariance matrix estimation. *Journal of American Statistical Association*, **494**, 672–684.

Campbell, J.Y., Lo, A. and MacKinlay, A. C. (1997). *The Econometrics of Financial Markets*. Princeton University Press, Princeton, NJ.

Campbell, J.Y. and Shiller, R.J. (1988a). The dividend-price ratio and expectations of future dividends and discount factors. *Review of Economic Studies*, **1**, 195–227.

Campbell, J.Y. and Shiller, R.J. (1988b). Stock prices, earnings and expected dividends. *Journal of Finance*, **43**, 661–676.

Carmona, R. (2004). *Statistical Analysis of Financial Data in S-Plus*. Springer, New York.

Carmona, R. (2013). *Statistical Analysis of Financial Data in R*. Springer, New York.

Carroll, R.J., Ruppert, D., Stefanski, L.A. and Crainiceanu, C.M. (2006). *Measurement Error in Nonlinear Models: a Modern Perspective* (2nd ed.). Chapman and Hall, London.

Chen, M. and An, H. (1998). A note on the stationarity and the existence of moments of the GARCH models. *Statistica Sinica*, **8**, 505–510.

Cheung, Y.-W. and Lai, K.S. (1995). Lag order and critical values of the augmented Dickey–Fuller test. *Journal of Business and Economic Statistics*, **13**, 227–280.

Christie, A.A. (1982). The stochastic behavior of common stock variances: value, leverage and interest rate effects. *Journal of Financial Economics*, **10**, 407–432.

Choi, B.S. (1992). *ARMA Model Identification*. Springer-Verlag, New York.

Cochrane, J.H. (2005). *Asset Pricing*. Princeton University Press, Princeton, NJ.

Connor, G. (1984). A unified beta pricing theory. *Journal of Economic Theory*, **34**, 13–31.

Copeland, T.E., Weston, J.F. and Shastri, K. (2005). *Financial Theory and Corporate Policy* (4th ed.). Pearson Addison Wesley, Boston.

Cryer, J.D. and Chan, K.S. (2010).*Time Series Analysis with Applications in R*. Springer, New York.

Davies, N., Triggs, C.M. and Newbold, P. (1977). Significance levels of the Box–Piece portmanteau statistic in finite samples. *Biometrika*, **64**, 517–522.

Davis, C. and Kahan, W. (1970). The rotation of eigenvectors by a perturbation III. *SIAM Journal on Numerical Analysis*, **7**, 1–46.

Davis, R.A. and Mikosch, T. (2009). Probabilistic properties of stochastic volatility models. In *Handbook in Financial Times Series* T.G. Anderson, R.A. Davis, J.-P. Kreiss, and T. Mikosch (eds). Springer, New York, 255–267.

De Bondt, W.F. and Thaler, R.H. (1985). Does the stock market overreact? *Journal of Finance*, **40**, 793–805.

DeMiguel, V., Garlappi, L., Nogales, F.J. and Uppal, R.(2008). A generalized approach to portfolio optimization: Improving performance by constraining portfolio norms. *Management Science*, **55**, 798–812.

Deo, R.S. (2000). Spectral tests of the martingale hypothesis under conditional heteroscedasticity. *Journal of Econometrics*, **99**, 291–315.

Dickey, D.A. and Fuller, W.A. (1979). Distribution of the estimators for autoregressive time series with a unit root. *Journal of American Statistical Association*, **74**, 427–431.

Ding, Z., Engle, R. and Granger, C. (1993). A long memory property of stock market returns and a new model. *Journal of Empirical Finance*, **1**, 83–106.

Durbin, J. and Koopman, S.J. (2012). *Time Series Analysis by State Space Methods* (2nd edition). Oxford University Press, Oxford.

Durlauf, S.N. (1991). Spectral based testing of the martingale hypothesis. *Journal of Econometrics*, **50**, 355–376.

Engle, R.F. (1982). Autoregressive conditional heteroscedasticity with estimates of the variance of U.K. inflation. *Econometrica*, **50**, 987–1008.

Engle, R.F. and Bollerslev, T. (1986). Modelling the persistence of conditional variances. *Econometric Reviews*, **5**, 1–50.

Engle, R.F. and Granger, C.W.J. (1987). Co-integration and error correction: representation, estimation, and testing. *Econometrica*, **55**, 251–276.

Engle, R.F., Lilien, D.M. and Robins, R.P. (1987). Estimating time varying risk premia in the term structure: the ARCH-M model. *Econometrica*, **55**, 391–407.

Fama, E.F., and MacBeth, J.D. (1973). Risk, return, and equilibrium: Empirical tests. *Journal of Political Economy*, **71**, 607–636.

Fama, E. and French, K. (1992). The cross-section of expected stock returns. *Journal of Finance*, **47**, 427–465.

Fama, E. and French, K. (1993). Common risk factors in the returns on stocks and bonds. *Journal of Financial Economics*, **33**, 3–56.

Fan, J., Fan, Y. and Lv, J. (2008). Large dimensional covariance matrix estimation via a factor model. *Journal of Econometrics*, **147**, 186–197.

Fan, J. and Li, R. (2001). Variable selection via nonconcave penalized likelihood and its oracle properties. *Journal of American Statistical Association*, **96**, 1348–1360.

Fan, J., Li, Y. and Yu, K. (2012). Vast volatility matrix estimation using high frequency data for portfolio selection. *Journal of American Statistical Association*, **107**, 412–428.

Fan, J., Liao, Y. and Mincheva, M. (2011). High dimensional covariance matrix estimation in approximate factor models. *Annals of Statistics*, **39**, 3320–3356.

Fan, J., Liao, Y. and Mincheva, M. (2013). Large covariance estimation by thresholding principal orthogonal complements (with discussion). *Journal of the Royal Statistical Society, Series B*, **75**, 603–680.

Fan, J., Liao, Y. and Shi, X. (2015). Risks of large portfolios. *Journal of Econometrics*, **186**, 367–387.

Fan, J., Liao, Y. and Yao, J. (2013). Power enhancement in high dimensional cross-sectional tests. *Econometrica*, **83**, 1497–1541.

Fan, J., Qi, L. and Xiu, D. (2014). Quasi maximum likelihood estimation of GARCH models with heavy-tailed likelihoods (with discussion). *Journal of Business and Economic Statistics*, **32(2)**, 178–191.

Fan, J. and Yao, Q. (2003). *Nonlinear Time Series: Nonparametric and Parametric Methods*. Springer, New York.

Fan, J., Zhang, J. and Yu, K. (2012). Vast portfolio selection with gross-exposure constraints. *Journal of American Statistical Association*, **107**, 592–606.

Fiorentini, G. and Sentana, E. (2013). Consistent non-Gaussian pseudo maximum likelihood estimators. Mimeo, CEMFI.

Francq, C., Lepage, G. and Zakoäian, J.-M. (2011). Two-stage non Gaussian QML estimation of GARCH models and testing the efficiency of the Gaussian QMLE. *Journal of Econometrics*, **165**, 246–257.

Franke, J., Härdle, W. and Hafner, Ch. (2015). *Statistics of Financial Markets: An Introduction*. Springer-Verlag, Heidelberg.

Fuller, W.A. (1996). *Introduction to Statistical Time Series* (2nd edition). Wiley, New York.

Gagliardini, P., Ossola, E. and Scaillet, O. (2016). Time-varying risk premium in large cross-sectional equity datasets. *Econometrica*, **84**, 985–1046.

Gordon, M. (1962). *The Investment, Financing, and Valuation of the Corporation*, Irwin, Homewood, IL.

Gordon, N.J., Salmond, D.J. and Smith, A.F.M. (1993). A novel approach to no-linear and non-Gaussian Bayesian state estimation. *IEE Proceedings F (Radar and Signal Processing)*, **140**, 107–113.

Gourieroux, C. and Jasiak, J. (2001). *Financial Econometrics: Problems, Models, and Methods*. Princeton University Press, NJ.

Granger, C.W.J. (1969). Investigating causal relations by econometric models and cross-spectral methods. *Econometrics*, **37**, 424–438.

Granger, C.W.J. (1981). Some properties of time series data and their use in econometric models specification. *Journal of Econometrics*, **16**, 150–161.

Granger, C.W.J. and Newbold, P. (1974). Spurious regressions in Econometrics. *Journal of Econometrics*, **2**, 111–120.

Grossman, S.J. and Shiller, R. (1981). The determinants of the variability of stock market prices. *American Economic Review*, **71**, 222–227.

Hall, P. and Yao, Q. (2003). Inference in ARCH and GARCH models with heavy-tailed errors. *Econometrica*, **71(1)**, 285–317.

Hallin, M. and Puri, M.L. (1988). Optimal rank-based procedures for time series analysis: testing an ARAM model against other ARMA models. *Annals of Statistics*, **16**, 402–432.

Hamilton, J.D. (1994). *Time Series Analysis*. Princeton University Press, Princeton, NJ.

Hannan, E.J. (1986). Remembrance of things past. In *The Craft of Probability Modeling*, J. Gani (ed). Springer-Verlag, New York.

Hansen, L.P. and Jagannathan, R. (1991). Implications of security market data for models of dynamic economies. *Journal of Political Economy*, **99**, 225–262.

Hansen, L.P. and Singleton, K. (1983). Stochastic consumption, risk aversion and the temporal behavior of asset returns. *Journal of Political Economy*, **91**, 249–268.

Harvey, A.C. (1989). *Forecasting, Structural Time Series Models and the Kalman Filter*. Cambridge University Press, Cambridge.

Harvey, A.C., Ruiz, E. and Shephard, N. (1994). Multivariate stochastic variance models. *Review of Economic Studies*, **61**, 247–264.

Hatanaka, M. (1996). *Time-Series-Based Econometrics: Unit Roots and Cointegration*. Oxford University Press, Oxford.

Hong, Y. (1996). Consistent testing for serial correlation of unknown form. *Econometrica*, **64**, 837–864.

Hong, Y. and Lee, Y.J. (2003). Consistent testing for serial correlation of unknown form under general conditional heteroscedasticity. *Preprint*, Cornell University, Department of Economics.

Horowitz, J.L, Lobato, I.N., Nankervis, J.C. and Savin, N.E. (2006). Bootstrapping the Box–Pierce Q test: a robust test of uncorrelatedness. *Journal of Econometrics*, **133**, 841–862.

Hosking, J.R.M. (1981). Fractional differencing. *Biometrika*, **68**, 165–176.

Huang, C.F. and Litzenberger, R.H.(1988). *Foundations for Financial Economics*, North-Holland, NY.

Huang, D., Wang, H. and Yao, Q. (2008). Estimating GARCH models: when to use what? *Econometrics Journal*, **11**, 27–38.

Hull, J. (2014). *Options, Futures, and Other Derivatives* (Nine edition). Prentice Hall, Upper Saddle River, NJ.

Hurvich, C.M. and Tsai, C.L. (1989). Regression and time series model selection in small samples. *Biometrika*, **76**, 297–307.

Jagannathan, R. and Ma, T. (2003). Risk reduction in large portfolios: why imposing the wrong constraints helps. *Journal of Finance*, **58**, 1651–1683.

Jarque, C.M. and Bera, A.K. (1987). A test for normality of observations and regression residuals. *International Statistical Review*, **55**, 163–172.

Jegadeesh, N. and Titman, S. (1993). Returns to buying winners and selling losers: implications for stock market efficiency. *Journal of Finance*, **48**, 65–91.

Jensen, M.C., Black, F. and Scholes, M.S. (1972). The capital asset pricing model: some empirical tests. In *Studies in the Theory of Capital Markets*, Michael C. Jensen (ed). Praeger Publishers Inc., New York.

Johansen, S. (1995). *Likelihood-Based Inference in Cointegration Vector Autoregressive Modes.* Oxford University Press, Oxford.

Kalman, R.E. (1960). A new approach to linear filtering and prediction problems. *Journal of Fluids Engineering*, **82**, 35–45.

Karatzas, I. and Shreve, S.E. (1998). *Methods of Mathematical Finance.* Springer, New York.

Kazakevičius V. and Leipus R. (2002). On stationarity in the ARCH(∞) model. *Econometric Theory*, **18**, 1–16.

Kim, J.-Y. (1998). Large sample properties of posterion densities, Bayesian information criterion and the likelihood principle in nonstationary time series models. *Econometrica*, **66**, 359–380.

Kitagawa, G. (1987). Non-Gaussian state space modeling of nonstationary time series (with discussion). *Journal of the Royal Statistical Society, Series B*, **82**, 1032–1063.

Kitagawa, G. (1996). Monte Carlo filter and smoother for non-Gaussian nonlinear state space models. *Journal of Computational and Graphical Statistics*, **5**, 1–25.

Kitagawa, G. (2010). *Introduction to Time Series Modelling.* CRC/Chapman & Hall, Boca Raton.

Konishi, S. and Kitagawa, G. (1996). Generalised information criteria in model selection. *Biometrika*, **83**, 875–90.

Lam, C. and Fan, J. (2009). Sparsistency and rates of convergence in large covariance matrices estimation. *Annals of Statistics*, **37**, 4254–4278.

Lam, C. and Yao, Q. (2012). Factor modeling for high-dimensional time series: inference for the number of factors. *Annals of Statistics*, **40**, 694–726.

Lee, A.W. and Hansen, B.E. (1994). Asymptotic theory for a GARCH(1,1) quasi-maximum lkelihood estimator. *Econometric Theory*, **10**, 29–52.

Lehmann, E.L. (1999). *Elements of Large-Sample Theory.* Springer, New York.

Lin, M.T., Zhang, J.L., Cheng, Q. and Chen, R. (2005). Independent particle filters. *Journal of American Statistical Association*, **100**, 1412–1421.

Lintner, J. (1965). The valuation of risky assets and the selection of risky investments in stock portfolios and capital budgets. *Review of Economics and Statistics*, **47**, 13–37.

Liu, J.S. and Chen, R. (1998). Sequential Monte Carlo methods for dynamic systems. *Journal of American Statistical Association*, **93**, 1032–1044.

Ljung, G.M. and Box, G.E.P. (1978). On a measure of a lack of fit in time series models. *Biometrika*, **65**, 297–303.

Lowenstein, R. (2000).*When Genius Failed: the Rise and Fall of Long-Term Capital Management.* Random House, New York.

Lumsdaine, R. (1996). Consistency and asymptotic normality of the quasi-maximum likelihood estimator for IGARCH(1,1) and covariance stationary GARCH(1,1) models. *Econometrica*, **16**, 575–596.

Lütkepohl, H. (2006). *New Introduction to Multiple Time Series Analysis.* Springer, Berlin.

Markowitz, H.M. (1952). Portfolio selection. *Journal of Finance*, **7**, 77–91.

Markowitz, H.M. (1959). *Portfolio Selection: Efficient Diversification of Investments.* John Wiley & Sons, New York.

Marron, J.S. and Nolan, D. (1988). Canonical kernels for density estimation. *Statist. Prob. Lett.*, **7**, 195–199.

Mehra, R. and Prescott, E.C. (1985). The equity premium: a puzzle. *Journal of Monetary Economics*, **15**, 145–161.

Merton, R.C. (1973). An intertemporal capital asset pricing model. *Econometrica*, **41**, 867–887.

Mikosch, T. (1998). *Elementary Stochastic Calculus with Finance in View*. World Scientific, Singapore.

Mikosch, T. and Straumann, T. (2006). Stable limits of martingale transforms with application to the estimation of GARCH parameters. *Annals of Statistics*, **34**, 493–522.

Nelson, D.B. (1990). Stationarity and persistence in the GARCH(1,1) model. *Econometric Theory*, **6**, 318–334.

Nelson, D.B. (1991). Conditional heteroscedasticity in asset pricing: a new approach. *Econometrica*, **59**, 347–370.

Newey, W. and West, K. (1987). A simple, positive semi-definite, heteroscedasticity and autocorrelation consistent covariance matrix. *Econometrica*, **55**, 703–708.

Onatski, A. (2010). Determining the number of factors from empirical distribution of eigenvalues. *Review of Economics and Statistics*, **92**, 1004–1016.

Onatski, A. (2012). Asymptotics of the principal components estimator of large factor models with weakly influential factors. *Journal of Econometrics*, 168, 244–258.

Peng, L. and Yao, Q. (2003). Least absolute deviations estimation for ARCH and GARCH models. *Biometrika*, **90**, 967–975.

Penzer, J., Wang, M. and Yao, Q. (2009). Approximating volatilities by asymmetric power GARCH functions. *Australian & New Zealand Journal of Statistics*, **51**, 201–225.

Pesaran, M.H. and Yamagata, T. (2012). Testing CAPM with a large number of assets. *Manuscript*.

Pfaff, B. (2006). *Analysis of Integrated and Cointegrated Time Series with R*. Springer, New York.

Phillips, P.C.B. (1986). Understanding spurious regressions in Econometrics. *Journal of Econometrics*, **33**, 311–340.

Phillips, P.C.B. (1991). Optimal inference in cointegrated systems. *Econometrica*, **59**, 283–306.

Phillips, P.C.B. and Perron, P. (1988). Testing for unit roots in time series regression. *Biometrika*, **75**, 335–346.

Phillips, P.C.B. and Yu, J. (2011). Dating the timeline of financial bubbles during the subprime crisis. *Quantitative Economics*, **2**, 455–491.

Pötscher, B.M. (1989). Model selection under nonstationarity: autoregressive models and stochastic linear regression models. *Annals of Statistics*, **17**, 1257–1274.

Qi, H. and Sun, D. (2006). A quadratically convergent Newton method for computing the nearest correlation matrix. *SIAM Journal on Matrix Analysis and Applications*, **28**, 360–385.

Rachev, S.T., Mittnik, S., Fabozzi, F.J., Focardi, S.M. and Jasic, T. (2013). *Financial Econometrics: from Basics to Advanced Modeling*. Wiley, New York.

Romano, J.P. and Thombs, L.A. (1996). Inference for autocorrelations under weak assumptions. *Journal of American Statistical Association*, **91**, 590–600.

Ross, S.A. (1976). The arbitrage theory of capital asset pricing. *Journal of Economic Theory*, **13**, 341–360.

Rothman, A.J., Bickel, P.J., Levina, E. and Zhu, J. (2008). Sparse permutation invariant covariance estimation. *Electron. J. of Stat.*, **2**, 494–515.

Rothman, A.J., Levina, E. and Zhu, J. (2009). Generalized thresholding of large covariance matrices. *Journal of American Statistical Association*, **104**, 177–186.

Ruppert, D. (2004). *Statistics and Finance: an Introduction*. Springer, New York.

Ruppert, D. (2010). *Statistics and Data Analysis for Financial Engineering*. Springer, New York.

Rydberg, T.H. (2000). Realistic statistical modelling of financial data. *International Statistical Review*, **68**, 233–258.

Schwarz, G.E. (1978). Estimating the dimension of a model. *Annals of Statistics*, **6**, 461–464.

Sentana, E. (2009). The econometrics of mean-variance efficiency tests: a survey. *Econometrics Journal*, **12**, C65–C101.

Shao, X. (2011). Testing for white noise under unknown dependence and its applications to diagnostic checking for time series models. *Econometric Theory*, **27**, 312–343.

Sharpe, W.F. (1964). Capital asset prices: a theory of market equilibrium under conditions of risks. *Journal of Finance*, **19**, 425–442.

Shephard, N. (1996). Statistical aspects of ARCH and stochastic volatility. In *Time Series Models in Econometrics, Finance and Other Fields*, D.R. Cox, D.V. Hinkley and O.E. Barndorff-Nielsen (eds). Chapman and Hall, London. pp. 1–67.

Shephard, N.G. and Anderson, T.G. (2009). Stochastic volatility: origins and overview. In *Handbook in Financial Times Series*, T.G. Anderson, R.A. Davis, J.-P. Kreiss and T. Mikosch (eds). Springer, New York, pp. 233–254.

Shibata, R. (1980). Asymptotically efficient selection of the order of the model for estimating parameters of a linear process. *The Annals of Statistics*, **8**, 147–164.

Shumway, R.H. and Stoffer, D.S. (2011). *Time Series Analysis and Its Applications* (3rd edition). Springer, New York.

Stock, J.H. (1987). Asymptotic properties of least squares estimators of cointegrating vectors. *Econometrica*, **55**, 1035–1056.

Stock, J.H. and Watson, M.W. (2002). Forecasting using principal components from a large number of predictors. *Journal of American Statistical Association*, **97**, 1167–1179.

Stock, J.H. and Watson, M.W. (2005). Implications of dynamic factor models for VAR analysis. *NBER Working Paper No. W11467.*

Straumann, D. and Mikosch, T. (2006). Quasi-MLE in heteroscedastic times series: a stochastic recurrence equations approach. *Annals of Statistics*, **34**, 2449–2495.

Taylor, S.J. (1986). *Modelling Financial Time Series*. Wiley, New York.

Teräsvirta, T., Tjostheim, D. and Granger, C.W.J. (2010). *Modelling Nonlinear Economic Time Series*. Oxford University Press, Oxford.

Tiao, G.C. and Box, G.E.P. (1977). A canonical analysis of multiple time series. *Biometrika*, **64**, 355–366.

Tiao, G.C. and Box, G.E.P. (1981). Modeling multiple time series with applications. *Journal of American Statistical Association*, **76**, 802–816.

Tibshirani, R. (1996). Regression shrinkage and selection via lasso. *Journal of the Royal Statistical Society, Series B*, **58**, 267–288.

Tsay, R. (2010). *Analysis of Financial Time Series* (3rd edition). Wiley.

Tsay, R. (2013). *Multivariate Time Series Analysis: with R and Financial Applications*, Wiley.

Tsay, R. and Tiao, G. (1984). Consistent estimates of autoregressive parameters and extended sample autocorrelation function for stationary and nonstationary ARMA models. *Journal of American Statistical Association*, **79**, 84–96.

Tsay, R. and Tiao, G. (1985). Use of canonical analysis in time series model identification. *Biometrika*, **72**, 299–315.

Vasicek, O.A. (1977). An equilibrium characterization of the term structure. *Journal of Financial Economics*, **5**, 177–188.

Weil, P. (1989). The equity premium puzzle and the risk-free rate puzzle. *Journal of Monetary Economics*, **24**, 401–421.

Woodroofe, M. (1982). On model selection and the arc sine laws. *The Annals of Statistics*, **10**, 1182–1194.

Xiao, H. and Wu, W.B. (2011). Asymptotic inference of autocovariances of stationary processes. *A Preprint.*

Zou, H. and Li, R. (2008). One-step sparse estimates in nonconcave penalized likelihood models (with discussion). *Annals of Statistics*, **36**, 1509–1533.

Author Index

Subject Index

Printed in the United States
By Bookmasters